Metal Vinylidenes and Allenylidenes in Catalysis

Edited by
Christian Bruneau and
Pierre H. Dixneuf

Related Titles

Diederich, F., Stang, P. J., Tykwinski, R. R. (eds.)

Modern Supramolecular Chemistry

Strategies for Macrocycle Synthesis

2008
ISBN: 978-3-527-31826-1

Sheldon, R. A., Arends, I., Hanefeld, U.

Green Chemistry and Catalysis

2007
ISBN: 978-3-527-30715-9

Hiersemann, M., Nubbemeyer, U. (eds.)

The Claisen Rearrangement

Methods and Applications

2007
ISBN: 978-3-527-30825-5

Cornils, B., Herrmann, W. A., Muhler, M., Wong, C.-H. (eds.)

Catalysis from A to Z

A Concise Encyclopedia

2007
ISBN: 978-3-527-31438-6

Tietze, L. F., Brasche, G., Gericke, K. M.

Domino Reactions in Organic Synthesis

2006
ISBN: 978-3-527-29060-4

Overman, LE

Organic Reactions V67

2006
ISBN: 978-0-470-04145-1

Haley, M. M., Tykwinski, R. R. (eds.)

Carbon-Rich Compounds

From Molecules to Materials

2006
ISBN: 978-3-527-31224-5

Yudin, A. K. (ed.)

Aziridines and Epoxides in Organic Synthesis

2006
Hardcover
ISBN: 978-3-527-31213-9

Tidwell, TT

Ketenes 2e

2006
E-Book
ISBN: 978-0-471-76766-4

Cornils, B., Herrmann, W. A., Horvath, I. T., Leitner, W., Mecking, S., Olivier-Bourbigou, H., Vogt, D. (eds.)

Multiphase Homogeneous Catalysis

2005
ISBN: 978-3-527-30721-0

Metal Vinylidenes and Allenylidenes in Catalysis

From Reactivity to Applications in Synthesis

Edited by
Christian Bruneau and Pierre Dixneuf

WILEY-VCH

WILEY-VCH Verlag GmbH & Co. KGaA

The Editors

Pierre Dixneuf
CNRS-Université de Rennes 1
Institut de Chimie, UMR 6509
Campus de Beaulieu
035042 Rennes Cedex
France

Christian Bruneau
CNRS-Université de Rennes 1
Organométalliques et Catalyse
Campus de Beaulieu
35042 Rennes Cedex
France

All books published by Wiley-VCH are carefully produced. Nevertheless, authors, editors, and publisher do not warrant the information contained in these books, including this book, to be free of errors. Readers are advised to keep in mind that statements, data, illustrations, procedural details or other items may inadvertently be inaccurate.

Library of Congress Card No.: applied for

British Library Cataloguing-in-Publication Data
A catalogue record for this book is available from the British Library.

Bibliographic information published by the Deutsche Nationalbibliothek
Die Deutsche Nationalbibliothek lists this publication in the Deutsche Nationalbibliografie; detailed bibliographic data are available in the Internet at http://dnb.d-nb.de.

© 2008 WILEY-VCH Verlag GmbH & Co. KGaA, Weinheim

All rights reserved (including those of translation into other languages). No part of this book may be reproduced in any form – by photoprinting, microfilm, or any other means – nor transmitted or translated into a machine language without written permission from the publishers. Registered names, trademarks, etc. used in this book, even when not specifically marked as such, are not to be considered unprotected by law.

Typesetting Thomson Digital, Noida, India
Printing   betz-druck GmbH, Darmstadt
Binding   Litges & Dopf GmbH, Heppenheim
Cover Design Adam-Design, Weinheim

Printed in the Federal Republic of Germany
Printed on acid-free paper

ISBN: 978-3-527-31892-6

Contents

Preface *XIII*
List of Contributors *XV*

1	**Preparation and Stoichiometric Reactivity of Mononuclear Metal Vinylidene Complexes** *1*	
	Michael I. Bruce	
1.1	Introduction *1*	
1.2	Preparative Methods *2*	
1.2.1	From 1-Alkynes *2*	
1.2.1.1	Migration of Other Groups (SiR$_3$, SnR$_3$, SR, SeR) *5*	
1.2.2	The η^2-Alkyne $\rightarrow$ Hydrido(η^1-Alkynyl) $\rightarrow$ Vinylidene Transformation *6*	
1.2.3	From Metal Alkynyls *6*	
1.2.3.1	Some Specific Examples *8*	
1.2.3.2	Redox Rearrangements of Metal Alkynyls and Vinylidenes *9*	
1.2.4	From Metal Allenylidenes via Metal Alkynyls *11*	
1.2.5	From Metal-Carbyne Complexes *11*	
1.2.6	From Metal-Carbon Complexes *14*	
1.2.7	From Acyl Complexes *15*	
1.2.8	From Vinyls *15*	
1.2.9	From Alkenes *16*	
1.2.10	Miscellaneous Reactions Affording Vinylidenes *16*	
1.2.11	Vinylvinylidene Complexes *17*	
1.3	Stoichiometric Reactions *19*	
1.3.1	Reactions at C$_\alpha$ *20*	
1.3.1.1	Deprotonation *20*	
1.3.1.2	Group 16 Nucleophiles. Oxygen *20*	
1.3.1.3	Alcohols *21*	
1.3.1.4	Sulfur *21*	
1.3.1.5	Group 15 Nucleophiles. Nitrogen *22*	
1.3.1.6	Phosphorus *22*	
1.3.1.7	Halogen Nucleophiles *22*

1.3.1.8	Carbon Nucleophiles	22
1.3.1.9	Hydride	22
1.3.2	Intramolecular Reactions	23
1.3.2.1	Formation of Cyclopropenes	23
1.3.2.2	Attack on Coordinated Phosphines	24
1.3.2.3	Coupling	24
1.3.2.4	Vinylidene/Alkyne Coupling	25
1.3.2.5	Formation of π-Bonded Ligands	25
1.3.3	Reactions at C_β	25
1.3.3.1	Protonation	26
1.3.3.2	Alkylation	27
1.3.3.3	Other Electrophiles	27
1.3.4	Cycloaddition Reactions	27
1.3.5	Adducts with Other Metal Fragments	28
1.3.6	Ligand Substitution	30
1.3.7	Miscellaneous Reactions	31
1.4	Chemistry of Specific Complexes	33
1.4.1	Reactions of Ti(=C=CH$_2$)Cp$_2^*$	33
1.4.2	Complexes Derived From Li[M(C≡CR)(CO)(NO)Cp] (M=Cr, W)	34
1.4.3	Reactions of M(=C=CRR′)(CO)$_5$ (M = Cr, Mo, W)	35
1.4.4	Reactions of M(=C=CRR′)(CO)(L) Cp (M = Mn, Re)	37
1.4.5	Reactions of [M(=C=CRR′)(L′)(P)Cp′]$^+$ (M = Fe, Ru, Os)	39
1.4.6	Reactions of [Ru{=C=C(SMe)$_2$}(PMe$_3$)$_2$Cp]$^+$	39
1.4.7	Reactions of *trans*-MCl(=C=CRR′)(L)$_2$ (M=Rh, Ir)	41
1.5	Reactions Supposed to Proceed via Metal Vinylidene Complexes	42
	Abbreviations	45
	References	46
2	**Preparation and Stoichiometric Reactivity of Metal Allenylidene Complexes**	**61**
	Victorio Cadierno, Pascale Crochet, and José Gimeno	
2.1	Introduction	61
2.2	Preparation of Allenylidene Complexes	62
2.2.1	General Methods of Synthesis	62
2.2.2	Group 6 Metals	63
2.2.3	Group 7 Metals	64
2.2.4	Group 8 Metals	65
2.2.4.1	Octahedral and Five-Coordinate Derivatives	66
2.2.4.2	Half-Sandwich Derivatives	66
2.2.4.3	Other Synthetic Methodologies	68
2.2.5	Group 9 Metals	68
2.3	Coordination Modes and Structural Features	69
2.4	Stoichiometric Reactivity of Allenylidenes	69
2.4.1	General Considerations of Reactivity	69
2.4.2	Electrophilic Additions	70

2.4.3	Nucleophilic Additions	71
2.4.3.1	Group 6 Metal-Allenylidenes	72
2.4.3.2	Group 7 Metal-Allenylidenes	73
2.4.3.3	Group 8 Metal-Allenylidenes	74
2.4.3.4	Group 9 Metal-Allenylidenes	78
2.4.4	C–C Couplings	79
2.4.5	Cycloaddition and Cyclization Reactions	81
2.4.5.1	Reactions Involving the $M=C_\alpha$ Bond	81
2.4.5.2	Reactions Involving the $C\alpha=C_\beta$ Bond	82
2.4.5.3	Reactions Involving the $C_\beta=C_\gamma$ Bond	84
2.4.5.4	Reactions Involving Both $C_\alpha=C_\beta$ and $C_\beta=C_\gamma$ Bonds (1,2,3-Heterocyclizations)	87
2.4.6	Other Reactions	89
2.5	Concluding Remarks	90
	References	91

3 Preparation and Reactivity of Higher Metal Cumulenes Longer than Allenylidenes 99

Helmut Fischer

3.1	Introduction	99
3.2	Steric and Electronic Structure	100
3.3	Synthesis of Cumulenylidene Complexes	103
3.3.1	Butatrienylidene Complex Synthesis	103
3.3.2	Pentatetraenylidene Complex Synthesis	108
3.3.3	Hexapentaenylidene Complex Synthesis	113
3.3.4	Heptahexaenylidene Complex Synthesis	113
3.4	Reactions of Higher Metal Cumulenes	114
3.4.1	Butatrienylidene Complexes	114
3.4.2	Pentatetraenylidene Complexes	119
3.4.3	Hexapentaenylidene Complexes	123
3.4.4	Heptahexaenylidene Complexes	123
3.5	Summary and Conclusion	124
	References	125

4 Theoretical Aspects of Metal Vinylidene and Allenylidene Complexes 129

Jun Zhu and Zhenyang Lin

4.1	Introduction	129
4.2	Electronic Structures of Metal Vinylidene and Allenylidene Complexes	130
4.2.1	Metal Vinylidene Complexes	130
4.2.2	Metal Allenylidene Complexes	132
4.3	Barrier of Rotation of Vinylidene Ligands	132
4.4	Tautomerization Between η^2-Acetylene and Vinylidene on Transition Metal Centers	134

4.4.1	η^2-Acetylene to Vinylidene *134*
4.4.2	Vinylidene to η^2-Acetylene *139*
4.5	Reversible C–C σ-bond Formation by Dimerization of Metal Vinylidene Complexes *141*
4.6	Metal Vinylidene Mediated Reactions *142*
4.6.1	Alkynol Cycloisomerization Promoted by Group 6 Metal Complexes *142*
4.6.2	Unusual Intramolecular [2 + 2] Cycloaddition of a Vinyl Group with a Vinylidene C=C Bond *148*
4.6.3	Intramolecular Methathesis of a Vinyl Group with a Vinylidene C=C Double Bond *149*
4.6.4	[2 + 2] Cycloaddition of Titanocene Vinylidene Complexes with Unsaturated Molecules *150*
4.7	Heavier Group 14 Analogs of Metal Vinylidene Complexes *150*
4.8	Allenylidene Complexes *151*
4.9	Summary *152*
	References *153*

5 Group 6 Metal Vinylidenes in Catalysis (Cr, Mo, W) *159*
Nobuharu Iwasawa

5.1	Introduction *159*
5.2	Preparation of Fischer-type Carbene Complexes through the Generation of the Vinylidene Complexes *159*
5.3	Utilization of Pentacarbonyl Vinylidene Complexes of Group 6 Metals for Synthetic Reactions *164*
5.3.1	Catalytic Addition of Hetero-Nucleophiles *165*
5.3.2	Catalytic Addition of Carbo-Nucleophiles *172*
5.3.3	Electrocyclization and Related Reactions *178*
5.4	Utilization of Vinylidene to Alkyne Conversion *184*
5.5	Synthetic Reactions Utilizing Other Kinds of Vinylidene Complexes of Group 6 Metals *186*
5.6	Conclusion *187*
	References *188*

6 Ruthenium Vinylidenes in the Catalysis of Carbocyclization *193*
Arjan Odedra and Rai-Shung Liu

6.1	Introduction *193*
6.2	Stoichiometric Carbocyclization via Ruthenium Vinylidene *193*
6.3	Catalytic Carbocyclization via Electrocyclization of Ruthenium-Vinylidene Intermediates *195*
6.3.1	Cyclization of *cis*-3-En-1-Ynes *195*
6.3.2	Cycloaromatization of 3,5-Dien-1-Ynes *196*
6.3.3	Ruthenium-Catalyzed Cyclization of 3-Azadienynes *202*
6.3.4	Cycloisomerization of *cis*-1-Ethynyl-2-Vinyloxiranes *203*

6.3.5	Catalytic Cyclization of Enynyl Epoxides *204*	
6.4	Catalytic Carbocyclization via Cycloaddition of Ruthenium Vinylidene Intermediates *208*	
6.4.1	Cyclocarbonylation of 1,1′-Bis(silylethynyl)ferrocene *208*	
6.4.2	Dimerization of 1-Arylethynes to 1-Aryl-Substituted Naphthalenes *209*	
6.4.3	Ruthenium-Catalyzed Cycloaddition Reaction between Enyne and Alkene *209*	
6.5	Catalyzed Cyclization of Alkynals to Cycloalkenes *211*	
6.6	Ruthenium-Catalyzed Hydrative Cyclization of 1,5-Enynes *211*	
6.7	Carbocyclization Initiated by Addition of *C*-Nucleophile to Ruthenium Vinylidene *213*	
6.8	Conclusion *214*	
	References *214*	
7	**Allenylidene Complexes in Catalysis** *217*	
	Yoshiaki Nishibayashi and Sakae Uemura	
7.1	Introduction *217*	
7.2	Propargylic Substitution Reactions *219*	
7.2.1	Propargylic Substitution Reactions with Heteroatom-Centered Nucleophiles *219*	
7.2.2	Propargylic Substitution Reactions with Carbon-Centered Nucleophiles *223*	
7.2.3	Reaction Pathway for Propargylic Substitution Reactions *224*	
7.2.4	Asymmetric Propargylic Alkylation with Acetone *227*	
7.2.5	Cycloaddition between Propargylic Alcohols and Cyclic 1,3-Dicarbonyl Compounds *231*	
7.3	Propargylation of Aromatic Compounds with Propargylic Alcohols *233*	
7.3.1	Propargylation of Heteroaromatic and Aromatic Compounds with Propargylic Alcohols *233*	
7.3.2	Cycloaddition between Propargylic Alcohols and Phenol and Naphthol Derivatives *234*	
7.4	Carbon–Carbon Bond Formation via Allenylidene-Ene Reactions *236*	
7.5	Reductive Coupling Reaction via Hydroboration of Allenylidene Intermediates *238*	
7.6	Selective Preparation of Conjugated Enynes *239*	
7.7	Preparation of Dicationic Chalcogenolate-Bridged Diruthenium Complexes and Their Dual Catalytic Activity *241*	
7.8	Other Catalytic Reactions via Allenylidene Complexes as Key Intermediates *243*	
7.9	Conclusion *246*	
	References *247*	

8	**Ruthenium Allenylidenes and Indenylidenes as Catalysts in Alkene Metathesis** *251*	
	Raluca Malacea and Pierre H. Dixneuf	
8.1	Introduction *251*	
8.2	Propargyl Derivatives as Alkene Metathesis Initiator Precursors: Allenylidenes, Indenylidenes and Alkenylalkylidenes *252*	
8.2.1	Allenylidene-Ruthenium Complexes as Alkene Metathesis Catalyst Precursors: the First Evidence *252*	
8.2.2	Allenylidene-Ruthenium Complexes in RCM, Enyne Metathesis and ROMP *254*	
8.2.2.1	RCM Reactions *254*	
8.2.2.2	Enyne Metathesis *254*	
8.2.2.3	ROMP Promoted by Allenylidene Complexes *255*	
8.2.3	Indenylidene-Ruthenium Complexes: the Alkene Metathesis Catalytic Species from Allenylidene Ruthenium Complexes *256*	
8.2.3.1	The First Evidence *256*	
8.2.3.2	The Intramolecular Allenylidene to Indenylidene Rearrangement Demonstration *259*	
8.2.3.3	Applications of Isolated Indenylidene-Ruthenium Complexes in ROMP *261*	
8.2.3.4	Indenylidene-Ruthenium(arene) Catalyst in Diene and Enyne RCM *262*	
8.2.4	Propargylic Ethers as Alkene Metathesis Initiator Precursors: Generation of Alkenyl Alkylidene-Ruthenium Catalysts *262*	
8.3	Indenylidene-Ruthenium Catalysts in Alkene Metathesis *265*	
8.3.1	Preparation of Indenylidene-Ruthenium Catalysts *265*	
8.3.2	Ruthenium Indenylidene Complexes in Alkene Metathesis *268*	
8.3.3	Polymerization with Ruthenium Indenylidene Complexes *271*	
8.3.4	Other Catalytic Reactions Promoted by Indenylidenes *273*	
8.4	Conclusion *274*	
	References *274*	
9	**Rhodium and Group 9–11 Metal Vinylidenes in Catalysis** *279*	
	Sean H. Wiedemann and Chulbom Lee	
9.1	Introduction *279*	
9.2	Rhodium and Iridium Vinylidenes in Catalysis *280*	
9.2.1	Introduction *280*	
9.2.2	Carbocyclization/Pericyclic Reactions *281*	
9.2.3	Anti-Markovnikov Hydrofunctionalization *288*	
9.2.4	Multi-Component Coupling *294*	
9.3	Rhodium Alkenylidenes in Catalysis *299*	
9.4	Group 10 and 11 Metal Vinylidenes in Catalysis *302*	
9.4.1	Introduction *302*	
9.4.2	Nickel Vinylidenes in Catalysis *302*	
9.4.3	Palladium Vinylidenes in Catalysis *303*	

9.4.4	Platinum Vinylidenes in Catalysis	*304*
9.4.5	Copper Vinylidenes in Catalysis	*306*
9.4.6	Gold Vinylidenes in Catalysis	*307*
9.5	Conclusion	*310*
9.6	Note Added in Proof	*310*
	References	*311*
10	**Anti-Markovnikov Additions of *O*-, *N*-, *P*-Nucleophiles to Triple Bonds with Ruthenium Catalysts**	***313***
	Christian Bruneau	
10.1	Introduction	*313*
10.2	C–O Bond Formation	*314*
10.2.1	Addition of Carbamic Acids: Synthesis of Vinylic Carbamates and Ureas	*314*
10.2.2	Addition of Carboxylic Acids: Synthesis of Enol Esters	*316*
10.2.3	Addition of Water: Synthesis of Aldehydes	*318*
10.2.4	Addition of Alcohols: Synthesis of Ethers and Ketones	*321*
10.2.4.1	Intermolecular Addition: Formation of Unsaturated Ethers and Furans	*321*
10.2.4.2	Intermolecular Addition with Rearrangement: Formation of Unsaturated Ketones	*321*
10.2.4.3	Intramolecular Addition: Formation of Cyclic Enol Ethers and Lactones from Pent-4-yn-1-ols and But-3-yn-1-ols	*323*
10.3	Formation of C–N Bonds via Anti-Markovnikov Addition to Terminal Alkynes	*325*
10.3.1	Addition of Amides to Terminal Alkynes	*325*
10.3.2	Formation of Nitriles *via* Addition of Hydrazines to Terminal Alkynes	*325*
10.4	Hydrophosphination: Synthesis of Vinylic Phosphine	*326*
10.5	C–C Bond Formation: Dimerization of Terminal Alkynes	*327*
10.6	Conclusion	*329*
	References	*330*

Index *333*

Preface

Catalytic transformations of alkynes have recently led to tremendous developments of synthetic methods with useful applications in the synthesis of natural products and molecular materials. Among them, the selective activations of terminal alkynes and propargylic alcohols via vinylidene- and allenylidene-metal intermediates play an important role, and have opened new catalytic routes toward *anti*-Markovnikov additions to terminal alkynes, carbocyclizations or propargylations, in parallel to the production of new types of molecular catalysts.

After the discovery of the first terminal vinylidene-metal complex in 1972, it was established that the stoichiometric activation of terminal alkynes by a variety of suitable metal complexes led to 1,2-hydrogen transfer and the formation of metal-vinylidene species, which is now a classical organometallic reaction. A metal-vinylidene intermediate was proposed for the first time in 1986 to explain a catalytic *anti*-Markovnikov addition to terminal alkynes. Since then, possible metal-vinylidene intermediate formation has been researched to achieve catalytic regioselective formation of carbon–heteroatom and carbon–carbon bonds involving the alkyne terminal carbon.

In parallel, since the first preparation of allenylidene-metal complexes in 1976, the formation of these carbon-rich complexes developed rapidly after the discovery, in 1982, that allenylidene-metal intermediates could be easily formed directly from terminal propargylic alcohols via vinylidene-metal intermediates. This decisive step has led to regioselective catalytic transformations of propargylic derivatives via carbon(1)–atom bond formation or alternately to propargylation. Due to their rearrangement into indenylidene complexes, metal-allenylidene complexes were also found to be catalyst precursors for olefin and enyne metathesis.

Higher cumulenyl moieties stabilized by organometallic fragments were introduced in the eighties and have recently received much attention. Such linear unsaturated carbon-rich cumulenyl-metal complexes have allowed access to new molecular architectures and have revealed interesting properties in the field of electronics and molecular wires.

The activation of alkynes to metal-vinylidenes with transition metal complexes of Groups 6–9, essentially, provides reactive intermediates with an electrophilic terminal carbon atom, whereas allenylidene species present two electrophilic carbon centers. Advantage has been taken of this property for the rational design of new catalytic transformations, and useful atom-economical catalytic transformations have been brought to light. The new trends aim at the use of multifunctional acetylenic substrates with the objective of performing unprecedented cascade catalytic reactions.

The content of this book gathers in the same volume all aspects of vinylidene- and allenylidene-metal complexes, including the preparation of these organometallic carbon-rich systems with a metal–carbon double bond, their stoichiometric reactivity and theoretical aspects, and their applications in catalysis for the production of fine chemicals, mainly in the field of selective transformations of functional terminal alkynes. It provides essential general information on catalytic transformations of alkynes and their use in synthesis.

This book should be of interest to academic and industrial researchers involved in the fields of organometallic, coordination and bioinorganic chemistry, transition metal catalysis, and organic synthesis.

We are grateful to the team from Wiley-VCH who made this project possible and to all contributors to this book for their enthusiasm in writing a chapter on their favorite selected topic.

Christian Bruneau
Pierre H. Dixneuf

List of Contributors

Michael I. Bruce
University of Adelaide
School of Chemistry & Physics
Adelaide 5005
South Australia
Australia

Christian Bruneau
UMR 6509 CNRS-Université de Rennes1
Institut Sciences Chimiques de Rennes
Campus de Beaulieu
Laboratoire Catalyse et Organométalliques
35042 Rennes Cedex
France

Victorio Cadierno
Universidad de Oviedo
Departamento de Química Orgánica e Inorgánica
Instituto Universitario de Química Organometálica "Enrique Moles"
(Unidad Asociada al CSIC)
Facultad de Química
c/ Julián Claveria 8
33071 Oviedo
Spain

Pascale Crochet
Universidad de Oviedo
Departamento de Química Orgánica e Inorgánica
Instituto Universitario de Química Organometálica "Enrique Moles"
(Unidad Asociada al CSIC)
Facultad de Química
c/ Julián Claveria 8
33071 Oviedo
Spain

Pierre H. Dixneuf
UMR 6226 CNRS-Université de Rennes1
Institut Sciences Chimiques de Rennes
Laboratoire Catalyse et Organométalliques
Campus de Beaulieu
35042 Rennes Cedex
France

Helmut Fischer
Universität Konstanz
Fachbereich Chemie
Universitätsstr. 1
78457 Konstanz
Germany

Metal Vinylidenes and Allenylidenes in Catalysis: From Reactivity to Applications in Synthesis
Edited by Christian Bruneau and Pierre Dixneuf
Copyright © 2008 WILEY-VCH Verlag GmbH & Co. KGaA, Weinheim
ISBN: 978-3-527-31892-6

Jos Gimeno
Universidad de Oviedo
Departamento de Química Orgánica e Inorgánica
Instituto Universitario de Química Organometálica "Enrique Moles"
(Unidad Asociada al CSIC)
Facultad de Química
33071 Oviedo
Spain

Nobuharu Iwasawa
Tokyo Institute of Technology
Department of Chemistry
O-okayama
Meguro-ku
Tokyo 152-8551
Japan

Chulbom Lee
Princeton University
Department of Chemistry
59 Frick Building
Princeton
NJ 08544
USA

Zhenyang Lin
The Hong Kong University of Science and Technology
Clear Water Bay
Kowloon
Hong Kong
The People's Republic of China

Rai-Shung Liu
National Tsing-Hua University
Department of Chemistry
Hsinchu
Taiwan

Raluca Malacea
UMR 6226 CNRS-Université de Rennes
Institut Sciences Chimiques de Rennes
Laboratoire "Catalyse et Organométalliques"
Campus of Beaulieu
35042 Rennes Cedex
France

Yoshiaki Nishibayashi
The University of Tokyo
Institute of Engineering Innovation
School of Engineering
Yayoi
Bunkyo-ku
Tokyo 113-8656
Japan

Arjan Odedra
National Tsing-Hua University
Department of Chemistry
101, Sec 2
Kuang Fu Road
Hsinchu 300
Taiwan

Sakae Uemura
Okayama University of Science
Faculty of Engineering
Okayama 700-0005
Japan

Sean H. Wiedemann
Princeton University
Department of Chemistry
60 Frick Building
Princeton
NJ 08544
USA

Jun Zhu
The Hong Kong University of Science
and Technology
Clear Water Bay
Kowloon
Hong Kong
The People's Republic of China

1
Preparation and Stoichiometric Reactivity of Mononuclear Metal Vinylidene Complexes

Michael I. Bruce

1.1
Introduction

Vinylidene, $:C=CH_2$, is a tautomer of ethyne, $HC\equiv CH$, to which it is related by a 1,2-H shift (Equation 1.1):

$$HC\equiv CH \rightarrow\, :C=CH_2 \qquad (1.1)$$

Attempts to produce vinylidene in the free state result in rapid reversion to ethyne, with a lifetime of 10^{-10} s [1]. As with many reactive organic intermediates, however, vinylidene can be stabilized by complexation to a metal center, using the lone pair for coordination and thus preventing the reversion to ethyne. Most 1-alkynes can be converted into the analogous vinylidene complexes by simple reactions with appropriate transition metal substrates (Equation 1.2):

$$ML_n + HC\equiv CR \rightarrow L_nM=C=CHR \qquad (1.2)$$

The first vinylidene complex, $Fe_2(\mu\text{-}C=CPh_2)(CO)_8$, was obtained by Mills and Redhouse by irradiation of diphenylketene with $Fe(CO)_5$ [2]. The first terminal vinylidene complex, $MoCl\{=C=C(CN)_2\}(PPh_3)_2Cp$, was prepared by King and Saran from the reaction between $Mo\{CCl=C(CN)_2\}(CO)_3Cp$ and PPh_3 [3]. Several groups reported synthetic approaches to metal vinylidene complexes during the late seventies, including platinum-stabilized carbenium ions by Clark and Chisholm [4], manganese and rhenium vinylidenes by Russian workers [5, 6], an iron vinylidene by Mays [7], and the present author's work on ruthenium and osmium systems [8]. Further studies, including Hughes's conversion of iron acyls to vinylidenes with Tf_2O [9] and Mansuy's serendipitous synthesis of a vinylidene-iron porphyrin system [10] followed soon after.

Much of the chemistry of metal-vinylidene complexes has been summarized in several reviews [11–14] and the following will merely summarize the main preparative methods and survey the reactions of many of the metal complexes so obtained. Complexes of most transition metals have been described, although most work has been developed using electron-rich ruthenium derivatives, which have been used in

Metal Vinylidenes and Allenylidenes in Catalysis: From Reactivity to Applications in Synthesis
Edited by Christian Bruneau and Pierre Dixneuf
Copyright © 2008 WILEY-VCH Verlag GmbH & Co. KGaA, Weinheim
ISBN: 978-3-527-31892-6

the multitude of catalytic reactions (either directly or prepared *in situ*) described in the main part of this volume. A special issue of *Coordination Chemistry Reviews* was devoted to the chemistry of vinylidene, allenylidene and cumulenylidene complexes [15]. Specific reviews of vinylidene complexes of elements of various Periodic groups have been published: Ti, [16] Mn [17, 18], W, [19] Fe, Ru, Os, [20–23] Rh, Ir, [24] and much chemistry is summarized in the particular Group volumes in the recently published COMC 3 [25]. Applications of metal vinylidene complexes to catalysis form the major part of this volume and have been previously reviewed [26–30].

There is not sufficient space to discuss all vinylidene complexes which have been reported, for example over 200 crystal structures are listed in the CCDC. Consequently, this article largely concentrates on the chemistry of metal vinylidene complexes which has been described since 1995. Vinylidene complexes are generally available for the metals of Groups 4–9, with several reactions of Group 10 alkynyls being supposed to proceed via intermediate vinylidenes. However, few of the latter compounds have yet been isolated. This chapter contains a summary of various preparative methods available, followed by a survey of stoichiometric reactions of vinylidene-metal complexes. A short section covers several non-catalytic reactions which are considered to proceed via vinylidene complexes. The latter, however, have been neither isolated nor detected under the prevailing conditions.

1.2
Preparative Methods

The main synthetic approaches to metal-vinylidene complexes will be discussed under the following headings:

1. From 1-alkynes via a 1,2-hydrogen shift.
2. The η^2-alkyne → hydrido(η^1-alkynyl) → vinylidene transformation.
3. From metal alkynyls.
4. From metal allenylidenes via metal alkynyls.
5. By deprotonation of metal-carbyne complexes.
6. From metal-carbon complexes.
7. From acyl complexes.
8. From metal-vinyl complexes.
9. From alkenes.
10. Miscellaneous methods involving the use of other precursors.
11. Vinylvinylidenes.

1.2.1
From 1-Alkynes (Equation 1.3)

$$ML_n + HC\equiv CR \rightarrow L_nM=C=CHRML_n \qquad (1.3)$$

This is the most common route to vinylidene complexes and occurs in reactions of the 1-alkynes with metal complexes, preferably with labile neutral or anionic ligands, which give neutral or cationic complexes, respectively. In the latter case, halide is commonly extracted, either by spontaneous displacement by a polar solvent, or by using sodium, silver or thallium salts.

Isomerization of the 1-alkyne to vinylidene generally occurs at d^6 metal centers which are coordinatively unsaturated. The relative stability of the vinylidene complex increases with rising electron density at the metal center. The kinetics and mechanisms of the various reactions have been studied extensively, commonly accepted mechanisms being initial formation of an η^2-alkyne complex which then undergoes either a 1,2-H migration with concomitant formation of the η^1-vinylidene directly, or an oxidative addition to form a hydrido(alkynyl) complex, from which the vinylidene ligand is formed by a 1,3 H-shift. A variety of theoretical treatments of the course of this reaction have appeared, a recent comprehensive summary by Wakatsuki [31] providing a useful starting point for discussion.

The reactions on Rh/Ir usually proceed via oxidative addition to give hydrido (alkynyl) complexes, which then undergo 1,3-H shifts to form the vinylidene complexes. In general, a unimolecular mechanism has been considered to be operative. Recent studies of $RhCl(PPr^i_2R)_2$ ($R = C \equiv NCBu^t = CHNMe$) complexes have shown a remarkable acceleration of the isomerization, with the =C=CHBu complex being formed within seconds [32]. Suitable cross-over experiments showed that a bimolecular mechanism, earlier suggested by some experimental and computational results [33], did not operate.

A range of metal-ligand centers has been shown to facilitate the formation of vinylidene complexes from 1-alkynes, a selection of recent examples being given in Table 1.1. In some cases, the preparation of the vinylidene is improved by deprotonation of the initial product to give the corresponding alkynyl compound, which can be reprotonated (see next section). Syntheses of both cationic and neutral vinylidene complexes can be achieved, the former by displacement of halide or triflate in a polar solvent, or of a labile neutral ligand, such as dinitrogen in $ReCl(N_2)(dppe)_2$. Neutral vinylidene complexes are obtained by the latter route and offer the opportunity of exchange of halide for other anionic ligands (for example).

Efficient syntheses of ruthenium complexes from readily available starting materials, such as $RuCl_2(=C=CH_2)(L)_2$ from $\{RuCl_2(cod)\}_n$, H_2 and PPr^i_3 (L) in 2-butanol with C_2H_2 [34] or $RuCl_2(=C=CHBu^t)(PCy_3)(Imes)$ from $\{RuCl_2(p\text{-cymene})\}_2$, [ImesH]Cl, $NaOBu^t$ and $HC \equiv CBu^t$ [35], have been described. Reactions of allyl-Ru complexes with 1-alkynes in the presence of HCl result in ready displacement of the allyl group and formation of neutral complexes $RuCl(=C=CHR)(PPh_3)Cp'$ [36, 37]. Alternatively, complexes containing hemi-labile ligands, such as $PPr^i_2\{CH_2C(O)OMe\}$, $PPr^i_2(CH_2CH_2OMe)$, and $PPr^i_2(CH_2CH_2NMe_2)$ ($=P \sim O$, $P \sim N$), can be used to generate a vacant coordination site for the incoming vinylidene [38]. Reactions of $[Ru(PPh_3)\{\kappa^2\text{-}PPh_2(CH_2CH=CH_2)\}(\eta^5\text{-}C_9H_7)]^+$ with $HC \equiv CAr$ give $[Ru(=C=CHAr)(PPh_3)\{\kappa^1\text{-}PPh_2(CH_2CH=CH_2)\}(\eta^5\text{-}C_9H_7)]^+$ [39]. In Group 9, other starting materials include $\{RhCl(L)_2\}_2$ and $IrH_2Cl(L)_2$ (the complex $IrCl(L)_2$ is inaccessible). Direct reaction of $Rh(C \equiv CH)(\eta\text{-}C_2H_4)(L)_2$ with $HC \equiv CH$ in

Table 1.1 Some metal-vinylidene complexes, $L_nM=C=CRR'$, prepared from 1-alkynes.

Metal	ML_n
Co	$Co(PBu^t_2CH_2CH_2\text{-}\eta\text{-}C_5H_4)$ [56]
Fe	$Fe(CO)_2(PEt_3)_2$ [57], $Fe(PP)Cp*$ [58], $FeCl(PP)_2$ [59]
Ir	$Ir\{N(SiMe_2CH_2PPh_2)_2\}$ [60]
Mn	$Mn(CO)_3(PP)$ [61], $Mn(CO)_2Cp$ [5, 6]
Mo	$Mo(PP)(\eta\text{-}C_7H_7)$ [62]
Os	$Os(CO)(NO)(PMeBu^t_2)_2$ [63], $Os(CO)(L)Cp$ [64, 65], $OsX(pcp)$ [66, 67], $OsXP_2$ [68], $OsClPCp$ [69], $OsHX_2P_2$ [70–72], $OsP_2(\eta\text{-arene})$ [73, 74], OsP_2Cp' [75–77]
Re	$Re(CO)_3P_2$ [78], $Re(CO)_2\{MeC(CH_2PPh_2)_3\}$ [79–81], $Re(CO)_2Cp$ [82], $ReCl(PP)_2$ [42]
Rh	$RhX(L)_2$ [40, 52]
Ru	$RuXLP_2$ [83], $RuHXP_2$ [84–86], RuP_2L^{OEt} [87], $RuXP(N_3)$ [88], $Ru(pnp)$ [89], $RuCl(NN)P_2$ [46, 47], $RuCl(P\sim O)_2$ [90–93], $RuClL_2$ [48–50, 94, 95], $RuCl(PP)_2$ [53, 54, 96–100], $Ru(CO)LCp'$ [101–104], $Ru(L)PCp$ [105, 106], RuL_2Cp' [39, 45, 107, 108], $RuClPCp'$ [109–111], RuP_2Cp' [112–120], RuP_2Cp* [44, 121, 122], RuL_2Cp* [123–125], $RuXPTp$ [111, 125], RuL_2Tp' [126, 127], RuP_2Tp' [127–130], $RuClP\{O_2CCH(dmpz)_2\}$ [47], $Ru(P\sim O)(\eta\text{-arene})$ [131], $RuClP(\eta\text{-arene})$ [132]
Tc	$TcCl(PP)_2$ [133]
W	$W(CO)_3(PP)$ [134]

the presence of NEt$_3$ affords trans-$Rh(C\equiv CH)(=C=CH_2)(L)_2$ [40] while the more basic $[Rh(PMe_3)_4]Cl$ reacts directly with 1-alkynes by C–H activation and oxidative addition to give $[RhH(C\equiv CR)(PMe_3)_4]Cl$, no migration of H to the metal being observed [41].

A variety of substituents can be tolerated (usually H, alkyl, aryl, SiMe$_3$, CO$_2$R), but in some cases, intramolecular nucleophilic attack on a presumed intermediate vinylidene complex results in the formation of carbene complexes. Ready replacement of SiMe$_3$ by H makes $HC\equiv CSiMe_3$ an attractive precursor for the parent complexes containing $=C=CH_2$ ligands. However, the strongly nucleophilic character of the OH group in hydroxyalkyl-alkynes often results in rapid addition to C$_\alpha$. If the alkyl chain is long enough, cyclic oxacarbenes can be formed; if not, then intermolecular attack on a second molecule results in binuclear derivatives, which may contain both carbene and vinylidene functions (see Section 1.5).

Some notable complexes which have been reported include $[Re(CN)(=C=CHPh)(dppe)_2]^+$ from aminocarbene $[ReCl(CNH_2)(dppe)_2]^+$ and $HC\equiv CR$ [42]; $[Ru\{=C=CHC\equiv C[RuCl(CO)_2(PPh_3)_2]\}(PPh_3)_2Cp]PF_6$, from $RuCl(C\equiv CC\equiv CH)(CO)_2(PPh_3)_2$ and $[Ru(thf)(PPh_3)_2Cp]PF_6$ [43]; the fullerene derivative $RuCl\{(R)\text{-prophos}\}(\eta^5\text{-}C_{60}Me_5)$ can be converted to $[Ru(=C=CHPh)\{(R)\text{-prophos}\}(\eta^5\text{-}C_{60}Me_5)]^+$ with good diastereoselectivity [44].

Although the majority of ruthenium complexes contain tertiary phosphines as co-ligands, N-donor ligands are present in complexes obtained with Ru(tmeda)Cp [45], $RuCl(Me_2bpy)(PPh_3)_2$ [46], $RuCl(L)$ [L = (dmpz)$_2$-acetate [47], 2,6-(dmpz)$_2$-

pyridine [48], 2,6-(NMe$_2$CH$_2$)$_2$C$_5$H$_3$N [49]] fragments. Structurally characterized macrocyclic ruthenium vinylidene complexes include RuCl{=C=CH(C$_6$H$_4$X-4)} (16-tmc) (R = H, Cl, Me, OMe; 16-tmc = tetramethyl-1,5,9,13-tetra-azacyclohexadecane) [50], while [Ru(N$_4$Me$_8$)(=C=CHR)]$^{2-}$ (N$_4$Me$_8$ = octamethylporphyrinogen; R = H, Ph) have been obtained from the anionic [Ru(N$_4$Me$_8$)]$^{2-}$ and ethyne or HC≡CPh, respectively [51]. In this case, formation of an intermediate η^2-alkyne complex is unlikely, the probable mechanism being deprotonation of the alkyne and coordination of the alkynyl anion followed by proton transfer.

Complexes containing several vinylidene-ruthenium fragments attached to branching organic centers are formed from suitable poly(ethynyl) precursors containing branching organic centers, such as HC≡C–X–C≡CH [X = 1, 4-HOC$_6$H$_4$OH, 1,4-C$_6$H$_4${CPh(OH)}$_2$] and {RhCl(L)$_2$}$_2$ [52], or from 1,3,5-tri(alkynyl)benzenes (triskela) [53], while convergent syntheses of polynuclear dendrimer complexes have also been described [54]. Reactions between HC≡CR (R = Bu^t, Ph) and {RuCl(η-C$_2$H$_4$)(PCy$_3$)}(µ-Cl)$_3${Ru(η^6-p-cymene)} afford {RuCl(=C=CHR)(PCy$_3$)}(µ-Cl)$_3${Ru(η^6-p-cymene)}; ethyne gives an unusual tetranuclear µ-carbido complex [55].

1.2.1.1 Migration of Other Groups (SiR$_3$, SnR$_3$, SR, SeR)

Although the vast majority of the reactions involving 1-alkynes proceed by 1,2-migration of the terminal H atom, other groups have been found to participate in this transfer. The nature of the other substituent on the 1-alkyne is often crucial, in some cases, for example, the presence of H providing a high kinetic barrier to the alkyne/vinylidene rearrangement. This barrier is lowered by the presence of Group 14 substituents, such as SiMe$_3$ or SnPh$_3$, with well-documented examples of facile 1,2-migration of the heavier groups. The Group 14 substituent may be replaced by H *in situ* by conventional means, such as treatment with [NBu$_4$]F.

An early example was provided by the reactions between {RhCl(L)$_2$}$_n$ and RC≡CSiMe$_3$ [R = Me, Ph, CO$_2$Et, CO$_2$SiMe$_3$, CH$_2$OH, C(O)CHPh$_2$] [135]. Kinetic studies carried out with FcC≡CSiMe$_3$ in the same reaction confirmed the 1,2-migration of the SiMe$_3$ group [136]. Similar silyl migration reactions have been found with C$_2$(SiMe$_3$)$_2$ and Ru(OTf)(NO)(L)$_2$ [137], Co(η-C$_5$H$_4$CH$_2$CH$_2$PBu$^t_2$) [56], IrCl(N$_2$)(PPh$_3$)$_2$ [138] and IrPh$_2$(N$_2$)Tp* [139]. For the former, Me$_3$SiC≡CC≡CSiMe$_3$ affords IrCl{=C=C(SiMe$_3$)C≡CSiMe$_3$}(PPh$_3$). The reaction of an excess of C$_2$(SiMe$_3$)$_2$ with Ru(NCMe)$_2${(C$_2$B$_{10}$H$_{10}$)CMe$_2$(η-C$_5$H$_4$)} afforded the first bis(vinylidene)ruthenium complex, Ru{=C=C(SiMe$_3$)$_2$}$_2${(C$_2$B$_{10}$H$_{10}$)CMe$_2$(η-C$_5$H$_4$)} [108].

A sub-set of these reactions is provided by the redox rearrangements of several complexes which have been extensively studied by Connelly and coworkers [140]. Oxidation of the η^2-alkyne complexes M(η^2-Me$_3$SiC$_2$SiMe$_3$)(CO)$_2$(η-arene) (M = Cr, Mo) results in formation of the vinylidene cations [M{=C=C(SiMe$_3$)$_2$}(CO)$_2$(η-arene)]$^+$.

Reactions of RC≡CSnMe$_3$ with MnCp'(η^6-C$_7$H$_8$) in the presence of dmpe give Mn{=C=CR(SnMe$_3$)}(dppe)Cp', while with Ph$_3$SnC≡CC≡CSnPh$_3$, the alkynylvinylidene Mn{=C=C(SnPh$_3$)C≡CSnPh$_3$}(dmpe)Cp' is first formed. Subsequent irradiation then affords butatrienylidene Mn{=C=C=C=C(SnPh$_3$)$_2$}(dmpe)Cp'

[141–143]. Stannylalkynes and {RhCl(L)$_2$}$_2$ afford many Rh{=C=CR(SnPh$_3$)}(L)$_2$ complexes directly, which react with H$^+$ to cleave the SnPh$_3$ group [144].

The reaction of C$_2$(SMe)$_2$ with RuCl(PMe$_3$)$_2$Cp gives [Ru{=C=C(SMe)$_2$}(PMe$_3$)$_2$Cp]$^+$ via the η^2-C$_2$(SMe)$_2$ complex [145]. A 1,2-shift of SeR occurs in the reaction between RuCl(PPh$_3$)$_2$Cp and PhC≡CSePr^i in the presence of Na[BPh$_4$], which affords [Ru{=C=CPh(SePr^i)}(PPh$_3$)$_2$Cp]$^+$ [146].

1,2-Halogen shifts have been found for tungsten, with assumed formation of iodovinylidenes in reactions of 1-iodo-1-alkynes with W(CO)$_5$(thf) en route to cyclization of 2-(iodoethynyl)styrenes to naphthalenes and of iodo-alkynyl silyl enol ethers [147], while more substantial confirmation is found in Mn{=C=C(I)CH(OR)$_2$}(CO)$_2$Cp [R = Me, Et; (OR)$_2$ = O(CH$_2$)$_3$O], of which the XRD structure of Mn{=C=C(I)CH(OMe)$_2$}(CO)$_2$Cp was determined [148].

1.2.2
The η^2-Alkyne → Hydrido(η^1-Alkynyl) → Vinylidene Transformation (Equation 1.4)

$$ML_n + CH\equiv CR \longrightarrow L_nM \longleftarrow \overset{H}{\underset{R}{\overset{|}{\underset{|}{\overset{C}{\underset{C}{|||}}}}}} \longrightarrow L_nM\overset{H}{\underset{\underset{R}{C}}{\overset{|}{\underset{|||}{C}}}} \longrightarrow ML_n=C=CHR \qquad (1.4)$$

Formation of the intermediate η^2-alkyne complex has been reported in some reactions of 1-alkynes with metal centers, followed by rearrangement to the η^1-vinylidene. This occurs but rarely in the ruthenium series, for example, with Ru(PMe$_2$Ph)$_2$Cp, where both η^2-alkyne and vinylidene isomers of the product from C$_2$H$_2$ have been structurally characterized [149], and Ru(dippe)Cp*, where metastable [RuH(C≡CR)(dippe)Cp*]$^+$ (R = Ph, CO$_2$Me, SiMe$_3$) transform into [Ru(=C=CHR)(dippe)Cp*]$^+$ in solution or the solid state [123]. Direct conversion of [Ru(η^2-HC$_2$Ph)(CO)(PMePr$^i_2$)Cp*]$^+$, prepared at −40 °C, to [Ru(=C=CHPh)(CO)(PMePr$^i_2$)Cp*]$^+$ occurs upon warming to 25 °C [104]. In contrast, the alkyne complex predominates in the room temperature solution equilibrium of the PPh$_3$ analog [150].

The transformation predominates in Group 9 (Rh, Ir) chemistry. Reactions of {RhCl(L)$_2$}$_2$ with 1-alkynes give the η^2-alkyne complexes which slowly convert to the hydrido(alkynyl)s at room temperature. The latter are sensitive to air and not often isolated. Addition of pyridine affords RhHCl(C≡CR)(py)(L)$_2$, which readily lose pyridine in hydrocarbon solvents to give square-planar trans-RhCl(=C=CHR)(L)$_2$. Alternatively, the Cp complexes Rh(=C=CHR)(L)Cp can be obtained by reaction of the chloro complexes with TlCp. In the iridium series, heating for 36 h in refluxing toluene afforded the vinylidenes in 80–90% yields. Table 1.2 lists several examples of reactions in which the η^2-alkyne complexes have been detected.

1.2.3
From Metal Alkynyls

In contrast to the alkynyl anion, coordination to a metal center results in C$_\alpha$ being electron-poor and subject to frontier-orbital controlled nucleophilic attack, while the

Table 1.2 Some vinylidene complexes, $L_nM=C=CRR'$, obtained by η^2-alkyne → hydrido(η^1-alkynyl) → vinylidene transformations.

Metal	ML_n
Co	Co{P(CH$_2$CH$_2$PPh$_2$)$_2$}[151]
Fe	Fe(CO)$_2$(PEt$_3$)$_2$ [152]
Ir	IrCl(PPr$^i_2$CH$_2$CH$_2$OMe/NMe$_2$)$_2$ [153], IrClP$_2$ [138, 154–156]
Mn	Mn(CO)$_2$Cp [157]
Os	Os(CO)(NO)P$_2$ [63], OsHX(L)P$_2$ [158], OsXPCp [159], OsP$_2$Cp [160]
Rh	RhCl(PPr$^i_2$CH$_2$CH$_2$OMe/NMe$_2$)$_2$ [161, 162], RhXP$_2$ [32, 40, 163–166]
Ru	Ru(tmeda)Cp [45], RuCl(P)Cp [36], RuHCl(CO)P$_2$ [158], RuP$_2$Cp [149, 167]
W	W(CO)$_5$ [168], W(CO)$_3$(PP)[169]

metal center and C_β are electron-rich and electrophilic attack is charge controlled. Consequently, a common route to vinylidene complexes is addition of electrophiles (E^+) to pre-formed neutral alkynyl-metal complexes, taking advantage of the polarization of the M−C≡C− fragment so that C_β is the preferred site of attack (Equation 1.5):

$$L_nM-C\equiv CR + E^+ \rightarrow [L_nM=C=C(E)R]^+ \quad (1.5)$$

The alkynyl-metal complexes are strong carbon bases, with measured pK_a values for M(C≡CBu^t)L$_2$Cp being 13.6 [ML$_2$ = Fe(CO)(PMe$_3$)] and 20.8 [ML$_2$ = Ru(PMe$_3$)$_2$] [170].

Table 1.3 lists several examples of ML_n groups supporting this reaction.

While protonation affords the vinylidene expected by H migration from the original 1-alkyne, use of other electrophiles provides a convenient route to disubstituted vinylidenes. The stereospecificity of this reaction with Re(C≡CR)(NO)(PPh$_3$)

Table 1.3 Some metal vinylidene complexes, $L_nM=C=CRR'$, obtained from alkynyl-metal systems.

Metal	ML_n
Cr, Mo, W	M(CO)(NO)Cp [173–177]
Fe	FeP$_4$ [178], Fe(PP)Cp [179–183]
Ir	IrCl(L)$_2$ [154]
Mn	Mn(CO)$_2$Cp [184], Mn(PP)Cp [185]
Mo	Mo(CO)(P)$_2$Cp [186], MoH$_3$(PP)$_2$ [187]
Nb	NbLCp$^{Si}_2$ [188]
Os	OsH(GePh$_3$)(L)Cp [189]
Pt	PtMeP$_2$ [190]
Re	Re(NO)(PPh$_3$)Cp* [191, 192]
Rh	RhCl{N(CH$_2$CH$_2$PPh$_2$)$_3$} [193, 194], RhClP$_2$ [195]
Ru	Ru(PO)$_4$ [196], RuP$_2$Cp' [171, 197–201], RuP$_2$Tp [202, 203]
W	W(CO)$_5$ [204, 205], W(CO)$_2$Cp [186], WI(O)Tp [206]

Cp has been discussed [170b]. Alkylation with haloalkanes (often iodoalkanes), triflates (alkyl, benzyl, cyclopropyl), or $[R_3O]^+$ (R = Me, Et) is often the best entry to vinylidenes of any particular system. Other common electrophiles, such as halogens (Cl, Br, I), acylium ($[RCO]^+$), azoarenes ($[ArN_2]^+$), tropylium ($[C_7H_7]^+$), triphenylcarbenium (trityl, $[CPh_3]^+$), arylthio (ArS) and arylseleno (ArSe) have also been used.

Several complexes have been obtained from reactions of alkynyl anions, such as [M(C≡CR)(CO)$_5$]$^-$ or [M(C≡CR)(CO)(NO)Cp]$^-$ (M = Cr, Mo, W), obtained from M(CO)$_2$(NO)Cp and LiC≡CR, or [Mn(C≡CR)(CO)$_2$Cp]$^-$, under charge-control.

Alkylation may sometimes afford unusual complexes as a result of subsequent reactions (see also below). Reactions of Ru(C≡CPh)(PPh$_3$){P(OMe)$_3$}Cp with halides XCH$_2$R (R = CN, Ph, C$_6$F$_5$, C$_6$H$_4$CN-4, C$_6$H$_4$CF$_3$-4, 1-nap, CO$_2$Me) give Ru{P(O)(OMe)$_2$}(=C=CPhCH$_2$R)(PPh$_3$)Cp via an initial cationic vinylidene which loses MeX in an Arbuzov-like reaction [171]. A similar reaction with RNCS gives Ru{P(O)(OMe)$_2$}{=C=CPh(SH=NR)}(PPh$_3$)Cp with low yields, which can be improved by working at higher temperatures [172].

1.2.3.1 Some Specific Examples

An interesting series of bimetallic vinylidene complexes is formed in reactions of [W(C≡CBu^t)(CO)$_5$]$^-$ with cationic hydrocarbon-metal carbonyls, such as [M(CO)$_3$(η-C$_7$H$_7$)]$^+$ (M = Cr, W), [Mn(CO)$_3$(η-C$_6$H$_6$)]$^+$, [Re(CO)$_4$(η-C$_2$H$_4$)]$^+$, [Fe(CO)$_3$(η^5-C$_6$H$_6$R)]$^+$ or [Fe(CO)$_3$(η^5-C$_7$H$_9$)]$^+$ [207]. Reaction of the heterocumulene CO$_2$ with [W(C≡CR)(CO)$_3$(dppe)]$^-$ gives Li[W(=C=CRCO$_2$)(CO)$_3$(dppe)] which can be alkylated with [Me$_3$O]$^+$ to neutral W{=C=CR(CO$_2$Me)}(CO)$_3$(dppe) [208].

Reactions of M(CO)$_2$(NO)Cp (M = Mo, W) with LiC≡CR [R = Ph, Bu^t, CH$_2$OCH$_2$CH=CH$_2$, (CH$_2$)$_2$OSiMe$_2$Bu^t] give Li[M(C≡CR)(CO)(NO)Cp] which react further with electrophiles to give either vinylidene or η^2-alkyne complexes. The former are obtained when aqueous acids (HCl) or MeOTf (hard electrophiles) are used [174]. The parent complex W(=C=CH$_2$)(CO)(NO)Cp was formed when the product from LiC≡CSiMe$_3$ was quenched with aqueous NaHCO$_3$. Treatment of the vinylidene with LiBu reforms the alkynyl anion [174, 175]. η^2-Alkyne complexes are formed with soft electrophiles, such as SiClMe$_3$, their formation resulting from the initial product by thermal isomerization [209]. In the case of Mo(=C=CHBu^t)(CO)(NO)Cp, depending on solvent, tautomerization may occur via either a 1,2-H shift (non-polar) or by a multi-step route involving deprotonation/protonation and reductive elimination (in EtOH).

Reactions of [Mn(C≡CR)(CO)(PPh$_3$)Cp^{Me}]$^-$ (R = Me, Pr) with electrophiles such as H$^+$, RI (R = Me, Et, Bu^t), MeOTf, [Et$_3$O]$^+$, RC(O)Cl (R = Me, Ph), (tol)NCO, Ph$_2$C=C=O, CO$_2$ and PhCH=CHCOMe give directly the neutral complexes Mn(=C=CEMe)(CO)(PPh$_3$)Cp^{Me} [E = H, alkyl, RC(O), C(O)NH(tol), C(O)CHPh$_2$, C(O)OMe (after treatment with MeOTf) and CHPhCH$_2$C(O)Me, respectively] (Scheme 1.1) [184, 210]. Aldehydes and ketones react with the propynyl anion to give vinylcarbyne cations after hydroxide elimination, which react with bulky nucleophiles (PPh$_3$) to give vinylidenes [211]. Similarly, the BF$_3$ adducts of epoxides react with [Mn(C≡CMe)(CO)(PPh$_3$)Cp^{Me}]$^-$ to afford anionic [Mn

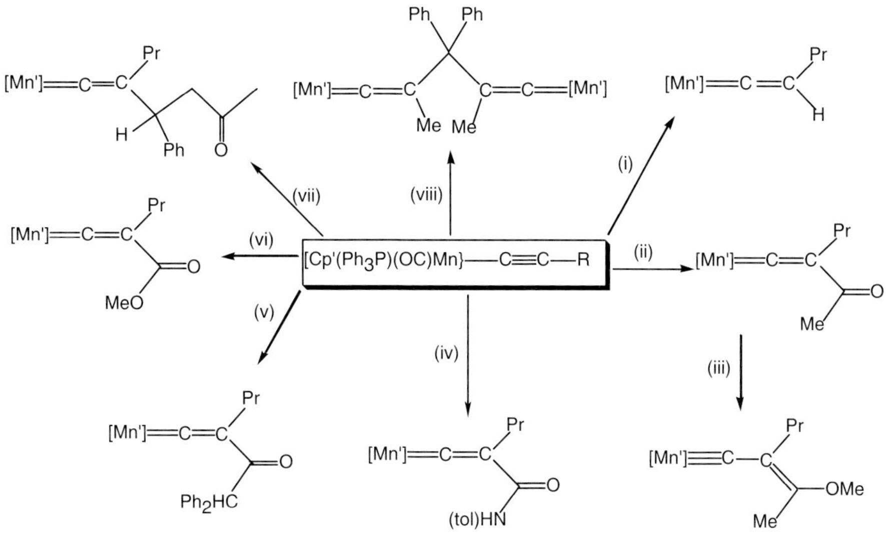

Scheme 1.1 [Mn'] = Mn(CO)(PPh$_3$)Cp'. Reagents: (for R = Pr, Cp' = Cp^{Me}) (i) H$^+$; (ii) MeC(O)Cl; (iii) MeOTf; (iv) NCO, then H$^+$; (v) Ph$_2$C=C=O, then H$^+$; (vi) CO$_2$, then MeOTf; (vii) PhCH=CHC(O)Me; (viii) (for R = Me, Cp' = Cp) [Mn(≡CCMe=CPh$_2$)(CO)(PPh$_3$)Cp]$^+$.

{=C=CMeCH$_2$CMe$_2$(CH$_2$)$_n$O(BF$_3$)}(CO)(PPh$_3$)Cp^{Me}]$^-$ (n = 0, 1, respectively), possibly via intermediate hydroxyalkyl-vinylidenes Mn{=C=CMeCH$_2$CMe$_2$(OH)}(CO)(PPh$_3$)Cp^{Me} which undergo intramolecular attack at C$_\alpha$ [210].

Protonation of Ru{C≡CCPh$_2$(C$_2$H[Co$_2$(CO)$_6$])}(PPh$_3$)$_2$(η^5-C$_9$H$_7$) and (E,Z)-Ru{C≡CCH=CH(C$_2$Ph[Co$_2$(CO)$_6$])}(PPh$_3$)$_2$(η^5-C$_9$H$_7$) gives the corresponding vinylidenes [212]. The complex trans-Rh(C≡CH)(η-C$_2$H$_4$)(L)$_2$ is protonated with [pyH]BF$_4$ to give trans-[Rh(=C=CH$_2$)(py)(L)$_2$]$^+$ and reacts with cyclopentadiene to give Rh(=C=CH$_2$)(L)Cp [40].

1.2.3.2 Redox Rearrangements of Metal Alkynyls and Vinylidenes

Oxidative coupling of metal alkynyls to give binuclear bis(vinylidene) complexes is exemplified by ML$_n$ = Nb(C≡CPh)Cp^{Si}$_2$ [213], Fe(PP)Cp*[214] or Ru(PP)Cp* (Equation 1.6): [215]

$$2L_nM-C\equiv CR \rightarrow [L_nM=C=CR-CR=C=ML_n]^{2+} \quad (1.6)$$

Suitable oxidizing agents are [FeCp$_2$]$^+$ or Ag$^+$, while the cationic species may be reduced back to the alkyne complexes using CoCp$_2$. Some of this chemistry has been reviewed [216].

Oxidation of trans-RuCl(C≡CCHPh$_2$)(dppe)$_2$ favors hydrogen atom transfer leading to trans-[RuCl(=C=CHCHPh$_2$)(dppe)$_2$]$^+$ [217]. Chemical oxidation of Ru(C≡CRc)(PPh$_3$)$_2$Cp (Rc = ruthenocenyl) gives the cyclopentadienylidene-ethylidene

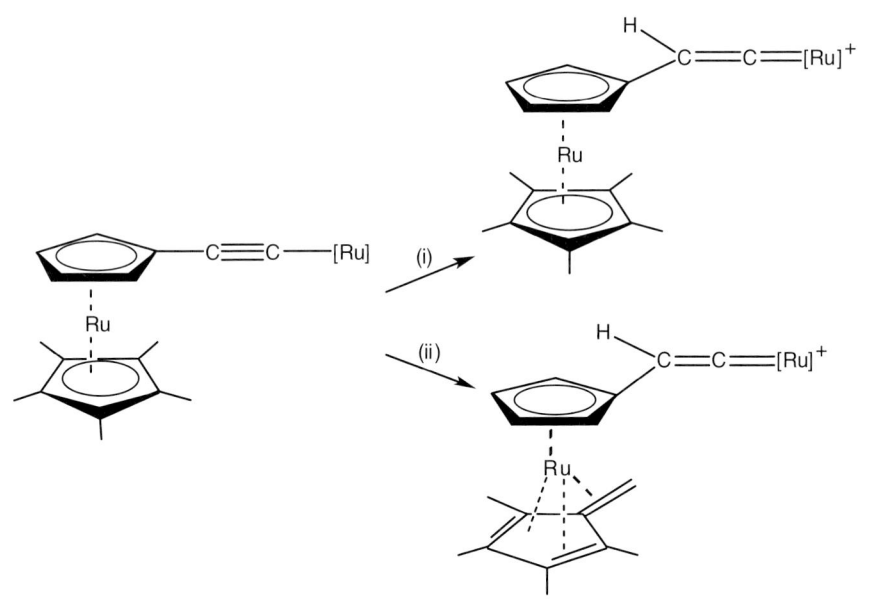

Scheme 1.2 [Ru] = Ru(PPh$_3$)Cp. Reagents: (i) [FeCp$_2$]$^+$; (ii) *p*-benzoquinone.

complex [Ru{η^1:η^6-(=C=C=C$_5$H$_4$)RuCp}(PPh$_3$)$_2$Cp]$^+$, while Ru{C≡C-η-C$_5$H$_4$)RuCp*}(PPh$_3$)$_2$Cp gives successively [Ru{=C=CH(η-C$_5$H$_4$)RuCp*}(PPh$_3$)$_2$Cp]$^+$ and the fulvene-vinylidene [Ru{=C=CH(η-C$_5$H$_4$)Ru(η^6-C$_5$Me$_4$CH$_2$)}(PPh$_3$)$_2$Cp]$^{2+}$ (Scheme 1.2) [218].

Examples of oxidative dehydro-dimerisation of vinylidenes are found with Mo(PP)(η-C$_7$H$_7$) [219], M(CO)$_2$Cp (M = Mn, Re) [220, 221] or Mn(PP)Cp^{Me} [185] centers (Equation 1.7):

$$2L_nM=C=CHR \rightarrow L_nM=C=CR-CR=C=ML_n \qquad (1.7)$$

For M(CO)$_2$Cp, this reaction proceeds via a 16-e alkynyl cation radical [M(C≡CPh)(CO)$_2$]$^+$, which, for M = Mn, couples at C$_\beta$ to give the bis(carbyne) dication [220]. For M = Re, a similar cation radical is formed, which with NEt$_3$ affords a mixture of {Cp(OC)$_2$Re}=C=CPhCPh=C={Re(CO)$_2$Cp} and the isomeric μ-vinylidene {Cp(OC)$_2$Re}$_2${μ-C=CPh(C≡CPh)} by competitive C$_\beta$–C$_\beta$ and C$_\beta$–M coupling [221]. With an alternative ligand set, the Mn{=C=CR(SnMe$_3$)} complexes can be destannylated with [NBu$_4$]F before oxidative coupling to the bis(vinylidene). Reductive uncoupling also occurs, making these systems of interest as potential energy sinks [142].

Oxidation [PhIO or Cu(OAc)$_2$] of [Fe(=C=CHMe)(dppe)Cp]$^+$ affords bis(vinylidene) [{Fe(dppe)Cp}$_2$(μ-C$_4$Me$_2$)]$^{2+}$, possibly via an intermediate radical cation [222]. Similar oxidative coupling of cyclopropenyl Ru{C=CPhCH(CN)}(PPh$_3$)$_2$Cp affords bis(vinylidene) [{Cp(Ph$_3$P)$_2$Ru}=C=CPhCH(CN)$_2$}$_2$]$^{2+}$ which, in turn, can be deprotonated to the bis(cyclopropenyl) [223]. Oxidation of [Ru(N$_4$Me$_8$)(=C=CH$_2$)]$^{2-}$ with PhN$_3$ or [FeCp$_2$]$^+$ affords [{Ru(N$_4$Me$_8$)}$_2$(μ-C=CHCH=C)]$^{4-}$ [51].

1.2.4
From Metal Allenylidenes via Metal Alkynyls

Ready addition of nucleophiles (Nu$^-$) to metal-allenylidene complexes affords alkynyl derivatives. Subsequent protonation or alkylation, as described in Section 1.2.3 above, then gives the corresponding vinylidene complexes (Equation 1.8):

$$[L_nM=C=C=CRR']^+ + Nu^- \rightarrow L_nM-C\equiv CC(Nu)RR'$$
$$\xrightarrow{H^+} [L_nM=C=CHC(Nu)RR']^+ \quad (1.8)$$

For example, addition of LiBu^t to Mn(=C=C=CPh$_2$)(CO)$_2$Cp, followed by protonation or methylation gives Mn{=C=CR(CPh$_2$Bu^t)}(CO)$_2$Cp^{Me} (R = H, Me) [224]. This is a method which has not been employed generally, and could perhaps be exploited to produce more unusual derivatives.

In the ruthenium series, more success has been reported, particularly with RuP$_2$Cp′ (Cp′ = Cp∗ [225], η^5-C$_9$H$_7$ [115, 226–232], Tp [130]). This can be rationalized in terms of a high contribution from the vinylidene to the structure of the alkenylcarbyne formed by protonation of the allenylidene (Equation 1.9):

$$[Ru]^+ = C = C = CR_2 \longrightarrow \left[[Ru]^{2+}\equiv C-C\begin{smallmatrix}CR_2\\H\end{smallmatrix} \longleftrightarrow [Ru]+=C=C\begin{smallmatrix}^+CR_2\\H\end{smallmatrix}\right]$$
$$(1.9)$$

Thus, protonation of the intermediate formed from RuCl(dippe)Cp∗ and HC≡CCHPh(OH), followed by addition of ketones, aldehydes, thiophene, dicarbonyl and related compounds affords [Ru{=C=CHCHPh(E)}(dippe)Cp∗]$^+$ [E = CH$_2$C(O)Me, CH(CH$_2$)$_3$CO, CH$_2$C(O)Ph, C=CHCH=CMeS, CH(COR)C(O)R (R = Me, OMe), CH(CN)$_2$, CH(COMe)C(O)Et] (Scheme 1.3) [225].

Regioselective Diels–Alder addition of 1,3-dienes (isoprene, cyclopentadiene, 1,3-cyclohexadiene) to C$_\beta$=C$_\gamma$ in [Ru(=C=C=CPh$_2$)(CO)(PPr^i$_3$)Cp′]$^+$ affords vinylidenes (Scheme 1.4) [232], while [Ru(=C=C=CHPh)(dippe)Cp∗]$^+$ adds nucleophiles such as pyrazole, PhSH, pyrrole or 2-methylfuran (the latter two in the presence of acid) to give [Ru(=C=CHCHPhR)(dippe)Cp∗]$^+$ (R = pz, SPh, pyr, fur) [228]. In the presence of base, acetone adds to C$_\gamma$ to give alkynyls which afford [Ru{=C=CHCRPhCH$_2$C(O)Me}(dippe)Cp∗]$^+$ with HBF$_4$ (Scheme 1.5). Regiospecific addition of [M{=CCH$_2$(OMe)}(CO)$_5$]$^-$ (M = Cr, Mo, W) to [Ru{(=C=C=CPh(R)}(PPh$_3$)$_2$(η^5-C$_9$H$_7$)]$^+$, followed by protonation, gives bimetallic [{(η^5-C$_9$H$_7$)(Ph$_3$P)$_2$Ru}=C=CHCPh(R)CH$_2$C(OMe)={M(CO)$_5$}]$^+$, containing vinylidene-carbene bridges [115].

1.2.5
From Metal-Carbyne Complexes

The deprotonation of carbyne complexes is the formal reverse of the addition of a proton to the vinylidene (Equation 1.10, Table 1.4):

$$[L_nM\equiv CCHRR']^+ + B \rightarrow L_nM=C=CRR' + [BH]^+ \quad (1.10)$$

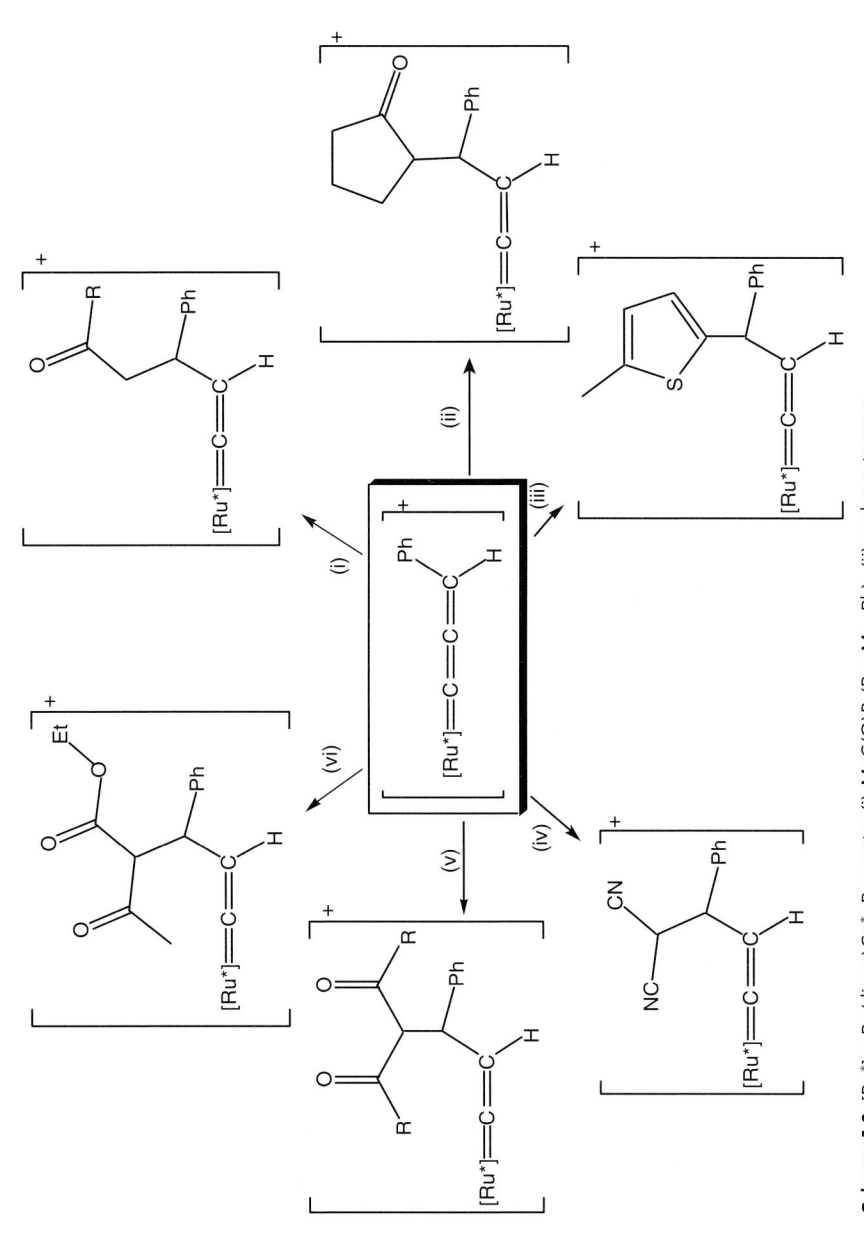

Scheme 1.3 [Ru*] = Ru(dippe)Cp*. Reagents: (i) MeC(O)R (R = Me, Ph); (ii) cyclopentanone; (iii) 2-Me-thiophene; (iv) CH$_2$(CN)$_2$; (v) RC(O)CH$_2$C(O)R (R = Me, OMe); (vi) MeC(O)CH$_2$CO$_2$Et. All reactions run in the presence of HBF$_4$.

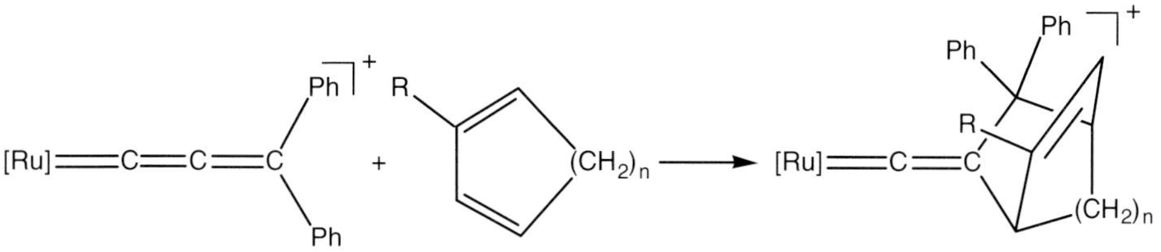

Scheme 1.4 [Ru] = Ru(CO)(PPr$^i_3$)Cp; R = Me, n = 0; R = H, n = 1,2.

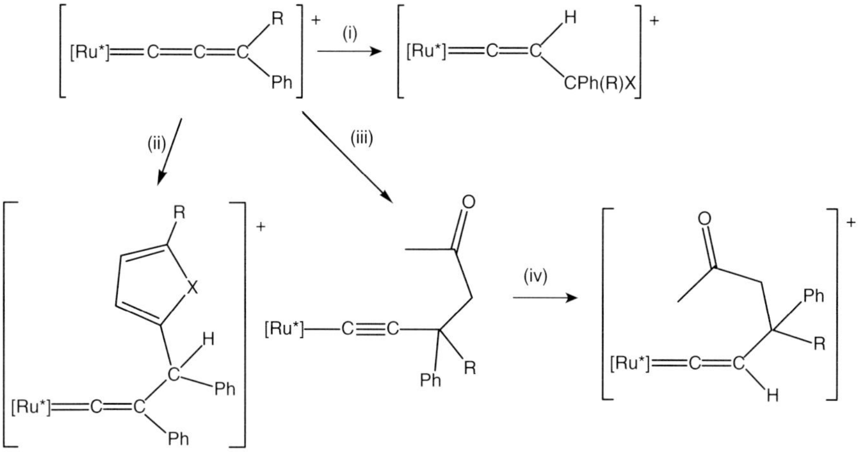

Scheme 1.5 [Ru*] = Ru(dippe)Cp*. Reagents: (i) (X = pzH, dmpzH, HSPh); (ii) pyrrole (X = NH, R = H), 2-Me-furan (X = O, R = Me); (iii) Me$_2$CO; (iv) H$^+$.

Deprotonation may be achieved by reaction with alkyllithiums, but has also been found in other circumstances. A reaction between CF$_3$I and Mo{≡C(CH$_2$Bu^t)}{P(OMe)$_3$}$_2$Cp afforded MoI(=C=CHBu^t){P(OMe)$_3$}$_2$Cp in a reaction suggested to proceed via electron transfer from the carbyne to CF$_3$I, followed by abstraction of H by the CF$_3$ radical. A related reaction occurs with [4-FC$_6$H$_4$N$_2$]BF$_4$ [233].

Table 1.4 Some vinylidene complexes, L$_n$M=C=CRR', obtained from carbynes.

Metal	ML$_n$
Mn	Mn(CO)(PPh$_3$)Cp [211, 234]
Mo	Mo(CO)(PP)$_2$ [235], Mo(CO)$_2$Tp* [236], MoP$_2$Cp [233, 237]
Os	Os(CO)$_2$P$_2$ [238], OsXPCp [239], OsXP$_2$ [240, 241], OsHLP$_2$ [242], OsHXLP$_2$ [71, 243–245]
Re	Re(CO)$_2$(p$_3$) [81], Re(CO)$_2$Cp [234], ReCl(PP)$_2$ [246]
W	W(CO)(PP)$_2$ [235, 247]

Highly nucleophilic anions [M(C≡CR)(CO)(PPh$_3$)Cp]$^-$ (M = Mn, Re) are obtained from M(CO)$_2$(PPh$_3$)Cp and LiC≡CR [211, 234]. These add to aldehydes and ketones to give vinylcarbynes [Mn(≡CCMe=CPh$_2$)(CO)(PPh$_3$)Cp]$^+$. Subsequent addition of nucleophiles to C$_\gamma$ gives neutral vinylidene complexes Mn{=C=CMeCPh$_2$(Nu)}(CO)(PPh$_3$)Cp for Nu = H or alkyl; with LiC≡CBu^t, only attack at C$_\alpha$ to give a carbene complex was observed [211]. Coupling of the carbyne with [Mn(C≡CMe)(CO)(PPh$_3$)Cp]$^-$ gave the binuclear bis(vinylidene) {[Mn(CO)(PPh$_3$)Cp]=C=CMe}$_2$CPh$_2$.

1.2.6
From Metal-Carbon Complexes

The reaction of Ru(≡C)Cl$_2$(PCy$_3$)(L′) (**1**, L′ = Imes; Scheme 1.6) with CH(C$_6$H$_4$NO$_2$-4)(CO$_2$Me)$_2$ in the presence of 4-Bu^{t}C$_5$H$_4$N (Bu^{t}py) gives Ru{=C=C(C$_6$H$_4$NO$_2$-4){CH(CO$_2$Me)$_2$}Cl$_2$(Imes)(Bu^{t}py)$_2$ [248]. Similar complexes were obtained from Feist's acid dimethyl ester in the presence of malonate, while Ru(=CHPh)Cl$_2$(Imes)(Bu^{t}py)$_2$ reacts with the ester to give Ru{=C=C(CO$_2$Me)(CHCO$_2$Me$_2$)}Cl$_2$(Imes)(Bu^{t}py)$_2$, probably via an intermediate carbide complex. A related reaction is that between Ru(≡C)Cl$_2$(PCy$_3$)$_2$ (**1**, L = PCy$_3$) and C$_2$(CO$_2$Me)$_2$ to give cyclopropenylidene Ru{=CC$_2$(CO$_2$Me)$_2$}Cl$_2$(PCy$_3$)$_2$, which ring-opens with HX to give Ru{=C=C(CO$_2$Me)$_2$}X$_2$(PCy$_3$)$_2$ [249].

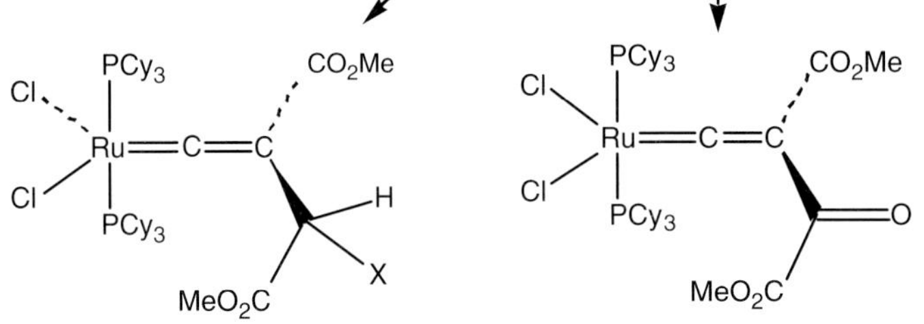

Scheme 1.6 Reagents: (i) C$_2$(CO$_2$Me)$_2$; (ii) HX [X = NHAr, OH, OPh, B(pin)]; (iii) py-N-oxide.

1.2.7
From Acyl Complexes (Equation 1.11)

$$L_nM-C(=O)R + H^+ \longrightarrow [L_nM=C(OH)R]^+ \xrightarrow{-H_2O} [L_nM=C=C(H)R]^+ \quad (1.11)$$

An early approach to vinylidenes was by the formal dehydration of metal acyls, which is best achieved by treatment with an electrophile, often the proton in the form of a non- or weakly-coordinating strong acid. The reaction appears to proceed stepwise via a hydroxycarbene formed by protonation of the acyl, subsequent dehydration of which affords the vinylidene. Occasionally, mixtures of the two complexes are obtained, again suggesting the intermediacy of the carbene.

This reaction was first described for the conversion of Fe{C(O)Me}(CO)(PPh$_3$)Cp to [Fe(=C=CH$_2$)(CO)(PPh$_3$)Cp]BF$_4$ by treatment with a HBF$_4$.OEt$_2$/triflic anhydride mixture [9a]. Transformations on Re(NO)PCp [170b] and Fe(PP)Cp [9b, 250] centers have also been reported. The complexes [Mn{C(O)CH$_2$R}(CO)$_2$Cp^{Me}]$^-$ and [M(SnPh$_3$){C(O)CH$_2$R}(CO)Cp]$^-$ (M = Fe, Ru), obtained from reactions of Mn(CO)$_3$Cp^{Me} or M(SnPh$_3$)(CO)$_2$Cp with LiCH$_2$R, could be converted to the corresponding vinylidene complexes by treatment with AcCl [251]. More recently, lithiated diarylmethanes and M(CO)$_6$ (M = Cr, W) have given anionic Li[M{C(O)CAr$_2$}(CO)$_5$] which with (CF$_3$CO)$_2$O gave reactive M(=C=CAr$_2$)(CO)$_5$ [252].

1.2.8
From Vinyls

Treatment of TaCl$_2$Cp*$_2$ with (CH$_2$=CH)MgBr afforded TaH(=C=CH$_2$)Cp*$_2$ directly in 75% yield, presumably by an α-H shift from an intermediate Ta(CH=CH$_2$)$_2$Cp*$_2$; the vinyl Ta(CH=CH$_2$)(CO)Cp*$_2$ was formed by a formal reversal of this shift on treatment with CO. UV irradiation of this vinyl reformed the vinylidene [253].

Although OsI(C≡CPh)(L)(η-C$_6$H$_6$) cannot be protonated to the corresponding vinylidene, reaction with NaBH$_4$ to give OsH(CH=CHPh)(L)(η-C$_6$H$_6$) followed by halogenation (CCl$_4$ or CH$_2$I$_2$) and reaction with LiBu^t gives Os(=C=CHPh)(L)(η-C$_6$H$_6$) [73, 74].

Several vinylidene complexes have been obtained from cyclopropenyl-ruthenium complexes Ru(C=CPhCR^{1}R^2)(PPh$_3$)$_2$Cp, themselves obtained by intramolecular cyclization of [Ru{=C=CPh(CH$_2$R^{1}R^2)}(PPh$_3$)$_2$Cp]$^+$ upon deprotonation [197, 223, 254]. Thus, ring-opening occurs when the neutral cyclopropenyls are treated with electrophiles such as CPh$_3^+$, HgCl$_2$, H$^+$ or tcnq (Section 1.3.2).

1.2.9
From Alkenes

Oxidative addition of α-chlorovinylsilanes to $Cr(CO)_3(\eta\text{-arene})$ or $Mn(CO)_3Cp^{Me}$ afforded mono- or di-substituted vinylidenes by elimination of Me_3SiX across the M−C bond: intermediate η^2-alkene complexes were detected in the Mn system, slow conversion of $Mn\{\eta^2\text{-}CBr_2{=}CMe(SiMe_3)\}(CO)_2Cp^{Me}$ to $Mn({=}C{=}CMe_2)(CO)_2Cp^{Me}$ occurring [255]. The reaction is solvent dependent, the major product from $Me_2C{=}CCl(SiMe_3)$ being the vinylidene in pentane, but butatriene $Mn(\eta^2\text{-}Me_2C{=}CC{=}CMe_2)(CO)_2Cp^{Me}$ in thf [256].

Irradiation of $M(\eta\text{-}C_2H_4)(L)Cp$ (M = Rh, Ir, L = PMe₃; M = Ir, L = $\eta\text{-}C_2H_4$) in a matrix initially affords $MH(CH{=}CH_2)(L)Cp$ and then the dihydride $MH_2({=}C{=}CH_2)Cp$ [257]. Treatment of $\{RhCl(L)_2\}_n$ or $IrCl(coe)(L)_2$ with $RR'C{=}CHCl$ [RR' = Me₂, MePh, Ph₂, (CH₂)₅] and Na in benzene gives trans-$MCl({=}C{=}CRR')(L)_2$, although the major product from $IrCl(coe)(L)_2$ and $Ph_2C{=}CHCl$ is trans-$IrCl(\eta\text{-}PhC_2Ph)(L)_2$ [258]. The reaction probably involves formation of an intermediate vinyl radical and Ph group migration. Addition of norbornadiene (nbd) to $\{IrCl(coe)_2\}_2$ gives a polymer which with PR_3 affords trans-$IrCl\{{=}C{=}CH(C_7H_9)\}(PR_3)_2$ (PR_3 = PPr^i_3, $PMeBu^t_2$) which, on heating, loses cyclopentadiene to give trans-$IrCl({=}C{=}CHR)(PR_3)_2$ (R = 5-norbornen-2-yl). It is thought that steric congestion afforded by the tertiary phosphines results in rearrangement of an initially-formed nbd-dimer complex **2** [259].

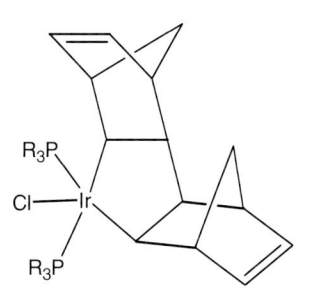

1.2.10
Miscellaneous Reactions Affording Vinylidenes

The following are some examples of reactions which have produced vinylidene complexes but are either not of general application or have not been further developed. Oxygen atom transfer occurs in reactions of $NbH(\eta^2 - OC - CPh_2)Cp'_2$ with nitriles or isonitriles to give isocyanates and $Nb({=}C{=}CPh_2)Cp'_2$ [260]. Metathesis of $Ph(R)C{=}C{=}NPh$ (R = Me, aryl) with $W({=}CHPh)(CO)_5$ proceeds via $W\{C(NPh{=}CHPh){=}C(R)Ph\}(CO)_5$, which is converted to $W\{{=}C{=}C(R)Ph\}(CO)_5$ by treatment with $BF_3.OMe_2$ [261].

Vinylidene **3** was obtained from Mn(CO)$_2$Cp and the cyclic alkynes SiMe$_2$ C≡CSiMe$_2$ (SiMe$_2$)$_n$SiMe$_2$ [262, 263]; formation of **4** from Fe$_2$(CO)$_9$ and (C≡CSiMe$_2$ OSiMe$_2$OSiMe$_2$)$_2$ may involve intramolecular addition of an iron-vinylidene (formed by SiMe$_3$ migration) to the second C≡C triple bond [262].

3

4

Unusual iron-porphyrin vinylidene complexes were obtained from DDT [1,1-bis(4-chlorophenyl)-2,2,2-trichloroethane] and Fe(tpp) [tpp = *meso*-tetraphenylporphinato (2-)] in the presence of a reducing agent [10a, 264]. The derived *N*,*N*′-vinylene-bridged porphyrin reacts with metal carbonyls [Fe$_3$(CO)$_{12}$, Ru$_3$(CO)$_{12}$] to break one or both N−C bonds with insertion of the vinylidene into an M−N bond. While the iron complex was formed in 90% yield, the reaction with Ru$_3$(CO)$_{12}$ afforded three products, the vinylidene being formed in only 40% yield [265].

Isomerization of η^2-vinyl ether complexes of OsHCl(L)$_2$ to carbenes OsHCl{=CMe(OR)}(L)$_2$ may be followed by C−OR bond cleavage to give either vinylidene OsHCl(=C=CH$_2$)(L)$_2$ (R = Et) or carbyne OsHCl(≡CMe)(OPh)(L)$_2$ (R = Ph) [266]. Rearrangement of [OsCl$_2$(CH=CHPh)(N=CR$_2$)(L)$_2$]$^+$ [CR$_2$ = CMe$_2$, C(CH$_2$)$_5$] with L′ = MeCN or H$_2$O gives [OsCl(=C=CHPh)(NH=CR$_2$)(L′)(L)$_2$]$^+$, or in the absence of another ligand, neutral OsCl$_2$(=C=CHPh)(NH=CR$_2$)(L)$_2$ [267].

1.2.11
Vinylvinylidene Complexes

An important sub-set of vinylidenes are those containing further unsaturation in conjugation with the metalla-allene system. Vinylvinylidene complexes have been obtained from methylated propargyl alcohols, where the normal dehydration to give allenylidene complexes does not occur. Thus, reactions of HC≡CCMe$_2$(OH) with metal halide complexes in the presence of [NH$_4$]PF$_6$ afford binuclear complexes such as **5** containing the vinylidene-alkylidene ligand, C$_{10}$H$_{12}$, which has been described for ML$_n$ = Ru(PPh$_3$)$_2$Cp [268, 269], Ru(dippe)Tp [127] and Os(PPh$_3$)$_2$Cp [77]. A related reaction is the formation of allenylidene [Ru{=C=C=C(C$_{13}$H$_{20}$)}(PPh$_3$)$_2$(η^5-C$_9$H$_7$)]$^+$ (**6**) from [Ru{=C=CH(C$_6$H$_9$)}(PPh$_3$)$_2$(η^5-C$_9$H$_7$)]$^+$ and HC≡C(C$_6$H$_9$) or HC≡C(C$_6$H$_{10}$)(OH); it can also be obtained directly from RuCl(PPh$_3$)$_2$(η^5-C$_9$H$_7$) and an excess of HC≡C(C$_6$H$_{10}$)OH [270].

5 [Ru] = Ru(PPh$_3$)$_2$Cp

6 [Ru] = Ru(PPh$_3$)$_2$(η^5-C$_9$H$_7$)

Z and E isomers of W{=C=C(CO$_2$Me)C(CO$_2$Me)=CHPh}(CO)$_3$(dppe) are formed from C$_2$(CO$_2$Me)$_2$ and W(=C=CHPh)(CO)$_5$ by [2+2]-cycloaddition of the alkyne to the vinylidene, followed by ring-opening of the cyclobutenyl so formed, corresponding to a formal insertion of the alkyne into the C$_\alpha$–C$_\beta$ bond. The first-formed Z isomer is photo-converted to the E isomer (Z/E = 1/8) [271]. Similar reactions are found with RuCl(=C=CHCO$_2$Me)(PEt$_3$)Tp and HC≡CCO$_2$Me [129], and with the neutral allenylidene OsCl(=C=C=CPh$_2$)(PPr$^i_3$)Cp and C$_2$(CO$_2$Me)$_2$ to give OsCl{=C=C(CO$_2$Me)C(CO$_2$Me)=C=CPh$_2$}(PPr$^i_3$)Cp [272].

Deprotonation of vinylcarbyne complexes affords vinylvinylidenes in a reaction reversed by addition of HOTf. Reactions of Mn(≡CCMe=CPh$_2$)(CO)(PPh$_3$)Cp with nucleophiles resulting in addition to C$_\gamma$ occur with larger groups such as PPh$_3$ [211]. If C$_\gamma$ bears a proton, however, facile deprotonation of the vinylcarbynes occurs to give Mn(=C=CMeCR=CR'$_2$)(CO)(PPh$_3$)Cp (R' = H, R = tol, Bu^t; R' = Me, R = H). The alkynylvinylidene [Re{=C=CHC≡CC≡C(tol)}(NO)(PPh$_3$)Cp*]$^+$ is obtained by protonation (HBF$_4$) of Re{(C≡C)$_3$(tol)}(NO)(PPh$_3$)Cp* [192].

Vinylvinylidenes FeC=C=CHCR^1=CH$_2$)(CO)$_2$(PEt$_3$)$_2$ (R^1 = Me, Et, Pr^i, Bu^t) are obtained by dehydration (silylated silica) of FeH{C≡CCMeR^1(OH)}(CO)$_2$(PEt$_3$)$_2$ [273]. Rapid dehydration occurs in the reactions of RuCl(cod)Tp with HC≡CCMePh(OH) (in the presence of PPr^{i}Ph$_2$) resulting in the formation of RuCl(=C=CHCPh=CH$_2$)(PPr^{i}Ph$_2$)Tp [130], or of OsCl$_2$(PPh$_3$)$_3$ with HC≡CCMe$_2$(OH) to give the η^2-alkyne-Os complex, which is dehydrated at 85 °C to OsCl(=C=CHCMe=CH$_2$)(PPh$_3$)Cp [159]. Reaction of OsCl(CO)(L)Cp with Ag[BF$_4$] followed by 1-ethynyl-1-cyclohexanol affords [Os{=C=CH(C$_6$H$_9$)}(CO)(L)Cp]$^+$ [64].

Alkylation of Ru{C≡C(C$_6$H$_9$)}(PPh$_3$)$_2$Cp with BrCH$_2$CO$_2$Me affords [Ru{=C=C(C$_6$H$_9$)(CH$_2$CO$_2$Me)}(PPh$_3$)Cp]$^+$ [199]. The reaction between Ru(NCMe)$_2$ {(C$_2$B$_{10}$H$_{10}$)CMe$_2$(η-C$_5$H$_4$)} and HC≡CSiMe$_3$ gives the binuclear vinylvinylidene {(C$_2$B$_{10}$H$_{10}$)CMe$_2$(η-C$_5$H$_4$)}Ru{=C=C(SiMe$_3$)CH=CH(SiMe$_3$)}{Ru(C$_2$B$_{10}$H$_{10}$)CMe$_2$(η-C$_5$H$_4$)} (**7**) [108].

7

[Ru] = Ru{η^5,η^1-(C_5H_4)CMe_2($C_2B_{10}H_{10}$)}

Enynes HC≡CCR=CHR′ (R = R′ = H; R = Me, R′ = H; R = H, R′ = OMe) react with {RhCl(coe)$_2$}$_2$ in the presence of PPr$^i_3$ to give the expected η^2-alkyne or hydrido (alkynyl) complexes, which rearrange in toluene at 25–45°C to give vinylvinylidene complexes trans-RhCl(=C=CHCR=CHR′)(L)$_2$ [274]. Similar reactions with HC≡CCHMe(X) (X = OH, NH$_2$) afford trans-RhCl{=C=CHCHMe(X)}(L)$_2$, which dehydrate or deaminate with acidic alumina or traces of acid to give the parent complex RhCl(=C=CHCH=CH$_2$)(L)$_2$ [274, 275]. Phenyl-substituted alkynols give allenylidenes [276]. While formation of hydrido(alkynyl)s occurs on treatment with py, carbonylation of the =C=CHCMe=CH$_2$ complex gives trans-RhCl(CO)(L)$_2$ and free enyne HC≡CCMe=CH$_2$. Protonation (HBF$_4$) gives carbyne trans-[Rh(≡CCH=CMe$_2$)(L)$_2$]$^+$, a reaction which is reversed with NaH. Addition of HOTf to the C$_\beta$–C$_\gamma$ bond of trans-IrCl(=C=C=C=CPh$_2$)(L)$_2$ affords trans-IrCl{=C=CHC(OTf)=CPh$_2$}(L)$_2$ [277].

1.3
Stoichiometric Reactions

The vinylidene ligand is polarized so that C$_\alpha$ is electron poor (δ^+) and the C$_\beta$ is electron rich (δ^-). The HOMO is located on the metal and C$_\beta$, while the LUMO is found on C$_\alpha$. This provides a ready distinction between the modes of reaction, although the presence of both electrophilic and nucleophilic centers in the same molecule can lead to the formation of cyclic products. Although C$_\beta$ is electron-rich, reactions with electrophiles are often limited by the positively charged metal complex and are more commonly found with the neutral species. Conversely, addition of nucleophiles to C$_\alpha$ is facilitated by the positive charge. Many vinylidenes are based on electron-rich metal-ligand fragments, one common feature being the presence of bulky phosphines. Many early studies used phenylphosphines, often PPh$_3$, and the kinetic stability afforded to the unsaturated ligand may result from steric shielding of C$_\beta$ by the phenyl groups. However, smaller tertiary phosphines, such as PMe$_3$ or dmpe, can also be used and result in increased reactivity.

Much of the chemistry of vinylidene complexes has been developed with catalytic applications in mind, as detailed later in this volume. Early examples had low activity for alkene metathesis, although complexes containing imidazolylidene ligands showed improved efficiencies [35]. However, in many cases, reactions of the vinylidene ligand have resulted in transformation to other carbon-based ligands which have not been released from the metal fragment.

The stoichiometric reactivity of metal-vinylidene complexes will be covered in the following sequence:

1. Addition to C_α – reactions of nucleophiles
2. Addition to C_β – reactions with electrophiles
3. Cycloaddition reactions.

To some extent, the reactivity varies with the metal-ligand combination. Consequently, extensive studies of vinylidenes attached to a few specific ML_n fragments have been reported, with those containing $TiCp_2$, $M(CO)_5$ (M = Cr, Mo, W), M(CO)(NO)Cp (M = Cr, Mo, W), $[M(L')(P)Cp]^+$ (M = Fe, Ru; L' = CO, P) $[Ru\{=C=C(SMe)_2\}(PMe_3)_2Cp]^+$, and $MCl(L)_2$ (M = Rh, Ir) fragments being summarized in Section 1.4.

1.3.1
Reactions at C_α

Addition of neutral nucleophiles to C_α affords the corresponding vinyl complexes, whereas anionic nucleophiles usually give carbene complexes.

1.3.1.1 Deprotonation
Nucleophilic attack resulting in deprotonation of C_β has been reported for a range of metal centers (Equation 1.12):

$$[L_nM=C=CHR]^+ + B \rightarrow L_nM-C\equiv CR + [BH]^+ \qquad (1.12)$$

ML_n is exemplified by Fe(dppm)Cp [250a], Fe(CO)(PPh$_3$)Cp [9], Fe(dppe)Cp* [278], Ru(CO)(PMePr$^i_2$)Cp* [104], trans-Ru(dppe)$_2$ [96], Ru(dippe)Tp [127], Ru(CO)(PR$_3$)(η^5-C$_9$H$_7$) [102], Ru(PMe$_2$Ph)$_2$Cp* [279], RuCl(PPh$_3$)Cp* [109, 110], OsX(L)(η^6-arene) [73, 74, 280].

Double deprotonation of Ru(=C=CH$_2$)(PPh$_3$)$_2$Cp with LiBu^t affords the lithioalkynyl Ru(C≡CLi)(PPh$_3$)$_2$Cp [281]. Reversal of the intramolecular reductive elimination leading to vinylidene is found on treatment of trans-IrCl(=C=CHR)(L)$_2$ with CO to give IrH(CO)(C≡CR)(L)$_2$ [154].

1.3.1.2 Group 16 Nucleophiles. Oxygen
A commonly used nucleophile has been water. Although initial attack affords a hydroxy-carbene derivative, ready cleavage of the $C_\alpha-C_\beta$ bond resulting from formal keto–enol tautomerism occurs to give either the acyl or the metal carbonyl (usually cationic) and the corresponding organic fragment (Equation 1.13):

$$L_nM{=}C{=}CHR \xrightarrow{H_2O} L_nM{=}C(OH)(CH_2R) \longrightarrow L_nM-C(O)(H)(R) \to RCH_2CHO$$
$$\to L_nM-C{\equiv}O \; (-RCH_3)$$

(1.13)

Examples include M = OsX(L)(η^6-arene) [73, 74, 280], PtMe(PPh$_3$)$_2$ (in situ) [190], Ru(dppe)Cp [197], Ru(CO)(PMePr$^i_2$)Cp* [104], RuCl(PPh$_3$)$_2$(Me$_2$bpy) [46], Ru(PPh$_3$)$_2$Cp [8b], RuCl{O$_2$CCH(dmpz)$_2$}(PPh$_3$) [47]. In the presence of acid, [Ru{=C=CPhCH$_2$(CN)}(PPh$_3$){P(OMe)$_3$}Cp]$^+$ gives Ru{C(O)CH$_2$Ph}(PPh$_3$){P(OMe)$_3$}Cp [171].

A study of the hydration of HC≡CPh in the presence of mer, trans-RuCl$_2$(PPh$_3$){NPr(CH$_2$CH$_2$PPh$_2$)$_2$} showed the intermediacy of vinylidene, aquo(alkynyl), acyl and benzyl(carbonyl) complexes [282].

1.3.1.3 Alcohols

Reactions with alcohols proceed similarly, but as the products cannot eliminate any fragment, formation of alkoxy-carbenes is observed (Equation 1.14):

$$L_nM{=}C{=}CRR' + R''OH \to L_nM{=}C(OR'')CHRR' \tag{1.14}$$

ML$_n$ = Fe(CO)(PPh$_3$)Cp/ROH [9b], [but H$_2$O and MeOH do not react with the Fe(=C=CMeR) complexes (R = Bu^t, Ph)] [283], RuCl$_2$(pnp) [118, 284], Ru(PR$_3$)$_2$Cp [8b], Ru{(PPh$_2$CH$_2$)$_2$CMe(η-C$_5$H$_4$)} [285], Ru(PR$_3$)$_2$Tp [127, 203], OsCl(L)(η^6-arene) [73, 74, 280].

As with vinylidenes, the cationic carbenes can be readily deprotonated to give the corresponding alkoxy-vinyl complexes, particularly with bulky bases such as Bu^{t}O$^-$ or NMe$_3$ (Equation 1.15):

$$[L_nM{=}C(OR)CHRR']^+ + B \to L_nM-C(OR){=}CRR' + [BH]^+ \tag{1.15}$$

1.3.1.4 Sulfur

Addition of RSH to vinylidene gives the thio-carbene (Equation 1.16):

$$L_nM{=}C{=}CRR' + RSH \to L_nM{=}C(SR)CHRR' \tag{1.16}$$

for example, ML$_n$ = Fe(CO)(PPh$_3$)Cp [9b, 286].

Addition of SPh$^-$ to [Fe(=C=CMe$_2$)(CO)(PPh$_3$)Cp]$^+$ gives Fe{C(SPh)=CMe$_2$}(CO)(PPh$_3$)Cp; a 93/7 Z/E mixture is formed with the =C=CMePh complex [287]. The reaction of [S$_2$CNEt$_2$]$^-$ with RuCl{=C=CH(tol)}(PPh$_3$)Tp gives Ru{C[=CH(tol)]SC(NEt$_2$)S]}(PPh$_3$)Tp [128, 287].

Treatment of TaH(=C=CH$_2$)Cp*$_2$ with Group 16 nucleophiles (H$_2$O, MeOH, RCH$_2$SH) results in elimination of C$_2$H$_4$ and formation of Cp*$_2$TaH(X) (X = O, η^2-CH$_2$O, η^2-CHRS, respectively) [280, 288].

1.3.1.5 Group 15 Nucleophiles. Nitrogen

Aminocarbenes are formed from NH_3, primary or secondary amines (Equation 1.17):

$$L_nM=C=CR^1R^2 + NHR^3R^4 \rightarrow L_nM=C(NR^3R^4)CHR^1R^2 \quad (1.17)$$

for example, $ML_n = Fe(CO)(PPh_3)Cp$ [9b], $RuCl_2(pnp)$ [289], $RuCl\{PPh_2(o\text{-tol})\}Cp$ [116], $W(CO)(NO)Cp$ [290, 291].

Pyridines and other heterocyclic nitrogen bases add to C_α (Equation 1.18)

$$L_nM=C=CRR' + py \rightarrow L_nM-C(py)=CRR' \quad (1.18)$$

for example, $ML_n = Mn/Re(CO)_2Cp$ [234], $Fe(CO)(PPh_3)Cp$ [9], $OsCl(PPr^i_3)(\eta^6\text{-arene})$ [73, 74, 280], pyrazole with $Ru(PPh_3)_2Tp$ [292].

1.3.1.6 Phosphorus

Phosphines add to C_α to give phosphonio-alkyl complexes (Equation 1.19):

$$L_nM=C=CRR' + PR''_3 \rightarrow L_nM-C(PR'_3)=CRR' \quad (1.19)$$

Products from apparent reaction at C_β are probably formed by attack on the η^2-alkyne, followed by migration induced by steric effects, for example, with Ru $(=C=CHPh)(CO)(PPh_3)(\eta^5\text{-}C_9H_7)]^+$ to give $[Ru\{E\text{-}CH=CPh(PPh_3)\}(CO)(PPh_3)(\eta^5\text{-}C_9H_7)]^+$ [102], $[Fe\{C(PPh_3)=CH_2\}(CO)(PPh_3)Cp]^+$ (migration on heating) [9], $Ru(=C=CHR)(PPh_3)_2Cp$ [293, 294].

1.3.1.7 Halogen Nucleophiles

The chloro-carbene complex $[Fe(=CClMe)(CO)(PPh_3)Cp]^+$ is formed from $[Fe(=C=CH_2)(CO)(PPh_3)Cp]^+$ and HCl [9]. Attack of fluoride on $[Ru\{=C=CPhCH_2(OMe)\}(PPh_3)_2Cp]^+$ affords vinyl $Ru\{CF=CPhCH_2(OMe)\}(PPh_3)_2Cp$ [295].

1.3.1.8 Carbon Nucleophiles

With cyanide, $[Ru(=C=CMePh)(PMe_3)_2Cp]^+$ gives >95% $Z\text{-}Ru\{C(CN)=CMePh\}(PMe_3)_2Cp$, which isomerizes in the presence of acid to >95% E isomer [296]. $Li[CuR_2(CN)]$ (R = Ph, $CH=CH_2$) and $[Fe(=C=CMe_2)(CO)(PPh_3)Cp]^+$ give $Fe(CR=CMe_2)(CO)(PPh_3)Cp$, the vinyl complex transforming to $Fe(\eta^3\text{-}CH_2CHC=CMe_2)(PPh_3)Cp$ [297].

1.3.1.9 Hydride

Borohydride reacts with vinylidene complexes to give σ-vinyls. Examples include $Fe(=C=CH_2)(CO)(PPh_3)Cp$ [9] and $Fe(=C=CMePh)(dppm)Cp$ [283]. A carbyne complex is formed by addition of $K[BHBu^s_3]$ to $MoBr(=C=CPh)(PEt_3)\{P(OMe)_3\}Cp$ [237].

1.3.2
Intramolecular Reactions

1.3.2.1 Formation of Cyclopropenes

Deprotonation of complexes containing Ru(=C=CPhCH$_2$R) fragments gives cyclopropenyl complexes by intramolecular attack on C$_\alpha$. Reactions of vinylidene complexes [Ru(=C=C(Ph)CH$_2$R)(P)$_2$Cp]$^+$ [P$_2$ = (PPh$_3$)$_2$, dppe; R = CN, C$_6$F$_5$, Ph, C$_6$H$_4$CN-4, C$_6$H$_4$CF$_3$-4, 1-nap, CH=CH$_2$, CH=CMe$_2$, C$_6$H$_9$] with [NBu$_4$]OH in acetone give Ru(C=CPhCHR)(P)$_2$Cp [Scheme 1.7, **8** (R = CN shown)], for which an extensive chemistry has been developed [171, 197, 202, 203, 223, 295, 298]. Addition of electrophiles (E = H$^+$, [CPh$_3$]$^+$, HgCl$_2$, tcnq) to **8** gives vinylidene complexes [Ru{=C=CPh(CN)(E)}(PPh$_3$)$_2$Cp]$^+$ by attack at the methine carbon.

Deprotonation of Ru{C=CPhCH(CO$_2$Me)}(PPh$_3$)$_2$Cp affords the furyl complex Ru{C=CPhCHC(OMe)O}(PPh$_3$)$_2$Cp (**9**) via the cyclopropenyl, which can be observed at low temperatures. Furan ring-opening occurs with electrophiles, for example, tcnq gives zwitterionic Ru$^+${=C=Ph[CH(CO$_2$Me)(tcnq$^-$)]}(PPh$_3$)$_2$Cp (**10**) [223]. Vinyl complex Ru{CF=CPh(CH$_2$OMe)}(PPh$_3$)$_2$Cp is obtained from the corresponding vinylidene and F$^-$; deprotonation of this complex does not occur. Several related ring-closing reactions occur with vinylidenes such as Ru$^+${=C=CPh[CH(CN)(tcnq$^-$)]}(PPh$_3$)$_2$Cp (**11**) which with [NBu$_4$]X (X = OH, CN) in MeOH give cyclopropenes Ru{C(OMe)CPh=C(CN)}(PPh$_3$)$_2$Cp and Ru{C=CPhC(CN)$_2$}(PPh$_3$)$_2$Cp, respectively; protonation of the latter gives Ru{=C=CPh[CH(CN)$_2$]}(PPh$_3$)$_2$Cp]$^+$ [223].

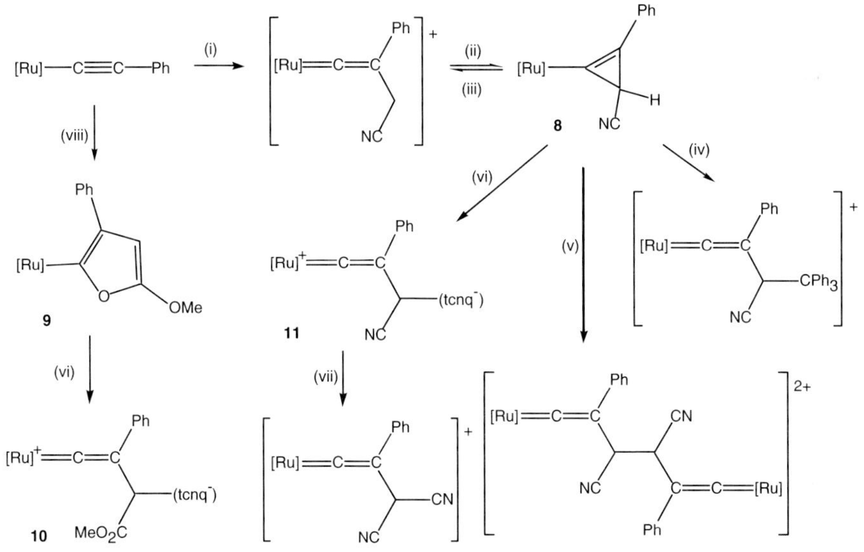

Scheme 1.7 [Ru] = Ru(PPh$_3$)$_2$Cp. Regents: (i) ICH$_2$CN; (ii) [NBu$_4$]OH; (iii) H$^+$ (iv) [CPh$_3$]$^+$, (v) CH$_2$=CHCH$_2$I; (vi) tcnq; (vii) [NBu$_4$]CN/MeOH, then H$^+$; (viii) BrCH$_2$CO$_2$Me.

Complexes [Ru{=C=CPh(CH$_2$R)}(dppp)Cp* (R = CN, CO$_2$Me), formed similarly, are deprotonated to cyclopropenyls which then undergo displacement of cyclopropenyl by azide, ring-opening with HgCl$_2$ or ring enlargement with acetone, the latter giving dihydrofuran Ru{C=CPhC(CN)CMe$_2$O}(dppp)Cp* [298].

Related chemistry of the Ru(PPh$_3$)$_2$Tp fragment has been described, although the enhanced lability of one of the PPh$_3$ ligands has allowed unusual reactions to take place [202]. Deprotonation of Ru{=C=CPh(CH$_2$CN)}(PPh$_3$)$_2$Tp affords cyclopropenyl Ru{C=CPhCH(CN)}(PPh$_3$)$_2$Tp, which ring-opens with [CPh$_3$]$^+$ or HgCl$_2$ (E) to give Ru{=C=CPh[CH(E)(CN)](PPh$_3$)$_2$Tp]$^+$. However, with CO in MeOH, vinyl Ru{C(OMe)=CPh(CH$_2$CN)}(CO)(PPh$_3$)Tp is formed, while pyrazole gives **12** by coordination and attack of the second N center at C$_\alpha$ [202].

12   **13**

1.3.2.2 Attack on Coordinated Phosphines

The related iron complexes [Fe{=C=CPh(CH$_2$R)}(dppe)Cp]$^+$ (R = CN, 4-CNC$_6$H$_4$, 4-CF$_3$C$_6$H$_4$) are also converted to cyclopropenyl complexes, but for R = CH=CH$_2$, Ph, intramolecular attack on the dppe ligand gives 1-ferra-2,5-diphospha-[2.2.1]-bicyclic complex **13** [299]. For R = C$_6$F$_5$, a mixture of the two types of complex is obtained. Intramolecular attack of deprotonated ligands such as dppm or dppe, has also been reported as resulting from base attack (KOH, Na[N(tms)$_2$]) on dppe in Fe(=C=CMe$_2$)(dppe)Cp [240, 299], Fe(=C=CMeBu^t)(dppm)Cp to give bicyclic Fe{C(=CMeBu^t)CH(PPh$_2$)$_2$}Cp [300], and [Ru(=C=CMeR)(dppm)(η^5-C$_9$H$_7$)]$^+$ (R = Bu^t, Ph) [301]. Deprotonation of bis(vinylidene)s {Ru(dppm)Cp*}$_2${μ-(=C=CRCR=C=)} (R = H, Me) affords the complexes Ru{C≡CCR=CCH(PPh$_2$)$_2$[RuCp*]}(dppm)Cp* [302].

1.3.2.3 Coupling

Attack of MeO$^-$ on the CO of Mo{=C=CPh(CH$_2$CH=CH$_2$)}(CO)(dppe)Cp gives Mo{η^1,η^2-C(CO$_2$Me)=CPh(CH$_2$CH=CH$_2$)}(dppe)Cp (**14**) as a result of intramolecular coupling of the resulting CO$_2$Me group with the vinylidene C$_\alpha$, while the benzylvinylidene affords Mo{η^3-CH(CO$_2$Me)CPhCHPh}(dppe)Cp[303]. On W, coupling of W{=C=CH(PPh$_2$)}(CO)(NO)Cp affords eight-membered cyclic dimers {W[μ-η^1:P-(C≡C)PPh$_2$]$_2$(CO)$_2$(NO)}$_2$ and **15** [176]. Spontaneous insertion of the vinylidene ligand of Mn(=C=CHCO$_2$Me)(CO)$_3$(dppm) [from Mn(C≡CCO$_2$Me)(CO)$_3$(dppm) and HBF$_4$ at 200 K] into a Mn–P bond gives Mn{OC(OMe)CH=CPPh$_2$CH$_2$PPh$_2$}(CO)$_3$ (**16**) [61].

14 [Mo] = Mo(dppe)Cp

Intramolecular coupling of the vinylidene from HC≡CPh and Ru(O$_2$CCHPh$_2$) (PPh$_3$)Tp affords Ru{C(=CHPh)OC(CHPh$_2$)=O}(PPh$_3$)Tp [304]. Treatment of OsCl (=C=CHPh)(PPr$^i_3$)Cp with RMgCl gives the osma-indene OsH(CR=CHC$_6$H$_4$)(L)Cp which isomerizes to the *exo*-allyl complex Os(η^3-CH$_2$CHCHPh)(L)Cp when R = Me, and when R = Ph protonation of the latter gives stilbene [305].

1.3.2.4 Vinylidene/Alkyne Coupling

Coupling of 1-alkynes via vinylidenes occurs on Fe(CO)$_2$(PR$_3$)$_2$ [57], while HC≡CCO$_2$Me couples with RuCl(=C=CHR)(PMePr$^i_2$)Tp (R = CO$_2$Me) to give vinylvinylidene [Ru{=C=C(CO$_2$Me)CH=CH(CO$_2$Me)}(PEt$_3$)$_2$Tp]$^+$ via [2 + 2]-cycloaddition [129]. Treatment of similar precursors with C$_2$Ph$_2$/LiNPr$^i_2$ gives the enynyl complexes Ru(CPh=CPhC≡CR)(PMePr$^i_2$)Tp (R = H, Bu^t, Ph, SiMe$_3$).

Intramolecular coupling of vinylidene and alkynyl occurs by migratory insertion upon carbonylation of *trans*-Rh(C≡CR)(=C=CHR')(PPr^{i}Ph$_2$)$_2$ to give Rh{C(C≡CR)=CHR'}(CO)(PPr^{i}Ph$_2$)$_2$ [306]. Binuclear complexes containing bridging C$_8$ or C$_4$–C$_6$H$_4$–C$_4$ skeletons, {FcCH=C[Rh(CO)(L)$_2$]C≡C–}$_2$ and 1,4-C$_6$H$_4${CH=C(C≡CPh)[Rh(CO)(L)$_2$]}$_2$, respectively, have been obtained from the corresponding vinylidene complexes under CO [52].

1.3.2.5 Formation of π-Bonded Ligands

Attack on the vinylidene ligand may be followed by rearrangement to give π-bonded ligands. For example, alkene complexes Mn{η^2-*trans*-PhCH=CHP(O)(OR)$_2$}(CO)$_2$Cp are formed in the reactions of Mn(=C=CHPh)(CO)$_2$Cp with P(OR)$_3$ [R = Et, Ph; (OR)$_2$ = O(CH$_2$)$_3$O] [307]. The η^2-thioketene complex RhCl(η^2-RCH=C=S)(L)$_2$ is formed from from S$_8$ and RhCl(=C=CHR)(L)$_2$ [308]. Allenes can be obtained from diazoalkanes and W(=C=CH$_2$)(CO)(NO)Cp: CH$_2$N$_2$ gives W(η^2-CH$_2$=C=CH$_2$)(CO)(NO)Cp, while an *E/Z* mixture of the η^2-CH$_2$=C=CHCO$_2$Et complex is obtained with CH(CO$_2$Et)N$_2$ [309].

1.3.3 Reactions at C$_\beta$

As in alkynyl complexes, C$_\beta$ is electron rich and thus reacts readily with electrophiles. These reactions proceed more readily with neutral rather than cationic vinylidenes, although many examples of both types of reaction are known. Addition of electrophiles is influenced by steric effects: whereas double protonation of [W(C≡CR)(CO)$_5$]$^-$ (R = Bu^t, Ph) in the presence of [NMe$_4$]I gives trans-WI(≡CCH$_2$R)(CO)$_4$, alkylation of [W(C≡CBu^t)(CO)$_5$]$^-$ with FSO$_3$Me or [Et$_3$O]$^+$ gives only the vinylidenes W(=C=CBu^{t}R)(CO)$_5$. However, protonation of the latter in the presence of iodide affords trans-WI(≡CCHRBu^t)(CO)$_4$ [310].

1.3.3.1 Protonation

The simplest reaction is addition of a proton (usually from a strong non-coordinating acid, such as HBF$_4$, HPF$_6$, HOTf, often as their ether adducts) to give the corresponding carbyne (Equation 1.20):

$$L_nM=C=CHR + H^+ \rightarrow [L_nM\equiv C-CH_2R]^+ \quad (1.20)$$

The reaction may be reversible by reaction with hydride [301]. Mechanistic studies of the protonation of Re complexes showed three routes to be operative. Regiospecific protonation of C$_\beta$ is most direct but under some conditions, protonation occurs more rapidly at Re. The resulting hydrido-complex then undergoes rearrangement by migration to C$_\beta$ or further protonation to give [ReH(≡CCH$_2$R)(dppe)$_2$]$^{2+}$, which is then deprotonated to [Re(≡CCH$_2$R)(dppe)$_2$]$^+$ (Scheme 1.8) [311].

Examples of protonation reactions include complexes containing MCl(L)$_2$ (M = Ir, Rh; R = H, Me, Ph) [33, 95], OsH(L)(η-C$_5$H$_4$SiPh$_3$), Os(GePh$_3$)(L)Cp′ (Cp′ = Cp, Cp^{Me}) [189], OsHCl(=C=CHPh)(L)$_2$, which gives [OsHCl(≡CCH$_2$Ph)(OH$_2$)(L)$_2$]$^+$ with HBF$_4$·H$_2$O [243], OsHCl(≡CCH$_2$Ph)(L)$_2$ with HBF$_4$·OEt$_2$ [72], OsHF(Hpz)(L) [241], ReCl(dppe)$_2$ [312], Rh{E(CH$_2$CH$_2$PPh$_2$)$_3$} (E = N, P) [194], RuCl(PPh$_3$)Cp* [313], RuHCl(P)$_2$ [314], RuCl$_2$(P)$_2$ (P = PPr$^i_3$, PCy$_3$) [315], RuCl$_2$(pnp) [95], TcCl(dppe)$_2$ [133] and WCl(CO)$_2$(dppe) fragments [134].

Mixtures of the carbyne with cyclic carbenes [RuCl{=C(CH$_2$Ph)OC(O)R}(L)$_2$]$^+$ were obtained by protonation of RuCl(O$_2$CR)(=C=CHPh)(L)$_2$ (R = alkyl, aryl), the

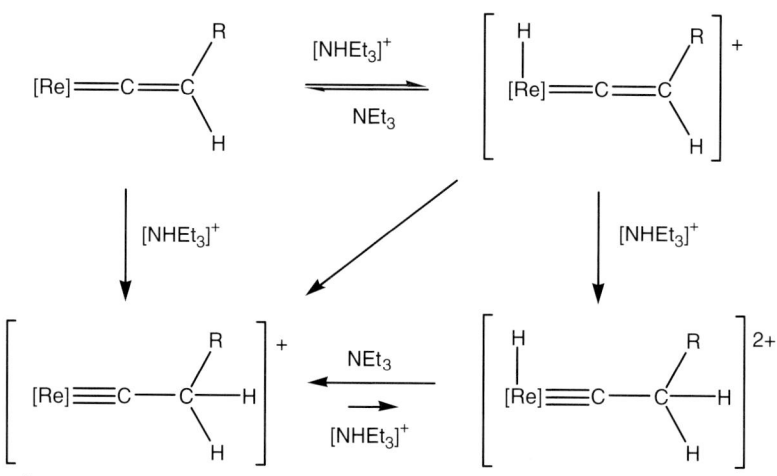

Scheme 1.8 [Re] = trans-ReCl(dppe)$_2$.

composition of which depends on the basicity of the RCO_2 ligand [315]. Comparison of RuCl(=C=CHR)(PPh$_3$)(L′) (L′ = Cp*, Tp) shows that the Cp* complexes are stable enough to be isolable, whereas the Tp analogs are labile, being detected by NMR methods [111]. Addition of HCl to Os(=C=C$_{10}$H$_6$)(CO)$_2$(PPh$_3$)$_2$ (C$_{10}$H$_6$ = 1- or 2-naphthylidene) gives Os(CH$_2$R)Cl(CO)$_2$(PPh$_3$)$_2$ (R = 1- or 2-naphthyl) [238]. At −40 °C, addition of HCl to trans-Rh(C≡CPh)(=C=CHPh)(L)$_2$ gives RhCl(C≡CPh)(Z-CH=CHPh)(L)$_2$, which couples at 25 °C to form the η^2-PhC$_2$(Z-CH=CHPh) complex; the enyne is displaced by CO [316]. Addition of HX (X = I, tfa) to Rh(=C=CHR)(L)Cp (R = H, Ph) gives kinetic products [Rh(Z-CH=CHR)(L)Cp]X which, for X = tfa, rearrange in polar solvents to the thermodynamic E isomers [317]. With HCl, the chlorocarbyne [Rh(≡C-CHClMe)(L)Cp]$^+$ is formed. Allylvinylidenes react with H$^+$ to give η^4-vinylallenes [205]; vinylvinylidenes add H$^+$ at C$_\delta$ to give cyclic carbenes [247].

1.3.3.2 Alkylation

Alkylation proceeds similarly in reactions commonly employing MeOTf. In some instances, the products obtained from neutral complexes containing anionic ligands X depend on the nature of X [315]. In several cases, X/OTf exchange occurs [111, 313], such as with RuCl(=C=CHR)(PPh$_3$)(L) (L = Cp*, Tp), MeCl being the detectable by-product. Alkylation of Mn{=C=CMeC(O)Me}(CO)(PPh$_3$)Cp^{Me} with MeOTf gives vinylcarbyne [Mn{≡CCMe=CMe(OMe)}(CO)(PPh$_3$)Cp^{Me}]OTf [210]. Oxidative addition of MeI to Ir(=C=CH$_2$){N(SiMe$_2$CH$_2$PPh$_2$)$_2$} gives IrMeI(=C=CH$_2$){N(SiMe$_2$CH$_2$PPh$_2$)$_2$}, which undergoes insertion of the vinylidene to give IrI(CMe=CH$_2$){N(SiMe$_2$CH$_2$PPh$_2$)$_2$}. Subsequent rearrangement forms IrI(η^3-allyl){N(SiMe$_2$CH$_2$PPh$_2$)$_2$} [60].

1.3.3.3 Other Electrophiles

Other electrophiles include [CPh$_3$]$^+$ to replace H, as in Mn(=C=CHPh)(CO)(PPh$_3$)Cp which gives Mn{=C=CPh(CPh$_3$)}(CO)(PPh$_3$)Cp [220a], addition of diazo compounds to [Ru(=C=CHR)(PP)Cp]$^+$ (R = Me, Ph) to give the methyl-vinylidenes [Ru(=C=CMeR)(PP)Cp]$^+$ [318], and BrCH$_2$CO$_2$Me to [Ru(=C=CHR)(PPh$_3$)$_2$Cp]$^+$ (R = C$_6$H$_9$, Ph) give the furyl derivative Ru(2-C$_4$HO-3-R-5-OMe)(PPh$_3$)$_2$Cp via [Ru{=C=CR(CH$_2$CO$_2$Me)}(PPh$_3$)$_2$Cp]$^+$ [199]. Addition of Sn(CH=CH$_2$)$_4$/CuCl to RuCl(=C=CHCO$_2$Me)(PPh$_3$)Cp gives Ru(η^3-CH$_2$CHC=CHCO$_2$Me)(PPh$_3$)Cp [36].

Reactions with dioxygen generally afford the corresponding metal carbonyl derivatives, with loss of organic aldehyde or acid. This reaction can be expressed as an analog of multiple bond metathesis and corresponds to oxidation of vinylidene to CO [46]. Oxidation of OsHCl(=C=CHPh)(L)$_2$ affords the styryl complex OsClO$_2$(E-CH=CHPh)(L)$_2$ [243].

1.3.4 Cycloaddition Reactions

The alternating electronic properties of the C$_\alpha$ and C$_\beta$ atoms in the vinylidene ligand enable dipolar molecules to enter into cycloaddition reactions. Intramolecular [2 + 2]-

cycloaddition has been studied most intensively with M(=C=CRR′)(CO)$_5$ (M = Cr, W) and M(=C=CRR′)(CO)$_2$Cp (M = Mn, Re) (Sections 1.4.3 and 1.4.4). Cycloaddition of RC≡CCO$_2$Me (R = H, CO$_2$Me) to vinylidene or allenylidene complexes to give vinylvinylidenes has been discussed above (Section 1.2.11).

Intramolecular metathesis of alkenylvinylidenes occurs with complexes containing a *gem*-CPh$_2$CH= group at the vinylidene ligand, such as [Ru(=C=CHCPh$_2$CH$_2$R)(PPh$_3$)$_2$Cp]$^+$ (R = CH=CH$_2$, C≡CH), which slowly and irreversibly transform to [Ru(=C=CHCH$_2$CPh$_2$R)(PPh$_3$)$_2$Cp]$^+$ in solution [319]. The methylated complex with R = CMe=CH$_2$ gives the cyclic η^2-allene complex **17**. Longer chains containing an extra CH$_2$ group are stable.

In **18** the functionalized indenyl group becomes η^6-bonded to Ru as a result of intramolecular cycloaddition in [Ru(=C=CHCR^{1}R^{2}CH$_2$CMe=CH$_2$)(PPh$_3$)$_2$(η^5-C$_9$H$_7$)]$^+$ [320]. Similar addition of vinylidene to allyl occurs in [Ru(=C=CHC$_6$H$_4$X-4)(PPh$_3$){PPh$_2$(CH$_2$CH=CH$_2$)}(η^5-C$_9$H$_7$)]$^+$ (X = H, Cl) to give bicyclic alkylidene complexes **19** [321].

17 [Ru] = Ru(PPh$_3$)$_2$Cp

18

19 [Ru] = Ru(PPh$_3$)(η^5-C$_9$H$_7$)
R = C$_6$H$_4$X

1.3.5
Adducts with Other Metal Fragments

Although the chemistry of bi- and poly-nuclear vinylidene complexes is beyond the scope of this chapter, several examples of the interaction of metal-vinylidene fragments with a second metal-ligand fragment have been described. This type of reaction usually produces complexes containing bridging vinylidene ligands (Equation 1.21):

$$L_nM=C=CRR' + M'L'_m \rightarrow L_nM(\mu\text{-}C=CRR')M'L'_m \quad (1.21)$$

Thus, Ti(=C=CH$_2$)Cp*$_2$ (prepared *in situ*) reacts with MX(PPh$_3$) (M = Cu, Au) to give Ti{C(=CH$_2$)M(L)X}Cp*$_2$ [XML = ClCu(PR$_3$) (R = Me, Ph), ClAu(PPh$_3$), PhC≡CCu(PMe$_3$)] [322]. Treatment of [M(=C=CH$_2$)(CO)$_2$Tp*]$^-$ (M = Mo, W) with WI(CO)(η^2-C$_2$Ph$_2$)Tp afforded binuclear carbyne Tp*(OC)$_2$M≡CCH$_2${W(CO)(η^2-C$_2$Ph$_2$)Tp}, which was deprotonated (KOBu^t) and coupled with I$_2$ to give the μ-C$_2$ complexes {Tp*(OC)$_2$M}C≡C{W(CO)(η^2-C$_2$Ph$_2$)Tp} [323]. Mn(=C=CHPh)(CO)$_2$Cp reacts with Re(CO)$_3$Cp to give Cp(OC)$_2$Mn(μ-C=CHPh)Re(CO)$_2$Cp [324].

A range of homo- and hetero-nuclear complexes has been obtained from Rh(=C=CHR)(L)Cp: addition of {RhCl(PPr^i$_3$)$_2$}$_2$ gives Cp(L)Rh(μ-C=CHR)RhCl

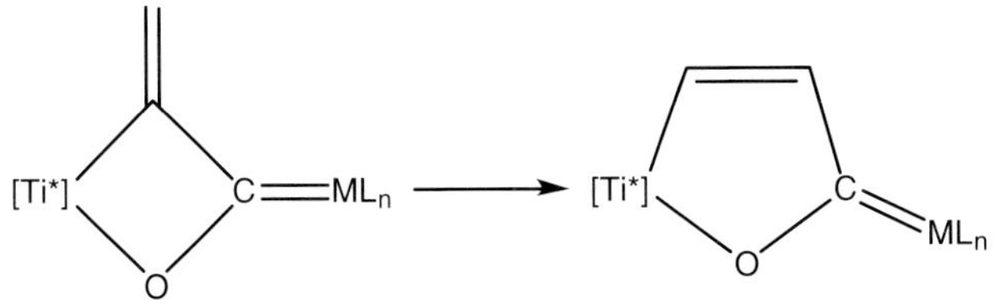

Scheme 1.9 [Ti*] = TiCp*$_2$.

(L)$_2$ [325]. Heterometallic complexes are formed with CuCl, Mn(CO)$_3$Cp and Fe$_2$(CO)$_9$, which afford Cp(L)Rh(μ-C=CHR){M'L$_m$'}, ({M'L$_m$'} = CuCl, Mn(CO)$_2$Cp) and Fe$_2$Rh(μ$_3$-CCHR)(CO)$_6$(L)Cp, respectively [326]. Alternatively, attack at the metal center may occur, as found with Rh(=C=CHR)(L)Cp and HgX$_2$ (X = Cl, I) or HgIMe/NEt$_3$, to give alkynyls Rh(HgX)(C≡CR)Cp [327]. With SnCl$_2$, insertion and protonation give Rh(SnCl$_3$)(CH=CHR)(L)Cp. Reactions between ER$_3$ (E = Al, R = Me, Et; Ga, R = Me) and Ir(=C=CH$_2$){N(SiMe$_2$CH$_2$PPh$_2$)$_2$} afford Ir(ER$_2$)(CR=CH$_2$){N(SiMe$_2$CH$_2$PPh$_2$)$_2$} by oxidative addition and migration of R to the vinylidene [60].

Cycloaddition of a CO ligand in metal carbonyls to Ti(=C=CH$_2$)Cp*$_2$ affords titana-oxetanes with a metal carbonyl fragment attached to the ring, Ti{C(=CH$_2$)C(=ML$_n$)O}Cp*$_2$ [20; ML$_n$ = M(CO)$_5$ (M = Cr, Mo, W), Re(CO)$_2$Cp, M$_2$(CO)$_9$ (M = Mn, Re), Fe(CO)$_4$, Rh(CO)(acac); Scheme 1.9] [328, 329]. These complexes rearrange between 70 and 110 °C to titana-oxacyclopentenes Ti{CH=CHC(=ML$_n$)O}Cp*$_2$ (21), a reaction reminiscent of the reverse of the alkyne/vinylidene rearrangement. Intramolecular vinylidene–carbene coupling occurs with Cr{=CMe(OMe)}(CO)$_5$ to give a mixture of Ti{CH$_2$C[=CMe(OMe)]C[=Cr(CO)$_5$]O}Cp*$_2$ which then undergoes coordination of OMe to Cr with concomitant CO loss [330]. If the reaction between Ti(=C=CH$_2$)Cp*$_2$ and isocyanide is carried out in the presence of W(CO)$_6$, oxametallacycle Ti{C(=NR)C(=CH$_2$)C[=W(CO)$_5$]O}Cp*$_2$ (R = 2,6-Me$_2$C$_6$H$_3$) is obtained.

Interaction of the vinylidene with a second C$_2$ fragment may also give a μ-cyclobutenylidinium complex (Scheme 1.10).

Heterometallic cyclobutenylidene complexes {(OC)$_5$Cr}(μ-C=CR'CCR$_2$){ML$_n$} have been obtained from reactions of Cr(=C=CR$_2$)(CO)$_5$ with alkynylmetal complexes M(C≡CR')L$_n$ [R = Me, Ph; ML$_n$ = Fe(CO)(L')Cp [L' = CO, PPh$_3$, P(OMe)$_3$],

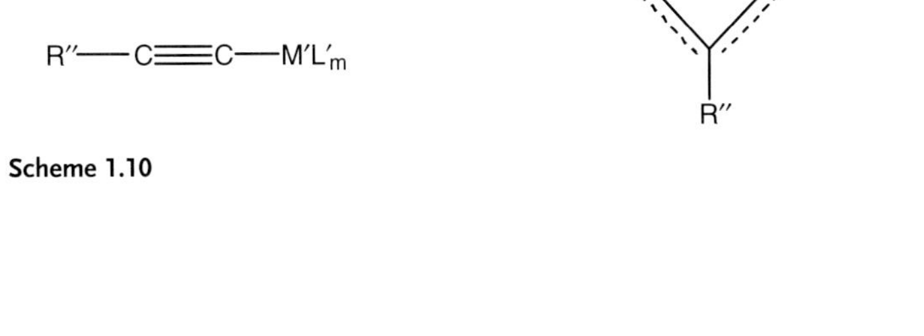

Scheme 1.10

Ni(PEt$_3$)Cp; R' = Bu, Me, Ph, C$_6$H$_4$NO$_2$-4, CO$_2$Me (not all combinations)] [331]. A spirocyclic bridging ligand is present in analogous products obtained from Cr{=C=C(CH$_2$)$_5$}(CO)$_5$. Similar chemistry with Cr(=C=CMe$_2$)(CO)$_5$ and diynyl complexes Fe(C≡CC≡CR)(CO)(L)Cp (L = CO, R = SiMe$_3$, Bu, Ph; L = PPh$_3$, R = SiMe$_3$) afforded {(OC)$_5$Cr}{μ-CC(C≡CR)CCMe$_2$}{Fe(CO)(L)Cp}, from which routes to tri- and tetra-metallic systems were developed [332]. This reaction has also been reported for Re(NO)(PPh$_3$)Cp* [191], Fe(CO)(L)Cp (L = CO [333], PPh$_3$ [9c], trans-RuCl(dppe)$_2$ [334, 335], and Ru(PPh$_3$)$_2$Cp [336]. Formation of [{Fe(CO)(PPh$_3$)Cp}$_2$(μ-C$_4$H$_3$)]$^+$ as a 3/1 mixture of diastereomers occurs spontaneously from the parent vinylidene in thf, but the PCy$_3$ complex gives the RR,SS pair of enantiomers [8].

Complexation of [Fe{=C=CH(nap)}(dppe)Cp*]$^+$ with the areneophile [RuCp*]$^+$ gives derivatives in which the RuCp* fragment is attached to the naphthyl group, the isomer in which it is attached to the unsubstituted C$_6$ ring being the major product (92%) [337].

1.3.6
Ligand Substitution

Displacement of the vinylidene ligand is obviously important when these complexes are used in synthesis. The product is usually the 1-alkyne, as found for trans-Ru(C≡CR)$_2$(dppe)$_2$ (R = Bu, Ph, SiMe$_3$, CH$_2$OMe, CH$_2$OSiMe$_3$) and [NH$_4$]PF$_6$, which afforded trans-Ru(C≡CR)(NH$_3$)(dppe)$_2$ and HC≡CR [96], [Ru(=C=CHR)(PPh$_3$)$_2$(η^5-C$_9$H$_7$)]$^+$ (R = Ph, C$_6$H$_4$NO$_2$-4, Fc) and nitriles R'CN (R' = Me, Et) [338], or RuHCl(=C=CH$_2$)(PR$_3$)$_2$ (R = Pr^i, Cy) and KPF$_6$ in MeCN, which resulted in migration of H to vinylidene to give Ru(CH=CH$_2$)(NCMe)$_2$(PR$_3$)$_2$ [339]. Addition of bidentate phosphines to hydrido complexes such as RuHCl(=C=CHR)(L)$_2$ may result in H migration to C$_\alpha$ to give vinyls, such as RuCl(CH=CHR)(PP) [340]. Labile vinylidenes in RuCl(=C=CHR)(PPh$_3$)Tp (R = Ph, SiMe$_3$, Bu, Bu^t, CO$_2$Et, C$_6$H$_9$) are replaced by other nucleophiles, L (PMe$_3$, PPh$_3$, MeCN, py, CO); phenylvinylidene can be displaced by HC≡CR (R = SiMe$_3$, Bu, Bu^t, CO$_2$Et, C$_6$H$_9$) [125]. Both vinylidene and PPh$_3$ in RuCl(=C=CHPh)(PPh$_3$)Cp* are displaced by P(OEt)$_3$ to give RuCl{P(OMe)$_3$}$_2$Cp* [341].

Photolytic demetallation of M(=C=CMeR)(CO)$_5$ (M = Cr, W; R = Pr, Bu^t) occurs with coupling of the vinylidene with CO to give (presumably) propadienones which dimerize to the isolated cyclobutane-1,3-diones [342]. Other examples include the reaction of [Ru(=C=CHPh)(ppfp)]$^+$ [ppfp = Fe{η-C$_5$H$_3$-1-CHMe(η-C$_5$H$_4$)-2-PPh$_2$}Cp] with CH$_2$=CHCHMe(OH) to give PhCH$_2$C(O)CHMeCH=CH$_2$ (ee 25%) [113].

An azavinylidene/vinylidene equilibrium can be established on Os [343]. Reactions of OsCl$_2$(=C=CHPh)(NH=CR$_2$)(L)$_2$ [22, CR$_2$ = CMe$_2$, C(CH$_2$)$_5$; Scheme 1.11] with NEt$_3$ or NH(CH$_2$CH=CH$_2$)$_2$ to give mixtures of 22 and OsCl(=C=CHPh)(=N=CR$_2$)(L)$_2$; the latter could be obtained pure with LiBu^t. With NH$_2$R' (R' = allyl, Ph), further reaction to form OsCl$_2$(=C=CHPh)(NH=CR$_2$)(NH$_2$R')(L) takes place. Complexes 22 lose imine on heating to give OsCl$_2$(=C=CHPh)(L)$_2$, while protonation with HBF$_4$.OEt$_2$ gives [Os(≡CCH$_2$Ph)(NH=CR$_2$)(L)$_2$]$^+$.

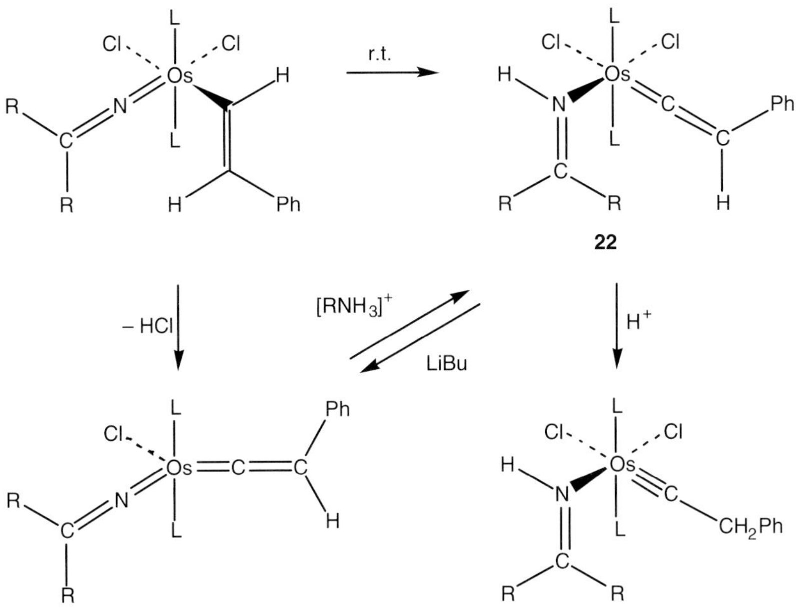

Scheme 1.11 L = PPr$^i_3$; CR$_2$ = CMe$_2$, (CH$_2$)$_5$.

There are many examples of reactions in which non-vinylidene ligands are exchanged. Neutral ligands such as CO can be exchanged for phosphines on Mn(CO)$_2$Cp [344], monodentate phosphines for dcype on RuCl$_2$(=C=CHR)(P)$_2$ (R = Bu^t, Ph) [340], while anionic ligands (usually Cl) exchange for OH, I or Me on IrCl(=C=CHR)(L)$_2$ (R = H, Me, Ph) [154], for acetate on OsCl(=C=CHPh){2,6-(PPh$_2$CH$_2$)$_2$C$_6$H$_3$} [67], for Cp on RhCl(=C=CHR)(L)$_2$ [308], OTf [111, 313] and N(PPh$_2$S)$_2$ on RuCl$_2$(=C=CHPh)(L)$_2$ [345] or RuHCl(=C=CHR)(L)$_2$ (I, NCO, OPh, OAc, tfa); Cl has been exchanged for CO, N$_2$ or [9]aneS$_3$ on the latter [312, 345]. Reactions of OsHCl(=C=CHR)(L)$_2$ with acetate, Na[acac] or P(OMe)$_3$ have given OsH(OAc)(=C=CHPh)(L)$_2$, OsH(acac)(=C=CHR)(L)$_2$ and OsHCl(=C=CHR)(L)$_2${P(OMe)$_3$}, respectively [243]. Exchange of F for OTf in Rh(OTf)(=C=CR)(L)$_2$ (R = Fc, C$_6$H$_4$) occurs with KF in acetone [52].

1.3.7
Miscellaneous Reactions

Insertion of the vinylidene into an Fe−N bond occurs when Fe(=C=CAr$_2$)(porph) (porph = tetra-4-tolylporphyrin, Ar = 4-ClC$_6$H$_4$) is irradiated [10]. Treatment of IrPh$_2${=C=C(SiMe$_3$)$_2$}Tp^{Me} with acids variously gives IrClPh(=CMePh)Tp^{Me} (with HCl), IrPh(tfa)(=CMePh)Tp^{Me} (with CF$_3$CO$_2$H) or [IrPh(OEt$_2$)(=CMePh)Tp^{Me}]$^+$ (with HBF$_4$) (Scheme 1.12) [139]. The reaction of the vinylidene with C$_2$(SiMe$_3$)$_2$ gives an iridacyclopentene, {IrCH$_2$CMe=CMeCH$_2$}{=C=C(SiMe$_3$)$_2$}Tp^{Me} (23, Scheme 1.13) which undergoes intramolecular insertion of the resulting C=C

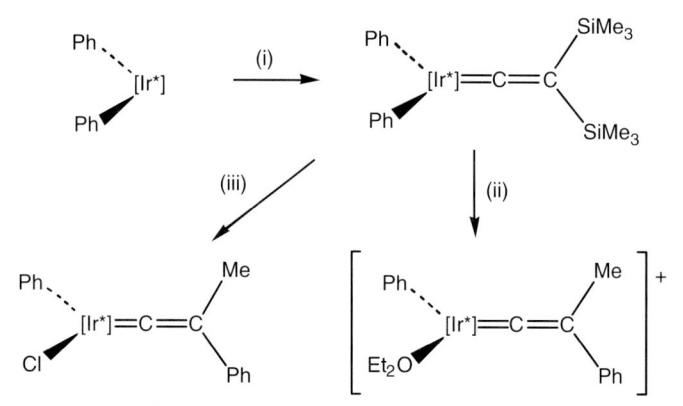

Scheme 1.12 [Ir*] = IrTp*. Reagents: (i) C$_2$(SiMe$_3$)$_2$; (ii) HBF$_4$; (iii) HCl/Et$_2$O.

Scheme 1.13 [Ir*] = IrTp*. Reagents: (i) C$_2$(SiMe$_3$)$_2$; (ii) CF$_3$CO$_2$H.

(SiMe$_3$)$_2$ into an Ir–C bond; protonation then gives the iridabenzene, {Ir(H)CHCMeCMeCHCMe}Tp^{Me} (**24**) [139].

Treatment of Ir{C$_4$(CO$_2$Me)$_4$}(C≡CPh)(CO)(PPh$_3$)$_2$ with HCl gives the bicyclic lactone Ir{C$_4$(CO$_2$Me)$_3$C(O)OC(=CHPh)}(CO)(PPh$_3$)$_2$ (**25**), also obtained from [Ir{C$_4$(CO$_2$Me)$_4$}(NCMe)(CO)(PPh$_3$)$_2$]$^+$ and HC≡CPh in the presence of water [346]. These reactions proceed by coupling with a CO$_2$Me substituent.

25 [Ir] = Ir(CO)(PPh$_3$)$_2$

1.4
Chemistry of Specific Complexes

Extensive studies of several readily available vinylidene complexes have led to much interesting chemistry, many reactions being specific to the metal-ligand combination used.

1.4.1
Reactions of Ti(=C=CH$_2$)Cp$_2^*$ (Scheme 1.14)

This complex is prepared *in situ* by intramolecular synchronous transfer of the α-H atom to R in TiR(CH=CH$_2$)Cp$^*_2$ (R = Me, CH=CH$_2$) to give methane or ethene, respectively. The corresponding η,N-C$_5$Me$_4$(CH$_2$CH$_2$NMe$_2$) complex has also been described [347]. Thermal decomposition results in the formation of fulvene Ti(CH=CH$_2$){η^1,η^5-CH$_2$C$_5$Me$_4$}Cp* (26) [348]. Addition of HC≡CR (R = Pr, Bu^t, Ph,

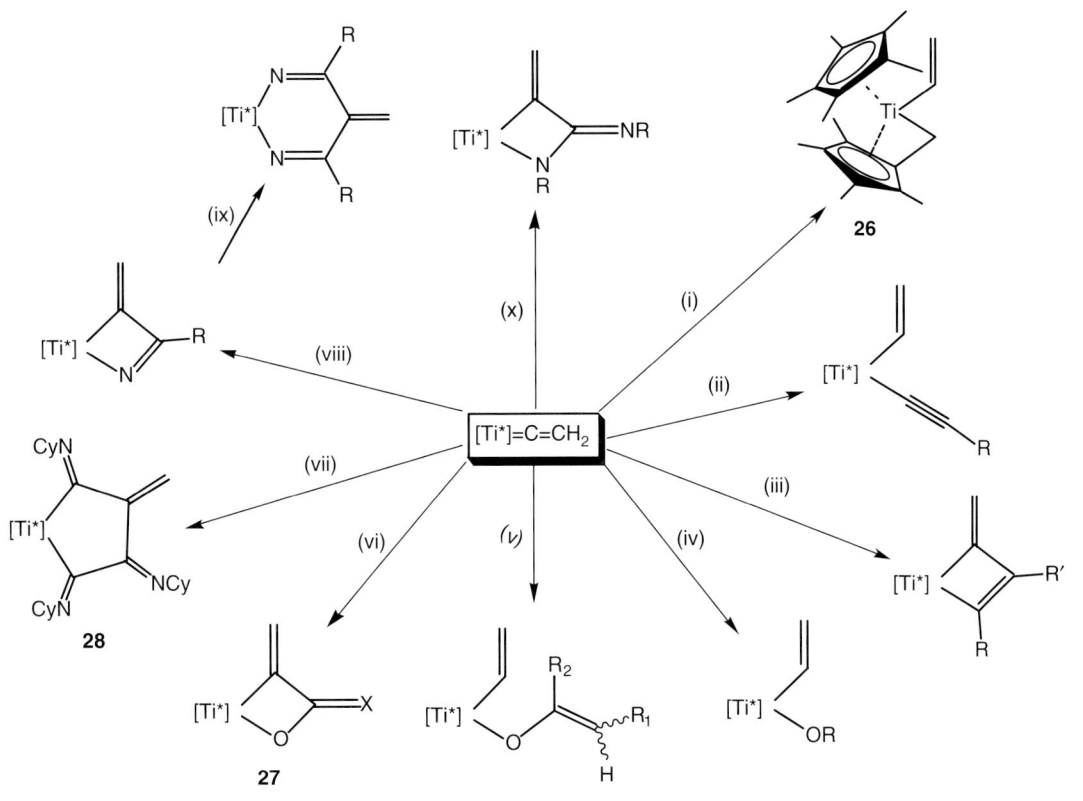

Scheme 1.14 Reagents: (i) r.t., spontaneous; (ii) HC≡CR; (iii) RC≡CR'; (iv) ROH; (v) R^{1}CH$_2$C(O)R^2; (vi) O=C=X(X=O, NR, CR$_2$); (vii) RNC; (viii) RCN; (ix) excess RCN; (x) RN=C=NR.

SiMe$_3$) or ROH (R = H, alkyl) gives Ti(X)(CH=CH$_2$)Cp$^*_2$ (X = C≡CR, OR). Enolizable ketones R^{1}CH$_2$C(O)R^2 give Ti(OCR^2=CHR^1)(CH=CH$_2$)Cp$^*_2$ [349], while heterocumulenes [O=C=X, X = O, NCy, CPh$_2$, C(CBu^t)$_2$] undergo cycloaddition to give titana-oxetanes Ti{C(=CH$_2$)C(X)O}Cp$^*_2$ (**27**) [350]. Similar cyclic complexes are formed from nitriles, carbodiimides and phospha-alkynes; disubstituted alkynes afford methylenetitanacyclobutenes. Isonitriles, for example CyNC, react to give intermediates Ti(η^2-CyN=CC=CH$_2$)Cp$^*_2$ which react further to give cyclic Ti{C(=NCy)C(=CH$_2$)C(=NCy)C(=NCy)}Cp$^*_2$ (**28**).

1.4.2
Complexes Derived From Li[M(C≡CR)(CO)(NO)Cp] (M = Cr, W)

Treatment of Li[M(C≡CR)(CO)(NO)Cp] [M = Cr, W; from LiC≡CR and M(CO)$_2$(NO)Cp] with electrophiles (E) gives M(=C=CER)(CO)(NO)Cp. Thermal isomerization of W{=C=CH(SiMe$_2$Bu^t)}(CO)(NO)Cp to the η^2-alkyne isomer occurs by a 1,2-silyl group shift [209]. Isomerization of Mo(=C=CHBu^t)(CO)(NO)Cp is solvent dependent. Addition of ICH$_2$CH$_2$I to Li[W(C≡CBu^t)(CO)(NO)Cp] followed by reaction with LiC≡CR (R = Bu^t, Ph, SiMe$_3$) affords metallafurans WI{η^2-OC(C≡CR)CH=CBu^t}(NO)Cp (**29**) [351].

29
[W] = WI(NO)Cp

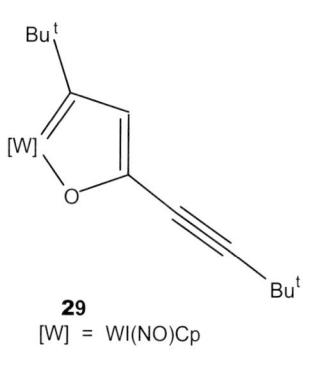

Aminocarbene complexes W{=C(NR^{1}R^2)CH$_2$R}(CO)(NO)Cp [R = Bu^t, SiMe$_2$Bu^t; R^1 = H, R^2 = Bu, Pr^i, Bu^t; R^{1}R^2 = (CH$_2$)$_4$] have been obtained from primary and secondary amines [290]. The intermediate η^2-carbamoyl-vinyl complexes W{(Z)-CH=CHBu^t}{η^2-C(O)NR^{1}R^2}(NO)Cp were obtained with shorter reaction times, and converted to the aminocarbenes on standing or with higher amine concentration. Transfer of H$^+$ from W(=C=CH$_2$)(CO)(NO)Cp in reactions with enamines CR^{1}R^2=C(NC$_4$H$_8$)CHR^{1}R^2 (NC$_4$H$_8$ = pyrrolidino, R^1,R^2 = alkyl) gives iminium cations, which eliminate pyrroline [352]. A Mannich reaction then affords β-aminoalkylated vinylidenes. Migration of N–H to C$_\alpha$ gives the vinylcarbenes W{=CHCH=C(CHR^{1}R^2)$_2$}(CO)(NO)Cp. The product from addition of ynamines R'C≡CNEt$_2$ to W(=C=CHR)(CO)(NO)Cp (R = Ph, Bu^t) depends upon R. For R = Ph, cyclobutenylidene complexes W{=CCPh=C(NEt$_2$)CHR'}(CO)(NO)Cp are formed, whereas for R = Bu^t, the aminocarbenes W{=C(NEt$_2$)CR'=C=CHBu^t}(CO)(NO)Cp are obtained [291]. For R = H, a mixture of the two products is formed. The primary step in all of these reactions is attack at C_α to give W{C(CHR)CR'=C=NEt$_2$}(CO)(NO)Cp (**A**), which may undergo attack of the W at C_α to give metallacyclobutene **B**, which transforms to allenyl-carbene **30**. Addition of the iminium group to C_β gives **31** via the cyclobutenylidene **C**, while attack by the carbonyl group at C_α gives cyclic acyl **D**, which can be trapped by the vinylidene or by ynamine to give **32** and **33**, respectively (Scheme 1.15).

Neutral phosphinovinyl complexes WCl{η^2-PR$_2$C=CHR'}(NO)Cp (R = Ph, R' H, Me, Ph; R = Bu^t, R = H) are obtained from PClR$_2$ and W(=C=CHR)(CO)(NO)Cp [353]. The mechanism probably involves nucleophilic attack by phosphorus on C_α, followed by migration of Cl to the metal center with elimination of CO induced by the lone pair on P. Phospha-allene complexes W(η^2-RP=C=CHR^2)(CO)(NO)Cp (R^2 = H, Ph; R = Bu^t, Cy) are formed from phospha-alkenes RP=C(NMe$_2$)$_2$ [354]. Arsa-alkenes react similarly to give η^2-arsa-allene complexes W{η^2-R^{1}As=C=CHR^2)(CO)(NO)Cp [R^1/R^2 = Bu^{t}C(O)/Bu^t; Fe(CO)$_2$Cp/Bu^t; Bu^{t}C(O)/Ph] [355].

1.4.3
Reactions of M(=C=CRR')(CO)$_5$ (M = Cr, Mo, W)

Thermolysis of Cr(=C=CMeR)(CO)$_5$ (R = Bu^t, Pr) results in the formation of Cr$_2${μ-(=C=CMeR)}(CO)$_9$; photolysis of this complex, or of M(=C=CMeR)(CO)$_5$ (M = Cr, W) directly, results in demetallation and formation of cyclobutane-1,3-diones, by formal dimerization of the vinylidene/CO-coupled product propadienone [342].

According to the nature of the reagent, imines react to give either zwitterionic or iminocarbene complexes, or azetidinylidene derivatives. Thus, for Cr(=C=CMe$_2$)(CO)$_5$, zwitterionic complexes {(OC)$_5$Cr}$^-${C(=CMe$_2$)(im)}$^+$ are obtained from MeN=CHPh, MeN=CMe(OMe), 2-methyl-4,5-dihydro-oxazole and -thiazole. Similar reactions are found for W(=C=CPh$_2$)(CO)$_5$. In contrast, xanthylideneimine, HN=C(C$_6$H$_4$)$_2$O, affords iminocarbene Cr{=CPr^{i}N=C(C$_6$H$_4$)$_2$O}(CO)$_5$. Amino(alkenyl)carbene complexes M{=C(NHPh)CAr=CHAr}(CO)$_5$ (M = Cr, W; Ar = C$_6$H$_4$R-4, R = H, Me, Br) are formed from M(=C=CAr$_2$)(CO)$_5$ and PhN=CHAr, together with lesser amounts of the 2-azetidin-1-ylidene complexes M{=CNPhCHArCHAr}(CO)$_5$ [261, 356]. The former are formed by cycloaddition to the alkyne complexes followed by H shift and ring-opening, whereas the latter are formed by cycloaddition to the vinylidene, probably by cyclization of the first-formed {(OC)$_5$Cr}$^-${C(=CMe$_2$)(NMe=CHPh)}$^+$ [357]. Complexes M(=C=CHPh)(CO)$_5$ (M Cr, Mo, W; by thermolytic isomerization of the HC≡CPh complexes) react with RN=CHPh (R = Et, Pr^i) to give the 2-azetidin-1-ylidene complexes, predominantly as the *syn* isomers (*syn/anti* > 9) [358]. Dialkylcarbodiimides RN=C=NR (R = Cy, Pr^i) similarly afford W{=CNRC(=NR)CHPh}(CO)$_5$. On warming, the O- and S-heterocycles give thiazole and oxazole complexes.

Regiospecific cycloaddition of the C≡C triple bond in alkynes MeC≡CX (R = alkyl, aryl; X = OEt, SMe, NMe$_2$, NEt$_2$) to the C=C double bond of the vinylidene in M

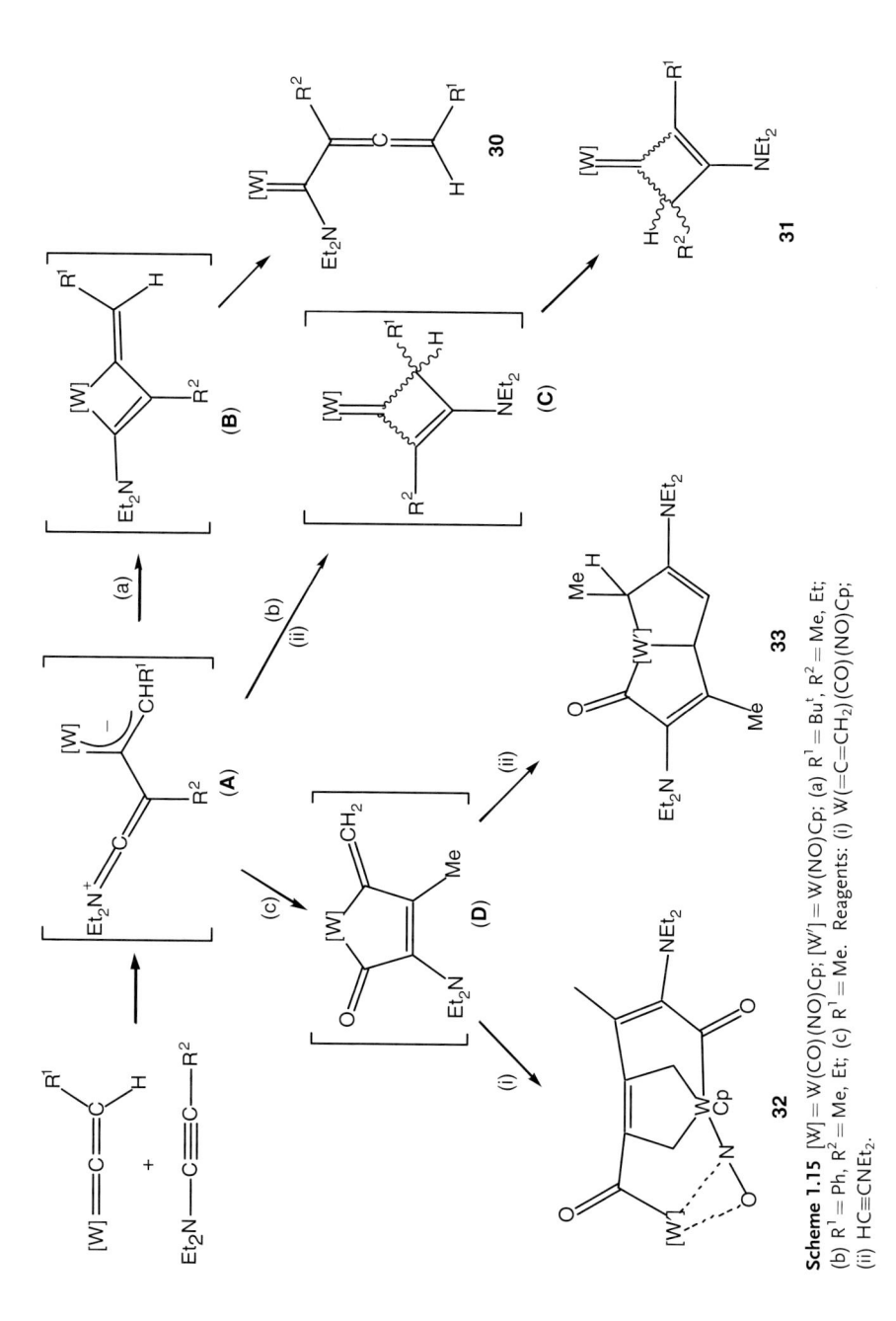

Scheme 1.15 [W] = W(CO)(NO)Cp; [W'] = W(NO)Cp; (a) R^1 = Bu^t, R^2 = Me, Et;
(b) R^1 = Ph, R^2 = Me, Et; (c) R^1 = Me. Reagents: (i) W(=C=CH$_2$)(CO)(NO)Cp;
(ii) HC≡CNEt$_2$.

1.4 Chemistry of Specific Complexes

(=C=CRR')(CO)$_5$ (M = Cr, RR' = MeBu^t; M = W, RR' = Ph$_2$, R = Me, Et, R' = Bu^t) or thermolabile W(η^2-HC$_2$R')(CO)$_5$ [R' = Ph, Bu^t, Cy, CO$_2$Me, CMe=CH$_2$, CH=CH(OMe)] give cyclobutenylidene complexes M{=CCR'=C(X)CHMe}(CO)$_5$ [357–360]. Oxidative cleavage of the products gives cyclobutenones [357].

1.4.4
Reactions of M(=C=CRR')(CO)(L) Cp (M = Mn, Re) (Scheme 1.16)

Thermally unstable M(=C=CHR)(CO)$_2$Cp (M = Mn, Re; R = H, Me, Ph) react with a variety of nitrogen bases [234]. Imines undergo [2 + 2]-cycloaddition reactions to give Mn(=CCH$_2$CHPhNR')(CO)$_2$Cp (R' = Me, Ph) which can be oxidized (K[MnO$_4$]) to β-lactams. For M = Re, R = H, Re{=CC(=CHPh)CHPhNR'}(CO)$_2$Cp (34; R' = Me, Ph) are formed. With Ph$_2$C=NH, iminocarbenes M{=C(CH$_2$R)N=CPh$_2$}(CO)$_2$Cp are obtained, together with Mn(=CCHPhCPh$_2$NH)(CO)$_2$Cp (35; for M = Mn, R = Ph). An organometallic analog of the Beckmann reaction is found in the formation of nitrile complexes Mn(NCCH$_2$R)(CO)$_2$Cp (R = H, Ph) from Ph$_2$C=NNH$_2$. Carbodiimides react with Re(=C=CH$_2$)(CO)$_2$Cp to give isocyanide complexes Re(CNR)(CO)$_2$Cp (R = Pr^i, Bu^t) by metathesis of the vinylidene C=C and C=N bonds. The manganese reaction affords Mn(CO)$_2${=C(NHPr^i)NPr^{i}C(=CH$_2$)-η-C$_5$H$_4$} (36), with some Mn{=C(NHPr^i)$_2$}(CO)$_2${η-C$_5$H$_4$C(O)Me} as a by-product. Benzalazine, PhCH=NN=CHPh, gives the bicyclic bis(carbene) complex {Re(CO)$_2$Cp}$_2${μ-(=CCH$_2$CHPhN)$_2$} (37) from Re(=C=CH$_2$)(CO)$_2$Cp [361, 362].

Reactions of Mn(=C=CHPh)(CO)$_2$Cp with P(OR)$_3$ (R = Et, Ph) afford Mn{η^2-PhCH=CHP(O)(OR)$_2$}(CO)$_2$Cp in an unusual variant of the Arbuzov reaction [307]. However, with PR$_3$ (R = Me, Et), Mn(η^2-C$_2$H$_2$)(CO)$_2$Cp' (Cp' = Cp, Cp^{Me}, Cp*) forms Mn{η^2-HC$_2$H(PR$_3$)}(CO)$_2$Cp' which transform to Mn{C(PR$_3$)=CH$_2$}(CO)$_2$Cp' [363].

The bis(vinylidene)s {Mn(CO)$_2$Cp'}$_2$(μ-C$_4$Ph$_2$) (Cp' = Cp, Cp*) are oxidized ([FeCp$_2$]$^+$, Ag$^+$) to give the bis(carbyne)s [{Cp'(OC)$_2$Mn}$_2$(≡CCPh=CPhC≡)]$^{2+}$, which react directly with H$_2$O to give cyclic bis(carbene)s {Mn(CO)$_2$Cp'}$_2$(μ-C$_4$Ph$_2$O) (38) [364].

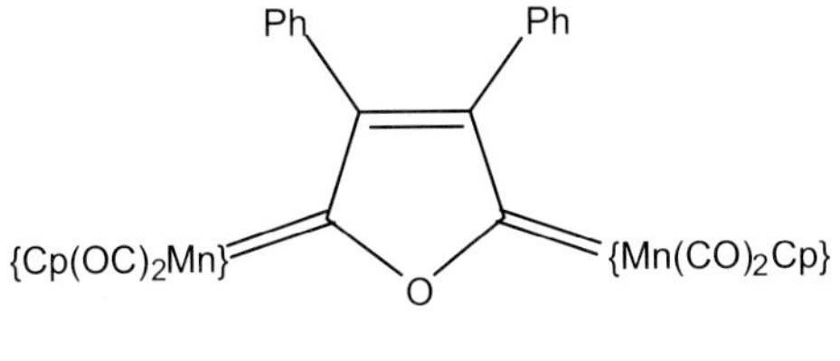

38

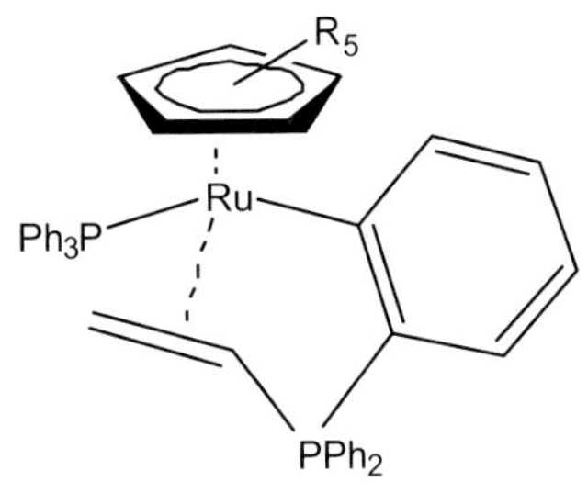

39 R = H, Me

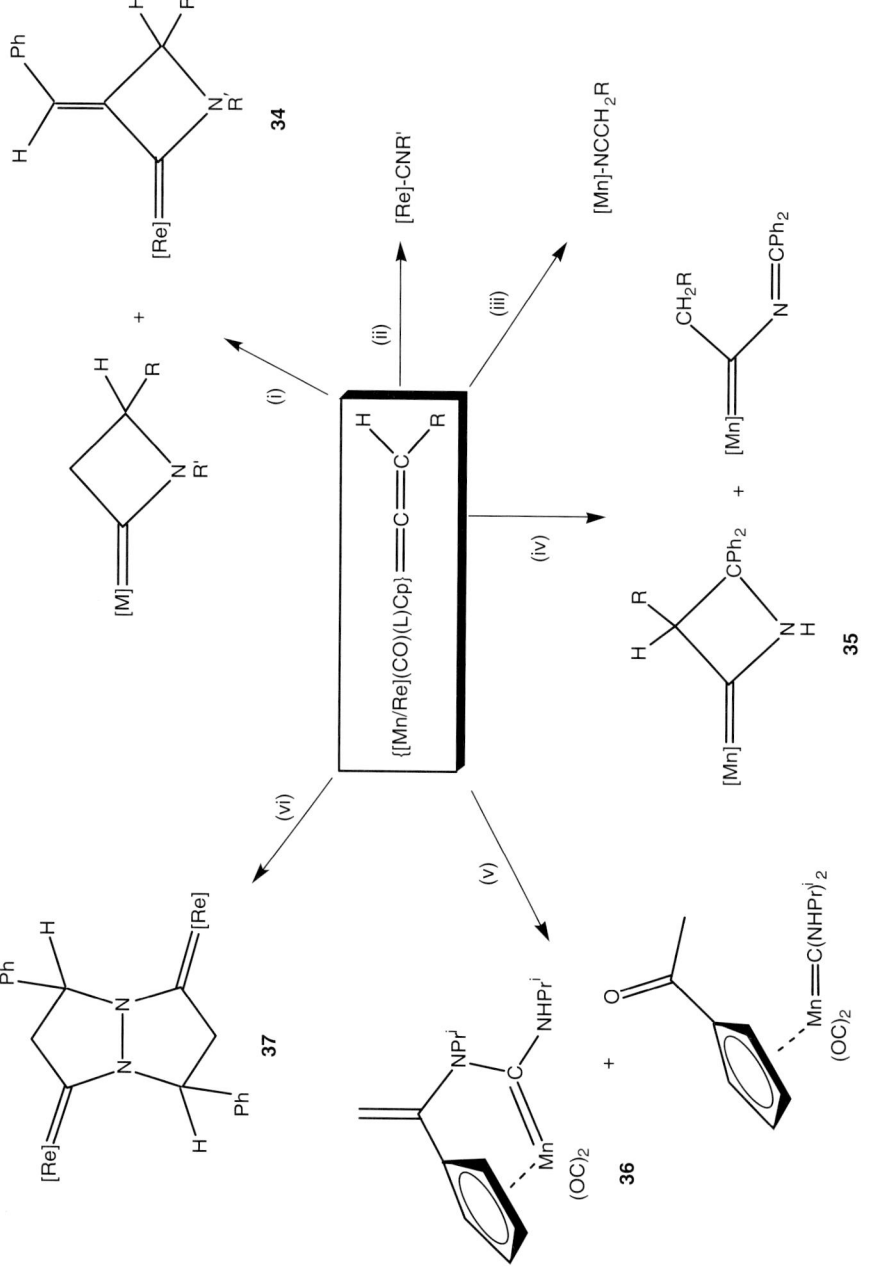

Scheme 1.16 [M] = {Mn/Re}(CO)(PPh₃)Cp. Reagents: (i) CHPh=NR′;
(ii) R′N=C=NR′ (R′ = Pr^i, Bu^t); (iii) Ph₂C=NNH₂; (iv) Ph₂C=NH;
(v) Pr^{i}N=C=NPr^i; (vi) PhCH=NN=CHPh.

1.4.5
Reactions of [M(=C=CRR′)(L′)(P)Cp′]$^+$ (M = Fe, Ru, Os)

Much of the chemistry of the vinylidene group was established with Group 8 complexes of the type [M(=C=CRR′)(L′)(P) Cp′]$^+$ and several of the ubiquitous reactions of vinylidenes with O-, S- and N-nucleophiles and with electrophiles have been mentioned above.

Thermal decomposition of the parent iron complex gives [Fe{C(PPh$_3$)=CH$_2$}(CO)(PPh$_3$)Cp]$^+$ [9]. Reactions of other Group 15 nucleophiles (Nu = py, 4-Mepy, PMe$_2$Ph) give similar products [Fe{C(Nu)=CH$_2$}(CO)(PPh$_3$)Cp]$^+$, while phosphonio-vinyls are also formed by heating Ru(=C=CH$_2$)(PPh$_3$)$_2$Cp′ (Cp′ = Cp, Cp*) [294]. Phosphonio-vinyl Ru{E-CH=CPh(PPh$_3$)}(CO)(PPh$_3$)(η^5-C$_9$H$_7$)]$^+$ results from attack on C$_\beta$ of an intermediate η^2-alkyne complex [102]. Reactions of [Fe(=C=CR$_2$)(CO) {P(OMe)$_3$}Cp]$^+$ with PhCH=NMe and 2-thiazolines give [2 + 2]-cycloadducts (azetidinylidenes) via iminium complexes, which upon oxidation (PhIO) afford β-lactams [365]. Thermolysis of [Ru(=C=CH$_2$)(PPh$_3$)$_2$Cp′]$^+$ (Cp′ = Cp, Cp*) in refluxing MeCN results in cyclometallation and insertion of the vinylidene into the Ru–P bond, giving the phosphonium complex **39** whereas the Cp/PMePh$_2$ complex, when heated in the presence of free PMePh$_2$, gives [Ru{C(=CH$_2$)PMePh$_2$}(PMePh$_2$)$_2$Cp]$^+$ [294].

The reaction of RuCl{=C=CH(CO$_2$Me)}(PPh$_3$)Cp with Sn(CH=CH$_2$)$_4$ gives Ru{η^3-CH$_2$CHC=CH(CO$_2$Me)}(PPh$_3$)Cp [36]. Addition of CH$_2$N$_2$ to [Ru(=C=CHR)(PP)Cp]$^+$ (R = Me, Ph; PP = PPh$_2$CHMeCHMePPh$_2$) gives the corresponding [Ru(=C=CMeR)(PP)Cp]$^+$ [318]. Initial displacement of PPh$_3$ from [Ru(=C=CHPh)(PPh$_3$)$_2$Cp]$^+$ by pyridine, followed by [2 + 2]-cycloaddition, gives [Ru{C(py)=CHPh)}(PPh$_3$)Cp]$^+$ [76]. This is followed by deprotonation to give aza-allyl Ru(η^3-NC$_5$H$_4$C=CHPh)(PPh$_3$)Cp, which is protonated to release E-2-PhCH=CHC$_5$H$_4$N. The reaction becomes catalytic if RC≡CSiMe$_3$ is used as the source of the vinylidene.

1.4.6
Reactions of [Ru{=C=C(SMe)$_2$}(PMe$_3$)$_2$Cp]$^+$ (Scheme 1.17)

Cleavage of a C–S bond in [Ru{=C=C(SMe)$_2$}(PMe$_3$)$_2$Cp]$^+$ occurs upon reduction (Na/Hg, Na[BHEt$_3$]) to give Ru(C≡CSMe)(PMe$_3$)$_2$Cp (**40**) and MeSSMe [145]. Nucleophilic addition of ROH (R = Me, Et) affords [Ru{=C(OR)CH$_2$(SMe)}(PMe$_3$)$_2$Cp]$^+$ and deprotonation of the former gives Ru{C(OMe)=CH(SMe)}(PMe$_3$)$_2$Cp. Addition of electrophiles (E = H$^+$, Me$^+$) to C$_\beta$ gives [Ru{=C=CE(SMe)}(PMe$_3$)$_2$Cp]$^+$ [E = H (**41**), Me (**42**)], while methylation ([OMe$_3$]$^+$) affords [Ru{=C=CR(SMe$_2$)}(PMe$_3$)$_2$Cp]$^+$ (R = H (**43**), Me). Reduction (Na/Hg or Na[BHEt$_3$]) of **42** and **43** give Ru(C≡CMe)(PPh$_3$)$_2$Cp, together with either MeSSMe or SMe$_2$. Deprotonation of **41** and **43** give **40** and [Ru{C≡C(SMe$_2$)}(PMe$_3$)$_2$Cp]$^+$ (**44**), respectively.

Addition of electrophiles (HBF$_4$, [MeSSMe$_2$]$^+$, [OMe$_3$]$^+$) to [Ru{=C=C(SMe)$_2$}(PMe$_3$)$_2$Cp]$^+$ gives [Ru{η^2-MeSC=CR(SMe)}(PMe$_3$)$_2$Cp]$^{2+}$ [R = H, SMe, SMe$_2$ (**45**), respectively] [293].

Tertiary phosphines abstract sulfur from **45** to give [Ru{C≡C(SMe$_2$)}(PMe$_3$)$_2$Cp]$^+$ and [PPh$_2$R(SMe)]$^+$ (R = Me, Ph). Nucleophiles (NaSR, 4-RC$_5$H$_4$N or SEt$_2$) displace

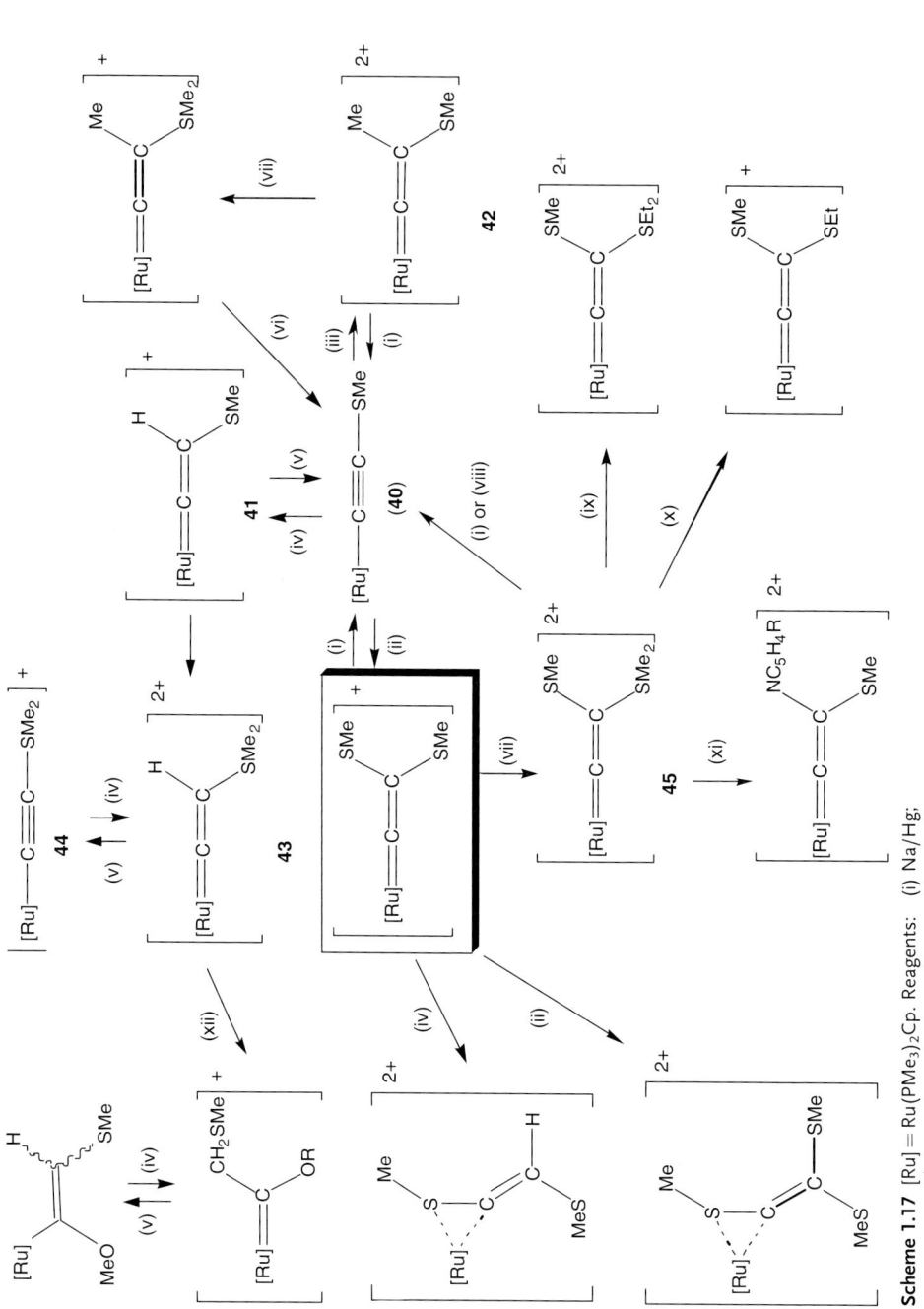

Scheme 1.17 [Ru] = Ru(PMe₃)₂Cp. Reagents: (i) Na/Hg; (ii) [Me₂SSMe]⁺; (iii) MeI; (iv) HBF₄·Et₂O; (v) NaOMe/MeOH; (vi) Na[BHEt₃]; (vii) [Me₃O]⁺; (viii) PRPh₂ (R = Me, Ph), -Ph₂RP=S; (ix) Et₂S; (x) NaSEt; (xi) NC₅H₄X-4 (X = H, Et, NMe₂); (xii) ROH (R = Me, Et).

SMe$_2$ from **45** to give [Ru{=C=C(SR)(SMe)}(PMe$_3$)$_2$Cp]$^+$ (R = Me, Et), [Ru{=C=C(4-RC$_5$H$_4$N)(SMe)}(PMe$_3$)$_2$Cp]$^+$ and [Ru{=C=C(SEt$_2$)(SMe)}(PMe$_3$)$_2$Cp]$^{2+}$, respectively.

1.4.7
Reactions of *trans*-MCl(=C=CRR′)(L)$_2$ (M=Rh, Ir)

Vinylidene transfer from Rh to Ir occurs on treatment of Rh(=C=CH$_2$)(L)Cp with {IrCl(L)$_2$}$_n$ to give *trans*-IrCl(=C=CH$_2$)(L)$_2$ [366]. Treatment of *trans*-RhCl(=C=CHR′)(L)$_2$ (R′=Ph, Bu^t) with RMgX (R = Me, Ph, CH=CH$_2$) gives *trans*-RhR(=C=CHR′)(L)$_2$, which with CO or Bu^{t}NC (L′) forms *trans*-Rh(CR=CHR′)(L′)(L)$_2$ as the Z isomers by stereoselective C–C coupling at C$_\alpha$ [367]. The free alkenes could be liberated by treatment of the carbonyls with acetic acid. Alkynes HC≡CR (R = Bu^t, Ph) react with Rh(η^3-benzyl)(L)$_2$ to give *trans*-Rh(C≡CR)(=C=CHR)(L)$_2$ which, with HCl for R = Ph, give the enyne complex *trans*-RhCl(η^2-PhC$_2$CH=CHPh)(L)$_2$; the enyne was displaced with CO [316]. Alternatively, treatment of the alkynyl(vinylidene) complex with CO gives *trans*-Rh{C(=CHR)C≡CR}(CO)(L)$_2$, which with CF$_3$CO$_2$H affords RCH=C=C=CHR.

The complex *trans*-RhF(=C=CHFc)(L)$_2$ reacts with Ph$_3$SnC≡CC≡CSnPh$_3$ to give *trans*-{(L)$_2$(FcCH=C=)Rh}$_2$(μ-C≡CC≡C) which, under CO, transforms into *trans*-{(L)$_2$(OC)Rh}$_2${μ-C(=CHFc)C≡CC≡C=CHFc} [52]. Similarly, 1,4-C$_6$H$_4${(CH=C=)RhF(L)$_2$}$_2$ reacts with RC≡CSnPh$_3$ (R = Me, Ph) to give 1,4-C$_6$H$_4${(CH=C=)Rh(C≡CR)}$_2$, which give 1,4-C$_6$H$_4${CH=C(CCR)Rh(CO)(L)$_2$}$_2$ under CO.

Octahedral RhH(C≡CR)$_2$(py)(L)$_2$ (R = H, Me, Bu^t, Ph) are formed from *trans*-Rh(C≡CR)(=C=CHR)(L)$_2$ and pyridine [40]. The complex *trans*-Rh(C≡CH)(η-C$_2$H$_4$)(L)$_2$ is protonated to *trans*-[Rh(=C=CH$_2$)(py)(L)$_2$]$^+$ by [pyH]BF$_4$, and converted to Rh(=C=CH$_2$)(L)Cp with cyclopentadiene. The ethene complex reacts with ethyne in the presence of NEt$_3$ to give *trans*-Rh(C≡CH)(=C=CH$_2$)(L)$_2$, while the reactions of *trans*-Rh(C≡CR)(η-C$_2$H$_4$)(L)$_2$ (R = Bu^t, Ph) with HC≡CCO$_2$Me afford *trans*-Rh(C≡CR){=C=CH(CO$_2$Me)}(L)$_2$.

Organolithiums LiR^3 (R^3 = Me, Ph, C≡CPh) react with *trans*-IrCl(=C=CR^{1}R^2)(L)^2 (R^1 = Me, R^2 = SiMe$_3$; R^1 = H, R^2 = Ph) to give *trans*-IrR^3(=C=CR^{1}R^2)(L)$_2$; use of Grignard reagents results only in halide metathesis [368]. The acid–base reactions between *trans*-Ir(OH)(=C=CR^{1}R^2)(L)$_2$ and HC≡CR^3 (R^3 = Ph, CO$_2$Me) give *trans*-Ir(C≡CR^3)(=C=CR^{1}R^2)(L)$_2$. Migratory insertion occurs in reactions of CO with *trans*-IrR^3(=C=CR^{1}R^2)(L)$_2$ and with *trans*-IrMe(=C=CHPh)(L)$_2$ to give vinyl complexes *trans*-Ir{(Z)-CR^3=CR^{1}R^2}(CO)(L)$_2$ and *trans*-Ir{(Z)-CMe=CHPh}(CO)(L)$_2$, respectively, from which the corresponding alkenes could be obtained by treatment with acid. Fluoride complex *trans*-IrF(=C=CPh)(L)$_2$ is obtained from *trans*-Ir(OH)(=C=CHPh)(L)$_2$ and NEt$_3$.3HF or from *trans*-Ir(O$_2$CCF$_3$)(=C=CHPh)(L)$_2$ and [NBu$_4$]F. The azido complexes *trans*-Ir(N$_3$)(=C=CHR)(L)$_2$ (R = Ph, CO$_2$Me), from the hydroxy compounds and NaN$_3$, convert to *trans*-Ir{CH(CN)R}(CO)(L)$_2$ under CO.

Vinylidenes *trans*-IrCl(=C=CHR)(L)$_2$ (R = H, Me, Ph) react with HBF$_4$ to give initially [IrHCl(=C=CHR)(L)$_2$]$^+$ which transform into carbynes *trans*-[IrCl(≡CCH$_2$R)(L)$_2$]$^+$ [154]. The vinylidenes are regenerated with NaH. Ligand displacement with X^- ($X = OH$, I, Me) gives trans-IrX(=C=CHR)(L)$_2$, while with CO, trans-IrI(=C=CH$_2$)(L)$_2$ affords IrHI(C≡CH)(CO)(L)$_2$. Conversion of trans-IrCl (=C=CHR)(L)$_2$ (R = H, Ph) to IrH$_2$(CH=CHR)(CO)(L)$_2$ and IrClI(CMe=CH$_2$)(L)$_2$ was achieved by treatment with NaOMe and MeI, respectively.

1.5
Reactions Supposed to Proceed via Metal Vinylidene Complexes

Many stoichiometric reactions, usually involving alkynes, have been reported to proceed via metal vinylidene complexes which have not been detected nor isolated. This section does not survey the extensive field of reactions *catalyzed* by metal vinylidene complexes.

The formation of carbene complexes from HC≡CR in alcohols (R'OH, R = Me, Et, Pr^i) such as Ru{=CCH$_2$R(OR')}(PR$_3$)(η-C$_6$Me$_6$) (PR$_3$ = PMe$_3$, PMe$_2$Ph, PPh$_3$), is attributed to the rapid attack of the initial vinylidene by the alcohol [369]. In the case of the Ru(PPh$_3$)$_2$Cp system, this reaction is slow, allowing isolation of the vinylidene, but prolonged heating results in transformation to the carbene. However, if the reaction is carried out in Bu^{t}OH, the vinylidene is stable and can be isolated in pure form. With RuCl(P~O)(η-C$_6$Me$_6$) [P~O = PPh$_2$(C$_6$H$_3$-2-O-OMe-6)] and HC≡CR, a mixture of the vinylidene and methoxy-carbene complexes is obtained, while HC≡CCMe$_2$(OH) reacts to form carbene complex Ru{=C(OMe)CH=CMe$_2$}(P~O)(η-C$_6$Me$_6$) [131].

The reactions are those of functionalized 1-alkynes, the first of which were described for alkynes bearing substituted hydroxymethyl groups, such as substituted propargyl alcohols, HC≡CCRR'(OH), and often proceed further to form metal allenylidene complexes, by spontaneous dehydration of a (usually unobserved) hydroxy-vinylidene complex (Equation 1.22):

$$ML_n + HC\equiv CCRR'(OH) \rightarrow L_nM=C=CHCRR'(OH)$$
$$\rightarrow L_nM=C=C=CRR' + H_2O \quad (1.22)$$

First described by Selegue for the Ru(PMe$_3$)$_2$Cp system [370], the generality of this approach has been established for many metal-ligand combinations, including Ru(EPh$_3$)$_2$Cp (E = P, Sb) [371], Os(L)$_2$Cp [272], Ru(PR$_3$)$_2$Tp (PR$_3$ = PPr^i$_3$, PPr^{i}Ph$_2$, PPh$_3$) [130], Ru(CO)(PMePr^i$_2$)Cp* [104], RuCl{O$_2$CCH(dmpz)$_2$}(PPh$_3$)[47] and Ru(PPh$_3$){PPh$_2$(C≡CPh)}(η^5-C$_9$H$_7$) [103].

ω-Hydroxyalkyl-1-alkynes containing longer chains react by intramolecular attack of the hydroxy group on C$_\alpha$ of the vinylidene, to give cyclic oxacarbene complexes (Equation 1.23):

$$ML_m + HC\equiv C(CH_2)_nOH \longrightarrow \underset{HO-(CH_2)_{n-1}}{L_mM=C=CH} \longrightarrow L_mM=C\underset{O-(CH_2)_{n-1}}{\diagup}$$

$$(1.23)$$

1.5 Reactions Supposed to Proceed via Metal Vinylidene Complexes

These reactions involve intramolecular nucleophilic attack on C_α. For example, $HC\equiv C(CH_2)_2OH$ and $RuCl_2(P)(\eta\text{-}C_6Me_6)$ give $[RuCl\{=CO(CH_2)_2CH_2\}(PR_3)(\eta\text{-}C_6Me_6)]^+$ ($PR_3 = PMe_3$, PMe_2Ph) [369]. The cations $[Ru\{=CO(CH_2)_nCH_2\}(P)_2(\eta^5\text{-}C_9H_7)]^+$ ($n = 5$–7) are obtained from $RuCl(P)_2(\eta^5\text{-}C_9H_7)$ and $HC\equiv C(CH_2)_n OH$, while $HC\equiv C(CH_2)_3CH_2OH$ also gives $[Ru\{=C=CH(CH_2)_4OH\}(P)_2(\eta^5\text{-}C_9H_7)$ [372]; $trans\text{-}RuCl\{=CO(CH_2)_2CHR\}(L)(L')]^+$ ($R = H$, Me; $L = L' = $ dppm, dppee; $L = $ dppm, $L' = $ bpy) have also been prepared [373]. A dicarbene complex is formed from $cis\text{-}RuCl_2(bpy)_2$, while $RuCl_2(PPh_3)_3$ and $HC\equiv C(CH_2)_2OH$ gave $[RuCl\{=CO(CH_2)_3\}(PPh_3)_2]^{2+}$ [373]. A variety of similar complexes was obtained from $RuCl(EPh_3)_2Cp$ (E = P, Sb) and $HC\equiv CCR^1R^2(OH)$ [$R^1 = R^2 = H$, Me; $R^1R^2 = $ MeEt, $(CH_2)_5$] [371], and $RuCl(cod)Tp$ with $HC\equiv CCR'_2(OH)$ and PR_3 gives hydroxyvinylidenes, allenylidenes or cyclic oxacarbene complexes, depending on substituents (Scheme 1.18) [130]. Related to these reactions is the formation of $[Ru(=COCH_2CH_2CHPh)(CO)(PPh_3)Cp]^+$ from $[Ru(=C=CHPh)(CO)(PPh_3)Cp]^+$ and ethylene oxide [150]. The oxacarbene can be deprotonated to give dihydrofuran $Ru(C=CPhCH_2CH_2O)(CO)(PPh_3)Cp$.

Reactions of propargylic alcohols $HC\equiv CCHR(OH)$ (R = H, Me) with $RuCl(L)_2Cp'$ (L = PMe_2Ph; PPh_3; $Cp' = Cp^* \eta^5\text{-}C_9H_7$) in MeOH apparently proceed by addition of MeOH across $C_\beta = C_\gamma$ of the allenylidene complexes formed by initial dehydration of the hydroxyvinylidenes [279, 375].

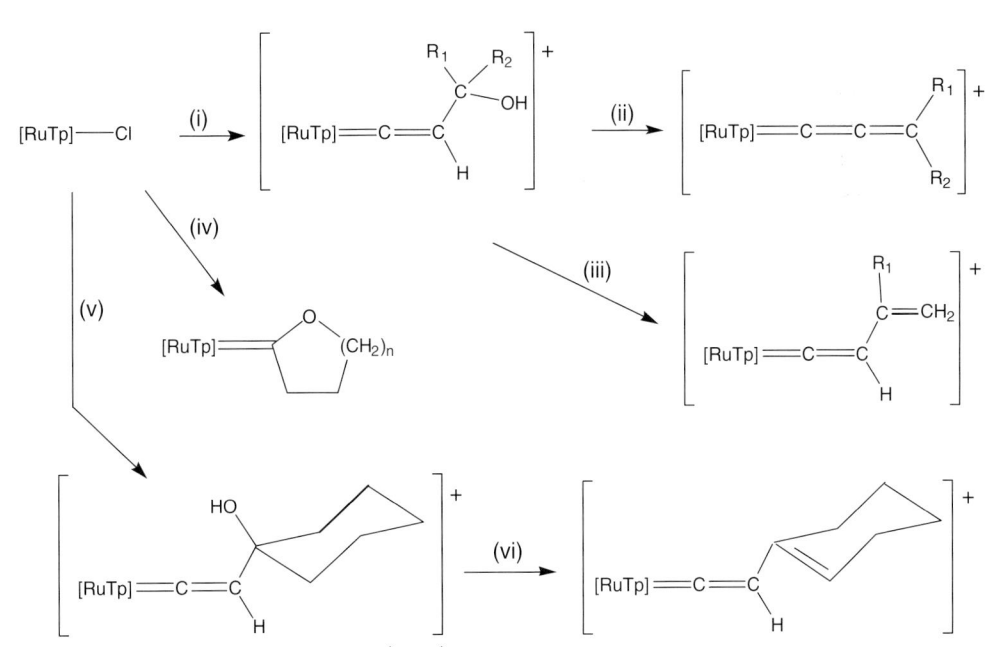

Scheme 1.18 [RuTp] = $Ru(PR_3)_2Tp$ ($PR_3 = P^iPr_3$, P^iPrPh_2, PPh_3).
Conditions: (i) $HC\equiv CCR^1R^2(OH)$ ($R^1 = R^2 = $ Ph, Fc; $R^1/R^2 = $ Me/Ph); (ii) $-H_2O$ ($R^1 = R^2 = $ Ph, Fc); (iii) $-H_2O$ ($R^1/R^2 = $ Me/Ph); (iv) $HC\equiv CCH_2)(CH_2)nOH$ ($n = 1, 2, 3$); (v) 1-$HC\equiv C$-l-OH-cyclohexane.

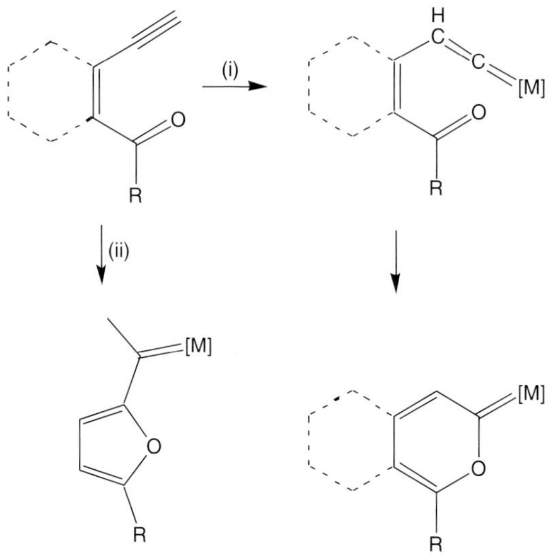

Scheme 1.19 [M] = M(CO)$_5$ (M = Cr, W). Reagents: (i) Cr(CO)$_5$(thf) (R = OMe, NEt$_2$); (ii) M(CO)$_5$(thf), R = Ph, 2-C$_4$H$_4$S.

Methylated propargyl alcohols react to give binuclear complexes, formed by attack of L$_n$M=C=C=CMe$_2$ or L$_n$MC≡CCMe=CH$_2$, which could be formed by dehydration/deprotonation of an initial vinylidene L$_n$M=C=CHCMe$_2$(OH). For example, RuCl(PPh$_3$)$_2$Cp reacts with HC≡CCMe$_2$(OH) to give [{Ru(PPh$_3$)$_2$Cp}$_2$(μ-C$_{10}$H$_{12}$)]$^{2+}$, in which the ligand contains a cyclohexenyl ligand formed by regioselective formation of two new C–C bonds, attached to the Ru atoms by a vinylidene and an alkylidene link [268].

On Cr, ene-carbonyl-vinylidene complexes undergo conversion to 2-pyranylidene complexes (Scheme 1.19) [376].

Intramolecular attack of acetate on a vinylidene is considered to be the explanation for formation of the cyclic vinyl ester Ru{C(=CHCO$_2$Me)OCMe=O}(PPh$_3$)Cp, obtained from the reaction between Ru(O$_2$CMe)(P)Cp and HC≡CCO$_2$Me [377]. Rapid Z/E isomerization occurs in solution. Similarly, cyclic aminocarbene complexes are obtained from reactions carried out in the presence of nitrogen ligands. For example, RuCl(cod)Tp reacts with HC≡CR in the presence of 2-aminopyridines, X, to afford RuCl{=CCH$_2$R(X)}Tp (X = 2-NH$_2$C$_5$H$_4$N, 2-NH$_2$-4-MeC$_5$H$_3$N, 2-NHMeC$_5$H$_4$N) [378].

Formation of metal acyl or carbonyl complexes from 1-alkynes in the presence of water is often assumed to proceed via attack on an intermediate vinylidene complex to give a hydroxycarbene complex (Equation 1.24):

$$M=C=CHR + H_2O \rightarrow M=C(OH)CH_2R \rightarrow M(CO) + CH_3R \qquad (1.24)$$

First reported for the Ru(PPh$_3$)$_2$Cp system [8], recent papers have described this reaction for RuCl$_2$(PPh$_3$)(pnp) with HC≡CPh, [Ru(C≡CH)(phen)(η-p-cymene)]$^+$ in TfOH to give Ru(COMe)(phen)(η-p-cymene), via characterized vinylidene, alkynyl and acyl derivatives, and from PtMe(C≡CR)(PPh$_3$)$_2$ and water to give [PtMe(CO)(PPh$_3$)$_2$]$^+$ [190].

Other reactions where vinylidenes have been invoked en route to products include the preparation of benzopyranylidene complexes from dienophiles and vinylidene-tungsten complexes formed from 2-HC≡CC$_6$H$_4$C(O)R [379], of Re{C(=CHPh)CBu^t=O}(CO)$_2$(PR$_3$)$_2$ from ReCl(CO)$_3$(PR$_3$)$_2$, LiBu^t and HC≡CPh [380], of pyrylium-iron complexes [Fe{CCHCRCPhOC(OMe)}(CO)$_2$Cp]$^+$ from PhC≡CR (R = H, Ph) and the product from HC≡CCO$_2$Me and [Fe(CO)$_2$Cp]$^+$ [381], of an iridafuran from irida-indene IrI(CH=CPhC$_6$H$_4$)(PPh$_3$)$_2$ and HC≡CCO$_2$Me [382] and of a [3.3.0]azairidabicyclo-octatriene from IrH$_2$Cl(PMe$_3$)$_3$ and 2-HC≡CC$_5$H$_4$N [383].

Coupling of alkynes with other unsaturated hydrocarbons at metal centers has also been demonstrated to occur via vinylidene complexes. Such reactions include dimerization of HC≡CBu^t with Ru(cod)(cot) to afford Z-Bu^{t}CH=C=C=CHBu^t, of RuX$_2$(=C=CHBu^t)(PPh$_3$)$_2$ with LiC≡CBu^t under CO to give RuCl{C(C≡CBu^t)=CHBu^t}(CO)(PPh$_3$)$_2$ [384], of [W(=C=CMeBu^t)(CO)(η^2-HC$_2$Bu^t)Cp*]$^+$ with Na[N(SiMe$_3$)$_2$] under CO to give W{η^1,η^2-C(=CMeBu^t)CMeCBu^t}(CO)$_2$Cp* [385], and dimerization of Me$_3$SiC≡CCO$_2$Et on RuCl{PPr$^i_2$CH$_2$C(O)OMe}Cp* to give RuCl{η^4-trans-C$_4$(SiMe$_3$)$_2$(CO$_2$Et)$_2$}Cp* [38]. The formation of η^2-dienyl-iridium complexes such as **46** from Ir(H)(CPh=CMe=O)(OCMe$_2$)(PPh$_3$)$_2$ and electron-rich alkynes also involves coupling of alkenyl and vinylidene ligands [374].

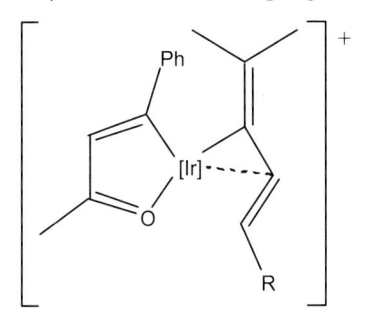

[Ir] = Ir(PPh$_3$)$_2$
R = PhCH$_2$, Ph, C$_6$H$_4$OMe-4

46

Abbreviations

C$_6$H$_9$	cyclohexenyl
cod	1,5-cyclo-octadiene
coe	cyclooctene
Cp^R, Cp^{Si}, Cp*	η-C$_5$H$_4$R, η-C$_5$H$_4$SiMe$_3$, η-C$_5$Me$_5$
dcype	1,2-bis(dicyclohexylphosphino)ethane
dmpz	3,5-dimethylpyrazolyl
dppee	cis-1,2-bis(diphenylphosphino)ethene
Fc	ferrocenyl

Imes	1,3-dimesitylimidazol-2-ylidene
L	PPr$^i_3$
L^{OEt}	CpCo{P(O)(OEt)$_2$}$_3$
NHC	N-heterocyclic carbene
P	tertiary phosphine
pnp	NPr(CH$_2$CH$_2$PPh$_2$)$_2$
PP	bidentate phosphine, usually dmpm, dmpe, dppm, dppe
(R)-chiraphos	(R)-PPh$_2$CHMeCHMePPh$_2$
(R)-prophos	(R)-PPh$_2$CH$_2$CHMePPh$_2$
Rc	ruthenocenyl
tcnq	tetracyanoquinodimethane
Tp, Tp*	[HB(pz)$_3$], [HB(dmpz)$_3$]
X	halide, acetate, OTf

References

1 Brown, R.F.C. and Eastwood, F. (1993) *Synlett*, **1**, 9.

2 (a) Mills, O.S. and Redhouse, A.D. (1966) *Chemical Communications*, 444. (b) Mills, O.S. and Redhouse, A.D. (1968) *Journal of the Chemical Society A*, 1282.

3 (a) King, R.B. and Saran, M.S. (1972) *Chemical Communications*, 1053. (b) King, R.B. (1974) *Annals of the New York Academy of Sciences*, **239**, 171. (c) King, R.B. (2004) *Coordination Chemistry Reviews*, **248**, 1533.

4 (a) Chisholm, M.H. and Clark, H.C. (1970) *J. Chem. Soc. D, Chemical Communications*, 763. (b) Chisholm, M.H. and Clark, H.C. (1972) *Journal of the American Chemical Society*, **94**, 1532. (c) Chisholm, M.H. and Clark, H.C. (1973) *Accounts of Chemical Research*, **6**, 202.

5 (a) Nesmeyanov, A.N., Antonova, A.B., Kolobova, N.E. and Anisimov, K.N. (1974) *Izvestiya Akademii Nauk SSSR-Seriya Khimicheskaya*, 2873. (b) Nesmeyanov, A.N., Aleksandrov, G.G., Antonova, A.B., Anisimov, K.N., Kolobova, N.E. and Struchkov, Yu.T. (1976) *Journal of Organometallic Chemistry*, **110**, C36. (c) Aleksandrov, G.G., Antonova, A.B., Kolobova, N.E. and Struchkov, Yu.T. (1976) *Koord. Khim.*, **2**, 1684.

6 Antonova, A.B., Kolobova, N.E., Petrovskii, P.V., Lokshin, B.V. and Obezyuk, N.S. (1977) *Journal of Organometallic Chemistry*, **137**, 55.

7 Bellerby, J.M. and Mays, M.J. (1976) *Journal of Organometallic Chemistry*, **117**, C21.

8 (a) Bruce, M.I. and Wallis, R.C. (1978) *Journal of Organometallic Chemistry*, **161**, C1. (b) Bruce, M.I., Swincer, A.G. and Wallis, R.C. (1979) *Journal of Organometallic Chemistry*, **171**, C5. (c) Bruce, M.I. and Wallis, R.C. (1979) *Australian Journal of Chemistry*, **32**, 1471. (d) Bruce, M.I. and Swincer, A.G. (1980) *Australian Journal of Chemistry*, **33**, 1471.

9 (a) Boland, B.E., Fam, S.A. and Hughes, R.P. (1979) *Journal of Organometallic Chemistry*, **172**, C29. (b) Boland-Lussier, B.E., Churchill, M.R., Hughes, R.P. and Rheingold, A.L. (1982) *Organometallics*, **1**, 628. (c) Boland-Lussier, B.E. and Hughes, R.P. (1982) *Organometallics*, **1**, 635.

10 (a) Mansuy, D., Lange, M. and Chottard, J.C. (1978) *Journal of the American Chemical Society*, **100**, 3213. (b) Mansuy, D., Lange, M. and Chottard, J.C. (1979) *Journal of the American Chemical Society*, **101**, 6437. (c) Chevrier, B., Weiss, R., Lange, M., Chottard, J.C. and Mansuy, D.

(1981) *Journal of the American Chemical Society*, **103**, 2899.
11 (a) Bruce, M.I., Swincer, A.G. (1983) *Advances in Organometallic Chemistry*, **22**, 59. (b) Bruce, M.I. (1991) *Chemical Reviews*, **91**, 197. (c) Bruce, M.I. (1998) *Chemical Reviews*, **98**, 2797.
12 Antonova, A.B. and Johansson, A.A. (1989) *Uspekhi Khimii*, **58**, 1197. (1989) *Russian Chemical Reviews*, **58**, 693 (Eng. trans.).
13 Werner, H. (1992) *Nachrichten*, **40**, 435.
14 (a) Cadierno, V., Gamasa, M.P. and Gimeno, J. (2001) *European Journal of Inorganic Chemistry*, 571. (b) Cadierno, V., Gamasa, M.P. and Gimeno, J. (2004) *Coordination Chemistry Reviews*, **248**, 1627.
15 King, R.B. (ed.) (2004) *Coordination Chemistry Reviews*, **248**, 1531–1715.
16 (a) Beckhaus, R. (1998) in *Metallocenes: Synthesis, Reactivity and Applications* (eds A. Togni and R.L. Halterman), Wiley-VCH, New York, vol. 1, p. 153.(b) Beckhaus, R. (1997) *Journal of the Chemical Society-Dalton Transactions*, 1991. (c) Beckhaus, R. (1997) *Angewandte Chemie-International Edition*, **36**, 686. (d) Beckhaus, R., Oster, J., Savy, J., Strauss, I. and Wagner, M. (1997) *Synlett*, 241. (e) Beckhaus, R. and Santamaria, C. (2001) *Journal of Organometallic Chemistry*, **617–618**, 81.
17 (a) Venkatesan, K., Blacque, O. and Berke, H. (2006) *Organometallics*, **25**, 5190. (b) Venkatesan, K., Blacque, O. and Berke, H. (2007) *Dalton Transactions*, 1091.
18 Antonova, A.B. (2007) *Coordination Chemistry Reviews*, **251**, 1521.
19 Fischer, H. and Szesmi, N. (2004) *Coordination Chemistry Reviews*, **248**, 1659.
20 Fe: Guerchais, V. (2002) *European Journal of Inorganic Chemistry*, 783.
21 Ru: (a) Le Bozec, H. and Dixneuf, P.H. (1995) *Izv. Akad. Nauk, Ser. Khim.*, 827. (1995) *Russian Chemical Bulletin*, **44**, 801 (Engl. trans). (b) Touchard, D. and Dixneuf, P.H. (1998) *Coordination Chemistry Reviews*, **178–180**, 409.

22 Ru, Os: Puerta, M.C. and Valerga, P. (1999) *Coordination Chemistry Reviews*, **193–195**, 977.
23 (a) Os: Esteruelas, M.A. and Oro, L.A. (2001) *Advances in Organometallic Chemistry*, **47**, 1 [OsHCl(P)$_2$ chemistry]. (b) Esteruelas, M.A. and López, A.M. (2005) *Organometallics*, **24**, 3584 [mainly Os(PPr$^i_3$)2Cp]. (c) Esteruelas, M.A., López, A.M. and Oliván, M. (2007) *Coordination Chemistry Reviews*, **251**, 795 [Os=C chemistry].
24 (a) Werner, H. (1994) *Journal of Organometallic Chemistry*, **475**, 45. (b) Werner, H., Ilg, K., Lass, R. and Wolf, J. (2002) *Journal of Organometallic Chemistry*, **661**, 137.
25 Crabtree, R.H. and Mingos, D.M.P. (eds) (2007) *Comprehensive Organometallic Chemistry III*, 13 volumes, Elsevier, Amsterdam, Ti: **4**, p. 556; Mo: **5**, p. 401; W, **5**, p. 671; Mn: **5**, p. 824; Re, **5**, p. 900; Fe, **6**, p. 111; Ru and Os: **6**, 409, 420, 587, 615; Ir: **7**, 352.
26 (a) Bruneau, C. and Dixneuf, P.H. (1999) *Accounts of Chemical Research*, **32**, 311. (b) Bruneau, C. (2004) *Topics in Organometallic Chemistry*, **11**, 125. (c) Bruneau, C. and Dixneuf, P.H. (2006) *Angewandte Chemie-International Edition*, **45**, 2176.
27 Katayama, H. and Ozawa, F. (2004) *Coordination Chemistry Reviews*, **248**, 1703.
28 Miki, K., Uemura, S. and Ohe, K. (2005) *Chemistry Letters*, **34**, 1068.
29 Varela, J.A. and Saa, C. (2006) *Chemistry – A European Journal*, **12**, 6450.
30 (a) Trost, B.M. (1966) *Chemische Berichte*, **129**, 1313. (b) Trost, B.M., Frederiksen, M.U. and Rudd, M.T. (2005) *Angewandte Chemie International Edition*, **44**, 6630.
31 Wakatsuki, Y. (2004) *Journal of Organometallic Chemistry*, **689**, 4092.
32 (a) Grotjahn, D.B., Zeng, X. and Cooksy, A.L. (2006) *Journal of the American Chemical Society*, **128**, 2798. (b) Grotjahn, D.B., Zeng, X., Cooksy, A.L., Kassel, W.S., Dipasquale, A.G., Zakharov, L.N., Rheingold, A.L. *Organometallics*, 2007, **26**, 3385.

33 Wakatsuki, Y., Koga, N., Werner, H. and Morokuma, K. (1997) *Journal of the American Chemical Society*, **119**, 360.

34 (a) Wolf, J., Stüer, W., Grunwald, C., Werner, H., Schwab, P. and Schulz, M. (1998) *Angewandte Chemie-International Edition*, **37**, 1124. (b) Stüer, W., Weberndörfer, B., Wolf, J. and Werner, H. (2005) *Dalton Transactions*, 1796.

35 (a) Louie, J. and Grubbs, R.H. (2001) *Angewandte Chemie-International Edition*, **40**, 247. (b) Trnka, T.M. and Grubbs, R.H. (2001) *Accounts of Chemical Research*, **34**, 18.

36 Braun, T., Meuer, P. and Werner, H. (1996) *Organometallics*, **15**, 4075.

37 Braun, T., Münch, G., Windmüller, B., Gevert, O., Laubender, M. and Werner, H. (2003) *Chemistry – A European Journal*, **9**, 2516.

38 Braun, T., Steinert, P. and Werner, H. (1995) *Journal of Organometallic Chemistry*, **488**, 169.

39 Bassetti, M., Alvarez, P., Gimeno, J. and Lastra, E. (2004) *Organometallics*, **23**, 5127.

40 Schaefer, M., Wolf, J. and Werner, H. (2004) *Organometallics*, **23**, 5713.

41 Marder, T.B., Zargarian, D., Calabrese, J.C., Herskowitz, T.H. and Milstein, D. (1987) *Journal of the Chemical Society, Chemical Communications*, 1484.

42 da Silva, M.F.C.G., Frausto da Silva, J.J.R., Pombeiro, A.J.L., Pellinghelli, M.A. and Tiripicchio, A. (1996) *Journal of the Chemical Society-Dalton Transactions*, 2763.

43 Bartlett, M.J., Hill, A.F. and Smith, M.K. (2005) *Organometallics*, **24**, 5795.

44 Matsuo, Y., Mitani, Y., Zhong, Y.-W. and Nakamura, E. (2006) *Organometallics*, **25**, 2826.

45 Gemel, C., Huffman, J.C., Caulton, K.G., Mauthner, K. and Kirchner, K. (2000) *Journal of Organometallic Chemistry*, **593–594**, 342.

46 Adams, C.J. and Pope, S.J.A. (2004) *Inorganic Chemistry*, **43**, 3492.

47 Kopf, H., Pietraszuk, C., Huebner, E. and Burzlaff, N. (2006) *Organometallics*, **25**, 2533.

48 Beach, N.J. and Spivak, G.J. (2003) *Inorganica Chimica Acta*, **343**, 244.

49 del Río, I., Gossage, R.A., Hannu, M.S., Lutz, M., Spek, A.L. and van Koten, G. (1999) *Organometallics*, **18**, 1097.

50 Wong, C.-Y., Che, C.-M., Chan, M.C.W., Leung, K.-H., Phillips, D.L. and Zhu, N. (2004) *Journal of the American Chemical Society*, **126**, 2501.

51 Bonomo, L., Stern, C., Solari, E., Scopelliti, R. and Floriani, C. (2001) *Angewandte Chemie-International Edition*, **49**, 1449.

52 Callejas-Gaspar, B., Laubender, M. and Werner, H. (2003) *Journal of Organometallic Chemistry*, **684**, 144.

53 Weiss, D. and Dixneuf, P.H. (2003) *Organometallics*, **22**, 2209.

54 McDonagh, A.M., Powell, C.E., Morrall, J.P., Cifuentes, M.P. and Humphrey, M.G. (2003) *Organometallics*, **22**, 1402.

55 Solari, E., Antonijevic, S., Gauthier, S., Scopelliti, R. and Severin, K. (2007) *European Journal of Inorganic Chemistry*, 367.

56 Foerstner, J., Kakoschke, A., Goddard, R., Rust, J., Wartchow, R. and Butenschön, H. (2001) *Journal of Organometallic Chemistry*, **617–618**, 412.

57 Janicke, M., Gauss, C., Koller, A. and Berke, H. (1997) *Journal of Organometallic Chemistry*, **543**, 171.

58 (a) Le Narvor, N. and Lapinte, C. (1995) *Organometallics*, **14**, 634. (b) Denis, R., Toupet, L., Paul, F. and Lapinte, C. (2000) *Organometallics*, **19**, 4240. (c) Weyland, T., Ledoux, I., Brasselet, S., Zyss, J. and Lapinte, C. (2000) *Organometallics*, **19**, 5235. (d) Roue, S., Le Stang, S., Toupet, L. and Lapinte, C. (2003) *Comptes Rendus Chimie*, **6**, 353. (e) Ghazala, S.I., Paul, F., Toupet, L., Toisnel, T., Hapiot, P. and Lapinte, C. (2006) *Journal of the American Chemical Society*, **128**, 2463.

59 Nakanishi, S., Goda, K., Uchiyama, S. and Otsuji, Y. (1992) *Bulletin of the Chemical Society of Japan*, **65**, 2560.

60 Fryzuk, M.D., Huang, L., McManus, N.T., Paglia, P., Rettig, S.J. and White, G.S. (1992) *Organometallics*, **11**, 2979.

61 Ruiz, J., Quesada, R., Vivanco, M., Castellano, E.E. and Piro, O.E. (2005) *Organometallics*, **24**, 2542.

62 Beddoes, R.L., Bitcon, C., Grime, R.W., Ricalton, A. and Whiteley, M.W. (1995) *Journal of the Chemical Society-Dalton Transactions*, 2873.

63 Renkema, K.B. and Caulton, K.G. (1999) *New Journal of Chemistry*, **23**, 1027.

64 Esteruelas, M.A., Gómez, A.V., López, A.M. and Oro, L.A. (1996) *Organometallics*, **15**, 878.

65 Asensio, A., Buil, M.L., Esteruelas, M.A. and Oñate, E. (2004) *Organometallics*, **23**, 5787.

66 Gusev, D.G., Maxwell, T., Dolgushin, F.M., Lyssenko, M. and Lough, A.J. (2002) *Organometallics*, **21**, 1095.

67 Wen, T.B., Zhou, Z.Y. and Jia, G. (2003) *Organometallics*, **22**, 4947.

68 (a) Wen, T.B., Yang, S.-Y., Zhou, Z.Y., Lin, Z., Lau, C.-P. and Jia, G. (2000) *Organometallics*, **19**, 3757. (b) Wen, T.B., Zhou, Z.Y., Lo, M.F., Williams, I.D. and Jia, G. (2003) *Organometallics*, **22**, 5217. (c) Wen, T.B., Hung, W.Y., Zhou, Z.Y., Lo, M.F., Williams, I.D. and Jia, G. (2004) *European Journal of Inorganic Chemistry*, 2837.

69 Baya, M., Buil, M.L., Esteruelas, M.A. and Oñate, E. (2005) *Organometallics*, **24**, 2030.

70 Esteruelas, M.A., Oro, L.A. and Valero, C. (1995) *Organometallics*, **14**, 3596.

71 Crochet, P., Esteruelas, M.A., López, A.M., Martínez, M.-P., Oliván, M., Oñate, E. and Ruiz, N. (1998) *Organometallics*, **17**, 4500.

72 Werner, H., Jung, S., Weberndörfer, B. and Wolf, J. (1999) *European Journal of Inorganic Chemistry*, 951.

73 Weinand, R. and Werner, H. (1985) *Journal of the Chemical Society. Chemical Communications*, 1145.

74 (a) Werner, H., Stahl, S. and Kohlmann, W. (1991) *Journal of Organometallic Chemistry*, **409**, 285. (b) Knaup, W. and Werner, H. (1991) *Journal of Organometallic Chemistry*, **411**, 471.

75 Pilar Gamasa, M., Gimeno, J., Gónzalez-Cueva, M. and Lastra, E. (1996) *Journal of the Chemical Society-Dalton Transactions*, 2547.

76 Gimeno, J., Gónzalez-Cueva, M., Lastra, E., Perrez-Carreño, E. and García-Granda, S. (2003) *Inorganica Chimica Acta*, **347**, 99.

77 Lalrempuia, R., Yennawar, H., Mozharivskyj, Y.A. and Kollipara, M.R. (2004) *Journal of Organometallic Chemistry*, **689**, 539.

78 Albertin, G., Antoniutti, S., Bordignon, E. and Bresolin, D. (2000) *Journal of Organometallic Chemistry*, **609**, 10.

79 Bianchini, C., Marchi, A., Marvelli, L., Peruzzini, M., Romerosa, A., Rossi, R. and Vacca, A. (1995) *Organometallics*, **14**, 3203.

80 Bianchini, C., Marchi, A., Marvelli, L., Peruzzini, M., Romerosa, A. and Rossi, R. (1996) *Organometallics*, **15**, 3804.

81 Bianchini, C., Mantovani, N., Marchi, A., Marvelli, L., Masi, D., Peruzzini, M., Rossi, R. and Romerosa, A. (1999) *Organometallics*, **18**, 4501.

82 Kolobova, N.E., Antonova, A.B., Khitrova, O.M., Antipin, M.Yu. and Struchkov, Yu.T. (1977) *Journal of Organometallic Chemistry*, **137**, 69.

83 Grunwald, C., Laubender, M., Wolf, J. and Werner, H. (1998) *Journal of the Chemical Society-Dalton Transactions*, 833.

84 Oliván, M., Eisenstein, O. and Caulton, K.G. (1997) *Organometallics*, **16**, 2227.

85 Jung, S., Ilg, K., Brandt, C.D., Wolf, J. and Werner, H. (2002) *Dalton Transactions*, 318.

86 Stuer, W., Wolf, J. and Werner, H. (2002) *Journal of Organometallic Chemistry*, **641**, 203.

87 Leung, W.-H., Chan, E.Y.Y., Williams, I.D. and Wong, W.-T. (1997) *Organometallics*, **16**, 3234.

88 Yang, S.-M., Chan, M.C.-W., Cheung, K.-K., Che, C.-M. and Peng, S.-M. (1997) *Organometallics*, **16**, 2819.

89 Bianchini, C., Purches, G., Zanobini, F. and Peruzzini, M. (1998) *Inorganica Chimica Acta*, **272**, 1.

90 Lindner, E., Geprägs, M., Gierling, K., Fawzi, R. and Steimann, M. (1995) *Inorganic Chemistry*, **34**, 6106.

91 Werner, H., Stark, A., Steinert, P., Grünwald, C. and Wolf, J. (1995) *Chemische Berichte*, **128**, 49.

92 Martin, M., Gevert, O. and Werner, H. (1996) *Journal of the Chemical Society-Dalton Transactions*, 2275.

93 Werner, H., Bachmann, P. and Martin, M. (2001) *Canadian Journal of Chemistry*, **79**, 519.

94 Bianchini, C., de los Ríos, I., López, C., Peruzzini, M. and Romerosa, A. (2000) *Journal of Organometallic Chemistry*, 593–594, 485.

95 Beach, N.J., Walker, J.M., Jenkins, H.A. and Spivak, G.J. (2006) *Journal of Organometallic Chemistry*, **691**, 4147.

96 Touchard, D., Haquette, P., Guesmi, S., Le Pichon, L., Daridor, A., Toupet, L. and Dixneuf, P.H. (1997) *Organometallics*, **16**, 3640.

97 Katayama, H. and Ozawa, F. (1998) *Chemistry Letters*, 67.

98 de los Ríos, I., Jiménez Tenorio, M., Puerta, M.C. and Valerga, P. (1998) *Organometallics*, **17**, 3356.

99 Hurst, S.K., Cifuentes, M.P., Morrall, J.P., Lucas, N.T., Whittall, I.R., Humphrey, M.G., Asselberghs, I., Persoons, A., Samoc, M., Luther-Davis, B. and Willis, A.C. (2001) *Organometallics*, **20**, 4664.

100 Bernechea, M., Lugan, N., Gil, B., Lalinde, E. and Lavigne, G. (2006) *Organometallics*, **25**, 684.

101 Lindner, E., Pautz, S. and Haustein, M. (1996) *Journal of Organometallic Chemistry*, **509**, 215.

102 Cadierno, V., Pilar Gamasa, M., Gimeno, J., Gónzalez-Bernardo, C., Perez-Carreño, E. and García-Granda, S. (2001) *Organometallics*, **20**, 5177.

103 Diez, J., Pilar Gamasa, M., Gimeno, J., Lastra, E. and Villar, A. (2006) *European Journal of Inorganic Chemistry*, 78.

104 Jiménez-Tenorio, M., Palacios, M.D., Puerta, M.C. and Valerga, P. (2004) *Journal of Organometallic Chemistry*, **689**, 2853.

105 Ruba, E., Mereiter, K., Schmid, R., Sapunov, V.N., Kirchner, K., Schottenberger, H., Calhorda, M.J. and Veiros, L.F. (2002) *Chemistry – A European Journal*, **8**, 3948.

106 Demerseman, B. and Toupet, L. (2006) *European Journal of Inorganic Chemistry* 1573.

107 Slugovc, C., Simanko, W., Mereiter, K., Schmid, R., Kirchner, K., Xiao, L. and Weissensteiner, W. (1999) *Organometallics*, **18**, 3865.

108 Sun, Y., Chan, H.-S., Dixneuf, P.H. and Xie, Z. (2006) *Organometallics*, **25**, 2719.

109 Bruce, M.I., Hall, B.C., Zaitseva, N.N., Skelton, B.W. and White, A.H. (1996) *Journal of Organometallic Chemistry*, **522**, 307.

110 Yi, C.S., Liu, N., Rheingold, A.L., Liable-Sands, L.M. and Guzei, I.A. (1997) *Organometallics*, **16**, 3729.

111 Beach, N.J., Williamson, A.E. and Spivak, G.J. (2005) *Journal of Organometallic Chemistry*, **690**, 4640.

112 Bruce, M.I., Hameister, C., Swincer, A.G. and Wallis, R.C. (1982) *Inorganic Syntheses*, **21**, 80.

113 Nishibayashi, Y., Takei, I. and Hidai, M. (1997) *Organometallics*, **16**, 3091.

114 Yeo, S.P., Henderson, W., Mak, T.C.W. and Hor, T.S.A. (1999) *Journal of Organometallic Chemistry*, **575**, 171.

115 Cadierno, V., Conejero, S., Pilar Gamasa, M. and Gimeno, J. (2000) *Dalton Transactions*, 451.

116 Barratta, W., Herdtweck, E., Vuano, S. and Rigo, P. (2001) *Journal of Organometallic Chemistry*, **617–618**, 511.

117 Hurst, S.K., Lucas, N.T., Cifuentes, M.P., Humphrey, M.G., Samoc, M., Luther-Davis, B., Asselberghs, I., Van Boxel, R. and Persoons, A. (2001) *Journal of Organometallic Chemistry*, **633**, 114.

118 Hamidov, H., Jeffery, J.C. and Lynam, J.M. (2004) *Chemical Communications*, 1364.

119 Enzmann, A. and Beck, W. (2004) *Zeitschrift fur Naturforschung*, **59b**, 865.
120 Nishibayashi, Y., Imajima, H., Onodera, G. and Uemura, S. (2005) *Organometallics*, **24**, 4106.
121 Bustelo, E., Carbo, J.J., Lledos, A., Mereiter, K., Puerta, M.C. and Valerga, P. (2003) *Journal of the American Chemical Society*, **125**, 3311.
122 Hansen, H.D. and Nelson, J.H. (2003) *Inorganica Chimica Acta*, **352**, 4.
123 de los Ríos, I., Jiménez Tenorio, M., Puerta, M.C. and Valerga, P. (1997) *Journal of the American Chemical Society*, **119**, 6529.
124 Barthel-Rosa, L.P., Maitra, K., Fischer, J. and Nelson, J.H. (1997) *Organometallics*, **16**, 1714.
125 (a) Mauthner, K., Slugovc, C., Mereiter, K., Schmid, R. and Kirchner, K. (1997) *Organometallics*, **16**, 1956. (b) Slugovc, C., Sapunov, V.N., Wiede, P., Mereiter, K., Schmid, R. and Kirchner, K. (1997) *Journal of the Chemical Society-Dalton Transactions*, 4209.
126 (a) Gemel, C., Wiede, P., Mereiter, K., Sapunov, V.N., Schmid, R. and Kirchner, K. (1996) *Journal of the Chemical Society-Dalton Transactions*, 4071. (b) Trimmel, G., Slugovc, C., Wiede, P., Mereiter, K., Sapunov, V.N., Schmid, R. and Kirchner, K. (1997) *Inorganic Chemistry*, **36**, 1076.
127 Jiménez Tenorio, M.A., Jiménez Tenorio, M., Puerta, M.C. and Valerga, P. (1997) *Organometallics*, **16**, 5528.
128 Buriez, B., Burns, I.D., Hill, A.F., White, A.J.P., Williams, D.J. and Wilton-Ely, J.D.E.T. (1999) *Organometallics*, **18**, 1504.
129 Jiménez Tenorio, M.A., Jiménez Tenorio, M., Puerta, M.C. and Valerga, P. (2000) *Organometallics*, **19**, 1333.
130 Pavlik, S., Mereiter, K., Puchberger, M. and Kirchner, K. (2005) *Journal of Organometallic Chemistry*, **690**, 5497.
131 Yamamoto, Y., Tanase, T., Sudoh, C. and Turuta, T. (1998) *Journal of Organometallic Chemistry*, **569**, 29.
132 Opstal, T. and Verpoort, F. (2003) *Journal of Molecular Catalysis A-Chemical*, **200**, 49.
133 Burrell, A.K., Bryan, J.C. and Kubas, G.J. (1994) *Organometallics*, **13**, 1067.
134 Birdwhistell, K.R., Burgmayer, S.J.N. and Templeton, J.L. (1983) *Journal of the American Chemical Society*, **105**, 7789.
135 Werner, H., Baum, M., Schneider, D. and Windmüller, B. (1994) *Organometallics*, **13**, 1089.
136 Katayama, H., Onitsuka, K. and Ozawa, F. (1996) *Organometallics*, **15**, 4642.
137 Huang, D., Streib, W.E., Eisenstein, O. and Caulton, K.G. (2000) *Organometallics*, **19**, 1967.
138 Konkol, M. and Steinborn, D. (2006) *Journal of Organometallic Chemistry*, **691**, 2839.
139 Ilg, K., Paneque, M., Poveda, M.L., Rendon, N., Santos, L.L., Carmona, E. and Mereiter, K. (2006) *Organometallics*, **25**, 2230.
140 (a) Connelly, N.G., Orpen, A.G., Rieger, A.L., Rieger, P.H., Scott, C.J. and Rosair, G.M. (1992) *Chemical Communications*, 1293. (b) Connelly, N.G., Geiger, W.E., Lagunas, C., Metz, B., Rieger, A.L., Rieger, P.H. and Shaw, M.J. (1995) *Journal of the American Chemical Society*, **117**, 12202.
141 Venkatesan, K., Fox, T., Schmalle, H.W. and Berke, H. (2005) *European Journal of Inorganic Chemistry*, 901.
142 Venkatesan, K., Blacque, O., Fox, T., Alfonso, M., Schmalle, H.W., Kheradmandan, S. and Berke, H. (2005) *Organometallics*, **24**, 920.
143 Venkatesan, K., Fernandez, F.J., Blacque, O., Fox, T., Alfonso, M., Schmalle, H.W. and Berke, H. (2003) *Chemical Communications*, 2006.
144 Baum, M., Mahr, N. and Werner, H. (1994) *Chemische Berichte*, **127**, 1877.
145 Miller, D.C. and Angelici, R.J. (1991) *Organometallics*, **10**, 79.
146 Hill, A.F., Hulkes, A.G., White, A.J.P. and Williams, D.J. (2000) *Organometallics*, **19**, 371.
147 Miura, T. and Iwasawa, N. (2002) *Journal of the American Chemical Society*, **124**, 518.

148 Löwe, C., Hund, H.-U. and Berke, H. (1989) *Journal of Organometallic Chemistry*, **371**, 311.

149 Lomprey, J.R. and Selegue, J.P. (1992) *Journal of the American Chemical Society*, **114**, 5518.

150 Nombel, P., Lugan, N. and Mathieu, R. (1995) *Journal of Organometallic Chemistry*, **503**, C22.

151 (a) Bianchini, C., Innocenti, P., Meli, A., Peruzzini, M., Zanobini, F. and Zanello, P. (1990) *Organometallics*, **9**, 2514. (b) Bianchini, C., Peruzzini, M. and Zanobini, F. (1991) *Organometallics*, **10**, 3415.

152 Gauss, C., Veghini, D. and Berke, H. (1997) *Chemische Berichte/Recueil*, **130**, 183.

153 Werner, H., Schulz, M. and Windmüller, B. (1995) *Organometallics*, **14**, 3659.

154 Höhn, A. and Werner, H. (1990) *Journal of Organometallic Chemistry*, **382**, 255.

155 Werner, H., Dirnberger, T. and Höhn, A. (1991) *Chemische Berichte*, **124**, 1957.

156 Werner, H., Höhn, A. and Schulz, M. (1991) *Journal of the Chemical Society-Dalton Transactions*, 777.

157 Kolobova, N.E., Ivanov, L.L. and Zhvanko, O.S. (1984) *Izv. Akad. Nauk, Ser. Khim.*, 1667.

158 Marchenko, A.V., Gerard, H., Eisenstein, O. and Caulton, K.G. (2001) *New Journal of Chemistry*, **25**, 1244.

159 Esteruelas, M.A., López, A.M., Ruiz, N. and Tolosa, J.I. (1997) *Organometallics*, **16**, 4657.

160 Baya, M., Crochet, P., Esteruelas, M.A., López, A.M., Modrevo, J. and Oñate, E. (2001) *Organometallics*, **20**, 4291.

161 Werner, H., Hampp, A., Peters, K., Peters, E.M., Walz, L. and von Schnering, H.G. (1990) *Zeitschrift für Naturforschung*, **45b**, 1548.

162 Windmüller, B., Wolf, J. and Werner, H. (1995) *Journal of Organometallic Chemistry*, **502**, 147.

163 Kukla, F. and Werner, H. (1995) *Inorganica Chimica Acta*, **235**, 253.

164 Werner, H., Bachmann, P., Laubender, M. and Gevert, O. (1998) *European Journal of Inorganic Chemistry*, 1217.

165 Gil-Rubio, J., Weberndörfer, B. and Werner, H. (1999) *Journal of the Chemical Society-Dalton Transactions*, 1437.

166 Canepa, G., Brandt, C.D. and Werner, H. (2001) *Organometallics*, **20**, 604.

167 Bustelo, E., de los Ríos, I., Jiménez Tenorio, M., Puerta, M.C. and Valerga, P. (2000) *Monatshefte für Chemie*, **131**, 1311.

168 Zayed, M.A. and Fischer, H. (2000) *Journal of Thermal Analysis and Calorimetry*, **61**, 897.

169 Birdwhistell, K.R., Tonker, T.L., Templeton, J.L. and Kenan, W.R., Jr. (1985) *Journal of the American Chemical Society*, **107**, 4474.

170 (a) Bullock, R.M. (1987) *Journal of the American Chemical Society*, **109**, 8087. (b) Wong A., Gladysz, J.A., (1982) *Journal of the American Chemical Society*, **104**, 4948.

171 Chang, C.-W., Lin, Y.-C., Lee, G.-H. and Wang, Y. (1999) *Journal of the Chemical Society-Dalton Transactions*, 4223.

172 Chang, C.-W., Lin, Y.-C., Lee, G.-H., Huang, S.-L. and Wang, Y. (1998) *Organometallics*, **17**, 2534.

173 Ipaktschi, J., Müller, B.G. and Glaum, R. (1994) *Organometallics*, **13**, 1044.

174 Ipaktschi, J., Demuth-Eberle, G.J., Mirzaei, F., Müller, B.G., Beck, J. and Serafin, M. (1995) *Organometallics*, **14**, 3335.

175 Ipaktschi, J., Mirzaei, F., Demuth-Eberle, G.J., Beck, J. and Serafin, M. (1997) *Organometallics*, **16**, 3965.

176 Ipaktschi, J., Munz, F. and Klotzbach, T. (2005) *Organometallics*, **24**, 206.

177 Dötz, K.H., Christoffers, C., Christoffers, J., Böttcher, D., Nieger, M. and Kotila, S. (1995) *Chemische Berichte*, **128**, 645.

178 Albertin, G., Antoniutti, S., Bordignon, E., del Ministro, E., Ianelli, S. and Pelizzi, G. (1995) *Journal of the Chemical Society-Dalton Transactions*, 1783.

179 Davison, A. and Solar, J.P. (1978) *Journal of Organometallic Chemistry*, **155**, C8.

180 Selegue, J.P. (1982) *Journal of the American Chemical Society*, **104**, 119.

181 Abbott, S., Davies, S.G. and Warner, P. (1983) *Journal of Organometallic Chemistry*, **246**, C65.

182 Pilar Gamasa, M., Gimeno, J., Lastra, E., Martín, B.M., Anillo, A. and Tiripicchio, A. (1992) *Organometallics*, **11**, 1373.

183 Pilar Gamasa, M., Gimeno, J., Godefroy, I., Lastra, E., Martín-Vaca, B.M., García-Granda, S. and Gutiérrez-Rodriguez, A. (1995) *Journal of the Chemical Society-Dalton Transactions*, 1901.

184 Lugan, N., Kelley, C., Terry, M.R., Geoffroy, G.L. and Rheingold, A.L. (1990) *Journal of the American Chemical Society*, **112**, 3220.

185 Unseld, D., Krivykh, V.V., Heinze, K., Wild, F., Artus, G., Schmalle, H. and Berke, H. (1999) *Organometallics*, **18**, 1525.

186 Nicklas, P.N., Selegue, J.P. and Young, B.A. (1988) *Organometallics*, **7**, 2248.

187 Henderson, R.A. and Oglieve, K.E. (1996) *Journal of the Chemical Society-Dalton Transactions*, 3397.

188 García-Yebra, C., López-Mardomingo, C., Fajardo, M., Antinolo, A., Otero, A., Rodriguez, A., Vallat, A., Lucas, D., Mugnier, Y., Carbo, J.J., Lledos, A. and Bo, C. (2000) *Organometallics*, **19**, 1749.

189 Baya, M., Esteruelas, M.A. and Oñate, E. (2001) *Organometallics*, **20**, 4875.

190 Belluco, U., Bertani, R., Meneghetti, F., Michelin, R.A. and Mozzon, M. (1999) *Journal of Organometallic Chemistry*, **583**, 131.

191 Weng, W., Bartik, T., Johnson, M.T., Arif, A.M. and Gladysz, J.A. (1995) *Organometallics*, **14**, 889.

192 Dembinski, R., Lis, T., Szafert, S., Mayne, C.L., Bartik, T. and Gladysz, J.A. (1999) *Journal of Organometallic Chemistry*, **578**, 229.

193 Bianchini, C., Laschi, F., Ottaviani, F., Peruzzini, M. and Zanello, P. (1988) *Organometallics*, **7**, 1660.

194 Bianchini, C., Meli, A., Peruzzini, M., Zanobini, F. and Zanello, P. (1990) *Organometallics*, **9**, 241.

195 Werner, H., Schneider, D. and Schulz, M. (1993) *Journal of Organometallic Chemistry*, **451**, 175.

196 Albertin, G., Antoniutti, S., Bordignon, E. and Granzotto, M. (1999) *Journal of Organometallic Chemistry*, **585**, 83.

197 Chang, C.-W., Ting, P.-C., Lin, Y.-C., Lee, G.-H. and Wang, Y. (1998) *Journal of Organometallic Chemistry*, **553**, 417.

198 Chang, C.-W. and Lin, Y.-C. (2003) *Journal of the Chinese Chemical Society (Taipei)*, **50**, 369.

199 Chang, K.-H., Sung, H.-L. and Lin, Y.-C. (2006) *European Journal of Inorganic Chemistry*, 649.

200 Cadierno, V., Pilar Gamasa, M. and Gimeno, J. (2001) *Journal of Organometallic Chemistry*, **621**, 39.

201 Cadierno, V., Conejero, S., Pilar Gamasa, M. and Gimeno, J. (2002) *Organometallics*, **21**, 3837.

202 Lo, Y.-H., Lin, Y.-C., Lee, G.-H. and Wang, Y. (1999) *Organometallics*, **18**, 982.

203 Lo, Y.-H., Lin, Y.-C., Lee, G.-II. and Wang, Y. (2004) *European Journal of Inorganic Chemistry*, 4616.

204 Mayr, A., Schaefer, K.C. and Huang, E.Y. (1984) *Journal of the American Chemical Society*, **106**, 1517.

205 Chen, D.-J., Ting, P.-C., Lin, Y.-C., Lee, G.-H. and Wang, Y. (1995) *Journal of the Chemical Society-Dalton Transactions*, 3561.

206 Crane, T.W., White, P.S. and Templeton, J.L. (1999) *Organometallics*, **18**, 1897.

207 Milke, J., Sünkel, K. and Beck, W. (1997) *Journal of Organometallic Chemistry*, **543**, 39.

208 Birdwhistell, K.R. and Templeton, J.L. (1985) *Organometallics*, **4**, 2062.

209 Ipaktschi, J., Mohsseni-Ala, J. and Uhlig, S. (2003) *European Journal of Inorganic Chemistry*, 4313.

210 Kelley, C., Lugan, N., Terry, M.R., Geoffroy, G.L., Haggerty, B.S. and Rheingold, A.L. (1992) *Journal of the American Chemical Society*, **114**, 6735.

211 Terry, M.R., Kelley, C., Lugan, N., Geoffroy, G.L., Haggerty, B.S. and

211 Rheingold, A.L. (1993) *Organometallics*, **12**, 3607.
212 Cadierno, V., Pilar Gamasa, M., Gimeno, J., Moreto, J.M., Ricart, S., Roig, A. and Molins, E. (1998) *Organometallics*, **17**, 697.
213 Antinolo, A., Otero, A., Fajardo, M., García-Yebra, C., López-Mardomingo, C., Martin, A. and Gómez-Sal, P. (1997) *Organometallics*, **16**, 2601.
214 Le Narvor, N., Toupet, L. and Lapinte, C. (1995) *Journal of the American Chemical Society*, **117**, 7129.
215 Bruce, M.I., Ellis, B.G., Low, P.J., Skelton, B.W. and White, A.H. (2003) *Organometallics*, **22**, 3184.
216 Valyev, D.A., Semeikin, O.V. and Ustynyuk, N.A. (2004) *Coordination Chemistry Reviews*, **248**, 1671.
217 Rigaut, S., Monnier, F., Mousset, F., Touchard, D. and Dixneuf, P.H. (2002) *Organometallics*, **21**, 2654.
218 Sato, M., Kawata, Y., Shintate, H., Habata, Y., Akabori, S. and Unoura, K. (1997) *Organometallics*, **16**, 1693.
219 Beddoes, R.L., Bitcon, C., Grime, R.W., Ricalton, A. and Whiteley, M.W. (1995) *Journal of the Chemical Society-Dalton Transactions*, 2873.
220 (a) Novikova, L.N., Peterleitner, M.G., Sevumyan, K.A., Semeikin, O.V., Valyaev, D.A., Ustynyuk, N.A., Khrustalev, V.N., Kuleshova, L.N. and Antipin, M.Yu. (2001) *Journal of Organometallic Chemistry*, **631**, 47. (b) Novikova, L.N., Peterleitner, M.G., Sevumyan, K.A., Semeikin, O.V., Valyaev, D.A. and Ustynyuk, N.A. (2002) *Applied Organometallic Chemistry*, **16**, 530.
221 Valyaev, D.A., Semeikin, O.V., Peterleitner, M.G., Borisov, Yu.A., Khrustalev, V.N., Mazhuga, A.M., Kremer, E.V. and Ustynyuk, N.A. (2004) *Journal of Organometallic Chemistry*, **689**, 3837.
222 Iyer, R.S. and Selegue, J.P. (1987) *Journal of the American Chemical Society*, **109**, 910.
223 Ting, P.-C., Lin, Y.-C., Lee, G.-H., Cheng, M.-C. and Wang, Y. (1996) *Journal of the American Chemical Society*, **118**, 6433.
224 Berke, H., Huttner, G. and von Seyerl, J. (1981) *Journal of Organometallic Chemistry*, **218**, 193.
225 Bustelo, E., Jiménez Tenorio, M., Puerta, M.C. and Valerga, P. (2006) *Organometallics*, **25**, 4019.
226 Cadierno, V., Pilar Gamasa, M., Gimeno, J., Borge, J. and García-Granda, S. (1994) *Chemical Communications*, 2495.
227 Cadierno, V., Conejero, S., Pilar Gamasa, M., Gimeno, J., Perez-Carreño, E. and García-Granda, S. (2001) *Organometallics*, **20**, 3175.
228 Bustelo, E., Jiménez Tenorio, M., Mereiter, K., Puerta, M.C. and Valerga, P. (2002) *Organometallics*, **21**, 1903.
229 Cadierno, V., Conejero, S., Pilar Gamasa, M., Gimeno, J., Falvello, L.R. and Llusar, R.M. (2002) *Organometallics*, **21**, 3716.
230 Cadierno, V., Pilar Gamasa, M., Gimeno, J., Perez-Carreño, E. and García-Granda, S. (2003) *Journal of Organometallic Chemistry*, **670**, 75.
231 Conejero, S., Diez, J., Pilar Gamasa, M. and Gimeno, J. (2004) *Organometallics*, **23**, 6299.
232 Baya, M., Buil, M.L., Esteruelas, M.A., López, A.M., Oñate, E. and Rodriguez, J.R. (2002) *Organometallics*, **21**, 1841.
233 Baker, P.K., Barker, G.K., Gill, D.S., Green, M., Orpen, A.G., Williams, I.D. and Welch, A.J. (1989) *Journal of the Chemical Society-Dalton Transactions*, 1321.
234 Terry, M.R., Mercando, L.A., Kelley, C., Geoffroy, G.L., Nombel, P., Lugan, N., Mathieu, R., Ostrander, R.L., Owens-Waltermire, B.E. and Rheingold, A.L. (1994) *Organometallics*, **13**, 843.
235 Nakamura, G., Harada, Y., Mizobe, Y. and Hidai, M. (1996) *Bulletin of the Chemical Society of Japan*, **69**, 3305.
236 Brower, D.C., Stoll, M. and Templeton, J.L. (1989) *Organometallics*, **8**, 2786.
237 Beevor, R.G., Green, M., Orpen, A.G. and Williams, I.D. (1987) *Journal of the Chemical Society-Dalton Transactions*, 1319.

238 Baker, L.-J., Rickard, C.E.F., Roper, W.R., Woodgate, S.D. and Wright, L.J. (1998) *Journal of Organometallic Chemistry*, **565**, 153.
239 Baya, M. and Esteruelas, M.A. (2002) *Organometallics*, **21**, 2332.
240 Roper, W.R., Waters, J.M., Wright, L.J. and van Meurs, F. (1980) *Journal of Organometallic Chemistry*, **201**, C27.
241 Esteruelas, M.A., Oliván, M., Oñate, E., Ruiz, N. and Tajada, M.A. (1999) *Organometallics*, **18**, 2953.
242 Barrio, P., Esteruelas, M.A. and Oñate, E. (2002) *Organometallics*, **21**, 2491.
243 Bourgault, M., Castillo, A., Esteruelas, M.A., Oñate, E. and Ruiz, N. (1997) *Organometallics*, **16**, 636.
244 Buil, M.L., Eisenstein, O., Esteruelas, M.A., García-Yebra, C., Gutiérrez-Puebla, E., Oliván, M., Oñate, E., Ruiz, N. and Tajada, M.A. (1999) *Organometallics*, **18**, 4949.
245 Esteruelas, M.A., García-Yebra, C., Oliván, M., Oñate, E. and Tajada, M.A. (2000) *Organometallics*, **19**, 5098.
246 (a) Almeida, S.S.P.R., Frausto da Silva, J.J.R. and Pombeiro, A.J.L. (1993) *Journal of Organometallic Chemistry*, **450**, C7. (b) Almeida, S.S.P.R. and Pombeiro, A.J.L. (1997) *Organometallics*, **16**, 4469.
247 Zhang, L., Pilar Gamasa, M., Gimeno, J., Cabajo, R.J., López-Ortiz, F., Lanfranchi, M. and Tiripicchio, A. (1996) *Organometallics*, **15**, 4274.
248 Heppert, J., Vilain, J.M., Gile, M.A., Carlson, R.G., Mason, M.H. and Powell, D. (Aug. 2004) *Abstr. 228th ACS Meeting, Philadelphia, PA*, INOR-690.
249 Caskey, S.R., Stewart, M.H., Johnson, M.J.A. and Kampff, J.W. (2006) *Angewandte Chemie-International Edition*, **45**, 7422.
250 Shaw, B.L., Smith, M.J. and Stretton, G.N. (1988) *Journal of the Chemical Society-Dalton Transactions* 725.
251 Adams, H., Broughton, S.G., Walters, S.J. and Winter, M.J. (1999) *Chemical Communications*, 1231.
252 Hagmayer, S., Früh, A., Haas, T., Drexler, M., Troll, C. and Fischer, H. (2007) *Organometallics*, **26**, 0000.
253 van Asselt, A., Burger, B.J., Gibson, V.C. and Bercaw, J.E. (1986) *Journal of the American Chemical Society*, **108**, 5347.
254 (a) Chang, K.-H. and Lin, Y.-C. (1998) *Chemical Communications*, 1441. (b) Chang, K.-H., Lin, Y.-C., Liu, Y.-H. and Wang, Y. (2001) *Journal of the Chemical Society-Dalton Transactions*, 3154.
255 (a) Schubert, U., Kirschgassner, U., Grönen, J. and Piana, H. (1989) *Polyhedron*, **8**, 1589. (b) Schubert, U. and Grönen, J. (1989) *Chemische Berichte*, **122**, 1237.
256 Schubert, U. and Grönen, J. (1987) *Organometallics*, **6**, 2458.
257 (a) Haddleton, D.M. and Perutz, R.N. (1986) *Chemical Communications*, 1734. (b) Bell, T.W., Haddleton, D.M., McCamley, A., Partridge, M.G., Perutz, R.N. and Willner, H. (1990) *Journal of the American Chemical Society*, **112**, 9212.
258 Wolf, J., Lass, R.W., Manger, M. and Werner, H. (1995) *Organometallics*, **14**, 2649.
259 Dziallas, M. and Werner, H. (1987) *Journal of Organometallic Chemistry*, **333**, C29.
260 Fermin, M.C. and Bruno, J.W. (1993) *Journal of the American Chemical Society*, **115**, 7511.
261 Fischer, H., Schlageter, A., Bidell, W. and Früh, A. (1991) *Organometallics*, **10**, 389.
262 Sakurai, H., Fujii, T. and Sakamoto, K. (1992) *Chemistry Letters*, 339.
263 Hojo, F., Fujiki, K. and Ando, W. (1996) *Organometallics*, **15**, 3606.
264 Battioni, J.-P., Dupré, D. and Mansuy, D. (1981) *Journal of Organometallic Chemistry*, **214**, 303.
265 (a) Chan, Y.W., Renner, M.W. and Balch, A.L. (1983) *Organometallics*, **2**, 1888. (b) Chan, Y.W., Wood, F.E., Renner, M.W., Hope, H. and Balch, A.L. (1984) *Journal of the American Chemical Society*, **106**, 3380. (c) Balch, A.L., Chan, Y.W., Olmstead, M.M., Renner, M.W. and Wood, F.E.

(1988) *Journal of the American Chemical Society*, **110**, 3897.

266 Ferrando, G., Gerard, H., Spivak, G.J., Coalter, J.N., III, Huffman, J.C., Eisenstein, O. and Caulton, K.G. (2001) *Inorganic Chemistry*, **40**, 6610.

267 Castarlenas, R., Esteruelas, M.A. and Oñate, E. (2000) *Organometallics*, **19**, 5454.

268 Selegue, J.P. (1983) *Journal of the American Chemical Society*, **105**, 5921.

269 Selegue, J.P., Young, B.A. and Logan, S.L. (1991) *Organometallics*, **10**, 1972.

270 Cadierno, V., Pilar Gamasa, M., Gimeno, J. and Lastra, E. (1999) *Journal of the Chemical Society-Dalton Transactions*, 3235.

271 Gamble, A.S., Birdwhistell, K.R. and Templeton, J.L. (1988) *Organometallics*, **7**, 1046.

272 Crochet, P., Esteruelas, M.A., López, A.M., Ruiz, N. and Tolose, J.I. (1998) *Organometallics*, **17**, 3479.

273 Gauss, C., Veghini, D., Orama, O. and Berke, H. (1997) *Journal of Organometallic Chemistry*, **541**, 19.

274 Rappert, T., Nürnberg, O., Mahr, N., Wolf, J. and Werner, H. (1992) *Organometallics*, **11**, 4156.

275 Werner, H. and Rappert, T. (1993) *Chemische Berichte*, **126**, 669.

276 Werner, H. (1997) *Chemical Communications*, 903.

277 Ilg, K. and Werner, H. (2002) *Chemistry – A European Journal*, **8**, 2812.

278 Akita, M., Tanaka, Y., Naitoh, C., Ozawa, T., Hayashi, N., Takeshita, M., Inagaki, A. and Chung, M.-C. (2006) *Organometallics*, **25**, 5261.

279 Le Lagadec, R., Roman, E., Toupet, L., Müller, U. and Dixneuf, P.H. (1994) *Organometallics*, **13**, 5030.

280 Werner, H., Weinand, R., Knaup, W., Peters, K. and von Schnering, H.G. (1991) *Organometallics*, **10**, 3967.

281 Kawata, Y. and Sato, M. (1997) *Organometallics*, **16**, 1093.

282 Bianchini, C., Casares, J.A., Peruzzini, M., Romerosa, A. and Zanobini, F. (1996) *Journal of the American Chemical Society*, **118**, 4585.

283 Pilar Gamasa, M., Gimeno, J., Lastra, E., Lanfranchi, M. and Tiripicchio, A. (1992) *Journal of Organometallic Chemistry*, **430**, C39.

284 Fillaut, J.-L., de los Ríos, I., Masi, D., Romerosa, A., Zanobini, F. and Peruzzini, M. (2002) *European Journal of Inorganic Chemistry*, 935.

285 Urtel, K., Frick, A., Huttner, G., Zsolnai, L., Kircher, P., Rutsch, P., Kaifer, E. and Jacobi, A. (2000) *European Journal of Inorganic Chemistry*, 33.

286 Knors, C., Kuo, G.H., Lauher, J.W., Eigenbrot, C. and Helquist, P. (1987) *Organometallics*, **6**, 988.

287 Buriez, B., Cook, D.J., Harlow, K.J., Hill, A.F., Welton, T., White, A.J.P., Williams, D.J. and Wilton-Ely, J.D.E.T. (1999) *Journal of Organometallic Chemistry*, **578**, 264.

288 Nelson, J.E., Parkin, G. and Bercaw, J.E. (1992) *Organometallics*, **11**, 2181.

289 Bianchini, C., Masi, D., Romerosa, A., Zanobini, F. and Peruzzini, M. (1999) *Organometallics*, **18**, 2376.

290 Ipaktschi, J., Uhlig, S. and Dülmer, A. (2001) *Organometallics*, **20**, 4840.

291 Ipaktschi, J., Mohseni-Ala, J., Dülmer, A., Steffens, S., Wittenburg, C. and Heck, J. (2004) *Organometallics*, **23**, 4902.

292 Murakami, M. and Hori, S. (2003) *Journal of the American Chemical Society*, **125**, 4720.

293 Miller, D.C. and Angelici, R.J. (1991) *Organometallics*, **10**, 89.

294 Onitsuka, K., Nishii, M., Matsushima, Y. and Takahashi, S. (2004) *Organometallics*, **23**, 5630.

295 Ting, P.-C., Lin, Y.-C., Cheng, M.-C. and Wang, Y. (1994) *Organometallics*, **13**, 2150.

296 Davies, S.G. and Smallridge, A.J. (1990) *Journal of Organometallic Chemistry*, **395**, C39.

297 Reger, D.L. and Swift, C.A. (1984) *Organometallics*, **3**, 876.

298 (a) Chang, C.-W., Lin, Y.-C., Lee, G.-H. and Wang, Y. (2000) *Organometallics*, **19**, 3211.

(b) Lin, Y.-C. (2001) *Journal of Organometallic Chemistry*, **617–618**, 141. (c) Huang, C.-C., Lin, Y.-C., Huang, S.-L., Liu, Y.-H. and Wang, Y. (2003) *Organometallics*, **22**, 1512.

299 Yen, Y.-S., Lin, Y.-C., Liu, Y.-H. and Wang, Y. (2007) *Organometallics*, **26**, 1250.

300 Pilar Gamasa, M., Gimeno, J., Lastra, E., Martín, B.M., Aguirre, A., García-Granda, S. and Pertierra, P. (1992) *Journal of Organometallic Chemistry*, **429**, C19.

301 Cadierno, V., Pilar Gamasa, M., Gimeno, J. and Martín-Vaca, B.M. (2001) *Journal of Organometallic Chemistry*, **617–618**, 261.

302 Bruce, M.I., Ellis, B.G., Skelton, B.W. and White, A.H. (2005) *Journal of Organometallic Chemistry*, **690**, 792.

303 Yang, J.-Y., Huang, S.-L., Lin, Y.-C., Liu, Y.-H. and Wang, Y. (2000) *Organometallics*, **19**, 269.

304 Sanford, M.S., Valdez, M.R. and Grubbs, R.H. (2001) *Organometallics*, **20**, 5455.

305 Esteruelas, M.A., Gónzalez, A.I., López, A.M., Oliván, M. and Oñate, E. (2006) *Organometallics*, **25**, 693.

306 Werner, H., Kukla, F. and Stenert, P. (2002) *European Journal of Inorganic Chemistry*, 1377.

307 (a) Antonova, A.B., Kovalenko, S.V., Korniets, E.D. and Ioganson, A.A. (1982) *Izvestiya Akademii Nauk SSSR-Seriya Khimicheskaya*, 1667. (b) Antonova, A.B., Kovalenko, S.V., Korniets, E.D., Ioganson, A.A., Struchkov, Yu.T., Akhmedov, A.I. and Yanovskii, A.I. (1983) *Journal of Organometallic Chemistry*, **244**, 35. (c) Antonova, A.B., Kovalenko, S.V., Cherkasov, R.A., Ovchinnikov, V.V., Ioganson, A.A., Korniyets, E.D. and Deykhina, N.A. (1987) *Zhurnal Obshchei Khimii*, **57**, 1030.

308 Werner, H. and Brekau, U. (1989) *Zeitschrift für Naturforschung*, **44b**, 1438.

309 Ipaktschi, J., Rooshenas, P. and Dülmer, A. (2006) *European Journal of Inorganic Chemistry*, 1456.

310 Carvalho, M.F.N.N., Henderson, R.A., Pombeiro, A.J.L. and Richards, R.L. (1989) *Journal of the Chemical Society, Chemical Communications*, 1796.

311 Carvalho, M.F.N.N., Almeida, S.S.P.R., Pombeiro, A.J.L. and Henderson, R.A. (1997) *Organometallics*, **16**, 5441.

312 Pombeiro, A.J.L., Hills, A., Hughes, D.L. and Richards, R.L. (1988) *Journal of Organometallic Chemistry*, **352**, C5.

313 Beach, N.J., Jenkins, H.A. and Spivak, G.J. (2003) *Organometallics*, **22**, 5179.

314 Jung, S., Ilg, K., Brandt, C.D., Wolf, J. and Werner, H. (2004) *European Journal of Inorganic Chemistry*, 469.

315 Gónzalez-Herrero, P., Weberndörfer, B., Ilg, K., Wolf, J. and Werner, H. (2001) *Organometallics*, **20**, 3672.

316 Schaefer, M., Mahr, N., Wolf, J. and Werner, H. (1993) *Angewandte Chemie-International Edition in English*, **32**, 1315.

317 Wolf, J. and Werner, H. (1987) *Journal of Organometallic Chemistry*, **336**, 413.

318 Consiglio, G., Schwab, R. and Morandini, F. (1988) *Journal of the Chemical Society, Chemical Communications*, 25.

319 Yen, Y.-S., Lin, Y.-C., Huang, S.-L., Liu, Y.-H., Sung, H.-L. and Wang, Y. (2005) *Journal of the American Chemical Society*, **127**, 18037.

320 Cadierno, V., Conejero, S., Diez, J., Pilar Gamasa, M., Gimeno, J. and García-Granda, S. (2003) *Chemical Communications*, 840.

321 Brana, P., Gimeno, J. and Sordo, J.A. (2004) *Journal of Organic Chemistry*, **69**, 2544.

322 Beckhaus, R., Oster, J., Wang, R. and Böhme, U. (1998) *Organometallics*, **17**, 2215.

323 Woodworth, B.E., White, P.S. and Templeton, J.L. (1998) *Journal of the American Chemical Society*, **120**, 9028.

324 Kolobova, N.E., Antonova, A.B. and Khitrova, O.M. (1978) *Journal of Organometallic Chemistry*, **146**, C17.

325 (a) Werner, H., Wolf, J., Müller, G. and Krüger, C. (1984) *Angewandte Chemie*, **96**, 421. (b) Werner, H., Wolf, J., Müller, G.

and Krüger, C. (1988) *Journal of Organometallic Chemistry*, **342**, 381.
326 Werner, H., García Alonso, F.J., Otto, H., Peters, K. and von Schnering, H.G. (1988) *Chemische Berichte*, **121**, 1565.
327 Brekau, U. and Werner, H. (1990) *Organometallics*, **9**, 1067.
328 Beckhaus, R., Oster, J. and Wagner, T. (1994) *Chemische Berichte*, **127**, 1003.
329 Beckhaus, R. and Oster, J. (1995) *Zeitschrift für Anorganische und Allgemeine Chemie*, **621**, 359.
330 Beckhaus, R., Oster, J., Kempe, R. and Spannenberg, A. (1996) *Angewandte Chemie-International Edition in English*, **35**, 1565.
331 Fischer, H., Leroux, F., Roth, G. and Stumpf, R. (1996) *Organometallics*, **15**, 3723.
332 Leroux, F., Stumpf, R. and Fischer, H. (1998) *European Journal of Inorganic Chemistry*, 1225.
333 Kolobova, N.E., Skripkin, V.V., Aleksandrov, G.G. and Struchkov, Yu.T. (1979) *Journal of Organometallic Chemistry*, **169**, 293.
334 Winter, R.F. (1999) *European Journal of Inorganic Chemistry*, 2121.
335 Rigaut, S., Olivier, C., Costuas, K., Choua, S., Fadhel, O., Massue, J., Turek, P., Saillard, J.-Y., Dixneuf, P.H. and Touchard, D. (2006) *Journal of the American Chemical Society*, **128**, 5859.
336 Bruce, M.I., Ellis, B.G., Skelton, B.W. and White, A.H. (2005) *Journal of Organometallic Chemistry*, **690**, 1772.
337 Shaw-Taberlet, J.A., Sinbandhit, S., Roisnel, T., Hamon, J.-R. and Lapinte, C. (2006) *Organometallics*, **25**, 5311.
338 Cadierno, V., Pilar Gamasa, M., Gimeno, J., Perez-Carreño, E. and García-Granda, S. (1999) *Organometallics*, **18**, 2821.
339 Jung, S., Ilg, K., Wolf, J. and Werner, H. (2001) *Organometallics*, **20**, 2121.
340 Werner, H., Jung, S., Gónzalez-Herrero, P., Ilg, K. and Wolf, J. (2001) *European Journal of Inorganic Chemistry*, 1957.
341 Bruce, M.I., Hall, B.C. and Tiekink, E.R.T. (1997) *Australian Journal of Chemistry*, **50**, 1097.
342 Abd-Elzaher, M.M., Weibert, B. and Fischer, H. (2005) *Organometallics*, **24**, 1050.
343 Castarlenas, R., Esteruelas, M.A., Gutiérrez-Puebla, E. and Oñate, E. (2001) *Organometallics*, **20**, 1545.
344 Nesmeyanov, A.N., Kolobova, N.E., Obezyuk, N.S. and Anisimov, K.N. (1976) *Izvestiya Akademii Nauk SSSR-Seriya Khimicheskaya*, 948.
345 Leung, W.-H., Lau, K.-K., Zhang, Q.-F., Wong, W.-T. and Tang, B. (2000) *Organometallics*, **19**, 2084.
346 O'Connor, J.M., Pu, L. and Chadha, R.K. (1990) *Journal of the American Chemical Society*, **112**, 9627.
347 O'Connor, J.M., Hiibner, K., Merwin, R., Pu, L. and Rheingold, A.L. (1995) *Journal of the American Chemical Society*, **117**, 8861.
348 Beckhaus, R., Sang, J., Oster, J. and Wagner, T. (1994) *Journal of Organometallic Chemistry*, **484**, 179.
349 Beckhaus, R., Strauss, I. and Wagner, T. (1994) *Journal of Organometallic Chemistry*, **464**, 155.
350 Beckhaus, R., Strauss, I., Wagner, T. and Kiprof, P. (1993) *Angewandte Chemie-International Edition in English*, **32**, 264.
351 Ipaktschi, J., Reimann, K., Serafin, M. and Dülmer, A. (2003) *Journal of Organometallic Chemistry*, **670**, 66.
352 (a) Ipaktschi, J., Mohsseni-Ala, J., Dülmer, A., Loschen, C. and Frenking, G. (2005) *Organometallics*, **24**, 977. (b) Ipaktschi, J., Rooshenas, P. and Dülmer, A. (2005) *Organometallics*, **24**, 6239.
353 Ipaktschi, J., Klotzbach, T. and Dülmer, A. (2000) *Organometallics*, **19**, 5281.
354 Weber, L., Noveski, G., Braun, T., Stammler, H.-G. and Neumann, B. (2007) *European Journal of Inorganic Chemistry*, 562.
355 Weber, L., Bayer, P., Noveski, G., Stammler, H.G. and Neumann, B. (2006)

European Journal of Inorganic Chemistry, 2299.
356 Abd-Elzaher, M.M., Froneck, T., Roth, G., Gvozdev, V. and Fischer, H. (2000) *Journal of Organometallic Chemistry*, **599**, 288.
357 Fischer, H., Podschadly, O., Früh, A., Troll, C., Stumpf, R. and Schlageter, A. (1992) *Chemische Berichte*, **125**, 2667.
358 Abd-Elzaher, M.M. and Fischer, H. (1999) *Journal of Organometallic Chemistry*, **588**, 235.
359 Fischer, H., Kirchbauer, F., Früh, A., Abd-Elzaher, M.M., Roth, G., Karl, C.C. and Dede, M. (2001) *Journal of Organometallic Chemistry*, **620**, 165.
360 (a) Fischer, H., Volkland, H.-P., Früh, A. and Stumpf, R. (1995) *Journal of Organometallic Chemistry*, **491**, 267. (b) Fischer, H., Karl, C.C. and Roth, G. (1996) *Chemische Berichte*, **129**, 615. (c) Fischer, H., Podschadly, O., Roth, G., Herminghaus, S., Klewitz, S., Heck, J., Houbrechts, S. and Meyer, T. (1997) *Journal of Organometallic Chemistry*, **541**, 321.
361 Kelley, C., Mercando, L.A., Lugan, N., Geoffroy, G.L., Xu, Z. and Rheingold, A.L. (1992) *Angewandte Chemie-International Edition in English*, **31**, 1053.
362 Alvarez, P., Lastra, E., Gimeno, J., Bassetti, M. and Falvello, L.R. (2003) *Journal of the American Chemical Society*, **125**, 2386.
363 Alt, H.G., Engelhardt, H.E. and Steinlein, E. (1988) *Journal of Organometallic Chemistry*, **344**, 227.
364 (a) Valyaev, D.A., Peterleitner, M.G., Leont'eva, L.I., Novikova, L.N., Semeikin, O.V., Khrustalev, V.N., Antipin, M., Yu., Ustynyuk, N.A., Skelton, B.W. and Whie, A.H. (2003) *Organometallics*, **22**, 5491. (b) Peterleitner, M.G., Valyaev, D.A., Novikova, L.N., Semeikin, O.V. and Ustynyuk, N.A. (2003) *Elektrokhimiya*, **39**, 1270.
365 (a) Barrett, A.G.M. and Sturgess, M.A. (1986) *Tetrahedron Letters*, **27**, 3811. (b) Barrett, A.G.M. and Sturgess, M.A. (1987) *Journal of Organic Chemistry*, **52**, 3940. (c) Barrett, A.G.M., Mortier, J., Sabat, M. and Sturgess, M.A. (1988) *Organometallics*, **7**, 2553.
366 Höhn, A., Otto, H., Dziallas, M. and Werner, H. (1987) *Chemical Communications*, 852.
367 Wiedemann, R., Steinert, P., Schaefer, M. and Werner, H. (1993) *Journal of the American Chemical Society*, **115**, 9864.
368 Werner, H., Ilg, K. and Weberndörfer, B. (2000) *Organometallics*, **19**, 3145.
369 (a) Ouzzine, K., Le Bozec, H. and Dixneuf, P.H. (1986) *Journal of Organometallic Chemistry*, **317**, C25. (b) Le Bozec, H., Ouzzine, K. and Dixneuf, P.H. (1991) *Organometallics*, **10**, 2768.
370 Selegue, J.P. (1982) *Organometallics*, **1**, 217.
371 Rao, K.M. and Agarwala, U.C. (1996) *Proc. Ind. Acad. Sci., Chem. Sci.*, **108**, 351.
372 Pilar Gamasa, M., Gimeno, J., Martín-Vaca, B.M., Isea, R. and Vegas, A. (2002) *Journal of Organometallic Chemistry*, **651**, 22.
373 Keller, A., Jasionka, B., Glowiak, T., Ershov, A. and Matusiak, R. (2003) *Inorganica Chimica Acta*, **344**, 49.
374 Li, X., Vogel, T., Incarvito, C.D. and Crabtree, R.H. (2005) *Organometallics*, **24**, 62.
375 Cadierno, V., Pilar Gamas, M., Gimeno, J., González-Cueva, M., Lastra, E., Borge, J., García-Granda, S. and Pérez-Carreño, E. (1996) *Organometallics*, **15**, 2137.
376 Miki, K., Yokoi, T., Nishino, F., Ohe, K. and Uemura, S. (2002) *Journal of Organometallic Chemistry*, **645**, 228.
377 Daniel, T., Mahr, N., Braun, T. and Werner, H. (1993) *Organometallics*, **12**, 1475.
378 Rüba, E., Hummel, A., Mereiter, K., Schmid, R. and Kirchner, K. (2002) *Organometallics*, **21**, 4955.
379 Iwasawa, N., Shido, M., Maeyama, K. and Kusama, H. (2000) *Journal of the American Chemical Society*, **122**, 10226.

380 Feracin, S., Hund, H.U., Bosch, H.W., Lippmann, E., Beck, W. and Berke, H. (1992) *Helvetica Chimica Acta*, **75**, 1305.

381 Shaw, M.J., Afridi, S.J., Light, S.L., Mertz, J.N. and Ripperda, S.E. (2004) *Organometallics*, **23**, 2778.

382 Bierstedt, A., Clark, G.R., Roper, W.R. and Wright, L.J. (2006) *Journal of Organometallic Chemistry*, **691**, 3846.

383 Le, T.X., Selnau, H.E. and Merola, J.S. (1994) *Journal of Organometallic Chemistry*, **468**, 257.

384 Wakatsuki, Y., Yamazaki, H., Kumegawa, N., Satoh, T. and Satoh, J.Y. (1991) *Journal of the American Chemical Society*, **113**, 9604.

385 McMullen, A.K., Selegue, J.P. and Wang, J.G. (1991) *Organometallics*, **10**, 3421.

2
Preparation and Stoichiometric Reactivity of Metal Allenylidene Complexes
Victorio Cadierno, Pascale Crochet, and José Gimeno

2.1
Introduction

Allenylidene (propadienylidene) complexes [M]=C=(C)$_n$=CR$_2$ ($n = 1$) belong to the series of cumulenylidene derivatives in which the unsaturated carbene unit is stabilized by a transition metal fragment. Although extremely reactive free allenylidene species:C=C=CH$_2$ can be generated experimentally and trapped in cold matrices (theoretically characterized as singlet ground state species) or used for identification via further reactions [1], these seminal studies provided little influence on the chemistry of allenylidene complexes. It was not until 1976 when the first examples [M{=C=C=CPh(NEt$_2$)}(CO)$_5$] (M = Cr, W) and [MnCp(=C=C=CtBu$_2$)(CO)$_2$] (Cp = η^5–C$_5$H$_5$) were discovered [2]. Since then, much progress has been made in the chemistry of metal-allenylidenes mainly due to their stoichiometric and catalytic applications in organic synthesis. The stabilization to a transition metal center is possible by the use of the lone pair of electrons on the carbenic carbon atom via formation of a dative metal–carbon bond. The electron back-donation from the metal fragment to the allenylidene chain may strengthen the bond (see Chapter 4 for theoretical studies). Although polynuclear derivatives are known, the most common allenylidenes are mononuclear complexes [3, 4]. A wide series of electron rich Group 6–9 metal fragments, including half-sandwich **A**, octahedral **B**, square-planar **C** and five-coordinate **D** (Figure 2.1), can stabilize allenylidene moieties =C=C=CR^{1}R^2. An unusual tetrahedral rhenium(VII) derivative [Re(NtBu)$_2$(SR)(=C=C=CPh$_2$)] (R = 1-adamantyl) has been isolated recently [5].

Very stable mononuclear metal d^6 rhenium(I), ruthenium(II) and osmium(II) allenylidenes dominate the scope of reported examples. In contrast, isoelectronic Group 6 metal(0) derivatives of the type [M(=C=C=CR^{1}R^2)(CO)$_5$] are generally thermally unstable. In this case, the stabilization of the carbene moiety requires the presence of electron-releasing substituents (R^1 = R^2 = 4-C$_6$H$_4$NR'$_2$, 4-C$_6$H$_4$OR', etc.). To date, no transition metal allenylidenes of Group 5, 10 and 11 have been described. Only one mononuclear example of Group 4 is known, namely [TiCp$_2${=C=C=CPh$_2$)

Metal Vinylidenes and Allenylidenes in Catalysis: From Reactivity to Applications in Synthesis
Edited by Christian Bruneau and Pierre Dixneuf
Copyright © 2008 WILEY-VCH Verlag GmbH & Co. KGaA, Weinheim
ISBN: 978-3-527-31892-6

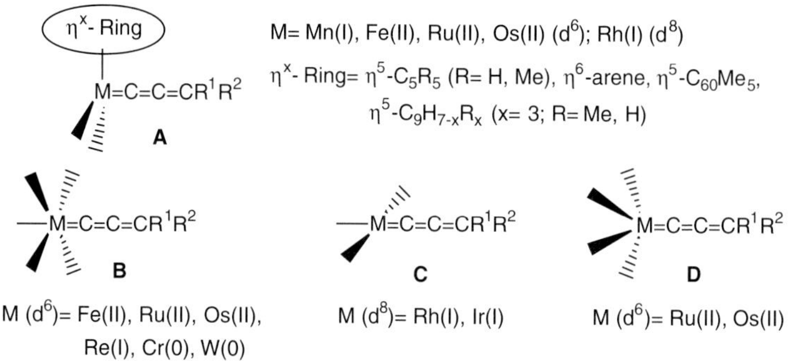

Figure 2.1 Allenylidenes of Group 6–9 metal fragments.

A: η^x-Ring—M=C=C=CR^{1}R^2; M= Mn(I), Fe(II), Ru(II), Os(II) (d^6); Rh(I) (d^8); η^x-Ring= η^5-C$_5$R$_5$ (R= H, Me), η^6-arene, η^5-C$_{60}$Me$_5$, η^5-C$_9$H$_{7-x}$R$_x$ (x= 3; R= Me, H)

B: M=C=C=CR^{1}R^2; M (d^6)= Fe(II), Ru(II), Os(II), Re(I), Cr(0), W(0)

C: M=C=C=CR^{1}R^2; M (d^8)= Rh(I), Ir(I)

D: M=C=C=CR^{1}R^2; M (d^6)= Ru(II), Os(II)

(PMe$_3$)] [6]. Nowadays, the accessibility of a number of efficient synthetic routes has facilitated the reactivity studies, disclosing a very rich and versatile chemistry. During the last decade the knowledge has grown to such an extent that several general reviews [3] and specialized accounts of chromium and tungsten [4a], Group 8 metals [4b], rhodium [4c], iridium [4d], ruthenium [4e–k], and osmium complexes [4l,m] are available. Recently, the chemistry of ruthenium and osmium complexes supported by macrocyclic ligands has been reviewed [4n]. In addition, brief accounts dealing with particular aspects of polynuclear and cluster species have also appeared [3a,b, 7].

In this chapter we have tried to give a general presentation of the most efficient synthetic routes, the main characteristic structural features and a discussion of the reactivity patterns. Special attention is devoted to the chemistry of mononuclear derivatives containing linear allenylidene groups as they are most frequently the active species in catalytic processes. Illustrative examples of the synthetic applications via stoichiometric reactions will also be covered. For more comprehensive information we refer the reader to the reviews and accounts mentioned above [3, 4, 7].

2.2
Preparation of Allenylidene Complexes

2.2.1
General Methods of Synthesis

The majority of synthetic approaches for allenylidene complexes use propargylic alcohols HC≡CCR^{1}R^2(OH) as sources of the allenylidene C$_3$ skeleton. In 1982, Selegue first introduced this synthetic strategy for the high yield preparation of the ruthenium(II) complex [RuCp(=C=C=CPh$_2$)(PMe$_3$)$_2$][PF$_6$] [8]. The alkynol is converted smoothly into the allenylidene unit via elimination of water (Equation 2.1).

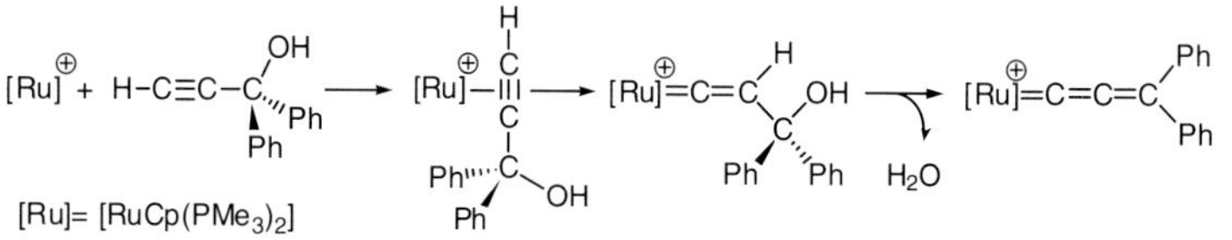

[Ru]= [RuCp(PMe$_3$)$_2$]

$$[Ru]^+ + HC\equiv C-C(OH)Ph_2 \longrightarrow [Ru]=C=C=C(Ph)_2 \quad -H_2O$$

(2.1)

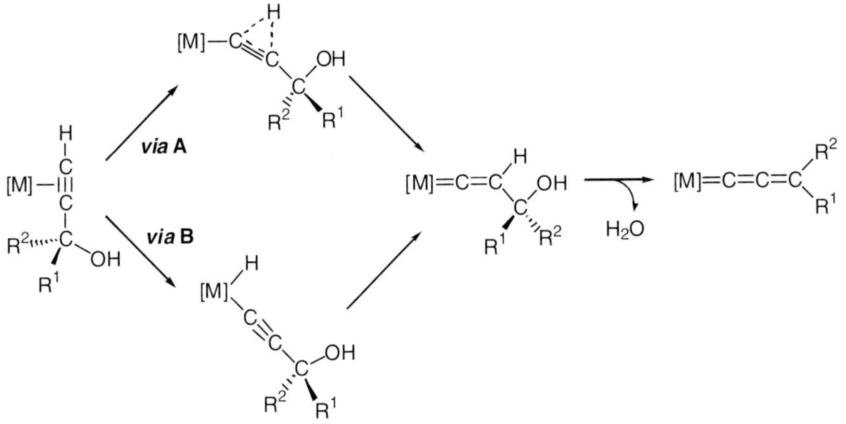

Scheme 2.1 Mechanism of the dehydration of 2-propyn-1-ols into allenylidene ligands.

The reaction is based on the spontaneous dehydration of the intermediate hydroxyvinylidene species formed, either via **A** or via **B** (Scheme 2.1), after the coordination of 2-propyn-1-ols at the metal center.

This synthetic strategy proved to be unsuitable for $[M(CO)_5]$ (M = Cr, W) metal fragments due to the thermal instability of the corresponding non-donor-substituted allenylidenes $[M(=C=C=CR^1R^2)(CO)_5]$ (R^1, R^2 = usually alkyl or aryl groups). An alternative general synthetic procedure using deprotonated functionalized acetylenes has been successfully applied (for example tris-amino or alkoxo prop-1-ynes; see Equation 2.2) [4a].

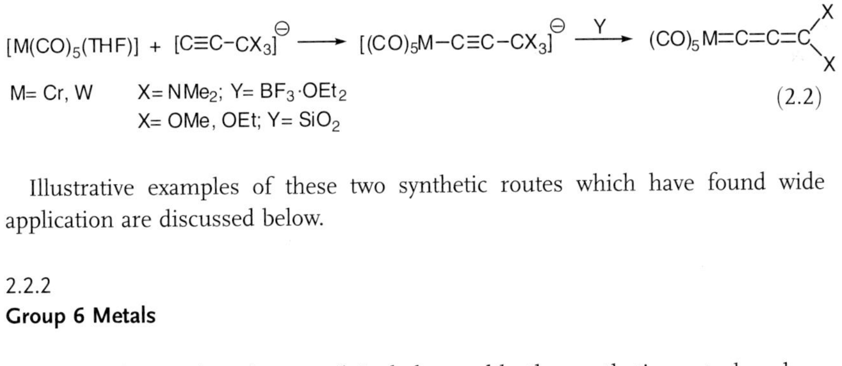

$$[M(CO)_5(THF)] + [C\equiv C-CX_3]^{\ominus} \longrightarrow [(CO)_5M-C\equiv C-CX_3]^{\ominus} \xrightarrow{Y} (CO)_5M=C=C=C\begin{matrix}X\\X\end{matrix}$$

M = Cr, W   X = NMe$_2$; Y = BF$_3\cdot$OEt$_2$
             X = OMe, OEt; Y = SiO$_2$ (2.2)

Illustrative examples of these two synthetic routes which have found wide application are discussed below.

2.2.2
Group 6 Metals

Fischer and coworkers have exploited thoroughly the synthetic route based on functionalized acetylides [4, 9]. Thus, by using deprotonated propynoic acid amides (alkynyl metallate) the reaction with $[M(CO)_5(THF)]$ followed by treatment with $[R_3O]BF_4$ affords N/O-substituted allenylidene complexes (Equation 2.3).

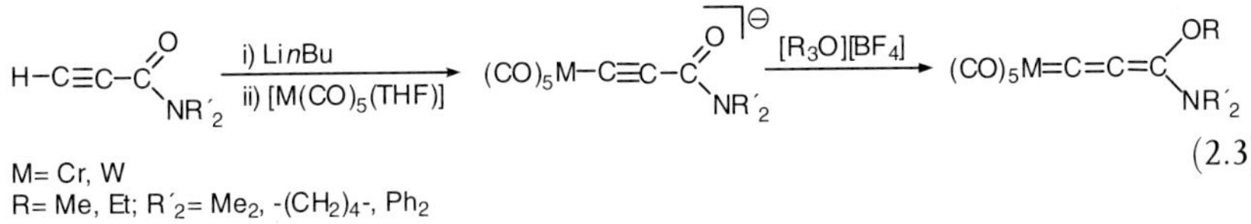

$$H-C\equiv C-C(=O)NR'_2 \xrightarrow[\text{ii) }[M(CO)_5(THF)]]{\text{i) Li}n\text{Bu}} (CO)_5M-C\equiv C-C(=O)^{\ominus}NR'_2 \xrightarrow{[R_3O][BF_4]} (CO)_5M=C=C=C(OR)NR'_2$$

M = Cr, W
R = Me, Et; R'$_2$ = Me$_2$, -(CH$_2$)$_4$-, Ph$_2$ (2.3)

Figure 2.2 Stable heteroatom-substituted Group 6 allenylidenes.

Analogous N/C substituted allenylidenes [M{=C=C=CPh(NMeR)}(CO)$_5$] (M = Cr, W; R = H, Me, Et) are obtained by using C-ethynylimines HC≡CC(=NMe)Ph instead of propynoic acid amides. O/C-, O/O-, and N/S-substituted allenylidenes are also accessible from ethynyl ketones HC≡CC(=O)R, propynoic acid esters HC≡CC(=O)OR and propynethioic acid amides HC≡CC(=S)NR$_2$, respectively, after sequential deprotonation and the corresponding alkylation (representative examples are shown in Figure 2.2) [4a, 9].

Some of these complexes have also been used as suitable precursors of related allenylidenes obtained through substitution or insertion reactions (see below). Other synthetic approaches to achieve selectively C/C substituted allenylidenes from reactions of dilithium derivatives Li$_2$[C≡CCR$_2$O], obtained by deprotonation of 2-propyn-1-ols, with [M(CO)$_5$(THF)] (M = Cr, W) and subsequent deoxygenation with phosgene are not always straightforward [10].

2.2.3
Group 7 Metals

In addition to the unusual tetrahedral Re(VII) complex [5], Re(I) derivatives are the only stable mononuclear allenylidenes of Group 7 metals after the preparation of the pioneer manganese derivatives [MnCp(=C=C=CR$_2$)(CO)$_2$] (R = tBu, Ph, CH$_2$Ph, cyclohexyl; obtained by reaction of the η^2-alkyne complex [MnCp(η^2-HC≡CCO$_2$Me)(CO)$_2$] with LiR and phosgene) [2b, 11]. Bianchini's group has studied extensively the chemistry of octahedral complexes [Re(=C=C=CR^{1}R^2)(CO)$_2$(triphos)][OTf] (triphos MeC(CH$_2$PPh$_2$)$_3$; TfO$^-$ = CF$_3$SO$_3^-$; **1** in Figure 2.3). Following Selegue's methodology rhenium(I) allenylidenes **1** can be prepared in CH$_2$Cl$_2$ from mono- and di-substituted propargylic alcohols HC≡CCR^{1}R^2(OH) (R^1 = Ph; R^2 = H, Me, Ph) and

R^1 = Ph; R^2 = H, Me, Ph

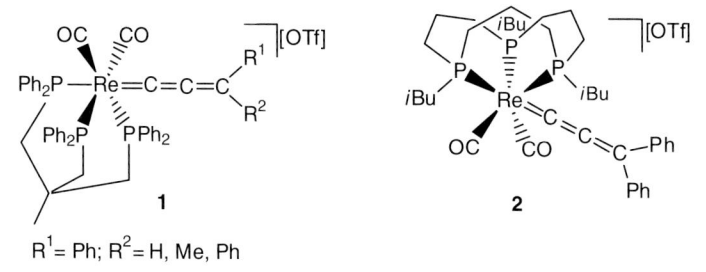

Figure 2.3 Allenylidene-rhenium(I) complexes.

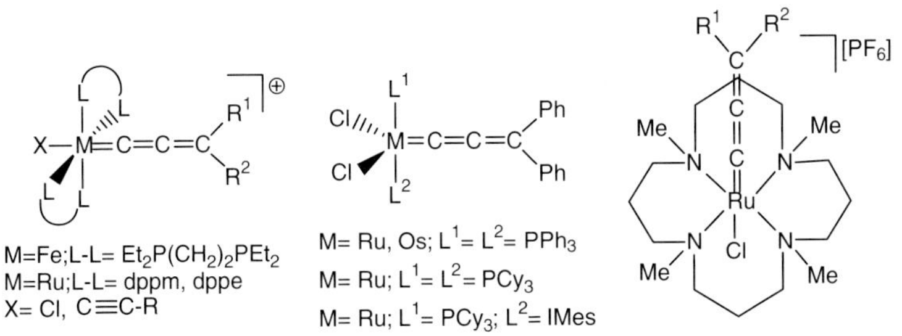

Figure 2.4 Example of octahedral and coordinate allenylidene complexes of Group 8 metals.

isolated as air-stable solids [12a–c]. A similar rhenium fragment containing the triphosphorus macrocyclic ligand 1,5,9-triisobutyl-1,5,9-triphosphacyclododecane has also been used to prepare the analogous octahedral allenylidene complex 2 (Figure 2.3) [12d]. Other Re(I) allenylidenes [Re(=C=C=CPh$_2$)(CO)$_2$(PR$_3$)$_3$][BF$_4$] (PR$_3$ = PPh(OEt)$_2$, PPh$_2$(OEt)) are also known [12e].

2.2.4
Group 8 Metals

Although a series of five-coordinated iron(0) derivatives [Fe(=C=C=CR$_2$)(CO)$_2$L$_2$] (L = CO, R = tBu; L = PEt$_3$, R = Ph, tBu) were known [13], only the application of Selegue's methodology (Equation 2.1) has given rise to the systematic synthesis of stable iron(II), ruthenium(II) and osmium(II) allenylidenes (some examples are included in Figures 2.4 and 2.5). The reactions usually proceed in a one-pot manner by reacting the precursor halide complex with the appropriate propargylic alcohol in the presence of a halide abstractor (NaBF$_4$, KPF$_6$, AgSbF$_6$, etc.). Limitations of this synthetic route mainly stem from: (i) the reluctance of the hydroxy-vinylidene intermediate **E** (Scheme 2.2) to undergo dehydration, especially when strong electron-releasing metal fragments are used, and (ii) the competitive formation of an alkenyl-vinylidene isomer arising from the activation of propargylic alcohols containing a C–H bond in the β-position with respect to the OH group (Scheme 2.2) [14]. Although spontaneous dehydration usually occurs, eventually

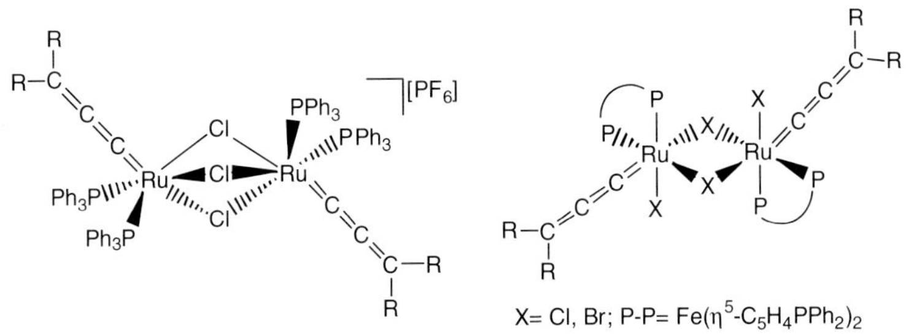

Figure 2.5 Dinuclear ruthenium (II) allenylidenes.

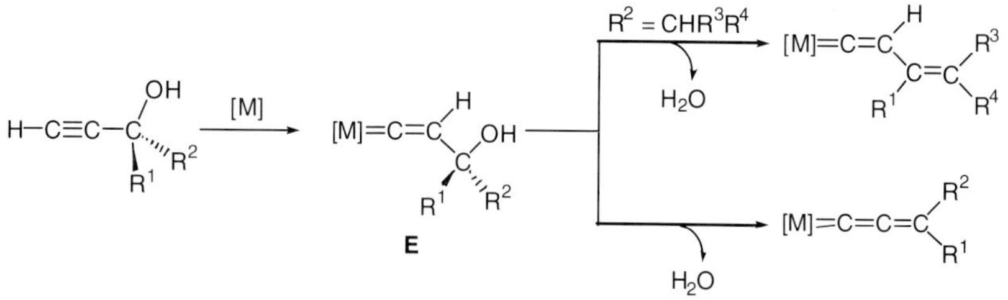

Scheme 2.2 Competitive formation of alkenyl-vinylidene vs. allenylidene group.

3-hydroxyvinylidene intermediates $[M]=C=C(H)C(OH)(R^1)(R^2)$ are stable and transformatios into the allenylidenes requires treatment with acidic Al_2O_3 [15].

The use of methanol or ethanol as solvent (or sometimes the molecule of water resulting from the spontaneous dehydration) often leads to the isolation of a Fischer-type alkoxy- or hydroxy-carbene $[M]=C(OR)CH=CR^1R^2$ instead of the desired allenylidene. Addition of nucleophiles to allenylidenes dominates the reactivity of these electrophilic groups (see below). Nevertheless, in some cases, the use of silver (I) salts Ag[X] ($X^- = PF_6^-$, TfO^-, BF_4^-) results in a more practical and flexible synthetic method since the use of nucleophilic polar solvents can be avoided.

2.2.4.1 Octahedral and Five-Coordinate Derivatives

Figure 2.4 shows illustrative complexes in which the allenylidene unit is generated from propargylic alcohols by reaction with a coordinatively unsaturated complex generated either by abstraction of a halide ligand or the dissociation of a labile ligand from the appropriate precursors. Ancillary ligands mainly include monodentate phosphines and phosphites [16], bidentate phosphines [4i–k,17], phosphinoester and phosphinoether P,O-donor ligands [18], bidentate pyrazolylphosphines [19a], tridentate aminodiphosphines [19b], poly(pyrazolyl)borates and related polydentate N-donor ligands including macrocycles [4n, 20].

The competitive formation of dinuclear complexes containing halide bridging systems has also been reported (representative examples are shown in Figure 2.5) [21].

Other synthetic methodologies have been used to prepare the unusual octahedral osmium(II) allenylidenes $[OsH(=C=C=CPh_2)(NCMe)_2(PiPr_3)_2][BF_4]$ and $[Os\{\kappa^2(C,O)\text{-}C(CO_2Me)=CH_2\}(=C=C=CPh_2)CO)(PiPr_3)_2][BF_4]$ from 1,1-diphenyl-2-propyn-1-ol [22].

2.2.4.2 Half-Sandwich Derivatives

Following the seminal preparation of $[RuCp(=C=C=CPh_2)(PMe_3)_2][PF_6]$ [8] an extensive series of analogous allenylidenes has been prepared using this methodology. Typical metal fragments include not only the classical η^5-cyclopentadienyl and η^6-arene chloride derivatives but also tethered-type ligands in which the rings are linked to the metal also through an ancillary κ^1-coordinated donor atom giving rise to $\eta^5{:}\kappa^1(L)$- or $\eta^6{:}\kappa^1(L)$-coordination modes (Figure 2.6).

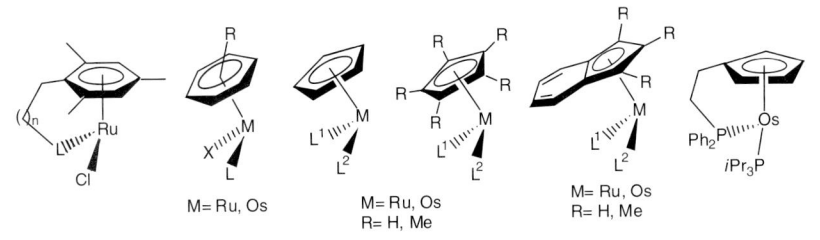

Figure 2.6 Half-sandwich metal fragments of Group 8 allenylidenes. Ancillary ligands include mono- and bi-dentate phosphines.

η^5-**Cyclopentadienyl and η^5-Indenyl Derivatives** The most common are cationic ruthenium(II) complexes [Ru(η^5-ring)(=C=C=CR^{1}R^2)(L^1)(L^2)][X] (X$^-$ = BF$_4^-$, BPh$_4^-$, PF$_6^-$, TfO$^-$, B(Ar$_F$)$_4^-$, and so on; Ar$_F$ = 3,5-C$_6$H$_3$(CF$_3$)$_2$) involving both η^5-C$_5$R$_5$ (R = H,Me) and η^5-C$_9$H$_7$ or η^5-1,2,3-C$_9$H$_4$Me$_3$ rings. The metal fragments mostly contain phosphine ligands. Illustrative examples of cyclopentadienyl derivatives are of the type: (i) L^1 = L^2 = PPh$_3$, PEt$_3$, PMe$_3$, PMeiPr$_2$, PPh$_2$NHR (R = Ph, nPr) [15b, 23], (ii) L^1 = CO, L^2 = PMeiPr$_2$, PiPr$_3$ [24], (iii) L^{1}L^2 = dippe (1,2-bis(diisopropylphosphino)ethane) [15a, 25] and (iv) L^1 = PPh$_3$, L^2 = Ph$_2$PCH$_2$C(=O)tBu [26]. Cyclopentadienyl-Os(II) and Fe(II) allenylidenes, that is, [OsCp(=C=C=CPh$_2$)(L^1)(L^2)][PF$_6$] (L^1 = L^2 = PPh$_3$, PiPr$_3$; L^1 = CO, L^2 = PiPr$_3$, PHPh$_2$) [27] and [Fe(η^5-C$_5$R$_5$)(=C=C=CPh$_2$)(dppe)][X] (dppe = 1,2-bis(diphenylphosphino)ethane; R = H, X$^-$ = BF$_4^-$; R = Me, X$^-$ = PF$_6^-$) [28], as well as an extensive series of analogous cationic allenylidenes of indenyl-Ru(II) and Os(II) fragments containing mono and bidentate phosphines [26, 29], including examples with chiral ligands [29g,i] or chiral substituents in the allenylidene chain [29d, e], have also been described. Atypical dinuclear ruthenium(III) complexes [Cp*RuCl(μ^2-ER)$_2$RuCp*(=C=C=CAr$_2$)]$^+$ (E = S, R = Me, iPr; E = S, Se, Te, R = Me), which have been synthesized by reaction of [Cp*RuCl(μ^2-SMe)$_2$RuCp*Cl] or [Cp*RuCl(μ^2-ER)$_2$RuCp*(H$_2$O)]$^+$ with the corresponding alkynol, are the only known ruthenium(III) allenylidenes [30]. They have been widely used in catalytic transformations of propargylic alcohols (see Chapter 7). Neutral Ru(II) and Os(II) derivatives are also known. Representative examples are: [RuClCp*(=C=C=CPh$_2$){κ^1(P)-iPr$_2$PCH$_2$CO$_2$Me}] [31a], [RuClCp(=C=C=CPh$_2$)(PPh$_3$)] [31b] and [OsClCp(=C=C=CPh$_2$)(PiPr$_3$)] [31c].

η^6-**Arene Derivatives** Most of these belong to the series of cationic complexes with general formula [RuCl(=C=C=CR^{1}R^2)(η^6-p-cymene)(L)][X], being prepared by using readily accessible dichloride precursors [RuCl$_2$(η^6-p-cymene)(L)] (R^1 and R^2 aryl groups; L = tertiary phosphine, such as PCy$_3$, PPh$_3$, PiPr$_3$ [32a–d], or N-heterocyclic carbene (NHC) ligands [32d–f]) and propargylic alcohols. Similarly, the use of complexes [MX$_2$(η^6–arene)(AR$_3$)] as starting materials has allowed the synthesis of the corresponding stable allenylidenes, that is, [RuCl(=C=C=CPh$_2$)(η^6-arene)(PR$_3$)][X] (arene = C$_6$Me$_6$, 1,2,4,5-C$_6$H$_2$Me$_4$; PR$_3$ = PMe$_3$, PPh$_3$, PCy$_3$ or Ph$_2$PCH$_2$P(=N-4-C$_5$F$_4$N)Ph$_2$; not all combinations) [32d,g,h] and [OsX

($=C=C=CR_2$)(η^6-arene)(AR_3)][PF_6] (X = Cl, I; R = Ph, 4-C_6H_4Me; arene = p-cymene, 1,3,5-$C_6H_3Me_3$, C_6H_6; AR_3 = PMe_3, PPh_3, PCy_3, AsiPr$_3$, SbiPr$_3$; not all combinations) [32i]. The dicationic species [Ru(=C=C=CMe$_2$)(η^6-p-cymene)(MeO-biphep)][SbF$_6$]$_2$ (MeO-biphep = (6,6'-dimethoxybiphenyl-2,2'-diyl)bis(diphenylphosphine)) and [Ru(=C=C=CPh$_2$)(η^6-p-cymene){κ^2(P,N)-Ph$_2$P-2-py}][BF$_4$]$_2$ are also known [32j,k].

η^5-Cyclopentadienyl and η^6-Arene Tethered Derivatives The use of Selegue's route using halide precursors has led to the synthesis of the following cationic allenylidenes: [Os{η^5:κ^1(P)-C$_5$H$_4$(CH$_2$)$_2$PPh$_2$}(=C=C=CPh$_2$)(PiPr$_3$)][PF$_6$] [33a], [Ru{η^5:κ^2-(P,P)-C$_5$H$_4$CH$_2$CMe(CH$_2$PPh$_2$)$_2$}(=C=C=CPh$_2$)][PF$_6$] [33b], [RuCl(=C=C=CPh$_2$){η^6:κ^1(P)-C$_6$H$_5$O(CH$_2$)$_2$PtBu$_2$}][PF$_6$] [33c] and [RuCl(=C=C=CPh$_2$){η^6:κ^1(P)-C$_6$H$_5$-(CH$_2$)$_3$PCy$_2$}][PF$_6$] [32a]. Related tethered η^6-arene complexes containing NHC-carbenes as side arms have also been described [33d].

2.2.4.3 Other Synthetic Methodologies

Besides Selegue's methodology several synthetic alternatives of Group 8 allenylidenes have been reported: (i) by trapping transient butatrienylidene complexes [M]$^+$=C=C=C=CH$_2$ in the presence of alcohols, thiols, selenothiols and secondary amines, which leads to heteroatom substituted allenylidenes [M]$^+$=C=C=C(ER$_n$)Me ([M] = [RuCp(PPh$_3$)$_2$], [RuCl(dppm)$_2$], [FeCp*(dppe)]; ER$_n$ = OR, NRMe, NR$_2$, SEt, SePh) [4e, 34a, b]. (ii) by trapping pentatetraenylidene intermediates [RuCl(=C=C=C=C=CPh$_2$)(η^6-C$_6$Me$_6$)(PR$_3$)]$^+$ with alcohols and primary amines which leads to heteroatom substituted alkenyl-allenylidenes [RuCl{=C=C=C(X)CH=CPh$_2$}(η^6-C$_6$Me$_6$)(PR$_3$)]$^+$ (X = OR, NR$_2$) [34c,d]. (iii) Starting from half sandwich acetylide cyclopentadienyl ruthenium(II) complexes which undergo chemical oxidation to give cationic allenylidene radicals [34e–g]. (iv) A unique case of an octahedral bis-allenylidene complex *trans*-[Ru(=C=C=CPh$_2$)$_2$(dppe)$_2$]$^{2+}$ prepared via oxidation with Ce(IV) ammonium nitrate of the corresponding *trans*-allenylidene-acetylide [34h].

2.2.5
Group 9 Metals

Typical square-planar rhodium(I) and iridium(I) allenylidenes *trans*-[MCl(=C=C=CR^{1}R^2)(PiPr$_3$)$_2$] have been prepared from substituted 2-propyn-1-ols via hydroxyvinylidenes *trans*-[MCl{=C=C(H)C(OH)R^{1}R^2}(PiPr$_3$)$_2$] by abstraction of water [35a–e]. The generation of iridium allenylidenes usually requires treatment with trace amounts of CF$_3$CO$_2$H and is also facilitated by using UV irradiation [35c,d]. In accord with the expected *trans*-influence of the π-acceptor allenylidene unit, substitution of the chloride ligand by different anionic nucleophiles is favored, affording new rhodium(I) and iridium(I) allenylidenes *trans*-[MX(=C=C=CR^{1}R^2)(PiPr$_3$)$_2$] (X = OH, I, N$_3$, OCN, SCN) [35a,d,f–i]. It is worth noting the interesting formation of the hydroxo derivatives which constitute the gateway for the synthesis of a variety of square-planar complexes, via OH$^-$/X$^-$ exchange reactions, not accessible through any other route [35d,g,h]. Other examples of Group 9 allenylidenes obtained from propargylic alcohols include *trans*-[RhCl(=C=C=CPh$_2$){κ^1(E)-iPr$_2$ECH$_2$CH$_2$

$$[M]{=}C{=}C{=}CR^1R^2 \longleftrightarrow [M]{-}C{\equiv}C{-}\overset{\oplus}{C}R^1R^2$$

Figure 2.7 Allenylidene/alkynl resonance.

OMe}$_2$] (E = P, As) [35j,k], and the cationic species trans-[Rh(=C=C=CPh$_2$)(L)(PiPr$_3$)$_2$][PF$_6$] (L = acetone, NH$_3$, py) [35l] and [Ir(=C=C=CPh$_2$)(η^4-COD)(PCy$_3$)][BF$_4$] (COD = 1,5-cyclooctadiene) [35m]. The half-sandwich Rh(I) complexes [RhCp(=C=C=CR^{1}R^2)(L)] (R^1 = Ph, R^2 = Ph, 2-C$_6$H$_4$Me; L = PiPr$_3$, iPr$_2$AsCH$_2$CH$_2$OMe) are also known, being obtained by reacting the corresponding square-planar precursor trans-[RhCl(=C=C=CR^{1}R^2)(L)$_2$] with NaCp [35e,j].

2.3
Coordination Modes and Structural Features

The allenylidene group usually acts as a terminal ligand featuring an almost linear coordination. Extensive structural studies from X-ray crystallography generally show that metal–carbon, C$_\alpha$–C$_\beta$ and C$_\beta$–C$_\gamma$ bond lengths are in accordance with the description of the bonding as a resonance of metal-carbene and metal-alkynyl mesomers (Figure 2.7), the latter (zwitterionic species [M]$^-$-C≡C-C$^+$R^{1}R^2 in neutral allenylidenes) being the dominant contribution to the observed structures.

Besides this most general terminal coordination, the allenylidene group can also act as a bridging ligand in binuclear complexes via two well-established μ-η^1:η^1 (end on) [36] and μ-η^1:η^2 (side on) [37] modes (**F** and **G**, respectively, in Figure 2.8). A number of atypical bridging modes, that is, **H** and **I**, in several cluster complexes show the electronic richness and coordinative versatility of the allenylidene chain [38].

2.4
Stoichiometric Reactivity of Allenylidenes

2.4.1
General Considerations of Reactivity

Based on a large number of stoichiometric studies, the main trends of allenylidene reactivity are presently well established, being governed by the electron-deficient

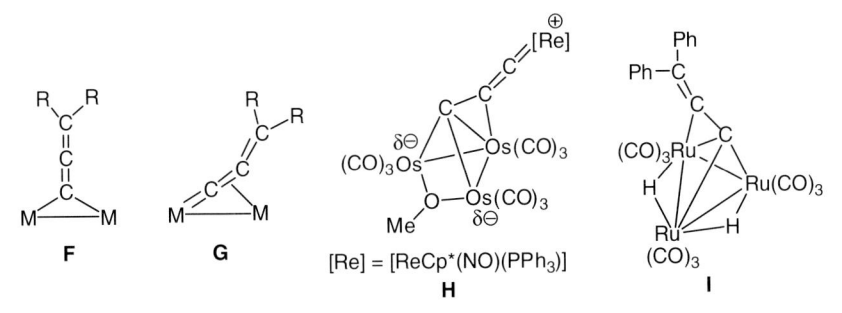

Figure 2.8 Structures **F**, **G**, **H** and **I**.

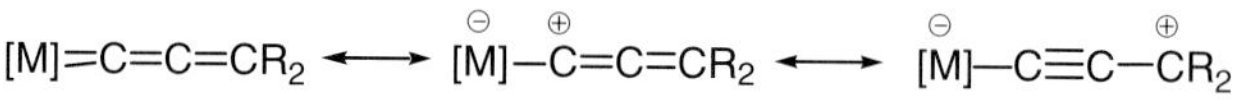

$[M]=C=C=CR_2 \longleftrightarrow [M]-\overset{\ominus}{C}=C=\overset{\oplus}{CR_2} \longleftrightarrow [M]-C\equiv C-\overset{\oplus}{CR_2}$

Figure 2.9 Canonical forms of allenylidene complexes.

character of the C_α and C_γ carbon atoms in the cumulenic chain, the C_β exhibiting nucleophilic character. This can also be rationalized by considering the mesomeric forms depicted in Figure 2.9. Theoretical calculations (see Chapter 4) are in accordance with these general reactivity patterns.

The alternating array of electrophilic/nucleophilic carbon sites makes allenylidene complexes unique organometallic reagents for C–C and C–heteroatom couplings via simple addition reactions. Thus, while electrophiles add selectively to C_β yielding alkenylcarbyne derivatives **J**, the nucleophilic attacks can take place both at the C_α or C_γ atoms, affording metal-allenyl **K** or metal-alkynyl **L** complexes, respectively (Scheme 2.3). Moreover, the unsaturated character of the allenylidene chain makes possible cyclization processes with a large variety of organic substrates, allowing the construction of original carbo- and heterocyclic compounds.

2.4.2
Electrophilic Additions

The nucleophilic character of the allenylidene C_β was experimentally demonstrated for the first time by Kolobova and coworkers in 1984. They obtained alkenyl-carbyne complexes [MnCp{≡CC(H)=CR$_2$}(CO)$_2$][X] (R = tBu, Ph; X$^-$ = Cl$^-$, BF$_4^-$, CF$_3$CO$_2^-$) by treatment of neutral manganese(I) allenylidenes [MnCp(=C=C=CR$_2$)(CO)$_2$] with Brønsted acids (HX) [39]. Since then, a large variety of neutral and cationic transition-metal allenylidenes have been selectively protonated or methylated at the C_β atom to afford stable alkenyl-carbyne species [M]≡CC(E)=CR^{1}R^2 (E = H, Me). Representative allenylidenes capable of being protonated are: [OsCpCl(=C=C=CPh$_2$)(PiPr$_3$)] [31c], [OsCp(=C=C=CPh$_2$)(PiPr$_3$)$_2$][PF$_6$] [27a], [OsH(=C=C=CPh$_2$)(NCMe)$_2$(PiPr$_3$)$_2$][BF$_4$] [40], [Os{η^5:κ^1(P)-C$_5$H$_4$CH$_2$CH$_2$PPh$_2$}(=C=C=CPh$_2$)(PiPr$_3$)][PF$_6$] [33a], [RuTpCl(=C=C=CR$_2$)(PR$_3$)] (Tp = tris(pyrazolyl)borate; PR$_3$ = PPh$_3$, PPh$_2$iPr; R = Ph, Fc) [20e], [{RuX(μ-X)(=C=C=CR$_2$)(dppf)}$_2$] (R = Ph, iPr; X = Cl, Br; dppf = 1,1′-bis(diphenylphosphino)ferrocene) [21b], [RuCl(=C=C=CPh$_2$){κ^2(P,O)-Cy$_2$PCH$_2$CH$_2$OMe}$_2$][PF$_6$] [18b], trans-[RuCl{=C=C=C(Me)R}(dppe)$_2$][BF$_4$] (R = Me, Ph) [41]

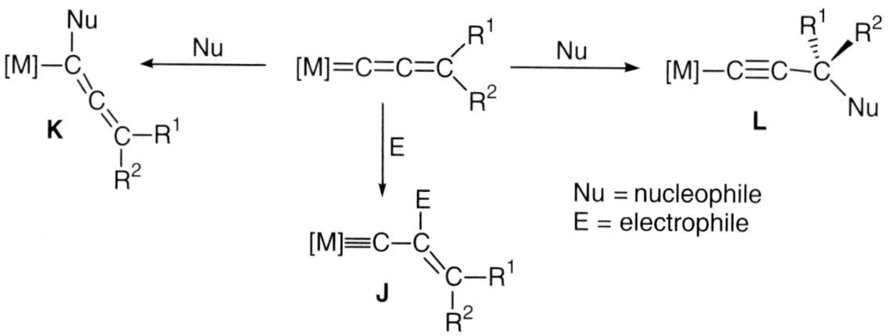

Scheme 2.3 Typical nucleophilic and electrophilic additions on metal allenylidenes.

Nu = nucleophile
E = electrophile

[RuCp*{=C=C=C(R)Ph}(dippe)][B(Ar$_F$)$_4$] (R = H, Ph) [42] and [Re(=C=C=CPh$_2$)(CO)$_2$(triphos)][OTf] [43]. In contrast, treatment of the neutral allenylidene complexes [RuCl$_2$-(=C=C=CPh$_2$){κ^2(P,O)-iPr$_2$PCH$_2$CO$_2$Me}{κ^1(P)-iPr$_2$PCH$_2$CO$_2$Me}] and trans-[IrCl{=C=C=C(R)Ph}(PiPr$_3$)$_2$] (R = Ph, tBu) with HX generates the α,β-unsaturated carbenes [RuCl$_2${=C(X)C(H)=CPh$_2$}{κ^2(P,O)-iPr$_2$PCH$_2$CO$_2$Me}{κ^1(P)-iPr$_2$PCH$_2$CO$_2$Me}] (X = Cl) [18a] and trans-[IrCl{=C(X)C(H)=C(R)Ph}(PiPr$_3$)$_2$] (R = Ph, tBu; X = OTf, CF$_3$CO$_2$) [35i], respectively, via HX-addition across the C$_α$=C$_β$ double bond, the Brønsted acids acting like classical NuH nucleophiles (see below). Nevertheless, it should be noted that the iridium derivatives (X = CF$_3$CO$_2$) readily evolve into the expected carbynes trans-[IrCl{≡CC(H)=C(R)Ph}(PiPr$_3$)$_2$][OTf] when dissolved in polar solvents [35i]. Interestingly, the Rh(I) allenylidenes trans-[RhCl{=C=C=C(R)Ph}(PiPr$_3$)$_2$] (R = Ph, 2–C$_6$H$_4$Me) have been found to react with HCl to afford the allenyl-Rh(III) species [RhCl$_2${C(H)=C=C(R)Ph}(PiPr$_3$)$_2$], via formal HCl addition across the Rh=C$_α$ bond [35e]. The course of this unexpected reaction involves an initial oxidative addition of HCl at the rhodium center followed by insertion of the allenylidene unit into the Rh–H bond. Probably, the most striking discovery related to this chemistry is the evidence that allenylidene-ruthenium complexes **3** rearrange, upon treatment with HOTf, into indenylidene complexes **5** which display extremely high catalytic activity in alkene metathesis (Scheme 2.4) [32d]. The process involves the initial formation of alkenyl-carbynes **4** which evolve into **5** through a formal electrophilic substitution of an *ortho*-proton of one of the phenyl groups by the C$_α$-atom of the carbyne moiety, with concomitant elimination of HOTf. A related acid-promoted allenylidene to indenylidene rearrangement has been observed for the osmium derivative [OsCp(=C=C=CPh$_2$)(CO)(PiPr$_3$)][PF$_6$] [27c]. Remarkably, the direct activation of HC≡CCPh$_2$(OH) by [RuCl$_2$(PPh$_3$)$_4$] leads also to the formation of an indenylidene complex; in his case an acidic medium is not required to promote the intramolecular allenylidene rearrangement. The chemistry and catalytic applications in olefin metathesis of ruthenium indenylidenes have been reviewed [44].

2.4.3
Nucleophilic Additions

The regioselectivity of nucleophilic additions on allenylidene complexes (C$_α$ versus C$_γ$) is subtly controlled by the electronic and steric properties of both the substituents on the unsaturated hydrocarbon chain and the ancillary ligands on the metal atom, as

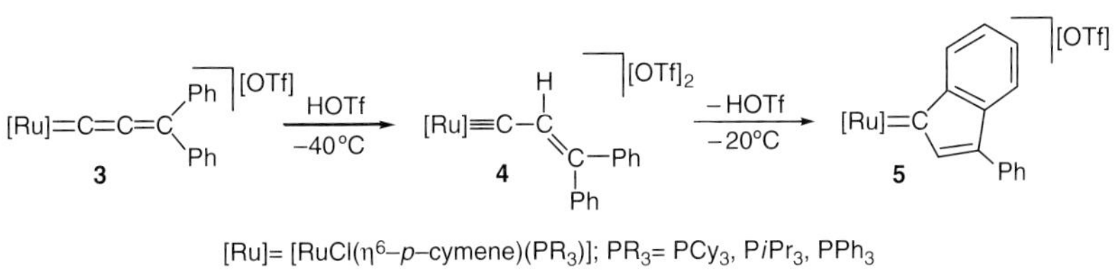

[Ru]= [RuCl(η^6–p–cymene)(PR$_3$)]; PR$_3$= PCy$_3$, PiPr$_3$, PPh$_3$

Scheme 2.4 Acid-promoted allenylidene to indenylidene rearrangement.

2.4.3.1 Group 6 Metal-Allenylidenes

The reactivity of Group 6 allenylidenes [M(=C=C=CR^{1}R^2)(CO)$_5$] (M = Cr, W; R^1 and R^2 = alkyl or aryl group) towards nucleophiles is clearly dominated by addition at the electrophilic α-carbon. Thus, it has been reported that compounds [M(=C=C=CiPr$_2$)(CO)$_5$] (M = Cr, W) react readily with PPh$_3$ to afford the stable zwitterionic phosphonio-allenyl derivatives [M{C(PPh$_3$)=C=CiPr$_2$}(CO)$_5$] [45]. The related species [Cr{C(PR$_3$)=C=C(4-C$_6$H$_4$NMe$_2$)$_2$}(CO)$_5$] (PR$_3$ = PMe$_3$, PHPh$_2$, PH$_2$Mes; Mes 2,4,6-C$_6$H$_2$Me$_3$) are also formed starting from [Cr{=C=C=C(4-C$_6$H$_4$NMe$_2$)$_2$}(CO)$_5$]. The PHPh$_2$ adduct evolves slowly into the allenylphosphine complex [Cr{κ^1(P)-PPh$_2$CH=C=C(4-C$_6$H$_4$NMe$_2$)$_2$}(CO)$_5$] by migration of the Cr(CO)$_5$ group from the allenyl C$_\alpha$ atom to P and synchronous H migration along the carbon chain. In contrast, the PH$_2$Mes adduct immediately isomerizes to [Cr{κ^1(P)-PH(Mes)C≡CCH(4-C$_6$H$_4$NMe$_2$)$_2$}(CO)$_5$] [46]. Undoubtedly, the most common reaction of these Group 6 allenylidenes (either isolated or generated *in situ*) is the addition of alcohols R^{3}OH across the C$_\alpha$=C$_\beta$ bond to afford Fischer-type α,β-unsaturated alkoxycarbene derivatives [M{=C(OR^3)CH=CR^{1}R^2}(CO)$_5$] (**M**) via nucleophilic attack of the alcohol at the electrophilic C$_\alpha$ and subsequent migration of the hydrogen atom to C$_\beta$ (Scheme 2.5) [47a–d]. Related N–H additions of amines (as well as imines and hydrazines) to afford α,β-unsaturated aminocarbenes **N** have also been reported [10a, 47e].

Similarly to bis(aryl)- or bis(alkyl)-substituted allenylidenes, amino-allenylidenes [M{=C=C=C(NR$'_2$)R}(CO)$_5$] (M = Cr, W; R = Ph, C(Me)$_2$OEt; R$'$ = Me, Et, iPr, Bn; not all combinations) add dimethylamine Me$_2$NH across the C$_\alpha$=C$_\beta$ to give alkenylaminocarbenes [M{=C(NMe$_2$)CH=C(NR$'_2$)R}(CO)$_5$] [9b,c, 48]. In contrast, when a solution of the alkoxy-substituted complex [Cr{=C=C=C(OMe)Ph}(CO)$_5$] is treated with one equivalent of Me$_2$NH the expected aminocarbene is not formed, the reaction leading instead to the allenylidene derivative [Cr{=C=C=C(NMe$_2$)Ph}(CO)$_5$] by substitution of the methoxy group [9c]. This unexpected substitution process, which is initiated by the nucleophilic attack of dimethylamine on the C$_\gamma$ atom of the allenylidene chain followed by elimination of methanol (Scheme 2.6), can be considered as the "allenylidene version" of the classical aminolysis of Fischer-type alkoxycarbene complexes. Exchange reactions of the alkoxy groups by primary and secondary amines in complexes [Cr{=C=C=C(NMe$_2$)OMe}(CO)$_5$] [9b, 49], and [M

Scheme 2.5 X-H (X = O, N) bond additions acrross the allenylidene C$_\alpha$=C$_\beta$ bond.

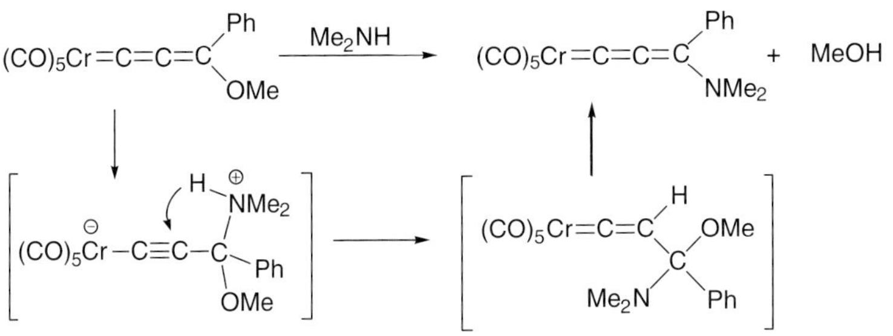

Scheme 2.6 Exchange of the methoxy group by dimethylamine.

{=C=C=C(OR)OEt}(CO)$_5$] (M = Cr, W; R = Et, (-)-menthyl, endo-bornyl) [9c] have also been described, allowing the preparation of a large variety of novel mono- and di-amino substituted Group 6 allenylidenes. Treatment of [Cr{=C=C=C(NMe$_2$)Ph}(CO)$_5$] with a large excess of ammonia or primary amines RNH$_2$ leads also to the substitution of the dimethylamino group, affording [Cr{=C=C=C(NHR)Ph}(CO)$_5$] (R = H, Ph or alkyl groups) [48b]. All these results seem to indicate a marked preference of the hetero-substituted Group 6 allenylidenes for C_γ vs. C_α additions in contrast to their alkyl or aryl-substituted counterparts.

2.4.3.2 Group 7 Metal-Allenylidenes

Only nucleophilic additions on complexes [Mn(η^5-C$_5$R$_5$)(=C=C=CPh$_2$)(CO)$_2$] (C$_5$R$_5$ = Cp or C$_5$H$_4$Me) and [Re(=C=C=CR^{1}R^2)(CO)$_2$(triphos)][OTf] (1 in Figure 2.3) have been described. Thus, the former react with MeO$^-$ and Me$_2$N$^-$ to generate anionic allenyl complexes [Mn(η^5-C$_5$R$_5$){C(X)=C=CPh$_2$}(CO)$_2$]$^-$, which on protonation can be transformed into the corresponding alkenylcarbenes [Mn(η^5-C$_5$R$_5$){=C(X)CH=CPh$_2$}(CO)$_2$] (X = OMe, NMe$_2$) [50a]. Related C_α-additions take place with PPh$_3$, affording stable phosphonio-allenyl derivatives [Mn(η^5-C$_5$R$_5$){C(PPh$_3$)=C=CPh$_2$}(CO)$_2$] [50a,c]. In contrast, addition of the carbanion tBu$^-$ to [Mn(η^5-C$_5$H$_4$Me)(=C=C=CPh$_2$)(CO)$_2$] occurs selectively at C_γ leading, after protonation, to the isolation of the vinylidene complex [Mn(η^5-C$_5$H$_4$Me){=C=C(H)CPh$_2$tBu}(CO)$_2$] [50b]. Mixtures of C_γ and C_α adducts are formed when tBuS$^-$ is used as nucleophile [50a]. On the basis of these results a rationalization of the C_γ versus C_α preference is not possible in this case. Concerning the reactivity of the cationic rhenium(I) complex [Re(=C=C=CPh$_2$)(CO)$_2$(triphos)][OTf], which is unreactive towards alcohols, it undergoes regioselective attacks of anionic nucleophiles (MeO$^-$, HO$^-$, Me$^-$, H$^-$, enolates, etc.) at the C_γ atom to afford stable neutral σ-alkynyl compounds [43]. Phosphines also attack the allenylidene-C_γ atom to give kinetic phosphonio-alkynyl products which are transformed thermally into thermodynamically more stable phosphonio-allenyl derivatives [51a]. Interestingly, when Ph$_2$PH is employed, the resulting phosphonio-allenyl species [Re{C(PHPh$_2$)=C=CPh$_2$}(CO)$_2$(triphos)][OTf] further evolves into the phosphonio-butadienyl derivative [Re{C(=PPh$_2$)C(H)=CPh$_2$}(CO)$_2$(triphos)][OTf] via a selective 1,3-P,C-H shift [51a]. This diphenylallenylidene readily reacts with thiols to give thiocarbenes [Re{=C(SR)CH=CPh$_2$}(CO)$_2$(triphos)][OTf] (R = Ph, 1-naphthyl, allyl) [51b]. Related N–H additions of primary amines and ammonia have been also described, the resulting complexes being formulated from X-ray data as azoniabutadienyl compounds [Re{C(=NHR)CH=CPh$_2$}(CO)$_2$(triphos)][OTf] (R = H, Ph, CH$_2$C≡CH) rather than Fischer-type aminocarbenes [Re{=C(NHR)CH=CPh$_2$}(CO)$_2$(triphos)][OTf] [51b]. Remarkably, in contrast to the diphenylallenylidene, the monosubstituted derivatives [Re(=C=C=CHR)(CO)$_2$(triphos)][OTf] (R = Me, Ph) undergo O–H addition of methanol or water across the C$_\alpha$=C$_\beta$ bond to afford the corresponding carbenes [Re{=C(OR′)CH=CHR}(CO)$_2$(triphos)][OTf] (R′ = Me, H) [12a,b]. On the basis of these results we can conclude that the fragment [Re(CO)$_2$(triphos)]$^+$ orients the addition of hard nucleophiles to the C$_\gamma$ atom, soft nucleophiles giving thermodynamically stable C$_\alpha$-adducts.

2.4.3.3 Group 8 Metal-Allenylidenes

Extensive studies on the reactivity of cationic half-sandwich ruthenium(II) allenylidenes have been reported, pointing out that the C$_\gamma$ versus C$_\alpha$ preference is strongly dependent on the steric and electronic properties of the ancillary ligands in the metal fragment. This is clearly exemplified by the behavior towards alcohols. Thus, while allenylidene ligands attached to the fragments [RuCl(η^6-arene)(L)]$^+$ (L = PR$_3$ or CNR) [32g, 47d, 52], [RuCp(CO)(PR$_3$)]$^+$ (PR$_3$ = PPh$_3$, PiPr$_3$) [47d, 24a], [Ru(η^5-1,2,3-Me$_3$C$_9$H$_4$)(CO)(PPh$_3$)]$^+$ [29b] and [Ru(η^5-C$_9$H$_7$)L$_2$]$^+$ (L$_2$ = dppe, dppm) [29a] are able to add alcohols across the C$_\alpha$=C$_\beta$ bond to yield Fischer-type α,β-unsaturated alkoxycarbenes, the more sterically demanding and/or electron-rich units [Ru(η^5-C$_9$H$_7$)(PPh$_3$)$_2$]$^+$ [29a], [RuCp(PPh$_3$)$_2$]$^+$ [23b], [RuCp*(dippe)]$^+$ [25a] and [RuCp*(PMePh$_2$)$_2$]$^+$ [53] make the allenylidene ligand resistant to alcohols or, in some cases [29a, 53], orient the addition across the C$_\beta$=C$_\gamma$ to generate vinylidene complexes [Ru]=C=C(H)CR$_2$(OR′). Of particular interest are the reactions of [RuCp(=C=C=CPh$_2$)(CO)(PiPr$_3$)][BF$_4$] (**6**) with allyl and propargyl alcohol (Scheme 2.7), the resulting alkoxy-carbenes **7** and **8** serving as intermediates for the construction, among others, of the polycyclic derivatives **9** and **10** via intramolecular cycloaddition processes [54]. Addition of water across the C$_\alpha$=C$_\beta$ of complex **6**, to afford the stable

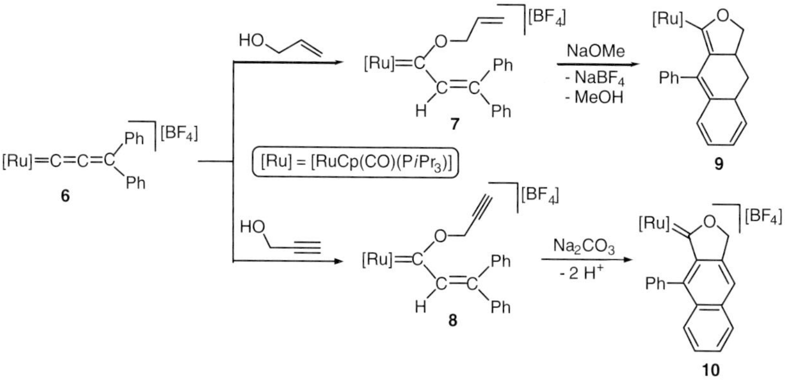

Scheme 2.7 Coupling of an allenylidene ligand with allyl propargyl alchohal.

hydroxycarbene [RuCp{=C(OH)CH=CPh$_2$}(CO)(PiPr$_3$)][BF$_4$], has also been described [24a]. In contrast, the hydroxycarbenes formed by addition of water to [RuCl(=C=C=CHPh)(η^6-p-cymene)(PR$_3$)][OTf] (PR$_3$ = PPh$_3$, PCy$_3$) are not stable, evolving into the carbonyl derivatives [RuCl(η^6-p-cymene)(CO)(PR$_3$)][OTf] by releasing styrene [55].

Similarly to alcohols and water, complex **6** and its related carbonyl counterpart [RuCp*(=C=C=CPh$_2$)(CO)(PMeiPr$_2$)][B(Ar$_F$)$_4$] (**11**) also add the N–H bond of primary and secondary amines across the C$_\alpha$=C$_\beta$, generating azoniabutadienyl species [RuCp{C(=NR^{1}R^2)CH=CPh$_2$}(CO)(PiPr$_3$)][BF$_4$] [56a–c] and [RuCp*{C(=NR^{1}R^2)CH=CPh$_2$}(CO)(PMeiPr$_2$)][B(Ar$_F$)$_4$] [56d], respectively. A similar reactivity was observed with benzophenone imine [24a, 56d]. The utility of this N–H addition reaction for the construction of complex molecular architectures is nicely illustrated in the behavior of **6** towards N,N-diallylamine and N-propargylamine, the reactions leading to the formation of the heterocyclic derivatives **12–13** and **14–15** (Figure 2.10), respectively, via base-promoted intramolecular cyclization of the corresponding azoniabutadienyl intermediates [56b,c]. It is also worth noting that the intramolecular version of this N–H addition process is known. Thus, it has been reported that activation of the propargylic alcohol HC≡CCPh$_2$(OH) by the complex [RuClCp(PPh$_2$NHnPr)$_2$], in the presence of AgOTf, affords the azaphosphacarbene **16** by intramolecular N–H addition of one of the phosphinoamine ligands to the C$_\alpha$=C$_\beta$ on the allenylidene intermediate [RuCp(=C=C=CPh$_2$)(PPh$_2$NHnPr)$_2$][OTf] [23e].

Thiols also react with allenylidenes **6** and **11** to afford α,β-unsaturated thiocarbenes, that is, [RuCp{=C(SnPr)CH=CPh$_2$}(CO)(PiPr$_3$)][BF$_4$] [24a] and [RuCp*{=C(SnPr)CH=CPh$_2$}(CO)(PMeiPr$_2$)][B(Ar$_F$)$_4$] [56d], via S–H addition across the C$_\alpha$=C$_\beta$ double bond of the cumulenic chain. Single-crystal X-ray diffraction studies on the latter indicate the existence of an important contribution of the tautomeric thiabutadienyl form [RuCp*{C(=SnPr)CH=CPh$_2$}(CO)(PMeiPr$_2$)][B(Ar$_F$)$_4$].

The crucial role of the ancillary ligands on the C$_\gamma$ versus C$_\alpha$ preference is also clearly reflected in the behavior of half-sandwich Ru(II) allenylidenes towards phosphines. Thus, while complexes **6** and **11** add phosphines at the C$_\alpha$ atom to yield cationic phosphonio-allenyl derivatives [RuCp{C(PR$_3$)=C=CPh$_2$}(CO)(PiPr$_3$)][BF$_4$] (PR$_3$ = PPh$_3$, PMePh$_2$, PHPh$_2$) [57] and [RuCp*{C(PR$_3$)=C=CPh$_2$}(CO)(PMeiPr$_2$)][B(Ar$_F$)$_4$] (PR$_3$ = PMe$_3$, PMeiPr$_2$) [56d], respectively, allenylidenes [Ru(η^5-C$_9$H$_7$)(=C=C=CR^{1}R^2)(PPh$_3$)$_2$][PF$_6$] containing the bulkier bis(triphenylphosphine)indenyl fragment react selectively at the C$_\gamma$, affording phosphonio-alkynyl species [Ru(η^5-C$_9$H$_7$){C≡CCR^{1}R^2(PR$_3$)}(PPh$_3$)$_2$][PF$_6$] (R^1, R^2 = alkyl, aryl or H; PR$_3$ = PPh$_3$,

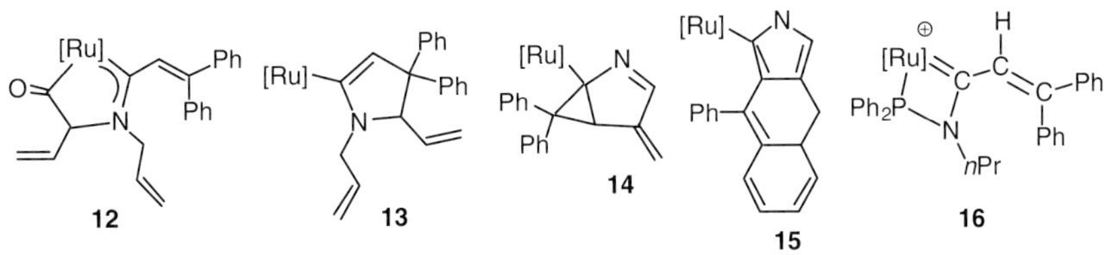

Figure 2.10 Structures of the ruthenium(II) complexes **12–16**.

PMePh$_2$, PMe$_2$Ph, PMe$_3$) [14b,d,58a,b]. A related C$_\gamma$-attack occurs in the reaction of [Ru(η^5-C$_9$H$_7$)(=C=C=CPh$_2$)(dppm)][PF$_6$] with PMe$_3$. Nevertheless, the resulting phosphonio-alkynyl derivative [Ru(η^5-C$_9$H$_7$){C≡CCPh$_2$(PMe$_3$)}(dppm)][PF$_6$] is not stable, evolving spontaneously into the allenyl isomer [Ru(η^5-C$_9$H$_7$){C(PMe$_3$)=C=CPh$_2$}(dppm)][PF$_6$], via PMe$_3$-migration from C$_\gamma$ to C$_\alpha$. This is favored by the presence of the less sterically demanding bis(diphenylphosphino)methane ligand and the steric congestion on the C$_\gamma$ atom due to the presence of phenyl groups [58a]. In this context, it should also be noted that the phosphonio-alkynyl derivatives [Ru(η^5-C$_9$H$_7$){C≡CCH(R^1)(PR$_3$)}(PPh$_3$)$_2$][PF$_6$] (R^1 = H, PR$_3$ = PPh$_3$; R^1 = Ph, PR$_3$ = PMe$_3$) have proven to be of particular synthetic interest, since they are excellent substrates for Wittig-type reactions, allowing the preparation of a large variety of neutral σ-enynyl complexes of general composition [Ru(η^5–C$_9$H$_7$){C≡CC(R^1)=CR^{2}R^3}(PPh$_3$)$_2$] [58b–f]. The addition of anionic heteroatom-centered nucleophiles (HO$^-$, MeO$^-$, pyrazolate, etc.) and carbanions (CN$^-$, enolates, alkyl or alkynyl reagents) to the cationic allenylidenes [Ru(η^5–C$_9$H$_7$)(=C=C=CR^{1}R^2)(PPh$_3$)$_2$][PF$_6$] [29c–e, 58a,e, 59], [Ru(η^5–C$_9$H$_7$)(=C=C=CHPh){(R)-BINAP}][PF$_6$] (BINAP 2,2′-bis(diphenylphosphino)-1,1′-binaphthyl) [29g], [Ru(η^5-1,2,3-Me$_3$C$_9$H$_4$)(=C=C=CPh$_2$)(CO)(PPh$_3$)][BF$_4$] [29b], [RuCp*(=C=C=CR^{1}R^2)(dippe)][BPh$_4$] [42], [RuCp(=C=C=CPh$_2$)(PPh$_3$)$_2$][PF$_6$] [23b], [RuCp(=C=C=CPh$_2$)(CO)(PiPr$_3$)][OTf] (6) [57, 60] and [Ru(η^5-C$_{60}$Me$_5$)(=C=C=CR^{1}R^2){(R)-prophos}][PF$_6$] (prophos = 1,2-bis(diphenylphosphino)propane) [61] has been extensively studied, the regioselectivity of the attack differing slightly from the trends observed with neutral soft nucleophiles. Thus, in all cases, regioselective C$_\gamma$-additions have been observed leading to the formation of neutral σ-alkynyl species [Ru]-C≡CCR^{1}R^2(Nu), except for complex 6 for which simultaneous C$_\alpha$ and C$_\gamma$ attacks take place [60]. Nevertheless, it should be noted that intramolecular C–C couplings involving the α-carbon of allenylidenes [Ru(η^5-C$_9$H$_7$)(=C=C=CPh$_2$)(dppm)][PF$_6$] and [Ru(η^5-Ring)(=C=C=CPh$_2$){κ^1(P)-Ph$_2$PCH$_2$C(=O)tBu}][PF$_6$] (Ring = Cp or C$_9$H$_7$) have been described. Thus, treatment of these species with K$_2$CO$_3$ or LitBu generates the allenylmetallacyclic species 17 and 18 formed by monodeprotonation of the methylenic unit of the phosphine ligand and subsequent attack of the initially generated carbanion to the electrophilic C$_\alpha$ (Figure 2.11) [26 58a].

Taking advantage of the regioselectivity shown by the indenyl-ruthenium(II) complexes [Ru(η^5-C$_9$H$_7$)(=C=C=CR^{1}R^2)(PPh$_3$)$_2$][PF$_6$] an efficient synthetic procedure for the propargylic substitution of 2-propyn-1-ols mediated by the metallic

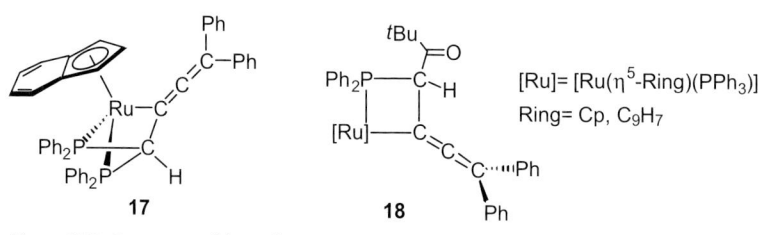

Figure 2.11 Structures of the ruthenium(II) complexes 17 and 18.

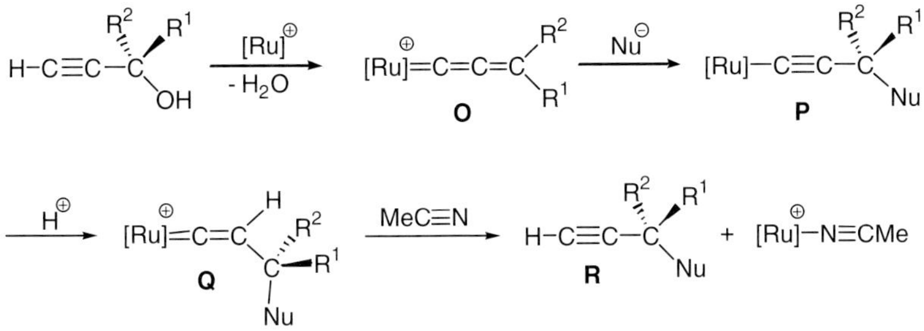

Scheme 2.8 [Ru(η^5-C$_9$H$_7$)(PPh$_3$)$_2$]$^+$-mediated propargylic substitution.

fragment [Ru(η^5–C$_9$H$_7$)(PPh$_3$)$_2$]$^+$ has been developed (Scheme 2.8). Thus, allenylidene complexes **O** are first formed and subsequently transformed into the corresponding σ-alkynyl derivatives **P** which undergo a selective C$_\beta$ protonation to afford the vinylidene complexes **Q**. Finally, demetallation of **Q** with acetonitrile leads to the functionalized terminal alkynes **R** in excellent yields. Following this route, a large variety of γ-ketoalkynes (including optically active representatives) [29d, 59b,c], 1,4-diynes [29e, 58e], and 1,5- and 1,6-enynes [59d,e] could be synthesized. Related processes have been described starting from the chiral allenylidene [Ru(η^5-C$_9$H$_7$)-(=C=C=CHPh){(R)-BINAP}][PF$_6$] allowing the preparation of propargylic-substituted compounds with complete enantioselectivity [29g]. In all cases the metal is recovered as the acetonitrile solvate.

The behaviour of osmium-allenylidenes towards nucleophiles follows similar trends to that observed for their analogous ruthenium counterparts. Thus, while the electron-rich bis-phosphine complexes [Os(η^5-Ring)(=C=C=CPh$_2$)(PR$_3$)$_2$][PF$_6$] (Ring = Cp, PR$_3$ = PiPr$_3$; Ring = C$_9$H$_7$, PR$_3$ = PPh$_3$) are inert towards alcohols and amines [27a, 29a], the more electrophilic carbonyl derivative [OsCp(=C=C=CPh$_2$)(CO)(PiPr$_3$)][PF$_6$] reacts readily with methanol and aniline to afford [OsCp{=C(OMe)CH=CPh$_2$}(CO)(PiPr$_3$)][PF$_6$] and [OsCp{C(=NHPh)CH=CPh$_2$}(CO)(PiPr$_3$)][PF$_6$], respectively [27c]. C$_\alpha$-additions of alcohols and phosphines to the (η^6-arene)-Os(II) allenylidene [OsCl(=C=C=CPh$_2$)(η^6-1,3,5-C$_6$H$_3$Me$_3$)(PMe$_3$)][PF$_6$] have also been described, allowing the preparation of [OsCl{=C(OR)CH=CPh$_2$}(η^6-1,3,5-C$_6$H$_3$-Me$_3$)(PMe$_3$)][PF$_6$] (R = Me, Et) and [OsCl{C(PR$_3$)=C=CPh$_2$}(η^6-1,3,5-C$_6$H$_3$Me$_3$)(PMe$_3$)][PF$_6$] (PR$_3$ = PMe$_3$, PPh$_3$), respectively [32i]. As expected, due to the presence of two coordinated bulky PiPr$_3$ ligands, the addition of anionic nucleophiles (Me$^-$, MeO$^-$, MeC(=O)CH$_2^-$) to [OsCp(=C=C=CPh$_2$)(PiPr$_3$)$_2$][PF$_6$] takes place selectively on the less sterically congested C$_\gamma$, generating neutral alkynyl species [OsCp{C≡CCPh$_2$(Nu)}(PiPr$_3$)$_2$] [27a]. Related C$_\gamma$-additions of anions have also been observed starting from the octahedral derivative [Os{κ^2(C,O)-C(CO$_2$Me)=CH$_2$}(=C=C=CPh$_2$)(CO)(PiPr$_3$)$_2$][BF$_4$] [22a]. It is also worth mentioning that the reduction of the C$_\beta$=C$_\gamma$ double bond of the neutral allenylidene [OsClCp(=C=C=CPh$_2$)(PiPr$_3$)] by action of NaBH$_4$ and methanol affords the vinylidene species [OsClCp{=C=C(H)-CHPh$_2$}(PiPr$_3$)] [31c]. In contrast to ruthenium and osmium, the reactivity of iron allenylidenes has been almost unexplored, only the behavior of the cationic diphenylallenylidene-Fe(II) derivative *trans*–[FeBr(=C=C=CPh$_2$)(depe)$_2$]

(depe = Et$_2$PCH$_2$CH$_2$PEt$_2$) being studied in detail. Thus, it has been found that this complex reacts exclusively at C$_\gamma$ with both neutral (amines, phosphines) and anionic (H$^-$, MeO$^-$, CN$^-$) nucleophiles [17b,d]. This behavior contrasts with that of the neutral Fe(0) derivative [Fe(=C=C=CtBu$_2$)(CO)$_5$] which undergoes PPh$_3$-attack at C$_\alpha$ to afford the zwitterionic phosphonio-allenyl species [Fe{C(PPh$_3$)=C=CtBu$_2$}(CO)$_5$] [62].

2.4.3.4 Group 9 Metal-Allenylidenes

As was commented previously, the reactivity of the square-planar Rh(I) and Ir(I) allenylidenes trans-[MCl(=C=C=CR^{1}R^2)(PiPr$_3$)$_2$] is marked by the strong trans-influence of the π-acceptor allenylidene unit, allowing the easy exchange of the chloride ligand by a large variety of anionic nucleophiles (I$^-$, HO$^-$, RO$^-$, RCO$_2^-$, N$_3^-$, SCN$^-$, etc.) [35]. Among the different products formed, compounds trans-[M(OPh)(=C=C=CR^{1}R^2)(PiPr$_3$)$_2$] (M = Rh, R^1 = Ph, R^2 = Ph, 2-C$_6$H$_4$Me; M = Ir, R^1 Ph, R^2 = tBu) and trans-[Rh{κ^1-(O)-O$_2$CMe}(=C=C=CR^{1}R^2)(PiPr$_3$)$_2$] (R^1 = Ph, R^2 Ph, 2-C$_6$H$_4$Me) are of particular interest since, upon treatment with carbon monoxide, they undergo migratory insertion of the allenylidene unit into the M–O bond to generate the σ-alkynyl complexes trans-[M{C≡CCR^{1}R^2(OPh)}(CO)(PiPr$_3$)$_2$] and trans–[Rh{C≡CCR^{1}R^2(O$_2$CMe)}(CO)(PiPr$_3$)$_2$], respectively [35d,h]. Similarly, the reactions of the hydroxo compounds trans-[Rh(OH)(=C=C=CR^{1}R^2)(PiPr$_3$)$_2$] (R^1 = R^2 = Ph, 4-C$_6$H$_4$OMe; R^1 = Ph, R^2 = tBu) with CH$_2$(CN)$_2$ and either CO or CNMe yield the carbonyl or the isocyanide complexes trans-[Rh{C≡CCR^{1}R^{2}CH(CN)$_2$}(L)(PiPr$_3$)$_2$] (L = CO, CNMe), via highly unstable allenylidene intermediates trans–[Rh{CH(CN)$_2$}(=C=C=CR^{1}R^2)(PiPr$_3$)$_2$] [35h].

Treatment of the azido complexes **19** with CO leads also to the migration of the N$_3^-$ ligand to the allenylidene unit (Scheme 2.9). Nevertheless, the initially formed azidoalkynyl compounds **20** are in this case thermally unstable, evolving slowly into the metallated acrylonitrile derivatives **23** via extrusion of N$_2$. The mechanism of formation of **23** involves the migration of the azido moiety from C$_\gamma$ to the C$_\alpha$ atom of the alkynyl ligand to generate the allenyl intermediates **21**, which by elimination of

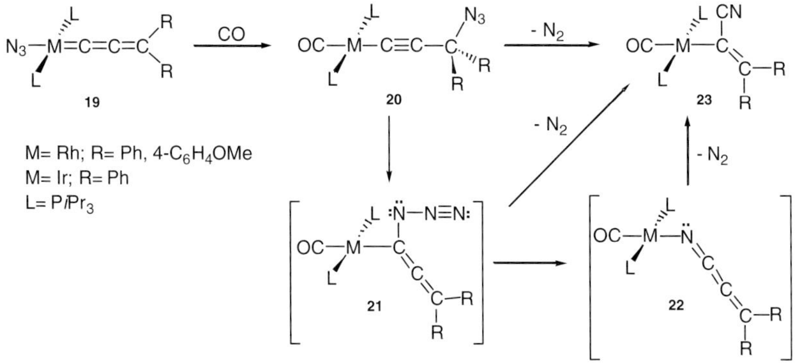

Scheme 2.9 CO-promoted allenylidene-azide couplings.

N$_2$ and shifting of the metal fragment (directly or via intermediate **22**) afford **23** [35d,f].

In this context, it should also be noted that an oxidatively induced C$_\alpha$–P coupling has been described in the oxidation reactions of complexes *trans*-[RhCl(=C=C=CR$_2$)(PiPr$_3$)$_2$] (R = Ph, 4-C$_6$H$_4$OMe) with Cl$_2$ or PhICl$_2$, affording phosphonio-allenyl products [RhCl$_3$|C(PiPr$_3$)=C=CR$_2$|(PiPr$_3$)]. They are formed by migration of one PiPr$_3$ group from the metal to the allenylidene α-carbon in the six-coordinate Rh(III) intermediates [RhCl$_3$(=C=C=CR$_2$)(PiPr$_3$)$_2$] [35e].

2.4.4
C–C Couplings

Besides the classical additions of carbon-centered nucleophiles to the electrophilic C$_\alpha$ and C$_\gamma$ atoms, transition-metal allenylidenes are able to promote a number of original carbon–carbon coupling processes. Thus, in the early days of this chemistry, it was reported that complex [MnCp(=C=C=CPh$_2$)(CO)$_2$] is able to insert *tert*-butyl isocyanide into the Mn=C bond giving rise to an intermediate species [MnCp(η^2-tBuN=C=C=C=CPh$_2$)(CO)$_2$] which readily reacts with water to afford [MnCp{η^2-Ph$_2$C=C=CHC(=O)NHtBu}(CO)$_2$] [63]. A related insertion of the methylene unit :CH$_2$ into the Rh=C bond of *trans*-[RhCl(=C=C=CR^{1}R^2)(PiPr$_3$)$_2$] (R^1 = R^2 = Ph, 4-C$_6$H$_4$OMe; R^1 = Ph, R^2 = CF$_3$, tBu) by treatment with diazomethane allowed the isolation of stable butatriene-Rh(I) compounds *trans*-[RhCl(η^2-H$_2$C=C=C=CR^{1}R^2)(PiPr$_3$)$_2$] [35e]. Remarkably, their iodide counterparts *trans*-[RhI(η^2-H$_2$C=C=C=CR$_2$)(PiPr$_3$)$_2$] (R = Ph, 4-C$_6$H$_4$OMe) were generated by reacting the corresponding allenylidene complexes *trans*-[RhI(=C=C=CR$_2$)(PiPr$_3$)$_2$] with MeI [35e]. This unusual C–C coupling reaction, in which MeI behaves as a :CH$_2$ source, involves oxidative addition of MeI at the rhodium center followed by insertion of the allenylidene unit into the Rh–Me bond. The resulting allenyl-Rh(III) intermediates [RhI$_2${η^1-C(Me)C=C=CR$_2$}(PiPr$_3$)$_2$] evolve through a β-H shift to give [RhHI$_2$(η^2-H$_2$C=C=C=CR$_2$)(PiPr$_3$)$_2$], which upon reductive elimination of HI generate the final butatriene-Rh(I) complexes. The same reactivity pattern has been observed in the reaction of *trans*-[IrI(=C=C=CPh$_2$)(PiPr$_3$)$_2$] with MeI [35i].

Treatment of one of these square-planar Rh(I)-allenylidenes, that is, **24**, with the Grignard reagent CH$_2$=CHMgBr has been reported to yield the η^3-pentatrienyl derivative **26** as the result of a C$_3$ + C$_2$ coupling process (Scheme 2.10) [35e]. With regard to the mechanism, an initial substitution of the chloride ligand takes place, leading to the vinyl-metal intermediate **25** which rearranges, by migratory insertion of the allenylidene unit into the Rh CH=CH$_2$ bond, to give the final product **26**. Compounds [OsClCp(=C=C=CPh$_2$)(PiPr$_3$)] and [RuClCp*(=C=C=CPh$_2$){κ^1(P)-iPr$_2$PCH$_2$CO$_2$Me}] undergo related C–C coupling processes on treatment with CH$_2$=CHMgBr [31c, 64]. In contrast, insertion of the allenylidene ligand into the Os–C(alkenyl) bond of **27** generates the five-membered metallacyclic compound **28** instead of the expected η^3-pentatrienyl isomer (Scheme 2.10) [22a].

Werner and coworkers have nicely exploited the ability of square-planar Rh(I) and Ir(I) allenylidenes to undergo C–C couplings with alkynes. For example, the Ir(I)

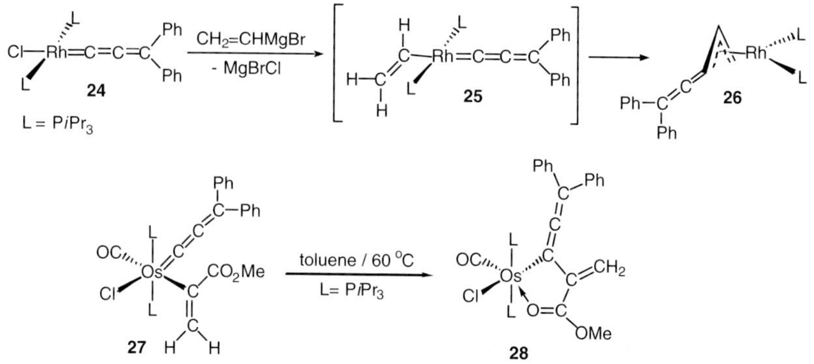

Scheme 2.10 Couplings of allenylidene ligands and vinyl group.

hydroxo-allenylidenes *trans*-[Ir(OH){=C=C=C(R)Ph}(P*i*Pr$_3$)$_2$] (R = Ph, *t*Bu) readily react with an excess of HC≡CR' (R' = Ph, CO$_2$Me) to afford *trans*-[Ir(C≡CR')$_2${η^1-(*E*)-CH=C(R')CH=C=C(R)Ph}(P*i*Pr$_3$)$_2$] at room temperature. The proposed mechanism involves an initial HO$^-$/R'C≡C$^-$ ligand exchange followed by the oxidative addition of one terminal alkyne molecule to generate the Ir(III) intermediate [IrH(C≡CR')$_2${C=C=C(R)Ph}(P*i*Pr$_3$)$_2$]. The rearrangement to an allenyl species followed by C–C coupling with a third alkyne molecule gives the final product [35d]. An intramolecular allenylidene–alkynyl coupling also takes place in the reaction of the rhodium complex **29** with carbon monoxide (Scheme 2.11). The initially formed thermally unstable allenyl derivative **30** evolves into the metallated cyclobutenone **31** when an excess of CO is present [35g].

Another unusual coupling reaction takes place after treatment of the bis(hydroxyalkynyl) rhodium(III) complex [RhH{C≡CCPh$_2$(OH)}$_2$(P*i*Pr$_3$)$_2$] with Al$_2$O$_3$ which, in the presence of chloride ions, yields the hexapentaene-Rh(I) derivative *trans*-[RhCl(η^2-Ph$_2$C=C=C=C=C=CPh$_2$)(P*i*Pr$_3$)$_2$] [65]. The formation of a bis(allenylidene) intermediate *trans*-[Rh(=C=C=CPh$_2$)$_2$(P*i*Pr$_3$)$_2$]$^+$ is proposed, which in the presence of Cl$^-$ undergoes the coupling of the two cumulenic groups. Alternatively, this hexapentaene complex can be generated by thermolysis of *trans*-[RhCl(=C=C=CPh$_2$)(P*i*Pr$_3$)$_2$] [35e]. Remarkably, the organic fragment Ph$_2$C=C=C=C=

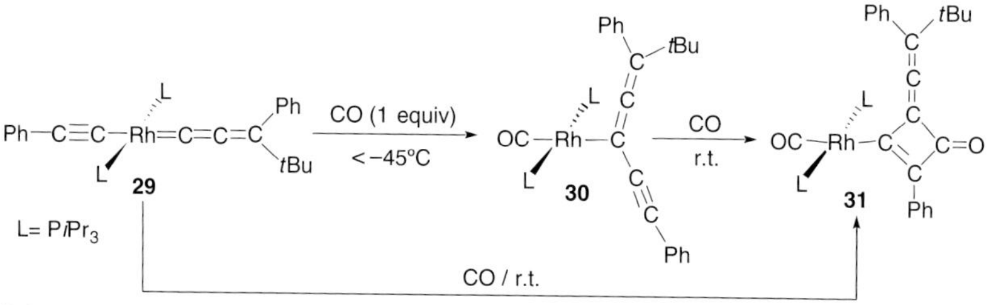

Scheme 2.11 CO-promoted alleynlidene–alkynyl coupling.

C=CPh$_2$ can be easily decoordinated by ligand displacement with CO, demonstrating the synthetic utility of transition-metal allenylidenes for the construction of elaborated organic compounds [65]. Linkage of two allenylidene moieties has also been observed in the thermal decomposition of [MnCp(=C=C=C*t*Bu$_2$)(CO)$_2$], which generates small quantities of the free tetrasubstituted hexapentaene *t*Bu$_2$C=C=C=C=C=C*t*Bu$_2$ [2b], as well as in the photolysis of [Cr(=C=C=CPh$_2$)(CO)$_5$] which affords Ph$_2$C=C=C=C=C=CPh$_2$ in high yield [66]. No detailed mechanisms have been proposed for these transformations.

2.4.5
Cycloaddition and Cyclization Reactions

The terms cycloaddition and cyclization generally refer to concerted and stepwise reactions, respectively. Nevertheless, in this section they will be used indiscriminately since, in most of the processes, the mechanisms are uncertain. Formation of non-cyclic products proceeding through a cyclization/cycloreversion pathway is also included in this section.

2.4.5.1 Reactions Involving the M=C$_\alpha$ Bond

Only a few cyclization processes of this type have been described up to now. Most of them result from inter- or intramolecular attacks of anionic nucleophiles containing at least two reactive heteroatoms. Thus, sodium dimethyldithiocarbamate reacts with the cationic allenylidene complex [RuTp(=C=C=CPh$_2$)(PPh$_3$)$_2$][PF$_6$] (32) to generate the allenyl-metallacycle 33 (Scheme 2.12), as the result of the nucleophilic addition of one of the sulfur atoms at the C$_\alpha$ carbon and subsequent coordination of the second sulfur to the ruthenium center, with concomitant release of a triphenylphosphine ligand [67]. Complex 33 is also accessible by treatment of the neutral derivative [RuTpCl(=C=C=CPh$_2$)(PPh$_3$)] with Na[S$_2$CNMe$_2$]. Under similar conditions, the allenylidene complex [Ru{κ^2(S,S)-S$_2$CNMe$_2$}(=C=C=CPh$_2$)(CO)(PPh$_3$)$_2$][PF$_6$] can be transformed into the analogous metalacycle [Ru{S=C(NMe$_2$)S-C=C=CPh$_2$}{κ^2(S,S)-S$_2$CNMe$_2$}(CO)(PPh$_3$)] [67].

A related intramolecular coupling between a monodentate acetate ligand and a transient diphenylallenylidene moiety occurs when the hydroxyalkynyl derivative 34 is treated with HPF$_6$, affording the ruthenacycle 36 (Scheme 2.13) [68]. This cyclization process is strongly dependent on the electronic properties of the organometallic

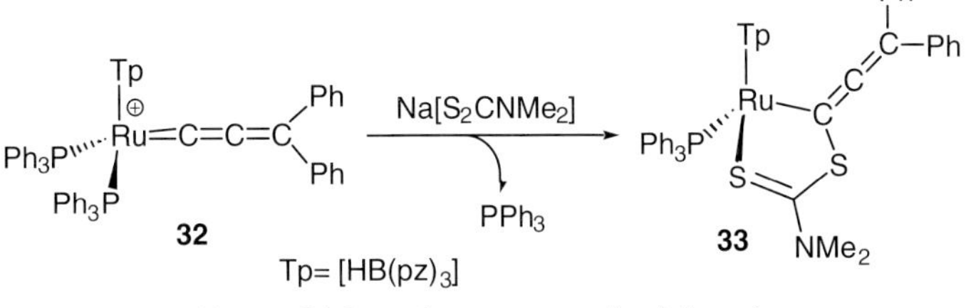

Scheme 2.12 Addition of dithiocarbamate to an allenylidene chain.

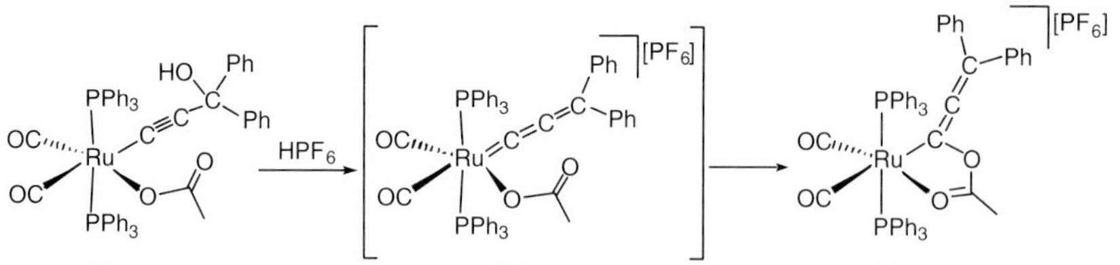

Scheme 2.13 Intramolecular coupling between acetate and allenylidene ligands.

fragment, as evidenced by the stability of the allenylidene species [Ru{κ^1(O)-OAc}(=C=C=CPh$_2$)(CNtBu)$_2$(PPh$_3$)$_2$][PF$_6$] closely related to the intermediate **35** [68].

More surprising is the coupling of the rhodium-allenylidene derivative **24** with terminal alkynes which results in the formation of the zwitterionic π-allyl-allenyl species **38** (Scheme 2.14) [35e]. The reaction is assumed to take place through an initial [2 + 2] cycloaddition between the carbon–carbon triple bond of the alkyne and the Rh=C bond, giving rise to the metallacyclobutene **37** which spontaneously evolves into **38** by migration of one PiPr$_3$ ligand from the metal to the hydrocarbon chain.

2.4.5.2 Reactions Involving the C$_\alpha$=C$_\beta$ Bond

The central C$_\alpha$=C$_\beta$ double bond of an allenylidene backbone can react with a wide range of unsaturated organic substrates to yield cyclic adducts. Most of the cyclization processes reported with dipolar substrates occur in a stepwise manner via an initial nucleophilic attack at the C$_\alpha$ atom and further rearrangement of the molecule involving a coupling with the C$_\beta$ carbon. Thus, the electron-poor ruthenium complex **6** readily adds 1,1-diethylpropargylamine to generate the unprecedented dihydropyridinium species **39** (Scheme 2.15) [56b]. Note that the course of the reaction is different when N-propargylamine is employed, the corresponding azoniabutadienyl complex being formed in this case (see above) [56b]. On the other hand, allenylidene **6** also reacts with ethyl diazoacetate affording the five-membered heterocyclic compound **40** [69].

Treatment of the tungsten complex [W(=C=C=CPh$_2$)(CO)$_5$] with benzylidene-isopropylimine gives rise to the formation of the azetidinylidene derivative **41** through a formal [2 + 2] cycloaddition between the C$_\alpha$=C$_\beta$ moiety and the iminic

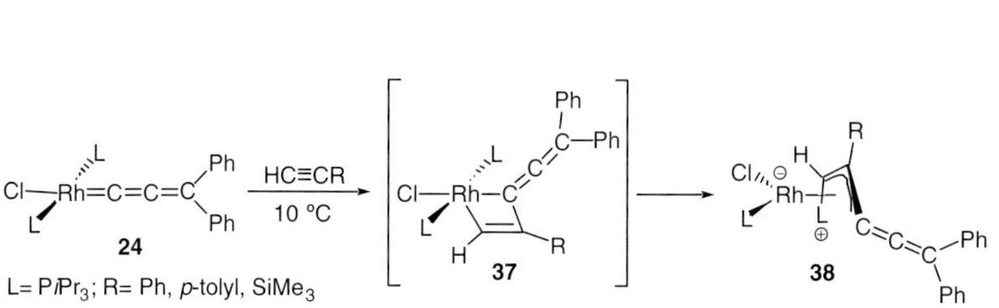

L= PiPr$_3$; R= Ph, p-tolyl, SiMe$_3$

Scheme 2.14 Formation of complexes **38** through cycloaddition between alkynes and the Rh=C moiety.

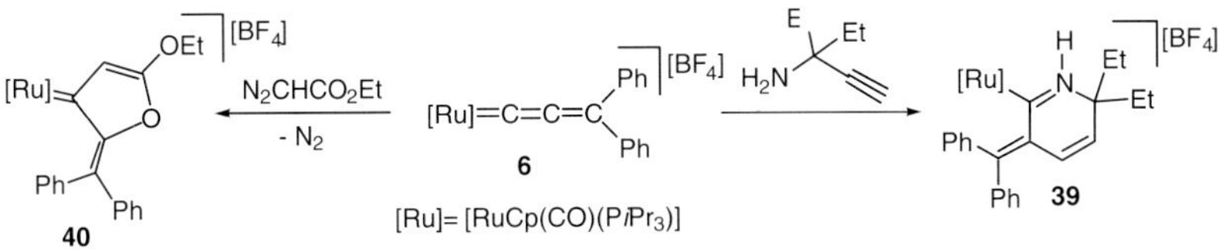

2.4 Stoichiometric Reactivity of Allenylidenes | 83

Scheme 2.15 Coupling of an allenylidene ligand with a propargylic amine or a diazo-compound.

C=N bond (Scheme 2.16) [10a]. A plausible mechanism consists in the initial attack of the nitrogen of the imine on the C_α carbon and subsequent cyclization of the resulting ylide.

Such a coupling process is also operative when C=C instead of C=N double bonds are implied. Thus, ruthenium allenylidene complexes bearing a $\kappa^1(P)$–allyldiphenylphosphine ligand have been shown to evolve smoothly into the bicyclic derivatives **42**, via an unusual intramolecular [2 + 2] cycloaddition of two C=C bonds (Scheme 2.17) [29f, 70]. This process has been carried out starting from isolated as well as in situ generated alkyl- or aryl-substituted allenylidenes, but is not observed with the related amino-allenylidene compound [Ru(η^5-C$_9$H$_7$){=C=C=C(NEt$_2$)C(Me)=CPh$_2$}{κ^1(P)-Ph$_2$PCH$_2$CH=CH$_2$}(PPh$_3$)][PF$_6$] [70].

Cycloaddition reactions of allenylidene ligands with alkynes have also been described. Thus, heating a toluene solution of the neutral osmium complex **43** in the presence of dimethylacetylenedicarboxylate leads selectively to the allenyl-vinylidene **45** (Scheme 2.18) [31c]. The formal insertion of the alkyne into the $C_\alpha=C_\beta$

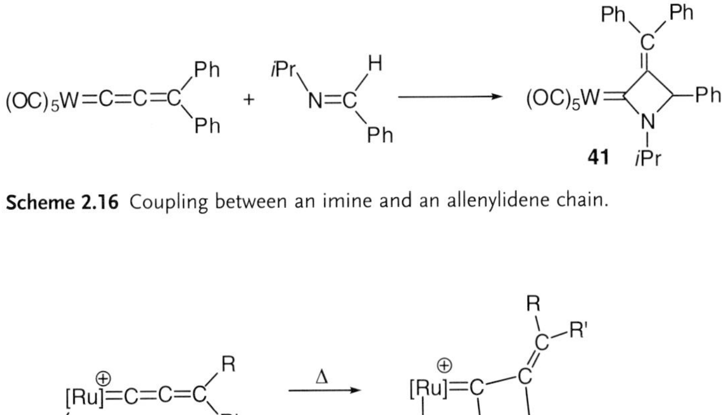

Scheme 2.16 Coupling between an imine and an allenylidene chain.

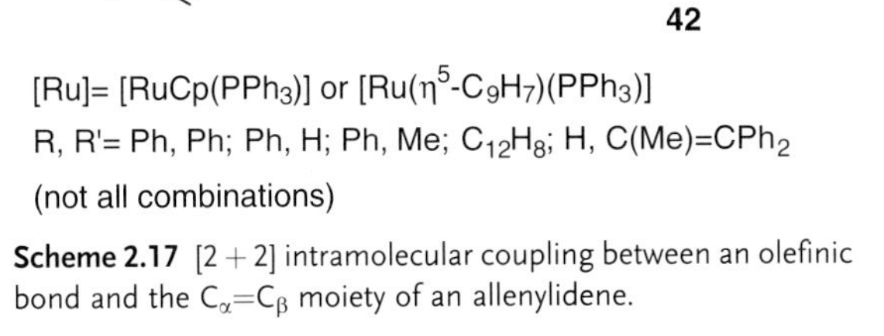

[Ru]= [RuCp(PPh$_3$)] or [Ru(η^5-C$_9$H$_7$)(PPh$_3$)]
R, R'= Ph, Ph; Ph, H; Ph, Me; C$_{12}$H$_8$; H, C(Me)=CPh$_2$
(not all combinations)

Scheme 2.17 [2 + 2] intramolecular coupling between an olefinic bond and the $C_\alpha=C_\beta$ moiety of an allenylidene.

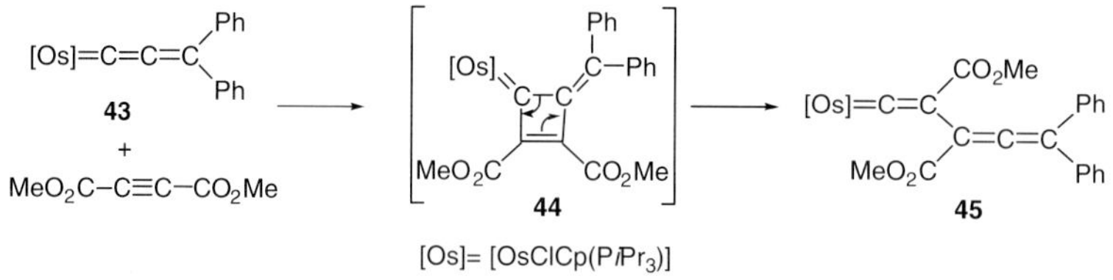

Scheme 2.18 Coupling of allenylidene **43** with an activated alkyne.

double bond can be rationalized through an initial cycloaddition followed by the ring-opening of the cyclobutenyl intermediate **44**.

Similar cyclizations also occur when chromium and tungsten diaryl-allenylidenes **46** are treated with organometallic alkynyl species (Scheme 2.19) [71]. Nevertheless, in contrast to intermediate **44**, the resulting 1,3-heterobinuclear cyclobutylidene complexes **47** are not prone to rearrange through cycloreversion. Structural analyses of these derivatives have evidenced strong electron-delocalization along the M–C–C (R′)–C–M′ bridge due to the significant contribution of a zwitterionic resonance form **47′**.

Allenylidene complexes **46** also react with the carbon–carbon triple bond of ynamines to yield similar mononuclear cyclobutenylidene derivatives **48**, although mixtures with the corresponding alkenyl-aminoallenylidene species **49** are formed (Scheme 2.20) [10c]. The former isomer results from the addition of the C≡C bond of ynamines across the $C_\alpha=C_\beta$ unsaturation, while the latter is provided by the formal [2 + 2] cycloaddition between C≡C and $C_\beta=C_\gamma$ bonds and subsequent cycloreversion. In both processes, stepwise cyclization initiated by the addition of the nucleophilic R′C≡CNEt$_2$ carbon at the C_α or C_γ position, respectively, is proposed. Relative proportions of **49** with respect to **48** increase with the electron-releasing capacity of the *para*-substituents of the diarylallenylidene skeleton (NMe$_2$ > OMe > Me > H). In contrast, the formation of **48** is favored when the reaction is carried out in low polarity solvents.

2.4.5.3 Reactions Involving the $C_\beta=C_\gamma$ Bond

Unlike the chromium and tungsten allenylidenes **46**, the coupling of the indenyl-ruthenium(II) allenylidene **50** with ynamine MeC≡CNEt$_2$ takes place regioselectively

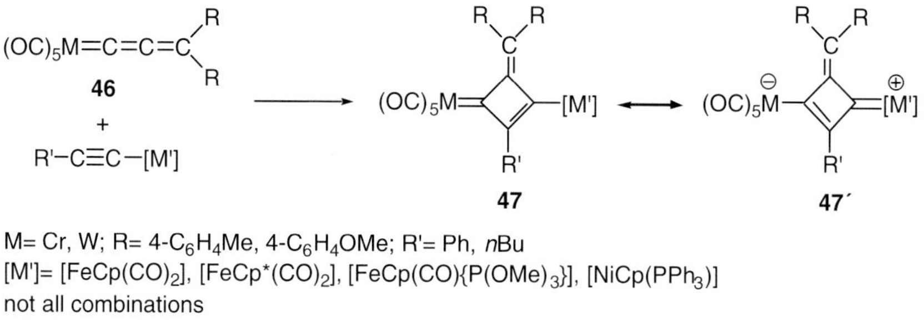

M= Cr, W; R= 4-C$_6$H$_4$Me, 4-C$_6$H$_4$OMe; R′= Ph, *n*Bu
[M′]= [FeCp(CO)$_2$], [FeCp*(CO)$_2$], [FeCp(CO){P(OMe)$_3$}], [NiCp(PPh$_3$)]
not all combinations

Scheme 2.19 Formation of heterobinuclear cyclobutenylidene complexes **47**.

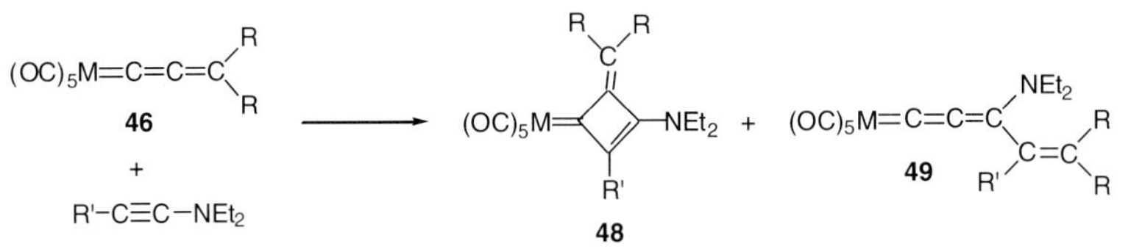

$(OC)_5M=C=C=C\begin{smallmatrix}R\\R\end{smallmatrix}$
46
+
$R'-C\equiv C-NEt_2$

$\longrightarrow$

$(OC)_5M=\underset{R'}{\overset{RR}{\underset{\|}{C}}}-NEt_2$ + $(OC)_5M=C=C=C\begin{smallmatrix}NEt_2\\C=C\\R'R\end{smallmatrix}$

48 **49**

M= Cr, W; R = Ph, 4-C$_6$H$_4$Me, 4-C$_6$H$_4$OMe, 4-C$_6$H$_4$NMe$_2$; R'= Me, Ph
not all combinations

Scheme 2.20 Coupling between chromium and tungsten allenylidene complexes and ynamines.

at the C$_\beta$=C$_\gamma$ unit. Then, spontaneous ring-opening of the resulting cyclobutylidene intermediate furnishes exclusively the alkenyl-aminoallenylidene complex **51**, structurally related to **49** (Scheme 2.21) [72]. This transformation has been extended to other indenyl ruthenium precursors such as [Ru(η^5-C$_9$H$_7$)(=C=C=CPh$_2$){κ^1(P)-Ph$_2$PCH$_2$CH=CH$_2$}(PPh$_3$)][PF$_6$] and [Ru(η^5-C$_9$H$_7$)(=C=C=CHPh)(PPh$_3$)$_2$][PF$_6$] and to the ynamine Me$_3$SiC≡CNEt$_2$ [70, 72]. Based on this reactivity, an original synthetic route to polyunsaturated allenylidene species has been developed (Scheme 2.21) [72]. Thus, after the first ynamine insertion, a formal substitution of amino group by hydrogen in **51** is performed by consecutive treatments with LiHBEt$_3$ and SiO$_2$. Like **50**, the resulting mono-substituted alkenyl-allenylidene **52** is able to insert ynamines via a cyclization/cycloreversion pathway to generate the corresponding dienyl-aminoallenylidene species. Further transformations in the presence of LiHBEt$_3$ and SiO$_2$ furnish the mono-substituted dienyl-allenylidene complex **53**. Finally, a third ynamine insertion provides the highly unsaturated trienyl-aminoallenylidene compound **54**. All the processes involved in this synthetic methodology are totally regio- and stereoselective giving rise to the formation of **54** as the *trans,trans* isomer exclusively. It is interesting to note that aminoallenylidene compounds, such as **51**, are not prone to insert ynamines even in the presence of a large excess of reagent.

The electron-deficient allenylidene moiety in complex [RuCp(=C=C=CPh$_2$)(CO)(PiPr$_3$)][BF$_4$] (**6**) can act as dienophile in Diels–Alder cycloadditions [73]. Thus,

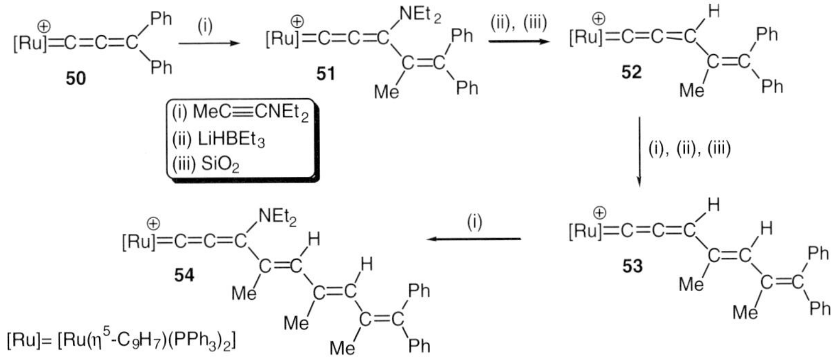

Scheme 2.21 Sequential insertion of ynamines through consecutive cyclization/cycloreversion processes.

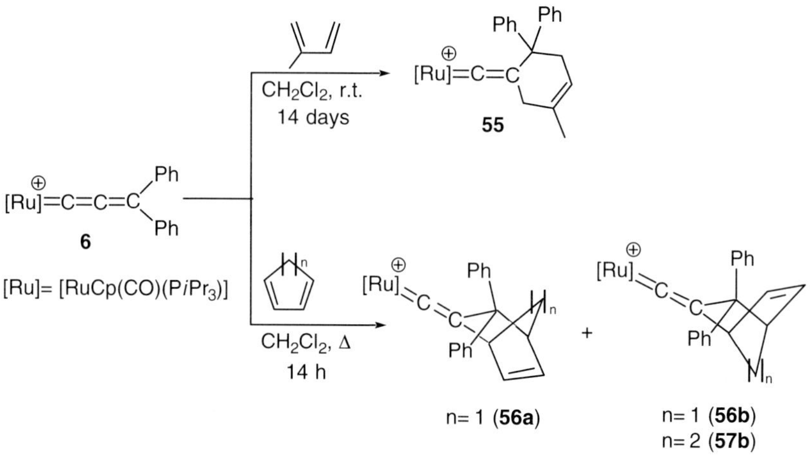

Scheme 2.22 Diels–Alder cycloaddition involving the $C_\beta=C_\gamma$ unsaturation.

treatment of dichloromethane solutions of **6** with a 20-fold excess of isoprene at room temperature slowly furnishes the cycloadduct **55** (Scheme 2.22). The process is regioselective with regard to the dienophile since only the $C_\beta=C_\gamma$ bond of the allenylidene skeleton is implied and also regioselective concerning the orientation of the diene with the exclusive attack of C(1) and C(4) carbons at the C_β and C_γ positions, respectively. Allenylidene **6** also undergoes Diels–Alder reactions with cyclopentadiene and cyclohexadiene in refluxing dichloromethane to afford the bicyclic products **56** and **57**, respectively, as a mixture of diastereomers for the former and as a sole diastereomer for the latter (Scheme 2.22).

The activation of the allenylidene group by an electron-deficient organometallic fragment is also evidenced when complex **6** and its osmium counterpart [OsCp(=C=C=CPh$_2$)(CO)(PiPr$_3$)][PF$_6$] are treated with carbodiimides in dichloromethane at room temperature. Under these conditions, the reactions yield Z- and E-iminiumazetidinylidenemethyl species **58** (Scheme 2.23), while the related bis(phosphine) complex [OsCp(=C=C=CPh$_2$)(PHPh$_2$)(PiPr$_3$)]PF$_6$] remains inert [27c, 74]. The formation of cycloadducts **58** has been rationalized in terms of a

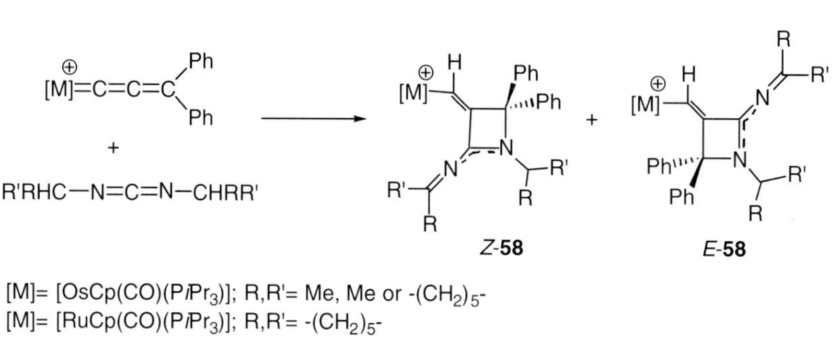

[M]= [OsCp(CO)(PiPr$_3$)]; R,R'= Me, Me or -(CH$_2$)$_5$-
[M]= [RuCp(CO)(PiPr$_3$)]; R,R'= -(CH$_2$)$_5$-

Scheme 2.23 Coupling of Ru(II) and Os(II) diphenylallenylidenes with carbodiimides.

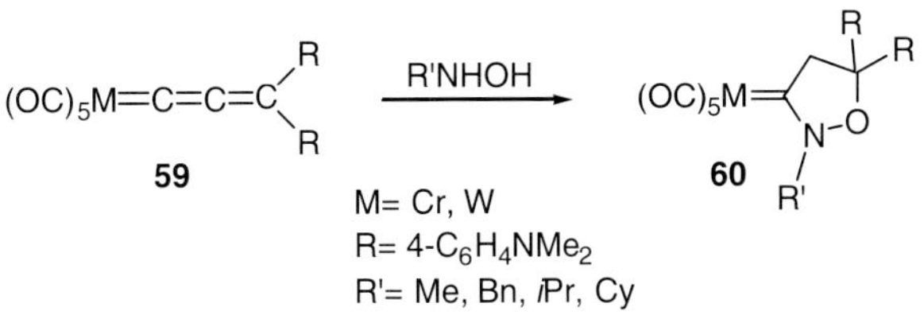

Scheme 2.24 Coupling of Group 6 allenylidenes with hydroxylamine.

stepwise [2 + 2] cycloaddition between the allenylidene $C_\beta=C_\gamma$ and one of the two C=N bonds of the carbodiimide, followed by an Alder-ene rearrangement.

2.4.5.4 Reactions Involving Both $C_\alpha=C_\beta$ and $C_\beta=C_\gamma$ Bonds (1,2,3-Heterocyclizations)

A wide range of dinucleophiles are prone to undergo cyclization processes by addition on both $C_\alpha=C_\beta$ and $C_\beta=C_\gamma$ bonds of an allenylidene moiety. The structure of the products generated depends on the nature of the substituents on the allenylidene chain (amino or hydrocarbon groups) and on the number of hydrogen atoms that the organic dinucleophile can deliver. In this way, saturated carbenes, alkenyl-carbenes and alkenyl species, all of them containing at least one heterocyclic unit, have been prepared selectively. Thus, reactions of chromium and tungsten diarylallenylidene compounds **59** with hydroxylamines lead to the cyclic aminocarbene derivatives **60** resulting formally from the addition of the nitrogen and oxygen atoms to C_α and C_γ, respectively, and the transfer of two hydrogen atoms to the C_β position (Scheme 2.24) [75].

When only one heteroatom of the dinucleophile possesses a hydrogen substituent, the reactions lead instead to alkenyl complexes rather than carbene compounds. Effectively, treatment of diphenylallenylidenes **1** and **6** with pyrazoles yields the heterocyclic derivatives **61** (Scheme 2.25) [76]. Interestingly, the dissymmetric 3-methylpyrazole (R = H, R' = Me) provides only one regioisomer, in which the methyl group points towards the metal. This process, which formally corresponds to the addition of two nitrogen nuclei at C_α and C_γ and a hydrogen atom at C_β, is assumed to take place through an initial nucleophilic attack at the C_α position.

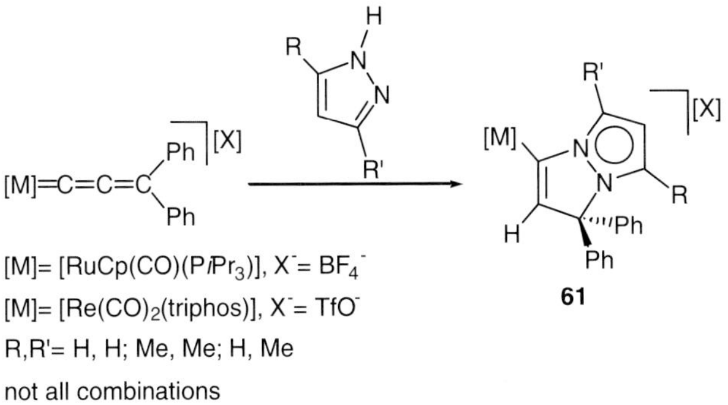

Scheme 2.25 1,2,3-diheterocyclizations with pyrazoles.

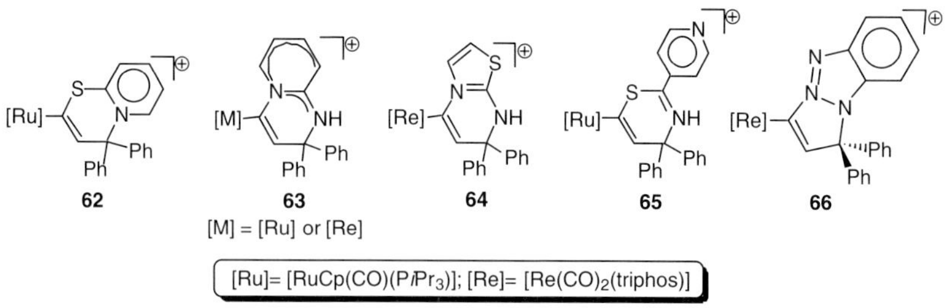

Figure 2.12 Products of 1,2,3-dihetrocyclizations of diarylallenylidene complexes.

Similar 1,2,3-diheterocyclizations have been performed by addition of other N,N- or N,S-dinucleophiles such as pyridine-2-thiol, 2-aminopyridine, 2-aminothiazole, thioisonicotinamide and 1H-benzotriazole giving rise to the formation of the five- and six-membered cyclic alkenyl derivatives **62–66** (Figure 2.12) [76a, 77].

Group 6 aminoallenylidene complexes display a slightly different reactivity towards N,N- or N,S-dinucleophiles. Thus, treatment of [Cr{=C=C=C(NMe$_2$)Ph}(CO)$_5$] with benzamidine, guanidine or thioacetamide yields selectively α,β-unsaturated carbenes **67** (Scheme 2.26) [49b], arising from nitrogen attack at C$_\gamma$, subsequent HNMe$_2$ elimination, and further reorganization of the molecule through a ring-closing process.

Starting from [Cr{=C=C=C(NMe$_2$)Ph}(CO)$_5$], the proper choice of other bifunctional nucleophiles also allows the preparation, in moderate to high yields, of the five-, seven- and eight-membered heterocyclic carbene species **68–71** (Figure 2.13) which, in some cases, are formed along with minor amounts of other derivatives [78]. Similar 1,2,3-diheterocyclizations involving (ethoxy)allenylidene complexes have been described, ethanol instead of HNMe$_2$ being released in these cases [49b].

Another type of coupling involving both C$_\alpha$=C$_\beta$ and C$_\beta$=C$_\gamma$ bonds of an allenylidene ligand is that observed by heating acetonitrile solutions of the osmium derivatives **72** which yields the 1-osma-4-hydrocyclopenta[c]pyrroles **73** (Scheme 2.27) [22b]. They are generated through the assembly of the alkenyl and allenylidene ligands with a molecule of acetonitrile. A plausible reaction pathway consists of the migratory insertion of the allenylidene chain into the Os-alkenyl bond, followed by the addition of the central carbon of the resultant allenyl group to the nitrile function and further electronic reorganization of the molecule.

X= NH; R= Ph, NH$_2$
X= S; R= Me

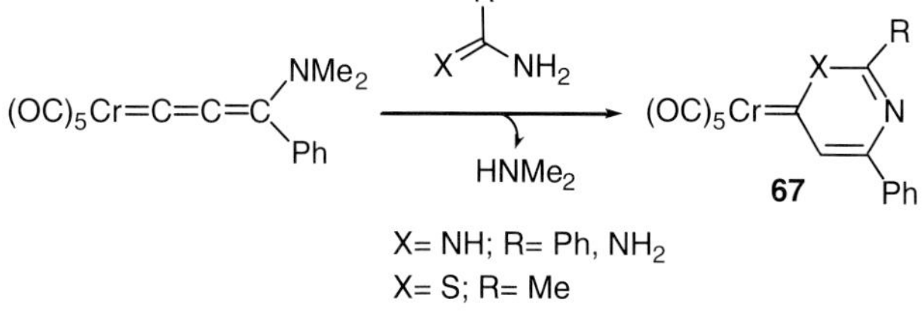

Scheme 2.26 1,2,3-diheterocyclizations of amino-allenylidene complexes.

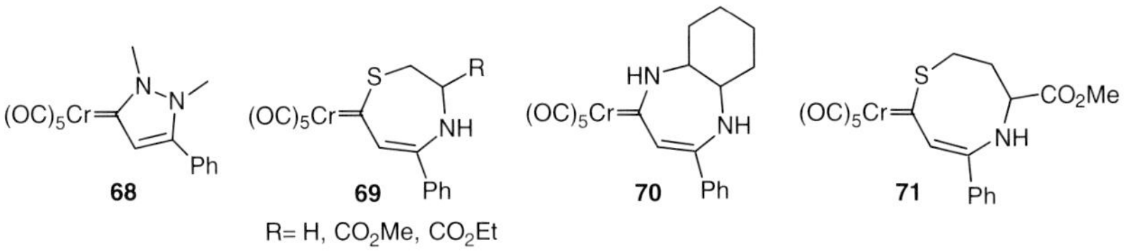

Figure 2.13 Structures of the chromium compound **68–71**.

The reaction of [RuClCp(PPh$_3$)$_2$] with 2-methyl-3-butyn-2-ol leads to the binuclear alkenyl-vinylidene-alkylidene complex **74** (Scheme 2.28) [79]. The suggested mechanism involves a cycloaddition process between a transient allenylidene and its alkenyl vinylidene tautomer, both generated *in situ* via the dehydration of a hydroxyvinylidene intermediate. Analogous compounds, containing other organometallic fragments such as [RuTp(dippe)]$^+$ [20a], [OsCp(PPh$_3$)$_2$]$^+$ [27b] and [Re(CO)$_2$(triphos)]$^+$ [12a], have also been synthesized. A related coupling between allenylidene and hydroxyvinylidene intermediates takes place when the rhenium(I) complex [Re(OTf)(CO)$_2$(triphos)] is treated with propargyl alcohol, affording a dinuclear vinylidene-carbene species [12a].

2.4.6
Other Reactions

The following types have been described:

1. The stoichiometric transfer of allenylidene ligands from one metal fragment to another metal center has been scarcely documented, the only examples known involving the allenylidene transfer from chromium compounds [Cr(=C=C=CR^{1}R^2)(CO)$_5$] (R^1, R^2 = aryl, amino or alkoxy groups) to [W(CO)$_5$(THF)] [9d]. DFT calculations indicate that the reaction proceeds by an associative pathway, the initial reaction step involving the coordination of W(CO)$_5$ to the C$_\alpha$=C$_\beta$ bond of the allenylidene ligand in the chromium precursor.

2. Cationic allenylidene complexes containing a hydrogen atom in the δ position, that is, [M]$^+$=C=C=C(R^1)CHR^{2}R^3, are known to undergo deprotonation processes upon treatment with bases affording neutral σ-enynyl derivatives [M]-C≡CC(R^1)=CR^{2}R^3. Representative examples include deprotonation of: [Ru(η^5-C$_9$H$_7$)(=C=

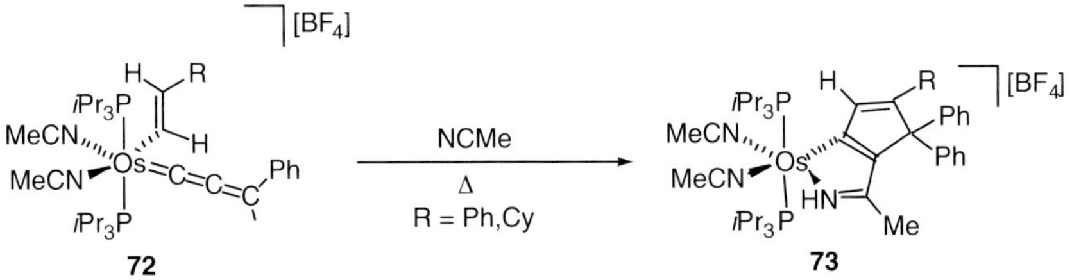

Scheme 2.27 Formation of osmacyclopentapyrrole derivatives **73**.

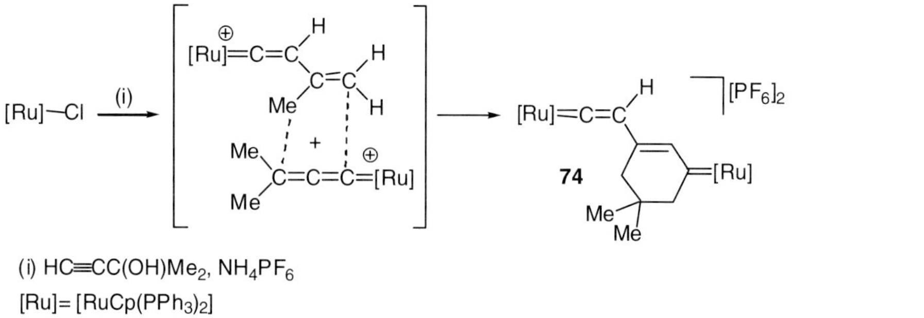

Scheme 2.28 Formation of the dinuclear ruthenium complex 74.

C=CMePh)(PPh$_3$)$_2$][PF$_6$] [29a], [RuCp(=C=C=CRMe)(PPh$_3$)$_2$][PF$_6$] (R=N-methylpyrrol-2-yl) [34b], [Re(=C=C=CMePh)(CO)$_2$(triphos)][OTf] [12b] and trans-[FeBr(=C=C=CMePh)(depe)$_2$][BPh$_4$] [17b]. In this context, it should be noted that ruthenium(II) allenylidenes trans-[RuCl{=C=C=C(R^1)CH$_2$R^2}(dppe)$_2$][BF$_4$], containing an acidic methylenic unit, readily react with the neutral diynyl complex trans-[RuCl{(C≡C)$_2$H}(dppe)$_2$] to afford the dinuclear C$_7$-bridged compounds 75 (see Figure 2.14). The process involves the initial protonation of the latter by the methylenic unit of the former and subsequent C–C coupling between the resulting organometallic species [80]. Related C$_9$-bridged species are formed using the triynyl derivative trans-[RuCl{(C≡C)$_3$SiMe$_3$}(dppe)$_2$] [80].

3. One-electron reduction of complexes trans-[RuCl(=C=C=CR$_2$)(dppe)$_2$][PF$_6$] (R = Ph, Me) with cobaltocene provides highly reactive radicals trans-[RuCl(C≡CC$^\bullet$R$_2$)(dppe)$_2$] which, in the presence of Ph$_3$SnH, can be trapped by H-transfer yielding alkynyl compounds trans-[RuCl(C≡CCHR$_2$)(dppe)$_2$] (Figure 2.14) [81].

2.5
Concluding Remarks

This chapter is intended to highlight the most important developments on the synthesis and reactivity of allenylidene complexes excluding catalytic processes.

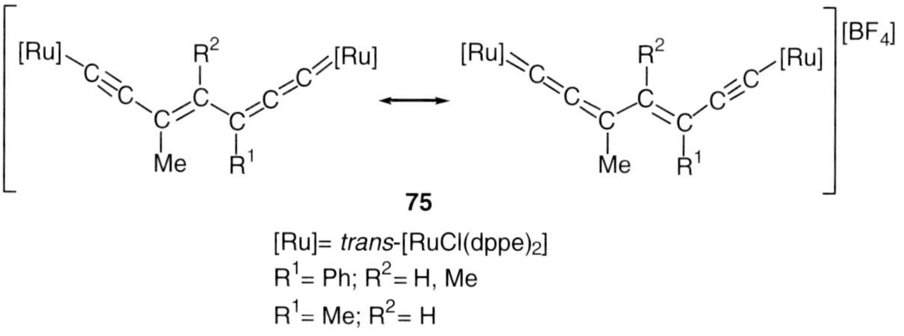

Figure 2.14 Structures of the dinuclear ruthenium complexes 75.

Since the discovery of the first allenylidene complexes in 1976 [2] and the first ruthenium derivative in 1982 [8], a large number of new derivatives, mainly of Group 6, 7, 8 and 9 metals, have been isolated and fully characterized [3, 4]. The ready accessibility of these derivatives from terminal alkynes and alkynols through efficient synthetic approaches has triggered reactivity studies mainly directed to stoichiometric and catalytic applications in organic synthesis. A wide scope of metal fragments [ML_n] have been used to stabilize the allenylidene chain =C=C=CR^{1}R^2 giving rise to a variety of structures ranging from typical half-sandwich to octahedral, five-coordinate and square-planar derivatives. Nowadays, both theoretical and experimental reactivity studies are well established, indicating that the electrophilic sites are located at C_α and C_γ while the C_β atom is a nucleophilic center. The extensive use of ancillary ligands with different steric and electronic properties has allowed the control of the reactivity sites on the cumulenic chain.

Much progress has been made in the chemistry of metal-allenylidenes during the last decade showing the potential utility in selective C–C and C–heteroatom bond formation. Relevant developments are: (i) Regio- and stereoselective additions at the C_γ atom which provide a valuable methodology for the synthesis of functionalized alkynyl complexes. Some of them are of interest for the synthesis of γ-ketoalkynes, 1,4-diynes, 1,6-enynes and propargylic substituted compounds. (ii) Inter- and intramolecular cycloaddition and cyclization reactions involving C_α=C_β and C_β=C_γ double bonds with unsaturated organic substrates. (iii) C–C coupling reactions with alkynes, alkenyl and alkynyl groups, leading to the selective formation of highly unsaturated hydrocarbon moieties.

In summary, allenylidene complexes have been the focus of extensive studies in modern organometallic chemistry due to the versatile reactivity of the unsaturated chain. Although the state of the art of this chemistry has achieved a high degree of fundamental knowledge, there is a growing interest in the use of metal-allenylidenes in synthetic applications. It is apparent that new achievements, extending the synthesis of new derivatives as well as its use in catalytic processes (see other chapters in this book), would enhance the rapid progress of this field in the near future.

References

1 Hartzler, H.D. (1961) *Journal of the American Chemical Society*, **83**, 4990–4996.

2 (a) Fischer, E.O., Kalder, H.J., Frank, A., Köhler, F.H. and Huttner, G. (1976) *Angewandte Chemie-International Edition in English*, **15**, 623–624. (b) Berke, H. (1976) *Angewandte Chemie-International Edition in English*, **15**, 624.

3 (a) Bruce, M.I. (1991) *Chemical Reviews*, **91**, 197–257. (b) Bruce, M.I. (1998) *Chemical Reviews*, **98**, 2797–2858. (c) Cadierno, V., Gamasa, M.P. and Gimeno, J. (2001) *European Journal of Inorganic Chemistry*, 571–591. (d) (2004) Vinylidene, Allenylidene and Metallacumulene Complexes *Coordination Chemistry Reviews*, **248**, 1531–1703.

4 (a) Fischer, H. and Szesni, N. (2004) *Coordination Chemistry Reviews*, **248**, 1659–1677. (b) Cadierno, V., Gamasa, M.P.

and Gimeno, J. (2004) *Coordination Chemistry Reviews*, **248**, 1627–1657. (c) Werner, H. (1997) *Chemical Communications*, 903–910. (d) Werner, H., Ilg, K., Lass, R. and Wolf, J. (2002) *Journal of Organometallic Chemistry*, **661**, 137–147. (e) Winter, R.F. and Záliš, S. (2004) *Coordination Chemistry Reviews*, **248**, 1565–1583. (f) Dragutan, I. and Dragutan, V. (2006) *Platinum Metals Review*, **50**, 81–94. (g) Cadierno, V., Conejero, S., Gamasa, M.P. and Gimeno, J. (2003) in *Perspectives in Organometallic Chemistry* (eds. C.G. Screttas and B.R. Steele), Royal Society of Chemistry, Cambridge, 285–296. (h) Le Bozec, H. and Dixneuf, P.H. (1995) *Russian Chemical Bulletin*, **44**, 801–812. (i) Touchard, D. and Dixneuf, P.H. (1998) *Coordination Chemistry Reviews*, **178–180**, 409–429. (j) Rigaut, S., Touchard, D. and Dixneuf, P.H. (2003) *Journal of Organometallic Chemistry*, **684**, 68–76. (k) Rigaut, S., Touchard, D. and Dixneuf, P.H. (2004) *Coordination Chemistry Reviews*, **248**, 1585–1601.
 (l) Esteruelas, M.A. and López, A.M. (2005) *Organometallics*, **24**, 3584–3613. (m) Esteruelas, M.A., López, A.M. and Oliván, M. (2007) *Coordination Chemistry Reviews*, **251**, 795–840. (n) Che, C.-M., Ho, C.-M. and Huang, J.-S. (2007) Coordination Chemistry Reviews, **251**, 2145–2166.

5 Li, X., Schopf, M., Stephan, J., Kippe, J., Harms, K. and Sundermeyer, J. (2004) *Journal of the American Chemical Society*, **126**, 8660–8661.

6 Binger, P., Müller, P., Wenz, R. and Mynott, R. (1990) *Angewandte Chemie-International Edition in English*, **29**, 1037–1038.

7 (a) Doherty, S., Corrigan, J.F., Carty, A.J. and Sappa, E. (1995) *Advances in Organometallic Chemistry*, **37**, 39–130. (b) El Amouri, H. and Gruselle, M. (1996) *Chemical Reviews*, **96**, 1077–1104. (c) Jia, G. and Lau, C.P. (1998) *Journal of Organometallic Chemistry*, **565**, 37–48. (d) Low, P.J. and Bruce, M.I. (2001) *Advances in Organometallic Chemistry*, **48**, 71–288.

8 Selegue, J.P. (1982) *Organometallics*, **1**, 217–218.

9 (a) Roth, G. and Fischer, H. (1996) *Organometallics*, **15**, 1139–1145. (b) Fischer, H., Szesni, N., Roth, G., Burzlaff, N. and Weibert, B. (2003) *Journal of Organometallic Chemistry*, **683**, 301–312. (c) Szesni, N., Weibert, B. and Fischer, H. (2006) *Inorganica Chimica Acta*, **359**, 617–632. (d) Szesni, N., Drexler, M. and Fischer, H. (2006) *Organometallics*, **25**, 3989–3995. (e) Szesni, N., Drexler, M., Maurer, J., Winter, R.F., de Montigny, F., Lapinte, C., Steffens, S., Heck, J., Weibert, B. and Fischer, H. (2006) *Organometallics*, **25**, 5774–5787.

10 (a) Fischer, H., Roth, G., Reindl, D. and Troll, C. (1993) *Journal of Organometallic Chemistry*, **454**, 133–149. (b) Fischer, H., Reindl, D. and Roth, G. (1994) *Zeitschrift für Naturforschung*, **49b**, 1207–1214. (c) Roth, G., Reindl, D., Gockel, M., Troll, C. and Fischer, H. (1998) *Organometallics*, **17**, 1393–1401.

11 (a) Berke, H. (1980) *Chemische Berichte*, **113**, 1370–1376. (b) Berke, H. (1986) *Organometallic Syntheses*, **3**, 239–240.

12 (a) Bianchini, C., Mantovani, N., Marchi, A., Marvelli, L., Masi, D., Peruzzini, M., Rossi, R. and Romerosa, A. (1999) *Organometallics*, **18**, 4501–4508. (b) Bianchini, C., Mantovani, N., Marvelli, L., Peruzzini, M., Rossi, R. and Romerosa, A. (2001) *Journal of Organometallic Chemistry*, **617–618**, 233–241. (c) Mantovani, N., Brugnati, M., Gonsalvi, L., Grigiotti, E., Laschi, F., Marvelli, L., Peruzzini, M., Reginato, G., Rossi, R. and Zanello, P. (2005) *Organometallics*, **25**, 405–418. (d) Baker, R.J., Edwards, P.G., Gracia-Mora, J., Ingold, F. and Malik, K.M.A. (2002) *Journal of the Chemical Society-Dalton Transactions*, 3985–3992. (e) Albertin, G., Antoniutti, S., Bordignon, E. and Bresolin, D. (2000) *Journal of Organometallic Chemistry*, **609**, 10–20.

13 (a) Berke, H., Grösmann, U., Huttner, G. and Zsolnai, L. (1984) *Chemische Berichte*, **117**, 3432–3442. (b) Gauss, C., Veghini, D.,

Orama, O. and Berke, H. (1997) *Journal of Organometallic Chemistry*, **541**, 19–38.

14 (a) Selegue, J.P., Young, B.A. and Logan, S.L. (1991) *Organometallics*, **10**, 1972–1980. (b) Cadierno, V., Gamasa, M.P., Gimeno, J., Borge, J. and García-Granda, S. (1997) *Organometallics*, **16**, 3178–3187. (c) Esteruelas, M.A., López, A.M., Ruiz, N. and Tolosa, J.I. (1997) *Organometallics*, **16**, 4657–4667. (d) Cadierno, V., Conejero, S., Gamasa, M.P., Gimeno, J. and Rodríguez, M.A. (2002) *Organometallics*, **21**, 203–209.

15 (a) Bustelo, E., Jiménez-Tenorio, M., Puerta, M.C. and Valerga, P. (2001) *European Journal of Inorganic Chemistry*, 2391–2398. (b) Aneetha, H., Jiménez-Tenorio, M., Puerta, M.C., Valerga, P. and Mereiter, K. (2003) *Organometallics*, **22**, 2001–2013.

16 (a) Harlow, K.J., Hill, A.F. and Wilton-Ely, J.D.E.T. (1999) *Journal of the Chemical Society-Dalton Transactions*, 285–292. (b) Schanz, H.-J., Jafarpour, L., Stevens, E.D. and Nolan, S.P. (1999) *Organometallics*, **18**, 5187–5190. (c) Albertin, G., Agnoletto, P. and Antoniutti, S. (2002) *Polyhedron*, **21**, 1755–1760. (d) Abdallaoui, I.A., Sémeril, D. and Dixneuf, P.H. (2002) *J. Mol. Catal. A: Chem.*, **182–183**, 577–583. (e) Wen, T.B., Zhou, Z.Y., Lo, M.F., Williams, I.D. and Jia, G. (2003) *Organometallics*, **22**, 5217–5225. (f) Albertin, G., Antoniutti, S., Bordignon, E. and Pegoraro, M. (2000) *Journal of the Chemical Society-Dalton Transactions*, 3575–3584.

17 (a) Venâncio, A.I.F., Martins, L.M.D.R.S., Fraústo da Silva, J.J.R. and Pombeiro, A.J.L. (2003) *Inorganic Chemistry Communications*, **6**, 94–96. (b) Venâncio, A.I.F., Martins, L.M.D.R.S. and Pombeiro, A.J.L. (2003) *Journal of Organometallic Chemistry*, **684**, 315–321. (c) Díez, J., Gamasa, M.P., Gimeno, J., Rodríguez, Y. and García-Granda, S. (2004) *European Journal of Inorganic Chemistry*, 2078–2085. (d) Venâncio, A.I.F., Guedes da Silva, M.F.C., Martins, L.M.D.R.S., Fraústo da Silva, J.J.R. and Pombeiro, A.J.L. (2005) *Organometallics*, **24**, 4654–4665. (e) Rigaut, S., Perruchon, J., Guesmi, S., Fave, C., Touchard, D. and Dixneuf, P.H. (2005) *European Journal of Inorganic Chemistry*, 447–460.

18 (a) Werner, H., Stark, A., Steinert, P., Grünwald, C. and Wolf, J. (1995) *Chemische Berichte*, **128**, 49–62. (b) Jung, S., Brandt, C.D. and Werner, H. (2001) *New Journal of Chemistry*, **25**, 1101–1103.

19 (a) Esquius, G., Pons, J., Yáñez, R., Ros, J., Mathieu, R., Lugan, N. and Donnadieu, B. (2003) *Journal of Organometallic Chemistry*, **667**, 126–134. (b) Bianchini, C., Peruzzini, M., Zanobini, F., Lopez, C., de los Rios, I. and Romerosa, A. (1999) *Chemical Communications*, 443–444.

20 (a) Jiménez-Tenorio, M.A., Jiménez-Tenorio, M., Puerta, M.C. and Valerga, P. (1997) *Organometallics*, **16**, 5528–5535. (b) Cadierno, V., Gamasa, M.P., Gimeno, J., Iglesias, L. and García-Granda, S. (1999) *Inorganic Chemistry*, **38**, 2874–2879. (c) Beach, N.J. and Spivak, G.J. (2003) *Inorganica Chimica Acta*, **343**, 244–252. (d) Albertin, G., Antoniutti, S., Bortoluzzi, M. and Zanardo, G. (2005) *Journal of Organometallic Chemistry*, **690**, 1726–1738. (e) Pavlik, S., Mereiter, K., Puchberger, M. and Kirchner, K. (2005) *Journal of Organometallic Chemistry*, **690**, 5497–5507. (f) Kopf, H., Pietraszuk, C., Hübner, E. and Burzlaff, N. (2006) *Organometallics*, **25**, 2533–2546. (g) Buriez, B., Burns, I.D., Hill, A.F., White, A.J.P., Williams, D.J. and Wilton-Ely, J.E.D.T. (1999) *Organometallics*, **18**, 1504–1516. (h) Hartmann, S., Winter, R.F., Brunner, B.M., Sarkar, B., Knödler, A. and Hartenbach, I. (2003) *European Journal of Inorganic Chemistry*, 876–891. (i) Wong, C.-Y., Che, C.-M., Chan, M.C.W., Leung, K.-H., Phillips, D.L. and Zhu, N. (2004) *Journal of the American Chemical Society*, **126**, 2501–2514. (j) Wong, C.-Y., Tong, G.S.M., Che, C.-M. and Zhu, N. (2006) *Angewandte Chemie-International Edition*, **45**, 2694–2698.

21 (a) Touchard, D., Guesmi, S., Bouchaib, M., Haquette, P., Daridor, A. and Dixneuf,

P.H. (1996) *Organometallics*, **15**, 2579–2581. (b) Cadierno, V., Díez, J., García-Garrido, S.E. and Gimeno, J. (2005) *Organometallics*, **24**, 3111–3117. (c) Saoud, M., Romerosa, A. and Peruzzini, M. (2000) *Organometallics*, **19**, 4005–4007.

22 (a) Bohanna, C., Callejas, B., Edwards, A.J., Esteruelas, M.A., Lahoz, F.J., Oro, L.A., Ruiz, N. and Valero, C. (1998) *Organometallics*, **17**, 373–381. (b) Bolaño, T., Castarlenas, R., Esteruelas, M.A. and Oñate, E. (2006) *Journal of the American Chemical Society*, **128**, 3965–3973.

23 (a) Tamm, M., Jentzsch, T. and Werncke, W. (1997) *Organometallics*, **16**, 1418–1424. (b) Bruce, M.I., Low, P.J. and Tiekink, E.R.T. (1999) *Journal of Organometallic Chemistry*, **572**, 3–10. (c) Bustelo, E., Tenorio, M.J., Puerta, M.C. and Valerga, P. (1999) *Organometallics*, **18**, 4563–4573. (d) Maddock, S.M. and Finn, M.G. (2001) *Angewandte Chemie-International Edition*, **40**, 2138–2141. (e) Pavlik, S., Mereiter, K., Puchberger, M. and Kirchner, K. (2005) *Organometallics*, **24**, 3561–3575.

24 (a) Esteruelas, M.A., Gómez, A.V., Lahoz, F.J., López, A.M., Oñate, E. and Oro, L.A. (1996) *Organometallics*, **15**, 3423–3435. (b) Jiménez-Tenorio, M., Palacios, M.D., Puerta, M.C. and Valerga, P. (2004) *Journal of Organometallic Chemistry*, **689**, 2853–2859.

25 (a) de los Ríos, I., Tenorio, M.J., Puerta, M.C. and Valerga, P. (1997) *Journal of Organometallic Chemistry*, **549**, 221–232. (b) Bustelo, E., Jiménez-Tenorio, M., Puerta, M.C. and Valerga, P. (2006) *Organometallics*, **25**, 4019–4025.

26 Crochet, P., Demerseman, B., Vallejo, M.I., Gamasa, M.P., Gimeno, J., Borge, J. and García-Granda, S. (1997) *Organometallics*, **16**, 5406–5415.

27 (a) Baya, M., Crochet, P., Esteruelas, M.A., Gutiérrez-Puebla, E., López, A.M., Modrego, J., Oñate, E. and Vela, N. (2000) *Organometallics*, **19**, 2585–2586. (b) Lalrempuia, R., Yennawar, H., Mozharivskyj, Y.A. and Kollipara, M.R. (2004) *Journal of Organometallic Chemistry*, **689**, 539–543. (c) Asensio, A., Buil, M.L., Esteruelas, M.A. and Oñate, E. (2004) *Organometallics*, **23**, 5787–5798.

28 (a) Nakanishi, S., Goda, K., Uchiyama, S. and Otsuji, Y. (1992) *Bulletin of the Chemical Society of Japan*, **65**, 2560–2561. (b) Argouarch, G., Thominot, P., Paul, F., Toupet, L. and Lapinte, C. (2003) *Comptes Rendus Chimie*, **6**, 209–222.

29 (a) Cadierno, V., Gamasa, M.P., Gimeno, J., González-Cueva, M., Lastra, E., Borge, J., García-Granda, S. and Pérez-Carreño, E. (1996) *Organometallics*, **15**, 2137–2147. (b) Gamasa, M.P., Gimeno, J., González-Bernardo, C., Borge, J. and García-Granda, S. (1997) *Organometallics*, **16**, 2483–2485. (c) Cadierno, V., Conejero, S., Gamasa, M.P. and Gimeno, J. (2000) *Journal of the Chemical Society-Dalton Transactions*, 451–457. (d) Cadierno, V., Conejero, S., Gamasa, M.P., Gimeno, J., Pérez-Carreño, E. and García-Granda, S. (2001) *Organometallics*, **20**, 3175–3189. (e) Cadierno, V., Conejero, S., Gamasa, M.P. and Gimeno, J. (2003) *Dalton Transactions*, 3060–3066. (f) Díez, J., Gamasa, M.P., Gimeno, J., Lastra, E. and Villar, A. (2005) *Organometallics*, **24**, 1410–1418. (g) Nishibayashi, Y., Imajima, H., Onodera, G. and Uemura, S. (2005) *Organometallics*, **24**, 4106–4109. (h) Díez, J., Gamasa, M.P., Gimeno, J., Lastra, E. and Villar, A. (2006) *European Journal of Inorganic Chemistry*, 78–87. (i) García-Fernández, A., Gimeno, J., Lastra, E., Madrigal, C.A., Graiff, C. and Tiripicchio, A. (2007) *European Journal of Inorganic Chemistry*, 732–741.

30 (a) Matsuzaka, H., Takagi, Y. and Hidai, M. (1994) *Organometallics*, **13**, 13–15. (b) Nishibayashi, Y., Wakiji, I. and Hidai, M. (2000) *Journal of the American Chemical Society*, **122**, 11019–11020. (c) Nishibayashi, Y., Imajima, H., Onodera, G., Hidai, M. and Uemura, S. (2004) *Organometallics*, **23**, 26–30.

31 (a) Braun, T., Steinert, P. and Werner, H. (1995) *Journal of Organometallic Chemistry*, **488**, 169–176. (b) Braun, T., Münch, G.,

Windmüller, B., Gevert, O., Laubender, M. and Werner, H. (2003) *Chemistry – A European Journal*, **9**, 2516–2530. (c) Crochet, P., Esteruelas, M.A., López, A.M., Ruiz, N. and Tolosa, J.I. (1998) *Organometallics*, **17**, 3479–3486.

32 (a) Fürstner, A., Liebl, M., Lehmann, C.W., Picquet, M., Kunz, R., Bruneau, C., Touchard, D. and Dixneuf, P.H. (2000) *Chemistry – A European Journal*, **6**, 1847–1857. (b) Le Gendre, P., Picquet, M., Richard, P. and Moïse, C. (2002) *Journal of Organometallic Chemistry*, **643–644**, 231–236. (c) Weiss, D. and Dixneuf, P.H. (2003) *Organometallics*, **22**, 2209–2216. (d) Castarlenas, R., Vovard, C., Fischmeister, C. and Dixneuf, P.H. (2006) *Journal of the American Chemical Society*, **128**, 4079–4089. (e) Jafarpour, L., Huang, J., Stevens, E.D. and Nolan, S.P. (1999) *Organometallics*, **18**, 3760–3763. (f) Opstal, T. and Verpoort, F. (2003) *Polymer Bulletin*, **50**, 17–23. (g) Pilette, D., Ouzzine, K., Le Bozec, H., Dixneuf, P.H., Rickard, C.E.F. and Roper, W.R. (1992) *Organometallics*, **11**, 809–817. (h) Cadierno, V., Díez, J., García-Álvarez, J. and Gimeno, J. (2004) *Chemical Communications*, 1820–1821. (i) Weberndörfer, B. and Werner, H. (2002) *Journal of the Chemical Society-Dalton Transactions*, 1479–1486. (j) den Reijer, C.J. Drago, D. and Pregosin, P.S. (2001) *Organometallics*, **20**, 2982–2989. (k) Govindaswamy, P., Mozharivskyj, Y.A. and Kollipara, M.R. (2004) *Polyhedron*, **23**, 3115–3123.

33 (a) Esteruelas, M.A., López, A.M., Oñate, E. and Royo, E. (2004) *Organometallics*, **23**, 3021–3030. (b) Urtel, K., Frick, A., Huttner, G., Zsolnai, L., Kircher, P., Rutsch, P., Kaifer, E. and Jacobi, A. (2000) *European Journal of Inorganic Chemistry*, 33–50. (c) Jung, S., Ilg, K., Brandt, C.D., Wolf, J. and Werner, H. (2002) *Journal of the Chemical Society-Dalton Transactions*, 318–327. (d) Çetinkaya, B., Demir, S., Özdemir, I., Toupet, L., Sémeril, D., Bruneau, C. and Dixneuf, P.H. (2003) *Chemistry – A European Journal*, **9**, 2323–2330.

34 (a) Guillaume, V., Thominot, P., Coat, F., Mari, A. and Lapinte, C. (1998) *Journal of Organometallic Chemistry*, **565**, 75–80. (b) Bruce, M.I., Hinterding, P., Low, P.J., Skelton, B.W. and White, A.H. (1998) *Journal of the Chemical Society-Dalton Transactions*, 467–473. (c) Péron, D., Romero, A. and Dixneuf, P.H. (1994) *Gazzetta Chimica Italiana*, **124**, 497–502. (d) Péron, D., Romero, A. and Dixneuf, P.H. (1995) *Organometallics*, **14**, 3319–3326. (e) Sato, M., Shintate, H., Kawata, Y., Sekino, M., Katada, M. and Kawata, S. (1994) *Organometallics*, **13**, 1956–1962. (f) Sato, M., Kawata, Y., Shintate, H., Habata, Y., Akabori, S. and Unoura, K. (1997) *Organometallics*, **16**, 1693–1701. (g) Sato, M., Iwai, A. and Watanabe, M. (1999) *Organometallics*, **18**, 3208–3219. (h) Rigaut, S., Costuas, K., Touchard, D., Saillard, J.Y., Golhen, S. and Dixneuf, P.H. (2004) *Journal of the American Chemical Society*, **126**, 4072–4073.

35 (a) Werner, H. and Rappert, T. (1993) *Chemische Berichte*, **126**, 669–678. (b) Werner, H., Rappert, T., Wiedemann, R., Wolf, J. and Mahr, N. (1994) *Organometallics*, **13**, 2721–2727. (c) Werner, H., Lass, R.W., Gevert, O. and Wolf, J. (1997) *Organometallics*, **16**, 4077–4088. (d) Ilg, K. and Werner, H. (2001) *Organometallics*, **20**, 3782–3794. (e) Werner, H., Weidemann, R., Laubender, M., Windmüller, B., Steinert, P., Gevert, O. and Wolf, J. (2002) *Journal of the American Chemical Society*, **124**, 6966–6980. (f) Laubender, M. and Werner, H. (1999) *Chemistry – A European Journal*, **5**, 2937–2946. (g) Gil-Rubio, J., Weberndörfer, B. and Werner, H. (2000) *Angewandte Chemie-International Edition*, **39**, 786–789. (h) Werner, H., Wiedemann, R., Laubender, M., Windmüller, B. and Wolf, J. (2001) *Chemistry – A European Journal*, **7**, 1959–1967. (i) Ilg, K. and Werner, H. (2001) *Chemistry – A European Journal*, **7**, 4633–4639. (j) Schwab, P. and Werner, H. (1994) *Journal of the Chemical*

Society-Dalton Transactions, 3415–3425.
(k) Windmüller, B., Wolf, J. and Werner, H. (1995) *Journal of Organometallic Chemistry*, **502**, 147–161. (l) Windmüller, B., Nürnberg, O., Wolf, J. and Werner, H. (1999) *European Journal of Inorganic Chemistry*, 613–619. (m) Esteruelas, M.A., Oro, L.A. and Schrickel, J. (1997) *Organometallics*, **16**, 796–799.

36 (a) Akita, M., Kato, S., Terada, M., Masaki, Y., Tanaka, M. and Moro-oka, Y. (1997) *Organometallics*, **16**, 2392–2412. (b) Edwards, A.J., Esteruelas, M.A., Lahoz, F.J., Modrego, J., Oro, L.A. and Schrickel, A.J. (1996) *Organometallics*, **15**, 3556–3562. (c) Etienne, M., Talarmin, J. and Toupet, L. (1992) *Organometallics*, **11**, 2058–2068. (d) Etienne, M. and Toupet, L. (1989) *Journal of the Chemical Society, Chemical Communications*, 1110–1111. (e) Berke, H., Härter, P., Huttner, G. and Zsolnai, L. (1982) *Chemische Berichte*, **115**, 695–705. (f) Berke, H. (1980) *Journal of Organometallic Chemistry*, **185**, 75–78. (g) Binger, P., Langhauser, F., Gabor, B., Mynott, R., Herrmann, A.T. and Krüger, C. (1992) *Journal of the Chemical Society, Chemical Communications*, 505–506.

37 (a) Capon, J.-F., Le Berre-Cosquer, N., Leblanc, B. and Kergoat, R. (1996) *Journal of Organometallic Chemistry*, **508**, 31–37. (b) Capon, J.F., Le Berre-Cosquer, N., Bernier, S., Pichon, R., Kergoat, R., L'Haridon, P. (1995) *Journal of Organometallic Chemistry*, **487**, 201–208. (c) Froom, S.F.T., Green, M., Mercer, R.J., Nagle, K.R., Orpen, A.G. and Rodrigues, R.A. (1991) *Journal of the Chemical Society-Dalton Transactions*, 3171–3183. (d) Ojo, W.-S., Pétillon, F.Y., Schollhammer, P., Talarmin, J. and Muir, K.W. (2006) *Organometallics*, **25**, 5503–5505.

38 (a) Bruce, M.I., Skelton, B.W., White, A.H. and Zaitseva, N.N. (2000) *Journal of the Chemical Society-Dalton Transactions*, 881–890. (b) Charmant, J.P.H., Crawford, P., King, P.J., Quesada-Pato, R. and Sappa, E. (2000) *Journal of the Chemical Society-Dalton Transactions*, 4390–4397.

39 Kolobova, N.E., Ivanov, L.L., Zhvanko, O.S., Khitrova, O.M., Batsanov, A.S. and Struchkov, Y.T. (1984) *Journal of Organometallic Chemistry*, **262**, 39–47.

40 Bolaño, T., Castarlenas, R., Esteruelas, M.A. and Oñate, E. (2007) *Organometallics*, **26**, 2037–2041.

41 Rigaut, S., Touchard, D. and Dixneuf, P.H. (2003) *Organometallics*, **22**, 3980–3984.

42 Bustelo, E., Jiménez-Tenorio, M., Mereiter, K., Puerta, M.C. and Valerga, P. (2002) *Organometallics*, **21**, 1903–1911.

43 Mantovani, N., Marvelli, L., Rossi, R., Bianchini, C., de los Rios, I., Romerosa, V. and Peruzzini, M. (2001) *Journal of the Chemical Society-Dalton Transactions*, 2353–2361.

44 Dragutan, V., Dragutan, I. and Verpoort, F. (2005) *Platinum Metals Review*, **49**, 33–40.

45 Berke, H., Härter, P., Huttner, G. and Zsolnai, L. (1981) *Zeitschrift für Naturforschung*, **36b**, 929–937.

46 Fischer, H., Reindl, D., Troll, C. and Leroux, F. (1995) *Journal of Organometallic Chemistry*, **490**, 221–227.

47 (a) Cosset, C., del Rio, I. and Le Bozec, H. (1995) *Organometallics*, **14**, 1938–1944. (b) Cosset, C., del Rio, I., Péron, V., Windmüller, B. and Le Bozec, H. (1996) *Synlett*, 435–436. (c) Dötz, K.H., Paetsch, D. and Le Bozec, H. (1999) *Journal of Organometallic Chemistry*, **589**, 11–20. (d) Ulrich, K., Porhiel, E., Péron, V., Ferrand, V. and Le Bozec, H. (2000) *Journal of Organometallic Chemistry*, **601**, 78–86. (e) Fischer, H. and Roth, G. (1995) *Journal of Organometallic Chemistry*, **490**, 229–237.

48 (a) Stein, F., Duetsch, M., Pohl, E., Herbst-Irmer, R. and de Meijere, A. (1993) *Organometallics*, **12**, 2556–2564. (b) Szesni, N., Weibert, B. and Fischer, H. (2005) *Inorganica Chimica Acta*, **358**, 1645–1656.

49 (a) Drexler, M., Hass, T., Yu, S.-M., Beckmann, H.S., Weibert, B. and Fischer, H. (2005) *Journal of Organometallic Chemistry*, **690**, 3700–3713. (b) Szesni, N., Drexler, M., Weibert, B. and Fischer, H. (2005) *Journal of Organometallic Chemistry*, **690**, 5597–5608.

50 (a) Berke, H., Huttner, G. and von Seyerl, J. (1981) *Zeitschrift für Naturforschung*, **36b**, 1277–1288. (b) Berke, H., Huttner, G. and von Seyerl, J. (1981) *Journal of Organometallic Chemistry*, **218**, 193–200. (c) Kolobova, N.E., Ivanov, L.L., Zhvanko, O.S., Khitrova, O.M., Batsanov, A.S. and Struchkov, Y.T. (1984) *Journal of Organometallic Chemistry*, **265**, 271–281.

51 (a) Peruzzini, M., Barbaro, P., Bertolasi, V., Bianchini, C., Rios de los, I., Mantovani, N., Marvelli, L. and Rossi, R. (2003) *Dalton Transactions*, 4121–4131. (b) Mantovani, N., Marvelli, L., Rossi, R., Bertolasi, V., Bianchini, C., de los Rios, I. and Peruzzini, M. (2002) *Organometallics*, **21**, 2382–2394.

52 (a) Devanne, D. and Dixneuf, P.H. (1990) *Journal of the Chemical Society. Chemical Communications*, 641–643. (b) Dussel, R., Pilette, D., Dixneuf, P.H. and Fehlhammer, W.P. (1991) *Organometallics*, **10**, 3287–3291. (c) Ruiz, N., Péron, D. and Dixneuf, P.H. (1995) *Organometallics*, **14**, 1095–1097. (d) Ruiz, N., Péron, D., Sinbandith, S., Dixneuf, P.H., Baldoli, C. and Maiorana, S. (1997) *Journal of Organometallic Chemistry*, **533**, 213–218. (e) Ulrich, K., Guerchais, V., Toupet, L. and Le Bozec, H. (2002) *Journal of Organometallic Chemistry*, **643–644**, 498–500. (f) Ghebreyessus, K.Y. and Nelson, J.H. (2003) *Inorganic Chemistry Communications*, **6**, 1044–1047.

53 Le Lagadec, R., Roman, E., Toupet, L., Müller, U. and Dixneuf, P.H. (1994) *Organometallics*, **13**, 5030–5039.

54 (a) Esteruelas, M.A., Gómez, A.V., López, A.M., Oñate, E. and Ruiz, N. (1998) *Organometallics*, **17**, 2297–2306. (b) Esteruelas, M.A., Gómez, A.V., López, A.M., Oliván, M., Oñate, E. and Ruiz, N. (2000) *Organometallics*, **19**, 4–14.

55 Bustelo, E. and Dixneuf, P.H. (2007) *Advanced Synthesis and Catalysis*, **349**, 933–942.

56 (a) Bernad, D.J., Esteruelas, M.A., López, A.M., Modrego, J., Puerta, M.C. and Valerga, P. (1999) *Organometallics*, **18**, 4995–5003. (b) Buil, M.L., Esteruelas, M.A., López, A.M. and Oñate, E. (2003) *Organometallics*, **22**, 162–171. (c) Buil, M.L., Esteruelas, M.A., López, A.M. and Oñate, E. (2003) *Organometallics*, **22**, 5274–5284. (d) Jiménez-Tenorio, M., Palacios, M.D., Puerta, M.C. and Valerga, P. (2004) *Journal of Organometallic Chemistry*, **689**, 2776–2785.

57 Esteruelas, M.A., Gómez, A.V., López, A.M., Modrego, J. and Oñate, E. (1998) *Organometallics*, **17**, 5434–5436.

58 (a) Cadierno, V., Gamasa, M.P., Gimeno, J., López-González, M.C., Borge, J. and García-Granda, S. (1997) *Organometallics*, **16**, 4453–4463. (b) Cadierno, V., Conejero, S., Gamasa, M.P., Gimeno, J., Asselberghs, I., Houbrechts, S., Clays, K., Persoons, A., Borge, J. and García-Granda, S. (1999) *Organometallics*, **18**, 582–597. (c) Cadierno, V., Gamasa, M.P., Gimeno, J., Moretó, J.M., Ricart, S., Roig, A. and Molins, E. (1998) *Organometallics*, **17**, 697–706. (d) Cadierno, V., Gamasa, M.P. and Gimeno, J. (1999) *Journal of the Chemical Society-Dalton Transactions*, 1857–1865. (e) Cadierno, V., Gamasa, M.P., Gimeno, J., Pérez-Carreño, E. and García-Granda, S. (1999) *Organometallics*, **18**, 2821–2832. (f) Cadierno, V., Gamasa, M.P. and Gimeno, J. (2001) *Journal of Organometallic Chemistry*, **621**, 39–45.

59 (a) Cadierno, V., Gamasa, M.P., Gimeno, J. and Lastra, E. (1999) *Journal of the Chemical Society-Dalton Transactions*, 3235–3243. (b) Cadierno, V., Gamasa, M.P., Gimeno, J., Pérez-Carreño, E. and García-Granda, S. (2003) *Journal of Organometallic Chemistry*, **670**, 75–83. (c) Cadierno, V., Conejero, S., Gamasa, M.P., Gimeno, J., Falvello, L.R. and Llusar, R.M. (2002) *Organometallics*, **21**, 3716–3726. (d) Cadierno, V., Conejero, S., Gamasa, M.P. and Gimeno, J. (2002) *Organometallics*, **21**, 3837–3840. (e) Cadierno, V., Conejero, S., Díez, J., Gamasa, M.P., Gimeno, J. and García-Granda, S. (2003) *Chemical Communications*, 840–841.

60 Esteruelas, M.A., Gómez, A.V., López, A.M., Modrego, J. and Oñate, E. (1997) *Organometallics*, **16**, 5826–5835.
61 Zhong, Y.-W., Matsuo, Y. and Nakamura, E. (2007) *Chemistry—An Asian Journal*, **2**, 358–366.
62 Berke, H., Groessmann, U., Huttner, G. and Orama, O. (1984) *Zeitschrift für Naturforschung*, **39b**, 1759–1766.
63 Kalinin, V.N., Derunov, V.V., Lusenkova, M.A., Petrovsky, P.V. and Kolobova, N.E. (1989) *Journal of Organometallic Chemistry*, **379**, 303–309.
64 Braun, T., Meuer, P. and Werner, H. (1996) *Organometallics*, **15**, 4075–4077.
65 Werner, H., Wiedemann, R., Mahr, N., Steinert, P. and Wolf, J. (1996) *Chemistry – A European Journal*, **2**, 561–569.
66 Abd-Elzaher, M.M., Weibert, B. and Fischer, H. (2005) *Organometallics*, **24**, 1050–1052.
67 Buriez, B., Cook, D.J., Harlow, K.J., Hill, A.F., Welton, T., White, A.J.P., Williams, D.J. and Wilton-Ely, J.D.E.T. (1999) *Journal of Organometallic Chemistry*, **578**, 264–267.
68 Harlow, K.J., Hill, A.F. and Welton, T. (1999) *Journal of the Chemical Society-Dalton Transactions*, 1911–1912.
69 Esteruelas, M.A., Gómez, A.V., López, A.M., Puerta, M.C. and Valerga, P. (1998) *Organometallics*, **17**, 4959–4965.
70 Díez, J., Gamasa, M.P., Gimeno, J., Lastra, E. and Villar, A. (2006) *Journal of Organometallic Chemistry*, **691**, 4092–4099.
71 Fischer, H., Leroux, F., Stumpf, R. and Roth, G. (1996) *Chemische Berichte*, **129**, 1475–1482.
72 (a) Conejero, S., Díez, J., Gamasa, M.P., Gimeno, J. and García-Granda, S. (2002) *Angewandte Chemie-International Edition*, **114**, 3589–3592. (b) Conejero, S., Díez, J., Gamasa, M.P. and Gimeno, J. (2004) *Organometallics*, **23**, 6299–6310.
73 Baya, M., Buil, M.L., Esteruelas, M.A., López, A.M., Oñate, E. and Rodríguez, J.R. (2002) *Organometallics*, **21**, 1841–1848.
74 Esteruelas, M.A., Gómez, A.V., López, A.M., Oñate, E. and Ruiz, N. (1999) *Organometallics*, **18**, 1606–1614.
75 Roth, G. and Fischer, H. (1996) *Journal of Organometallic Chemistry*, **507**, 125–136.
76 (a) Esteruelas, M.A., Gómez, A.V., López, A.M. and Oñate, E. (1998) *Organometallics*, **17**, 3567–3573. (b) Bertolasi, V., Mantovani, N., Marvelli, L., Rossi, R., Bianchini, C., de los Rios, I., Peruzzini, M. and Akbayeva, D.N. (2003) *Inorganica Chimica Acta*, **344**, 207–213.
77 (a) Bernad, D.J., Esteruelas, M.A., López, A.M., Oliván, M., Oñate, E., Puerta, M.C. and Valerga, P. (2000) *Organometallics*, **19**, 4327–4335. (b) Mantovani, N., Bergamini, P., Marchi, A., Marvelli, L., Rossi, R., Bertolasi, V., Ferretti, V., de los Rios, I. and Peruzzini, M. (2006) *Organometallics*, **25**, 416–426.
78 Szesni, N., Hohberger, C., Mohamed, G.G., Burzlaff, N., Weibert, B. and Fischer, H. (2006) *Journal of Organometallic Chemistry*, **691**, 5753–5766.
79 (a) Selegue, J.P. (1983) *Journal of the American Chemical Society*, **105**, 5921–5923. (b) Bruce, M.I., Hinterding, P., Tiekink, E.R.T., Skelton, B.W. and White, A.H. (1993) *Journal of Organometallic*, **450**, 209–218.
80 Rigaut, S., Olivier, C., Costuas, K., Choua, S., Fadhel, O., Massue, J., Turek, P., Saillard, J.-Y., Dixneuf, P.H. and Touchard, D. (2006) *Journal of the American Chemical Society*, **128**, 5859–5876.
81 Rigaut, S., Maury, O., Touchard, D. and Dixneuf, P.H. (2001) *Chemical Communications*, 373–374.

3
Preparation and Reactivity of Higher Metal Cumulenes Longer than Allenylidenes

Helmut Fischer

3.1
Introduction

Cumulenylidene complexes constitute a class of organometallic compounds in which a chain of sp-hybridized carbon atoms is terminated by a sp^2-hybridized carbon atom (CR_2 group) at one end and by a metal–ligand fragment (L_nM) at the other end.

$$L_nM=(C)_xR_2$$

Depending on the number of carbon atoms in the chain, several types of cumulenylidene complexes are conceivable (Chart 3.1). These cumulenylidene complexes may be regarded as formally derived from (i) carbene complexes ($x = 1$) by insertion of carbon atoms in between the metal–ligand fragment and the terminal CR_2 group or (ii) from organic carbonyl compounds such as ketones, ketenes, and their higher homologs by replacing the oxygen atom by a metal–ligand fragment.

The metal–ligand fragment L_nM, the number of carbon atoms x, and the substituents at the terminal sp^2-carbon may vary considerably and, correspondingly, the properties and reactivities. The early members of the series of cumulenylidene complexes ($x = 1, 2, 3$: carbene, vinylidene and allenylidene complexes) have established themselves as invaluable building blocks in stoichiometric synthesis and as highly potent catalyst precursors. The higher members might potentially be very useful candidates for application as one-dimensional wires and in opto-electronic devices.

The number of known, isolated and characterized complexes depends strongly on the length of the chain and drastically decreases with the number of carbon atoms in the chain. A great number of vinylidene complexes of many metals, with different terminal substituents R and various co-ligands have been synthesized and the reactivity has been studied extensively. At present, the solid-state structure of more than 230 vinylidene complexes has been determined by X-ray structure analyses. The number of isolated allenylidene complexes is somewhat smaller,

---

Metal Vinylidenes and Allenylidenes in Catalysis: From Reactivity to Applications in Synthesis
Edited by Christian Bruneau and Pierre Dixneuf
Copyright © 2008 WILEY-VCH Verlag GmbH & Co. KGaA, Weinheim
ISBN: 978-3-527-31892-6

x			
1	$L_nM=CR_2$	Carbene complexes	$O=CR_2$
2	$L_nM=C=CR_2$	Vinylidene complexes	$O=C=CR_2$
3	$L_nM=C=C=CR_2$	Allenylidene complexes	$O=C=C=CR_2$
4	$L_nM=C=C=C=CR_2$	Butatrienylidene complexes	$O=C=C=C=CR_2$
5	$L_nM=C=C=C=C=CR_2$	Pentatetraenylidene complexes	$O=C=C=C=C=CR_2$

Chart 3.1 Comparison of metallacumulenes and the corresponding carbonyl compounds.

however, there are still several hundreds of allenylidene complexes known and the structure of more than 110 complexes has been established by X-ray structure analyses. In contrast, only very few butatrienylidene complexes ($x=4$) have been synthesized. Until now (March 2007) 15 complexes have been isolated in a pure form and four more have been characterized in solution by spectroscopic means. A few other butatrienylidene complexes could be generated and be trapped by reactions with nucleophiles.

The number of pentatetraenylidene complexes ($x=5$) reported in the literature is even less. All in all, 11 complexes have been described. Complexes with an even larger number of carbon atoms in the chain are unknown. Therefore, cumulenylidene complexes with more than three carbon atoms in the chain are still rather elusive classes of metallacumulenes [1] and their involvement in a catalytic process has not yet been observed.

3.2
Steric and Electronic Structure

Until now, the structures of only three butatrienylidene [2–5] and four pentatetraenylidene complexes [6–9] (Chart 3.2) have been established by X-ray structure analysis.

$L_nM=C=C=C=CR_2$: $L_nM = Cl(PPr^i_3)_2Ir$, R = Ph [2, 3]
$L_nM = (C_5H_4Me)(Me_2PC_2H_4PMe_2)Mn$, R = $SnPh_3$ [4]
$L_nM = (C_5H_4Me)(Me_2PC_2H_4PMe_2)Mn$, R = $SnMe_3$ [5]

$L_nM=C=C=C=C=CR_2$: $L_nM = [Cl(Ph_2PC_2H_4PPh_2)Ru]+$, R = Ph [6]
$L_nM = Cl(PPr^i_3)_2Ir$, R = Ph [7]
$L_nM = (CO)_5W$, R = NMe_2 [8]
$L_nM = (CO)_5Cr$, $R_2 = (NMe_2)[CMe=C(NMe_2)_2]$ [9]

Chart 3.2 Mononuclear butatrienylidene and pentatetraenylidene complexes characterized by X-ray structure analysis.

The metal–carbon chain in these complexes is either linear or deviates only slightly from linearity. The M=C=C angle varies between 173.5° and 180°, the various C=C=C angles between 172.3° and 180°. The deviation from linearity is most pronounced with the butatrienylidene manganese complexes [4, 5].

As expected, the terminal C=C bond in all complexes is significantly longer than the internal double bonds. Whereas in the butatrienylidene iridium complex [2, 3] both internal double bonds are, within the error limit, equal in length, in the manganese complexes the central C=C bond is shorter than the (Mn)C=C bond.

The bond lengths in pentatetraenylidene complexes show a pronounced dependence on the π-donor potential of the terminal substituents and the type of the metal–ligand fragment. With all complexes a short–long–short–long pattern of the CC distances along the chain, starting from the metal, is observed. The bond length variation is most pronounced in the pentacarbonyl pentatetraenylidene complexes [8, 9] having strong π-donor substituents at one terminus of the chain and the acceptor "$(CO)_5M$" at the other end. This arrangement leads to strong π-interaction and to strong dipolarity of the complexes (see Scheme 3.18). In contrast, the internal C=C bonds in the neutral d^8 complex [6] are almost equal in length. The bond length alternation in the cationic d^6 ruthenium complex [7] is intermediate between those in the iridium complex and both pentacarbonyl complexes.

Based on these experimental data, DFT calculations have been carried out on several series of cumulenylidene complexes. The influence on bonding and the electronic structure of various factors such as

1. The number of carbon atoms in the chain in $[(CO)_5Cr=(C)_xH_2]$ ($x = 2$–9 [10])
2. The terminal substituents in $[(CO)_5Cr=(C)_xR_2]$ ($R = F$, SiH_3, $CH=CH_2$, NH_2, NO_2; $x = 2$–8 [11])
3. The charge in $[Cl(PH_3)_4Ru=(C)_xH_2]^z$ ($z = +1, 0, -1$; $x = 1$–8) [12]
4. The metal–ligand fragment: $[L_nM=(C)_xH_2]$ ($L_nM = (Cp)_2(PH_3)Ti$ (d^2), Cp (PH$_3$)$_2$Mo$^+$ (d^4), (CO)$_5$Cr, (CO)$_5$Mo, (CO)$_5$W, Cp(dppe)Fe$^+$, trans-Cl(dppe)$_2$Ru$^+$, Cp(PMe$_3$)$_2$Ru$^+$, BzCl(PH$_3$)Ru$^+$ (all d^6), trans-Cl(PH$_3$)$_2$Rh, and trans-Cl(PH$_3$)$_2$Ir (both d^8); $x = 4, 5$) [13].

has been analyzed. Certain trends have been identified. The most important results of these calculations may be summarized as follows.

1. The dissociation energy of the metal–cumulenylidene bond in $[(CO)_5Cr=(C)_xR_2]$ is essentially independent of the chain length but is significantly affected by π-donor and π-acceptor substituents R [10]. A π-donor group such as NH_2 causes a decrease in the bond dissociation energy (BDE) relative to that of the unsubstituted complex (R = H). The decrease in the BDE is more evident for odd-chain cumulenylidene ligands. π-Acceptor substituents like NO_2 lead to an increase in the BDE that, in turn, is more evident in even-chain metallacumulenes [11]. An increase in the electron richness of the d^6 metal fragment in $[L_nM=(C)_xH_2]$ causes a slight decrease in the BDE. The bond energies in d^8 complexes are larger than those in the d^6 analogs [13].

2. The atomic charges on the various carbon atoms of the cumulenylidene ligand show no significant difference except for the first and last atom of the chain bearing higher negative charges [10, 11, 13]. Deviating, the last atom of the chain in difluoro-substituted complexes [$(CO)_5Cr=(C)_xF_2$] is positively charged [11]. Charge distribution obviously is not important in determining the regioselectivity of either an electrophilic or a nucleophilic attack.

3. In complexes with an even number of carbon atoms in the chain the HOMO is oriented perpendicularly to the plane of the cumulenylidene ligand and the LUMO is an in-plane orbital. Conversely, in complexes with an odd number of carbon atoms the HOMO is an in-plane orbital and the LUMO is perpendicular to the plane [10, 12].

4. The energy of the LUMO in unsubstituted cumulenylidene complexes (R = H) with acceptor co-ligands (CO) at the metal (such as $Cr(CO)_5$, $Mo(CO)_5$, and $W(CO)_5$) in general decreases with increasing length of the carbon chain, indicating an increase in the electrophilicity. The decrease is slightly more pronounced in complexes with an odd number of carbon atoms than in even-numbered complexes. Contrary to the energy of the LUMO, the energy of the HOMO (and thus the nucleophilicity of the complexes) is essentially unaffected by the length of the chain [10, 13].

5. In contrast to d^6 $M(CO)_5$ complexes, the energy of the LUMO in cationic d^6 ruthenium complexes having donor co-ligands only changes marginally when x increases by 2. However, the LUMO (when $x > 2$) in odd-numbered complexes is about 0.3 eV lower in energy than in even-numbered ones. The energy of the HOMO increases almost evenly with increasing x [12, 13].

6. Replacing H in [$(CO)_5Cr=(C)_xH_2$] by π-donor (NH_2) or π-acceptor (NO_2) groups affects the energy of the HOMO and LUMO in different ways. Introducing NH_2 raises the energy of the HOMO as well as of the LUMO. Conversely, introducing NO_2 lowers the energy of both. The influence on the HOMO (LUMO) is much more pronounced in complexes with an even (odd) number of carbon atoms in the chain. The HOMO–LUMO gap in general decreases with increasing x, however, the value of the gap oscillates, being larger for complexes with an even-numbered carbon chain (R = H and NO_2). The reverse is true in the case of π-donor-substituted complexes (R = NH_2) [11].

7. Independent of the chain length, the substituents R and the metal–ligand fragment L_nM the LUMO in d^6 and d^8 complexes is predominantly localized on the odd carbon atoms. Therefore, nucleophiles are expected to add to the odd carbon atoms in the chain. The HOMO in d^6 and d^8 complexes has contributions mainly from the metal and the carbon atoms in even positions of the chain. As a consequence, electrophiles are expected to add either to the metal or to even-numbered carbon atoms. Deviating from these more general results, the HOMO in [$(CO)_5Cr=(C)_4H_2$] has been calculated to be more evenly distributed along the chain with major components from the odd-numbered carbon atoms C1 and C3 [10–13].

From these theoretical analyses of the electronic structure of cumulenylidene complexes it follows that different strategies are required for the stabilization of cumulenylidene complexes with an even and with an odd number of carbon atoms. For instance, π-donor substituents R in [(CO)$_5$Cr=(C)$_x$R$_2$] should stabilize complexes with an odd number of carbon atoms x and, conversely, complexes with an even number of carbon atoms should be stabilized by π-acceptor substituents [11].

3.3
Synthesis of Cumulenylidene Complexes

The cumulenylidene ligand can be introduced into transition metal complexes by coordinating precursors that already contain all needed carbon atoms x and subsequently transform the precursor ligand into the cumulenylidene ligand (route a). Alternatively, the cumulenylidene ligand can be constructed in the coordination sphere of the metal from fragments containing less than x carbon atoms (route b). In addition, modification of already existing cumulenylidene complexes (route c) through, for instance, substitution processes either at the metal or at the terminal carbon atom of the chain offers an approach to cumulenylidene complexes that are inaccessible by these direct routes.

3.3.1
Butatrienylidene Complex Synthesis

The majority of butatrienylidene complexes synthesized or generated so far were obtained by following route a and using butadiyne or a butadiyne derivative as the source of the C4 fragment. A few complexes of iridium or manganese were synthesized by the substitution route c. Until now route b has not been (successfully) employed.

The first butatrienylidene complexes were generated by Bruce et al. [14] and Lomprey and Selegue [15, 16] in 1993.

Lomprey and Selegue's method starts from an acylethynyl complex (**1**) obtained by reaction of [Cp(PPh$_3$)$_3$Ru–Cl] with Me$_3$Si–C≡CC(O)CMe$_2$H and KF in refluxing methanol. Treatment of complex **1** with trifluoroacetic anhydride afforded **2** in 76% yield. Based on the relatively large P–C$_\alpha$ coupling constant, complex **2** was suggested to partially dissociate to the ion pair **2a**. The ionized form **2a** was not observable, but the trifluoroacetate was readily displaced from **2/2a** by a variety of nucleophiles (Scheme 3.1) [16].

The synthesis introduced by Bruce et al. starts from butadiynyl lithium [14]. The addition of HBF$_4$ to solutions of buta-1,3-diynyl ruthenium complex **3** was proposed to afford the butatrienylidene cation **4** by protonation of the terminal carbon atom of the butadiynyl ligand. Complex **4** could neither be isolated nor spectroscopically detected. It readily decomposed by reaction with even traces of water in the air by nucleophilic attack of H$_2$O on the cationic center (Scheme 3.2).

The nucleophilicity of the terminal carbon atom of butadiynyl ligands was also used later by Dixneuf et al. for the generation of a related phenylbutatrienylidene

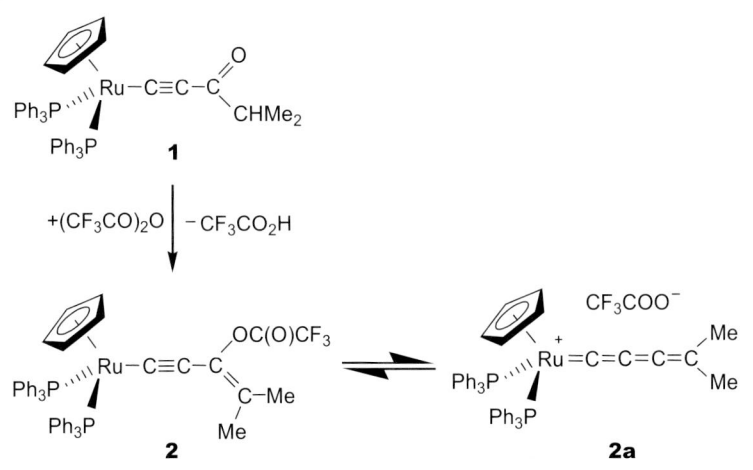

Scheme 3.1 Generation of a dimethylbutatrienylidene ruthenium complex by reaction of an acylethynyl complex with trifluoroacetic anhydride.

complex (**6**) by protonation of the phenyl buta-1,3-diynyl complex **5** (Scheme 3.3) [17].

The first isolable, albeit binuclear, butatrienylidene complexes, the cationic diiron complexes **8**, were likewise prepared by addition of an electrophile E^+ to neutral butadiynyl complexes. Instead of mononuclear butadiynyl complexes, binuclear C4-bridged butadiyndiyl complexes **7** were used as the starting complexes by Lapinte et al. (Scheme 3.4) [18]. Complexes **8** were characterized by multinuclear NMR, IR, UV–vis, and Mössbauer spectroscopies, mass spectrometry and cyclic voltammetry.

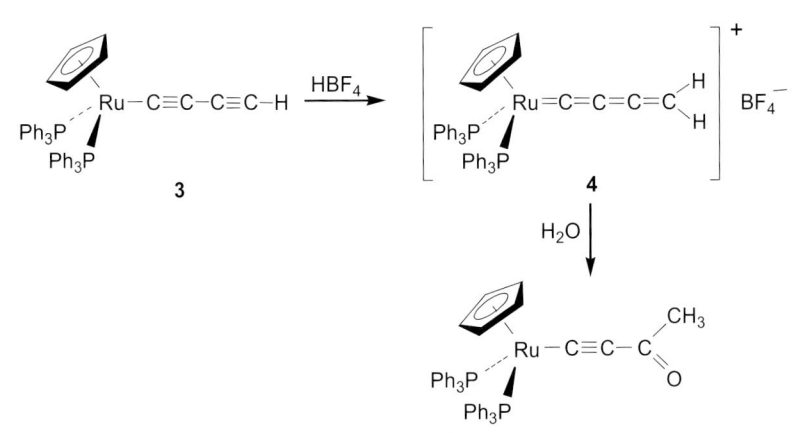

Scheme 3.2 Generation of $[Cp(PPh_3)_2Ru=(C)_4H_2]^+$ by protonation of a butadiynyl complex with trifluoroboric acid and subsequent reaction with water.

3.3 Synthesis of Cumulenylidene Complexes

Scheme 3.3 Generation of [Cl(dppe)$_2$Ru=(C)$_4$HPh]$^+$ by protonation of a phenylbutadiynyl complex with trifluoromethyl sulfonic acid.

Scheme 3.4 Synthesis of the first binuclear butatrienylidene complexes by addition of electrophiles to butadiyndiyl diiron complexes.

An alternative route to cationic butatrienylidene complex **4** involves 1,4-H shift in butadiyne complex **9**. The formation of **4** as an intermediate in the reaction of [Cp(PPh$_3$)$_3$Ru-Cl] with AgPF$_6$ and buta-1,3-diyne was deduced from trapping experiments. Complex **4** thus generated gave with diphenylamine the corresponding diphenylamino(methyl)allenylidene complex (Scheme 3.5) [19].

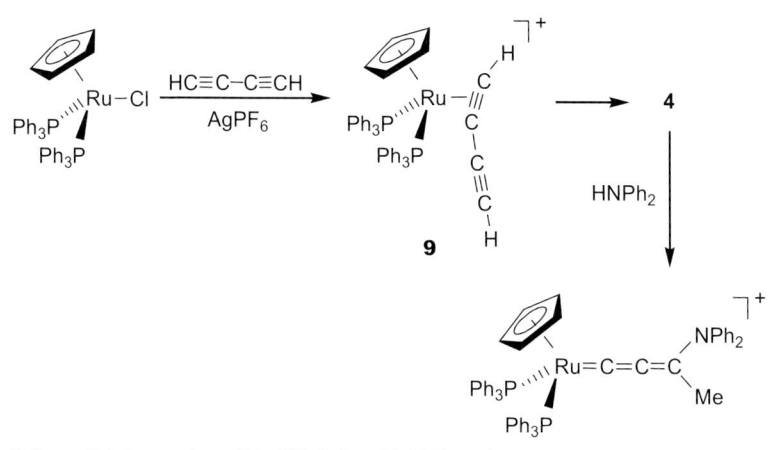

Scheme 3.5 Generation of [Cp(PPh$_3$)$_2$Ru=(C)$_4$H$_2$]$^+$ (**4**) by reaction of [RuCl(Cp)(PPh$_3$)$_2$] with butadiyne and addition of HNPh$_2$ across the C3–C4 bond of the chain.

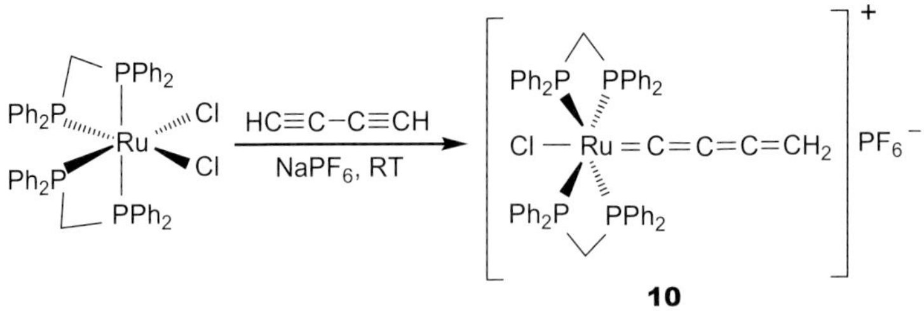

Scheme 3.6 Formation of the butatrienylidene complex **10** by reaction of [RuCl$_2$(dppm)$_2$] with butadiyne.

1,4-Rearrangement reactions have turned out to be very convenient and the most used methods for the synthesis of butatrienylidene complexes. 1,4-H shift reactions have been used by Bruce et al. [19, 20] and Winter et al. (e.g., Scheme 3.6) [21, 22] for generating butatrienylidene ruthenium complexes. Analogously, a butatrienylidene iron complex was obtained from [Cp*(dppe)Fe–Cl] (dppe = Ph$_2$PCH$_2$CH$_2$PPh$_2$), Me$_3$Si-C≡CC≡CH, and NaBPh$_4$ in methanol via 1,4-H shift (Scheme 3.7) [23, 24].

Neither of these complexes was isolable. Nevertheless, they were stable enough to be used in reactions, predominantly with nucleophiles (see Section 3.4.1).

The synthesis of the first neutral isolable butatrienylidene complex was reported by Werner et al. in 2000 [2] using an activated alkyne as the C4 source. Reaction of [IrCl(H)$_2$(PPr^i$_3$)$_2$] with HC≡C–C(OTf)=CPh$_2$ at −100 °C in the presence of one equivalent of triethylamine gave complex **11** in 77% yield (Scheme 3.8) [2, 3]. The reaction was accompanied by the evolution of H$_2$ and is, in part, reminiscent of the generation of a butatrienylidene ruthenium complex by Lomprey and Selegue [15]. Complex **11** was fully characterized, including an X-ray structure analysis, and could be transformed into several other butatrienylidene complexes via replacement of the *trans*-chloro ligand by other anionic groups (see Section 3.4.1).

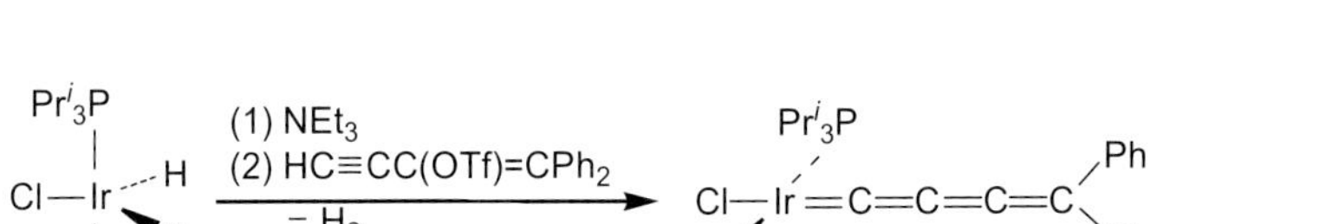

Scheme 3.7 Formation of a silyl-substituted butatrienylidene iron complex by reaction of [FeCl(Cp*)(dppe)] with trimethylsilylbutadiyne.

Scheme 3.8 Synthesis of the first isolable neutral butatrienylidene complex.

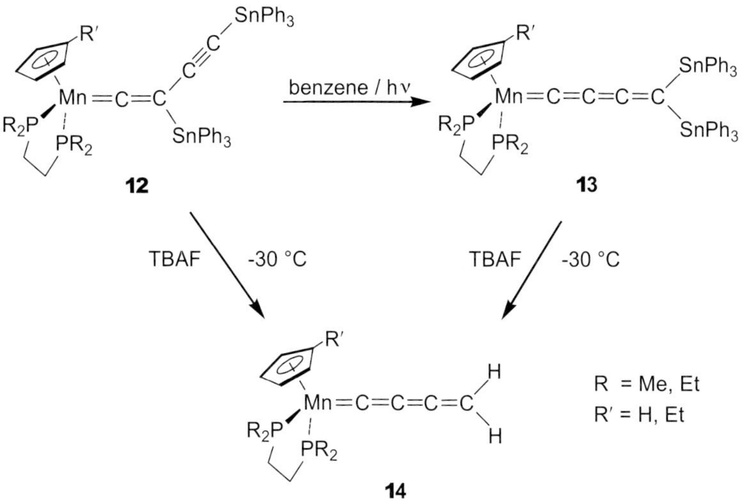

3.3 Synthesis of Cumulenylidene Complexes

Scheme 3.9 Synthetic routes to butatrienylidene manganese complexes.

Recently, isolable bis(triphenylstannyl)-substituted butatrienylidene complexes of manganese (**13**) were obtained by photolysis of alkynyl(triphenylstannyl)vinylidene complexes **12** (Scheme 3.9) [4, 5]. Treatment of the resulting bis(stannyl)butatrienylidene complexes **13** with tetrabutylammonium fluoride and water afforded the first characterizable butatrienylidene complexes (**14**) containing an unsubstituted [M=C=C=C=CH$_2$] moiety (Scheme 3.9). In contrast to **13**, complexes **14** were unstable above −5 °C and were therefore characterized in solution only by NMR spectroscopy at −40 °C. Complexes **14** were also formed instantaneously when solutions of **12** were treated at −30 °C with one equivalent of tetrabutylammonium fluoride.

Analogously to **12**, the mixed trimethylsilyl(trimethylstannyl)vinylidene complex **15** reacted rapidly with one equivalent of tetrabutylammonium fluoride at −30 °C to form **14** (R = R′ = Me) (Scheme 3.10) [5].

A bis(trimethylstannyl)-substituted butatrienylidene complex (**17**) related to **13** (R = R′ = Me) was formed in the reaction of cycloheptatrienyl(methylcyclopentadienyl)manganese with Me$_3$SnC≡C-C≡CSnMe$_3$ and dmpe. Complex **17** was obtained

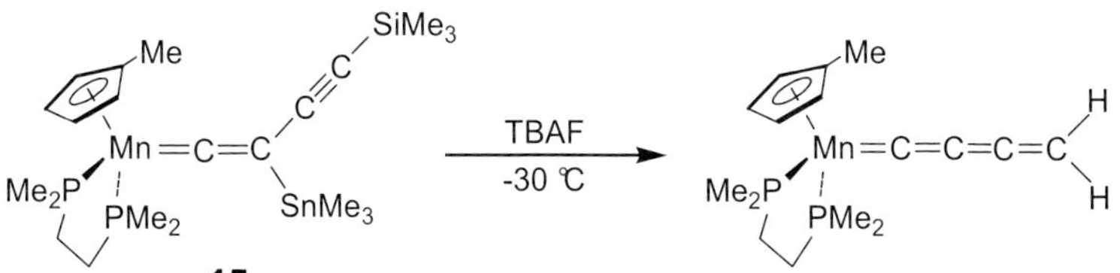

Scheme 3.10 Synthesis of [Cp′(dmpe)Mn=(C)$_4$H$_2$] by reaction of a stannyl (silylethynyl)vinylidene complex with tetrabutylammonium fluoride.

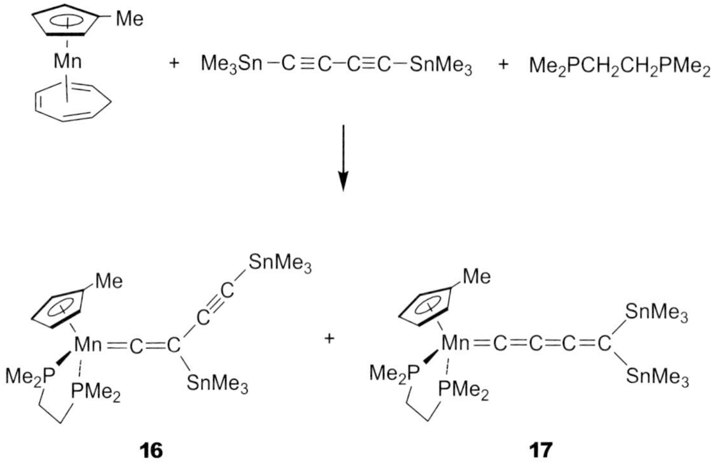

Scheme 3.11 Reaction of cycloheptatrienyl-(methylcyclopentadienyl)manganese with bis(trimethylstannyl)-butadiyne and bis(dimethylphosphino)ethane.

together with the vinylidene complex **16** related to **12** (Scheme 3.11). The mixture could not be separated, however, it was possible to grow crystals of butatrienylidene complex **17** [5].

All butatrienylidene complexes that have turned out to be isolable or, at least, spectroscopically detectable species are derived from electron-rich metal–ligand fragments. The butatrienylidene ligand is either unsubstituted or carries an aromatic or a SnR_3 substituent. π-Donor-substituted butatrienylidene complexes are unknown until now. Based on the results of the calculations mentioned above, π-donor-substituted butatrienylidene complexes are expected to be rather unstable, especially so when the metal–ligand fragment carries acceptor co-ligands. The relatively high stability of the tin-substituted butatrienylidene manganese complexes has been explained by the energetically high lying and thus strongly electron donating σ orbitals of the SnR_3 groups [5].

3.3.2
Pentatetraenylidene Complex Synthesis

In the majority of pentatetraenylidene complexes prepared or generated so far, the pentatetraenylidene ligand is derived from suitable C_5 precursors. Usually penta-1,3-diynyl derivatives like the alcohol $HC\equiv C-C\equiv C-CPh_2OH$, its trimethylsilyl ether, or the 5,5,5-tris(dimethylamino)-substituted penta-1,3-diyne are employed.

Analogously to butatrienylidene complexes, the first pentatetraenylidene complexes to be generated by Dixneuf *et al.* were ruthenium compounds. In 1990, the cationic pentatetraenylidene complexes $[(C_6Me_6)Cl(PR_3)Ru=C=C=C=C=CPh_2]^+$ ($PR_3 = PMe_3$, PMe_2Ph, $PMePh_2$) were proposed as intermediates in the reaction of $[(C_6Me_6)(PR_3)RuCl_2]$ with penta-1,3-diynes $HC\equiv C-C\equiv C-CPh_2OR'$ ($R' = H$;

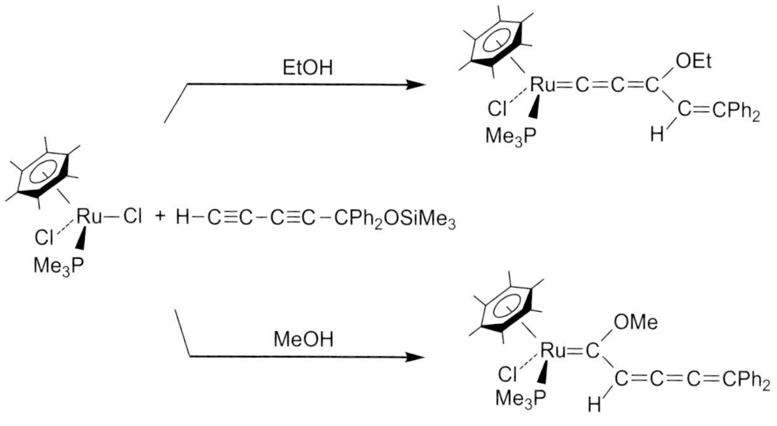

Scheme 3.12 Differing reactivity of [(C$_6$Me$_6$)(PMe$_3$)Ru= (C)$_5$Ph$_2$]$^+$ generated *in situ* towards methanol and ethanol.

SiMe$_3$) in the presence of ethanol, *i*-propanol, or diphenylamine to account for the formation of alkoxy- and amino(alkenyl)allenylidene complexes [25] or of a butatrienyl(methoxy)carbene complex in the presence of methanol [26]. Two representative examples are depicted in Scheme 3.12.

The formation of other mono- [27–29] or even bis[alkoxy(alkenyl)allenylidene] ruthenium complexes [28, 30] from the corresponding ruthenium chlorides and 5,5-diphenyl-penta-1,3-diynyl alcohol or trimethylsilyl ether in the presence of methanol (Scheme 3.13) and of the allenylidene complex **18** in the absence of methanol (Scheme 3.13) [30, 31] was also suggested to proceed via pentatetraenylidene intermediates. Neither one of these pentatetraenylidene complexes could be isolated or spectroscopically detected although their formation as an intermediate was very likely.

The first isolable pentatetraenylidene complex was obtained by Dixneuf *et al.* in 1994 when Me$_3$Si–C≡C–C≡C–C(Ph)$_2$OSiMe$_3$ was replaced by Me$_3$Si–C≡C–C≡C–C(C$_6$H$_4$NMe$_2$)$_2$OSiMe$_3$ in the reaction with [(C$_6$Me$_6$)(PMe$_3$)RuCl$_2$] (Scheme 3.14) [32]. Complex **19** was obtained in 27% yield and was fully characterized. The stability of **19** was suggested to be due to the π-electron-donating amino groups in position 4 of the aryl-substituent.

Another cationic pentatetraenylidene complex was synthesized in a stepwise fashion. Reaction of [RuCl$_2$(dppe)$_2$] with 5,5-diphenylpentadiynyl trimethylsilylether in the presence of NaPF$_6$/NEt$_3$ gave the corresponding butadiynyl complex. Subsequent [Ph$_3$C]$^+$PF$_6^-$ induced elimination of [OSiMe$_3$]$^-$ afforded complex **20** (Scheme 3.15) whose structure was established by spectroscopic means and additionally by an X-ray structure analysis [7]. It is noteworthy that the corresponding complex containing two dppm (dppm = Ph$_2$PCH$_2$PPh$_2$) instead of the dppe ligands could never be isolated. Obviously, the co-ligand sphere plays an important role in stabilizing pentatetraenylidene complexes.

The first neutral pentatetraenylidene complex was obtained by Werner *et al.* Treatment of a solution of the butadiynyl(hydrido) complex **21** at −78 °C with an equimolar amount of (CF$_3$SO$_2$)$_2$O followed by addition of two equivalents of NEt$_3$ at

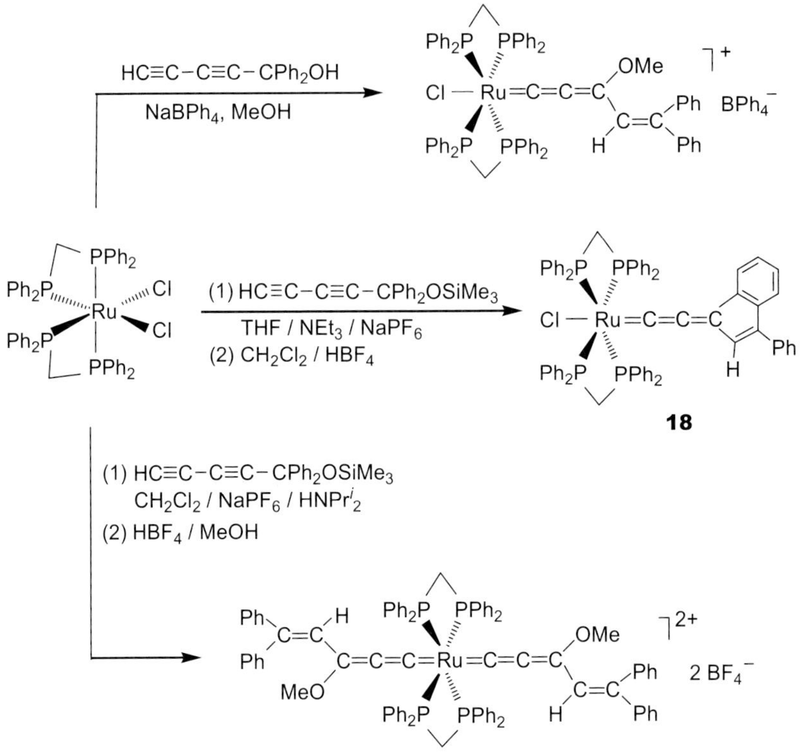

Scheme 3.13 Allenylidene complexes from pentatetraenylidene ruthenium complexes formed *in situ* by reaction of [RuCl$_2$(dppm)$_2$] and penta-1,3-diyne derivatives.

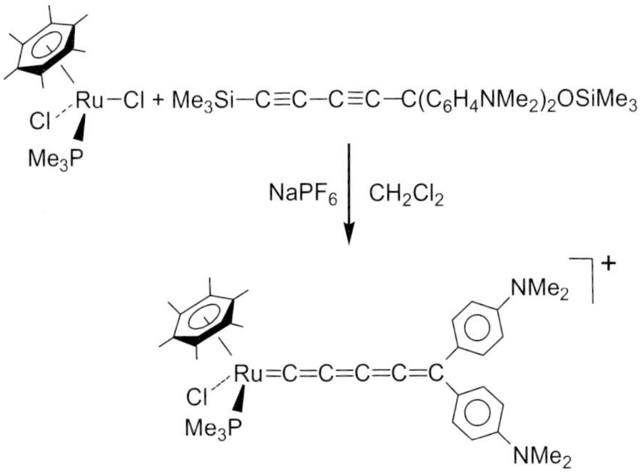

Scheme 3.14 Synthesis of the first isolable cationic pentatetraenylidene complex.

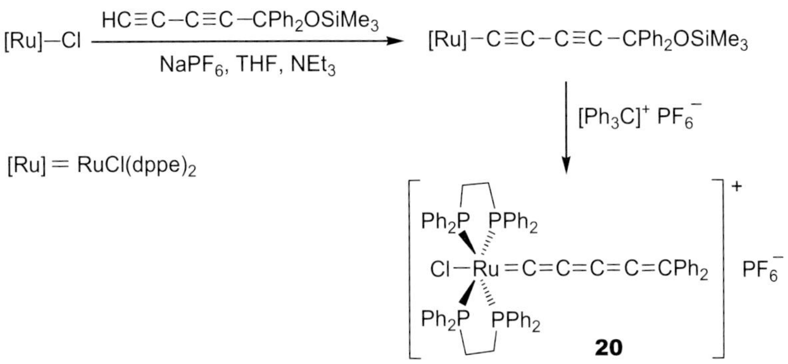

3.3 Synthesis of Cumulenylidene Complexes

$$[Ru]-Cl \xrightarrow[NaPF_6, THF, NEt_3]{HC\equiv C-C\equiv C-CPh_2OSiMe_3} [Ru]-C\equiv C-C\equiv C-CPh_2OSiMe_3$$

[Ru] = RuCl(dppe)$_2$

$\downarrow$ [Ph$_3$C]$^+$ PF$_6^-$

$$\left[\begin{array}{c} Ph_2P \overset{\frown}{} PPh_2 \\ Cl-Ru=C=C=C=C=CPh_2 \\ Ph_2P \underset{\smile}{} PPh_2 \end{array} \right]^+ PF_6^-$$

20

Scheme 3.15 Synthesis of [Cl(dppe)$_2$Ru=(C)$_5$Ph$_2$]PF$_6$.

room temperature afforded the pentatetraenylidene complex **22** in 80% yield (Scheme 3.16) [6]. Like **11**, the structure of complex **22** was established by an X-ray structural analysis. The analogous rhodium complex was later synthesized by the same methodology [33].

Subsequently, the first pentatetraenylidene complexes carrying non-aryl substituents at C5 of the chain were synthesized by Roth and Fischer [8]. They used 5,5,5-tris(drimethylamino)penta-1,3-diyne instead of 5,5-diarylpenta-1,3-diynes as the source of the C$_5$ fragment and the electron-rich metal–ligand fragments were replaced by pentacarbonylchromium and -tungsten. The resulting strongly dipolar complexes **23** (Scheme 3.17) turned out to be stable in air at room temperature. The high thermal stability is due to the strongly dipolar nature of these complexes (see Scheme 3.18).

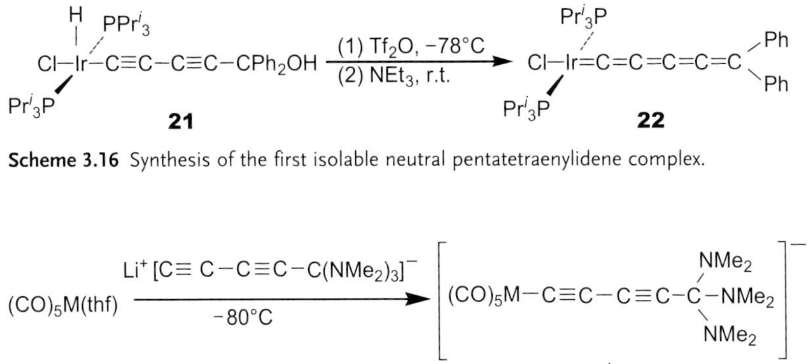

$$\begin{array}{c} H PPr^i_3 \\ | / \\ Cl-Ir-C\equiv C-C\equiv C-CPh_2OH \\ / \\ Pr^i_3P \end{array} \xrightarrow[(2)\ NEt_3,\ r.t.]{(1)\ Tf_2O,\ -78°C} \begin{array}{c} Pr^i_3P \\ / \\ Cl-Ir=C=C=C=C=C \overset{Ph}{\underset{Ph}{\diagdown}} \\ / \\ Pr^i_3P \end{array}$$

21 **22**

Scheme 3.16 Synthesis of the first isolable neutral pentatetraenylidene complex.

$$(CO)_5M(thf) \xrightarrow[-80°C]{Li^+ [C\equiv C-C\equiv C-C(NMe_2)_3]^-} \left[(CO)_5M-C\equiv C-C\equiv C-\underset{NMe_2}{\overset{NMe_2}{C}}-NMe_2 \right]^-$$

$\downarrow$ BF$_3$·Et$_2$O, -40°C

M = Cr, W

$$(CO)_5M=C=C=C=C=C \overset{NMe_2}{\underset{NMe_2}{\diagdown}}$$

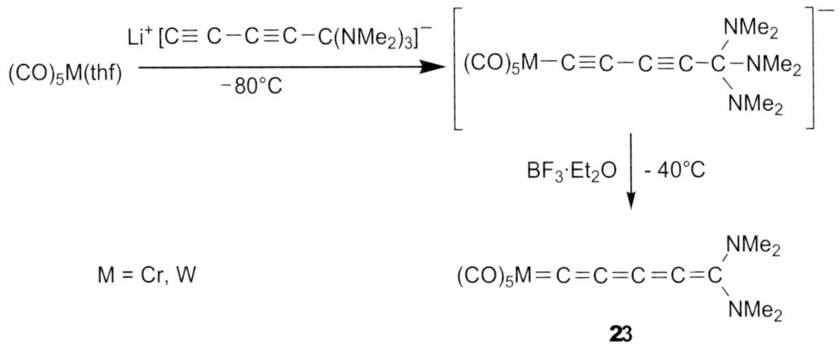

23

Scheme 3.17 Synthesis of π-donor substituted pentatetraenylidene complexes.

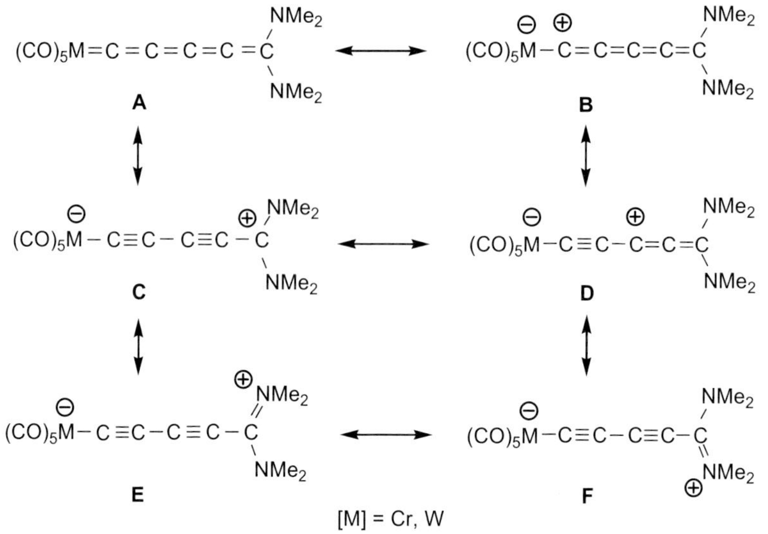

Scheme 3.18 Resonance structures of π-donor substituted pentatetraenylidene complexes.

The reaction of these complexes with ynamine allowed the modification of the pentatetraenylidene ligand. Insertion of the C≡C bond of the alkyne into the C4=C5 bond of the chain afforded alkenyl(amino)pentatetraenylidene complexes [9].

Another new approach to pentatetraenylidene complexes (construction of the cumulenylidene ligand in the coordination sphere of the metal from fragments containing less than x C atoms) was introduced by Gladysz et al. [34]. Coupling of the lithiated butadiynyl complex **24** with 9-fluorenone derivatives and subsequently with 3 equivalents of $[Me_3O]^+BF_4^-$ and $BF_3 \cdot OEt_2$ gave thermally labile pentatetraenylidene complexes **25** (Scheme 3.19).

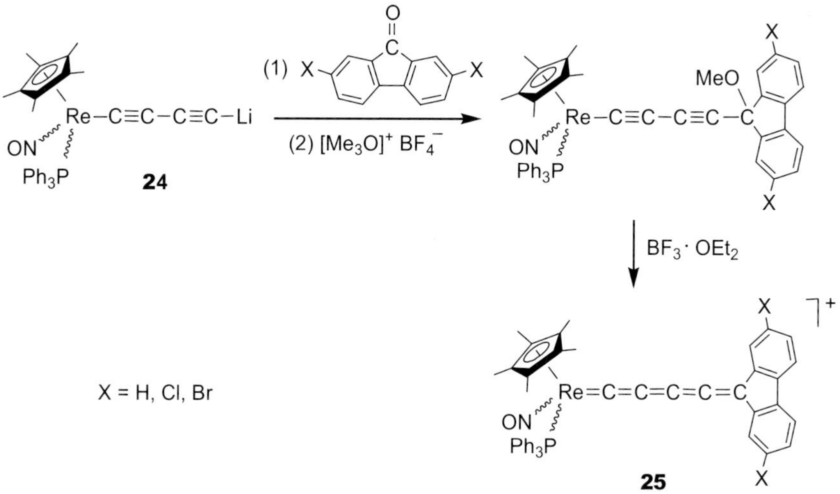

Scheme 3.19 Synthesis of pentatetraenylidene rhenium complexes.

3.3.3
Hexapentaenylidene Complex Synthesis

Until now, hexapentaenylidene complexes are unknown, however, a hexapentaenylidene complex has been proposed as an intermediate in the formation of a C_9-bridged diruthenium complex (Scheme 3.20) [35].

Attempts to generate a hexapentaenylidene complex by protonation of [Cp*(NO)(PPh$_3$)Re-C≡C-C≡C-C≡C-C$_6$H$_4$Me-4] have failed. Protonation occurs at C2 to give the corresponding cationic vinylidene complex [36].

3.3.4
Heptahexaenylidene Complex Synthesis

An example of the next higher homolog in the series of odd-numbered cumulenylidene complexes, heptahexaenylidene complexes, was already proposed in 1996 as an intermediate in the formation of the alkenyl(dimethylamino)pentatetraenylidene tungsten complex **27** (Scheme 3.21) in the sequential reaction of triyne Me$_3$Si-(C≡C)$_3$-C(NMe$_2$)$_3$ with LiMe·LiBr, [(CO)$_5$W(thf)], and BF$_3$·OEt$_2$ [37]. The putative heptahexaenylidene complex **26** could neither be isolated nor spectroscopically

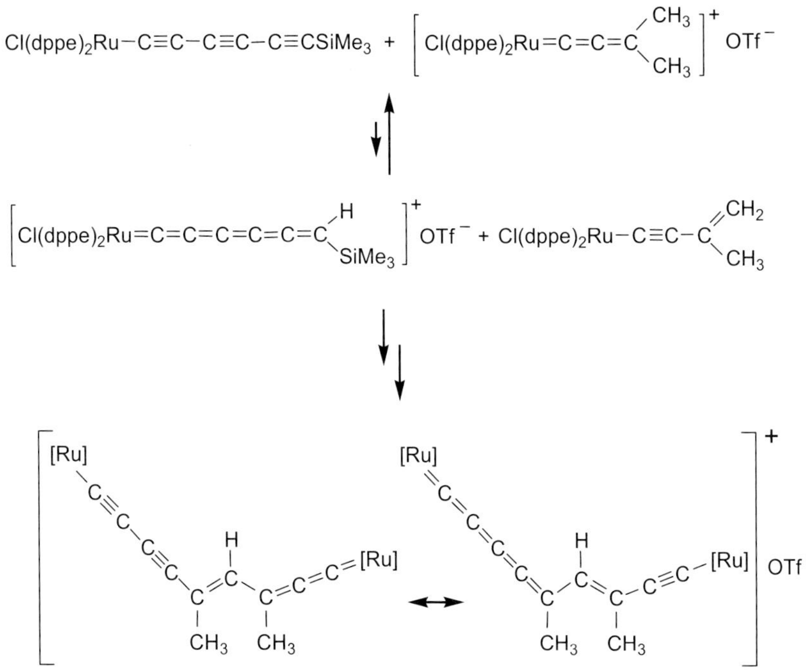

Scheme 3.20 Postulated hexapentaenylidene complex as intermediate in the synthesis of a C_9-bridged diruthenium complex.

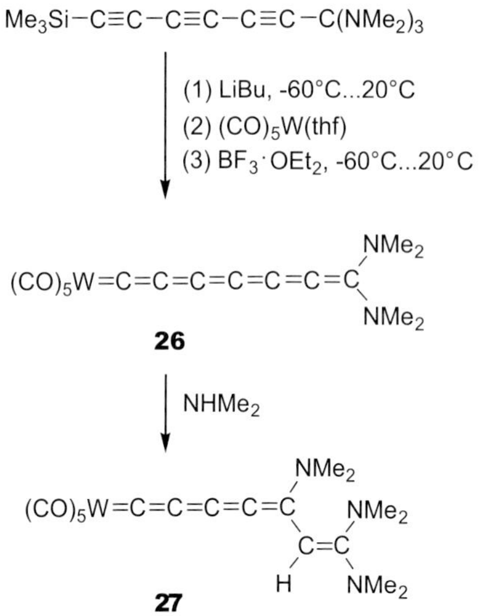

$Me_3Si-C\equiv C-C\equiv C-C\equiv C-C(NMe_2)_3$

(1) LiBu, -60°C...20°C
(2) $(CO)_5W(thf)$
(3) $BF_3 \cdot OEt_2$, -60°C...20°C

$(CO)_5W=C=C=C=C=C=C=C(NMe_2)_2$

26

$\downarrow$ $NHMe_2$

$(CO)_5W=C=C=C=C=C(NMe_2)-C(H)=C(NMe_2)_2$

27

Scheme 3.21 Synthesis of the first heptahexaenylidene complex and its reaction with dimethylamine.

detected. Once formed, it immediately added $HNMe_2$, generated in the course of the reactions as a by-product, and complex **27** was finally isolated.

However, by very careful control of the reaction conditions and the purity of the triyne, using lower temperatures and substituting SiO_2 for $BF_3 \cdot OEt_2$ it was recently possible to synthesize, isolate and characterize the first heptahexaenylidene tungsten complex **26** and its chromium analog [38].

3.4
Reactions of Higher Metal Cumulenes

DFT calculations indicate that the LUMO in d^6 and d^8 complexes is predominantly localized on the odd carbon atoms independent of the chain length, the substituents R and the metal–ligand fragment L_nM. Therefore, nucleophiles are expected to add to the odd carbon atoms of the chain in butatrienylidene as well as in pentatetraenylidene complexes.

3.4.1
Butatrienylidene Complexes

These expectations are confirmed by the addition of PPh_3 or CF_3COO^- to C3 of *in situ* generated $[Cp(PPh_3)_2Ru=(C)_4R_2]^+$ (R = H [19], Me [15, 16]), of CF_3COO^- to C3 of isolable *trans*-$[Cl(PPr^i_3)_2Ir=(C)_4R_2]$ [2, 39, 40], or of tertiary amines to C3 of $[Cl(dppm)_2Ru=(C)_4H_2]^+$ [21, 22] to form the corresponding adducts. Usually addition is followed by various rearrangements. Secondary amines, methanol, thiols, and phenylselenol typically add across the C3=C4 bond to give amino- [19, 20, 25, 41],

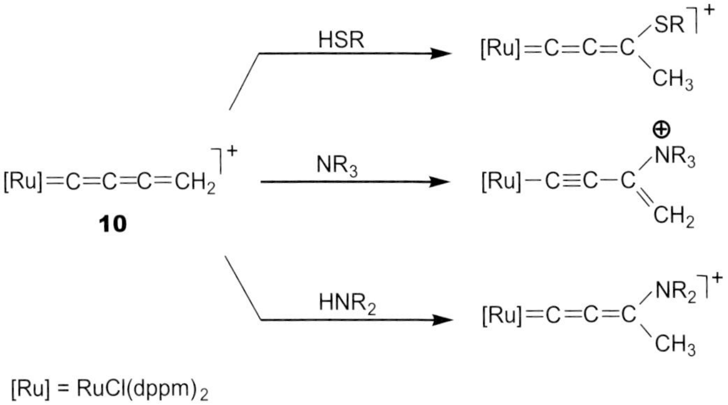

3.4 Reactions of Higher Metal Cumulenes

$$[Ru]=C=C=C=CH_2]^+ \quad \xrightarrow{HSR} \quad [Ru]=C=C=C{\overset{SR}{\underset{CH_3}{\big<}}}\Big]^+$$

$$\mathbf{10} \quad \xrightarrow{NR_3} \quad [Ru]-C\equiv C-C{\overset{\overset{\oplus}{NR_3}}{\underset{CH_2}{\big<}}}$$

$$\xrightarrow{HNR_2} \quad [Ru]=C=C=C{\overset{NR_2}{\underset{CH_3}{\big<}}}\Big]^+$$

[Ru] = RuCl(dppm)$_2$

Scheme 3.22 Reactivity of [Cl(dppm)$_2$Ru=(C)$_4$H$_2$]$^+$ towards various nucleophiles.

alkoxy- [17, 24], thio- [42], and phenylselenoallenylidene complexes [1c]. Some examples of the reactions of **10** with nucleophiles are shown in Scheme 3.22.

Allyl amines, allyl thioethers, and allylferrocenylselenide react analogously with butatrienylidene complex **10** by initial addition to C3 and subsequent hetero-Cope or hetero-Claisen rearrangement (Scheme 3.23) [21, 42–45].

The related reaction of **10** with ferrocenylmethylamine affords, in addition to the C3 adduct (as the minor product), a 2-ferrocenylethyl(dimethylamino)allenylidene complex as the major product, formed by migration of the resonance-stabilized [FcCH$_2$]$^+$ carbenium ion to the terminal carbon atom of the chain (Scheme 3.24) [46].

The formation of pyrrolyl- and indolyl-substituted allenylidene complexes by reaction of complex **10** with various pyrroles and *N*-methylindole [47] has also been rationalized as involving initial attack of the electron-rich heterocycle on C3 of **10** followed by proton migration to the terminal =CH$_2$ entity of the intermediate butenynyl-substituted σ-complex (Scheme 3.25).

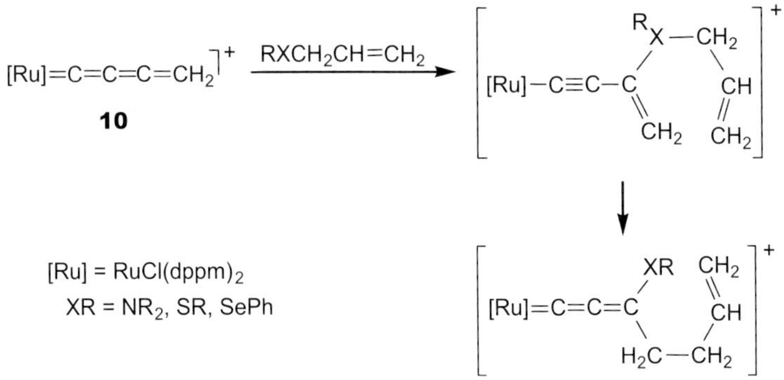

$$[Ru]=C=C=C=CH_2]^+ \quad \xrightarrow{RXCH_2CH=CH_2} \quad \left[[Ru]-C\equiv C-C{\overset{\overset{R}{\underset{CH_2}{X-CH_2}}}{\underset{CH_2}{\big<}}}{\overset{CH}{\underset{CH_2}{\big\|}}} \right]^+$$

10

$\downarrow$

[Ru] = RuCl(dppm)$_2$
XR = NR$_2$, SR, SePh

$$\left[[Ru]=C=C=C{\overset{XR}{\underset{H_2C-CH_2}{\big<}}}{\overset{CH_2}{\underset{CH}{\big\|}}} \right]^+$$

Scheme 3.23 Hetero-Cope (hetero-Claisen) rearrangement in vinylalkynyl ruthenium complexes formed by addition of heteroallyl compounds to C3 of complex **10**.

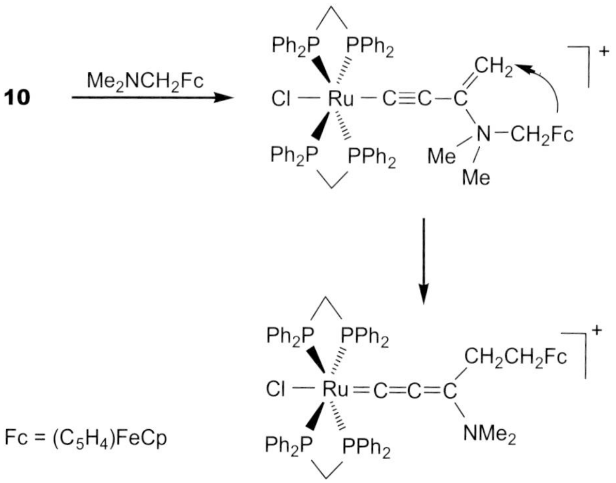

Scheme 3.24 Ferrocenylethyl(dimethylamino)allenylidene ruthenium by nucleophilic addition of ferrocenylmethylamine to C3 of **10** and subsequent rearrangement.

The reactions of various butatrienylidene complexes with water [14, 19, 20, 48] giving acylethynyl complexes very likely proceed by initial attack of water (or OH$^-$) at C3 of the butatrienylidene ligand.

The coupling of cationic allenylidene complexes with a neutral butadiynyl ruthenium complex to form bimetallic ruthenium systems with a W-shaped C_7 conjugated bridge was also suggested to proceed via a butatrienylidene complex as an intermediate [35, 49, 50]. In the reaction sequence a nucleophilic attack of deprotonated allenylidene complexes at C3 of the butatrienylidene complex (generated *in situ* by protonation at C4 of the butadiynyl complex) was proposed to be one of the key reaction steps (for the closely related coupling of an allenylidene complex with a hexatriynyl complex to give the corresponding binuclear ruthenium complex containing a W-shaped C_9 bridge see Scheme 3.20). Analogous intermediates were also proposed to account for the formation of a similar C_7 bridged diruthenium complex as a minor product in the reaction of [Cp*(dppe)Ru-Cl] with HC≡C-C≡CSiMe$_3$ and NH$_4$PF$_6$ in methanol [51].

The reaction of aromatic imines with [Cp(PPh$_3$)$_2$Ru=C=C=C=CH$_2$]$^+$ gave two types of products, either 1-azabuta-1,3-diene-2-ethynyl complexes or 4-ethynylquinoline complexes [52, 53]. The azabutadiene-2-ethynyl complexes were thought to be

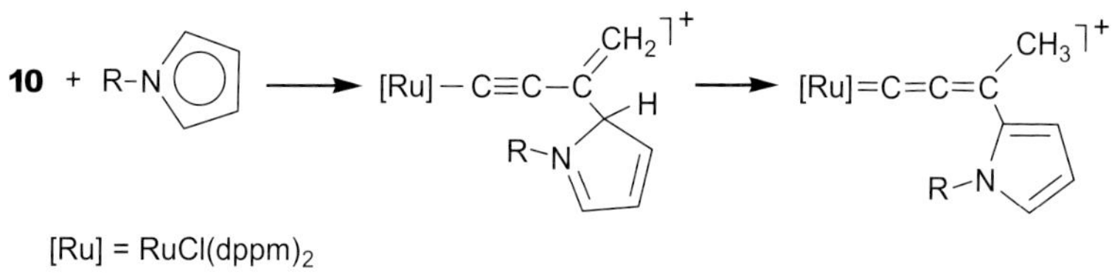

Scheme 3.25 Pyrrolyl-substituted allenylidene complexes by reaction of **10** with pyrroles.

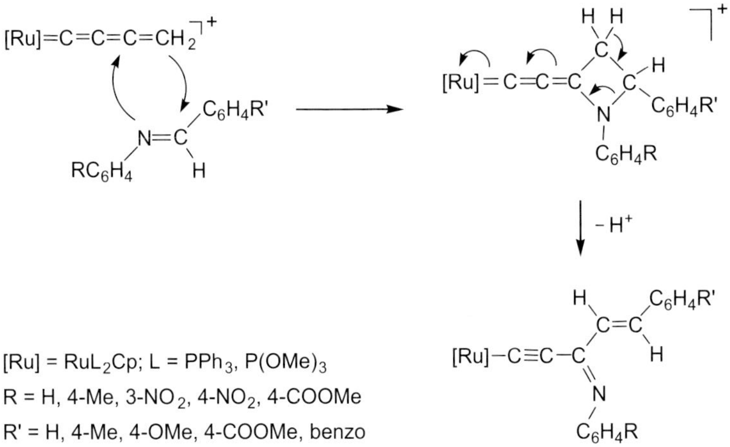

Scheme 3.26 Formation of azabutadiene-2-ethynyl complexes in the reaction of butatrienylidene complexes with imines.

[Ru] = RuL$_2$Cp; L = PPh$_3$, P(OMe)$_3$
R = H, 4-Me, 3-NO$_2$, 4-NO$_2$, 4-COOMe
R' = H, 4-Me, 4-OMe, 4-COOMe, benzo

formed by cycloaddition of the N=CH group to C3=C4 of the butatrienylidene ligand, followed by deprotonation and opening of the resulting four-membered ring (Scheme 3.26).

For the formation of the 4-ethynylquinoline complexes a mechanism was proposed involving nucleophilic attack of the terminal carbon of the butatrienylidene ligand at the imine carbon, followed by C–C bond formation between the *ortho* carbon of the N-aryl group and C3 of the butatrienylidene ligand. Deprotonation finally affords 4-ethynylquinoline complexes (Scheme 3.27). Some preference was observed for quinoline formation with the more electron-rich metal centers, whereas

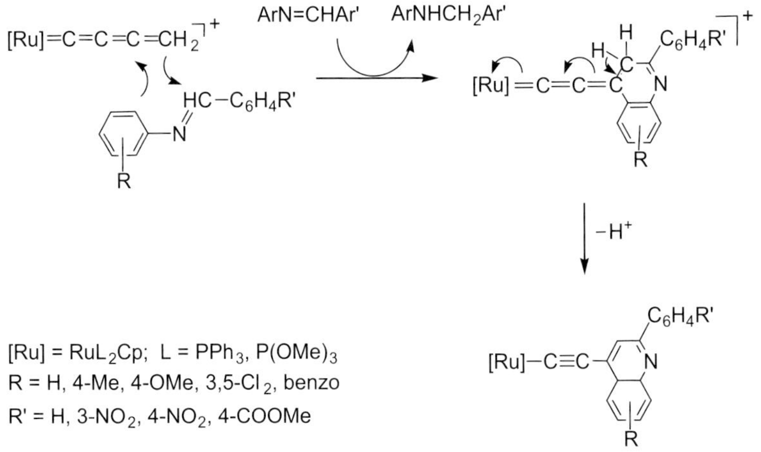

[Ru] = RuL$_2$Cp; L = PPh$_3$, P(OMe)$_3$
R = H, 4-Me, 4-OMe, 3,5-Cl$_2$, benzo
R' = H, 3-NO$_2$, 4-NO$_2$, 4-COOMe

Scheme 3.27 Formation of ethynylquinoline complexes in the reaction of butatrienylidene complexes with imines.

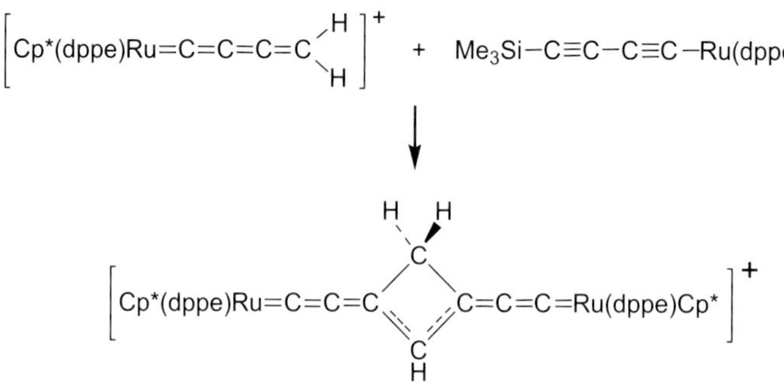

$$[Cp^*(dppe)Ru=C=C=C=C{\overset{H}{\underset{H}{\diagdown}}}]^+ + Me_3Si-C{\equiv}C-C{\equiv}C-Ru(dppe)Cp^*$$

$$\downarrow$$

$$\left[Cp^*(dppe)Ru=C=C{\diagup}{\overset{H\;\;H}{\underset{\underset{H}{C}}{C}}}{\diagdown}C=C=C=Ru(dppe)Cp^*\right]^+$$

Scheme 3.28 Formation of a C7-bridged diruthenium complex with a central cyclobutenylidene unit.

azabutadienylethynyl complexes were formed with $(4\text{-}RH_4C_6)HC=N(C_6H_4R\text{-}4)$ (R = Me, OMe) [53].

Cycloaddition of the terminal C≡C bond in $[Cp^*(dppe)Ru\text{-}C{\equiv}CC{\equiv}C\text{-}SiMe_3]$ to the C3=C4 bond in cationic $[Cp^*(dppe)Ru=C=C=C=CH_2]^+$ gave a diruthenium-complex containing a C7-bridge with a central cyclobutenylidene unit (Scheme 3.28) [51]. The complex was characterized by an X-ray structure analysis. A similar bis-[Ru(dppe)$_2$Cl] complex was already obtained earlier by oxidation of the diynyl complex $[RuCl_2(dppe)_2]$ with $[FeCp_2]PF_6$ [35, 50, 54].

The reactivity of d^6 butatrienylidene complexes towards simple electrophiles is rather unexplored. The manganese complexes $[(C_5H_4R)(P\text{-}P)Mn=C=C=C=C(SnPh_3)_2]$ (R = H, Me; P-P = dmpe, depe) were found to react with $[NBu_4]F/H_2O$ to give the unsubstituted butatrienylidene complexes $[(C_5H_4R)(P\text{-}P)Mn=C=C=C=CH_2]$ (see Scheme 3.9) [4, 5].

The reactivity of neutral square-planar d^8 butatrienylidene complex **11** (Scheme 3.8) strongly deviates from that of cationic d^6 ruthenium complexes. The deviation is readily understood when considering the orbital contributions of the metal and the carbon atoms of the chain to the LUMO. In d^6 and d^8 complexes the LUMO is predominantly localized at the metal, at C1 and C3. However, the relative contribution of the metal in d^6 and d^8 complexes is significantly different. In d^6 complexes the metal contributes considerably less than C1 and C3, in d^8 complexes its contribution is approximately equal to that of C1 and C3.

Whereas d^6 ruthenium complexes add protic nucleophiles to the C3=C4 of the butatrienylidene ligand to give alkenylallenylidene complexes (see Scheme 3.22), iridium complex **11** adds trifluoroacetic acid to the C2=C3 bond to form an alkenylvinylidene complex [2, 3]. The corresponding reaction of **11** with two equivalents of HCl yields, presumably again via an alkenylvinylidene complex, a five-coordinate alkenyl(dichloro) complex with a trigonal-bipyramidal coordination geometry (Scheme 3.29) [3].

Anionic nucleophiles such as I^-, OH^-, N_3^-, or CH_3^- did not add to C3 of the butatrienylidene ligand of **11** but rather replaced the *trans* chloro ligand. When the *trans* hydroxo butatrienylidene **28** was treated with PhOH, instead of an allenylidene complex (3,4 addition product) or an alkenylvinylidene complex (2,3 addition

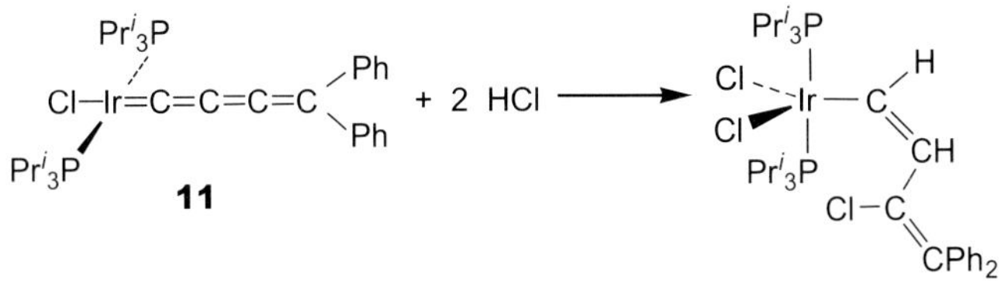

Scheme 3.29 Reaction of the butatrienylidene iridium complex **11** with HCl.

product), the *trans* phenoxybutatrienylidene complex **29** by substitution of PhO for Cl was produced. In contrast, from the corresponding reaction of **28** with MeOH, neither an addition product nor a *trans* methoxy-substituted complex was obtained. Instead, a butatrienyl(dihydrido) complex (**30**) was formed (Scheme 3.30) [3].

The reactions of butatrienylidene iridium complexes with CO depend strongly on the *trans* ligand. The *trans* chloro complex **11** reversibly adds CO to form a five-coordinate butatrienylidene complex **31** whereas the *trans* azido complex yields with CO an alkenyl(azido)ethynyl complex (**32**) and the *trans* methyl complex the alkynylalkenyl complex **33** (Scheme 3.31) [3].

3.4.2
Pentatetraenylidene Complexes

The LUMO in d^6 pentatetraenylidene complexes is predominantly localized on the odd carbon atoms and to a lesser extent on the metal. The coefficients on C1 and C3 are very similar, independent of the metal–ligand fragment and the terminal substituent. The coefficient at C5 is somewhat larger. In square-planar d^8 rhodium and iridium complexes the coefficient at the metal is comparable to that on C5 and is larger than those on C1 and C3. Thus, a nucleophilic attack at the metal of d^8 complexes has also to be taken into account.

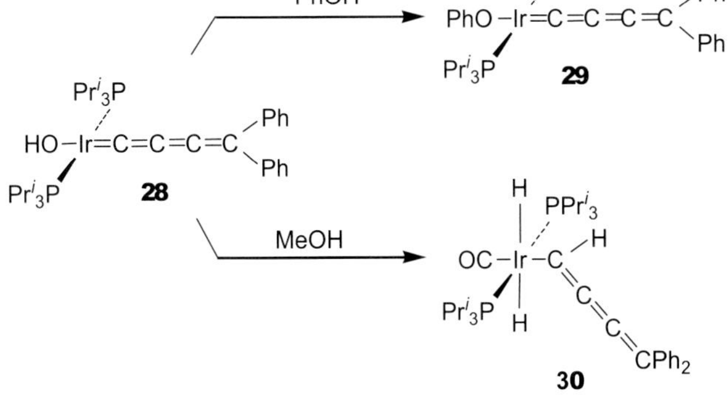

Scheme 3.30 Differing reactivity of butatrienylidene iridium complex **28** with methanol and phenol.

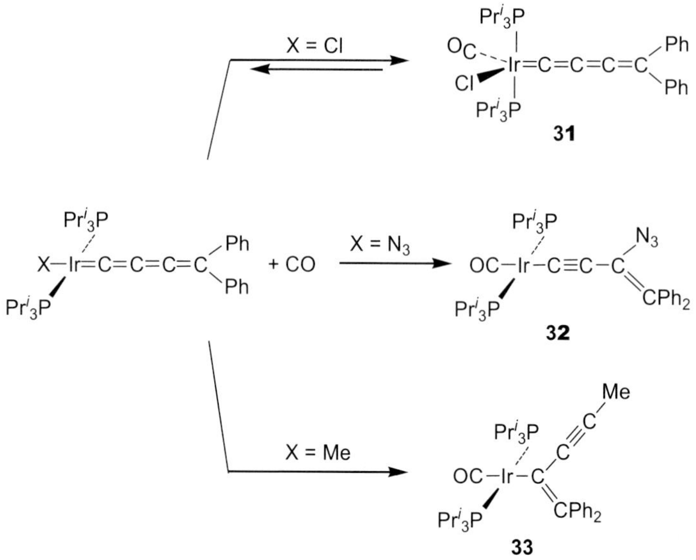

Scheme 3.31 Dependence on the *trans* ligand of the reactions of butatrienylidene iridium complexes with CO.

Experimental information on the reactivity of pentatetraenylidene complexes is still rather rare. The isolated pentatetraenylidene complexes [Cl(dppe)$_2$Ru=C=C=C=C=CPh$_2$]$^+$ (**20**) [7] and [(CO)$_5$M=C=C=C=C=C(NMe$_2$)$_2$] (M = Cr, W) (**22**) [8] were found to react with secondary amines by addition of the amine across the C3=C4 bond, very likely via an initial nucleophilic attack at C3, to give alkenyl(amino)allenylidene complexes (Scheme 3.32). Analogously, methanol

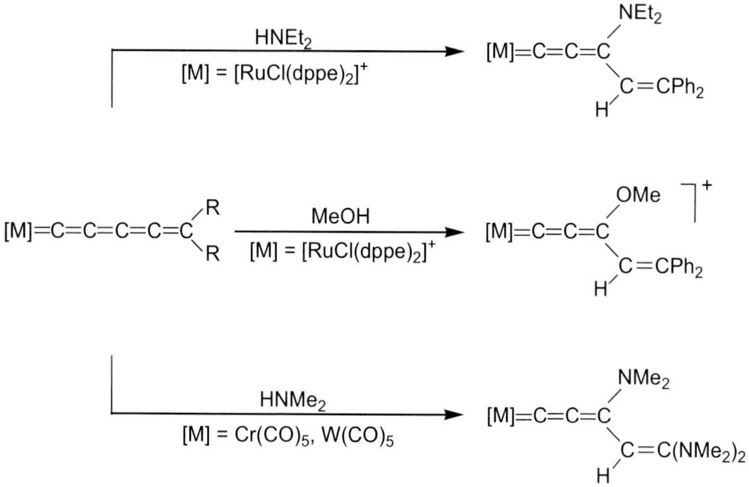

Scheme 3.32 Addition of amines and methanol to the C3–C4 bond of pentatetraenylidene complexes.

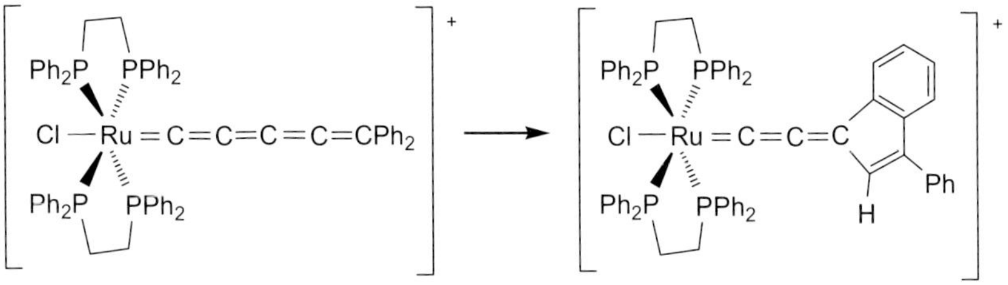

$$[\text{Ph}_2\text{P}\,\text{PPh}_2,\ \text{Cl}-\text{Ru}=\text{C}=\text{C}=\text{C}=\text{C}=\text{CPh}_2,\ \text{Ph}_2\text{P}\,\text{PPh}_2]^+ \longrightarrow [\text{Ph}_2\text{P}\,\text{PPh}_2,\ \text{Cl}-\text{Ru}=\text{C}=\text{C}=\text{C}(\text{indenyl-Ph,H}),\ \text{Ph}_2\text{P}\,\text{PPh}_2]^+$$

20

Scheme 3.33 Formation of an allenylidene complex by the intramolecular addition of an *ortho* C−H bond of a CPh$_2$ phenyl group at C3 of the chain in **20**.

adds across the C3=C4 bond of **21** [7] whereas both pentacarbonyl complexes **21** as well as $[(\text{C}_6\text{Me}_6)\text{Cl}(\text{PMe}_3)\text{Ru}=\text{C}=\text{C}=\text{C}=\text{C}=\text{C}(\text{C}_6\text{H}_4\text{NMe}_2\text{-}4)_2]^+$ (**19**) are inert towards alcohols [8, 32].

An unusual intramolecular addition of one *ortho* C-H bond of a phenyl substituent across the C3=C4 bond in **20** (Scheme 3.33) is observed on thermolysis of **20** in CHCl$_3$ [7]. A similar complex is also formed when the corresponding dppm complex is generated *in situ* from [Cl(dppm)$_2$Ru-C≡C-C≡C-C(Ph)$_2$OSiMe$_3$] and HBF$_4$ in CH$_2$Cl$_2$ [30, 31]. This transformation is actually the parent of the later commonly observed allenylidene to indenylidene intramolecular rearrangement.

The site of hydride addition (from NaBH$_4$) and thus reduction of cationic **20** affording an allenylethynyl complex is also C3 of the chain [55]. In contrast, reduction with one equivalent of cobaltocene, Cp$_2$Co, and subsequent trapping of the resulting radical with Ph$_3$SnH, which is known as a specific radical quencher by H transfer, yields the corresponding 5,5-diphenylpenta-1,3-diynyl complex [55]. Its formation suggests that the unpaired electron is predominantly localized on the C5 atom.

In contrast, soft carbon nucleophiles attack at C5. The reaction of **23** with diethylaminopropyne yields alkenyl(amino)pentatetraenylidene complexes (**34**) by insertion of the C≡C bond of the alkyne into the C4=C5 bond of the pentatetraenylidene ligand [9]. The reaction is initiated by a nucleophilic attack of the ynamine at C5 followed by ring closure and electrocyclic ring opening (Scheme 3.34). Complexes **34** are obtained as mixtures of *s-cis/s-trans* isomers.

Other C-nucleophiles such as Me$^-$, Ph$^-$, [C≡CPh]$^-$, likewise add to C5 of **23** forming diynyl metallates. Subsequent SiO$_2$-induced elimination of [NMe$_2$]$^-$ gives new dimethylamino(organyl)pentatetraenylidene complexes (Scheme 3.35) [56]. This substitution route affords a variety of pentatetraenylidene complexes so far not accessible by other routes.

The attack of a nucleophile at C1 of the chain in an *isolated* pentatetraenylidene complex has been observed only once. The reaction of CH$_2$N$_2$ with *trans*-[Cl(PPr^i$_3$)$_2$Rh=C=C=C=C=CPh$_2$] proceeded by addition of "CH$_2$" to C1, N$_2$ elimination and rearrangement of the thus-formed diphenylhexapentaene ligand to form a C2,C3-bonded 6,6-diphenylhexapentaene complex (Scheme 3.36) [33].

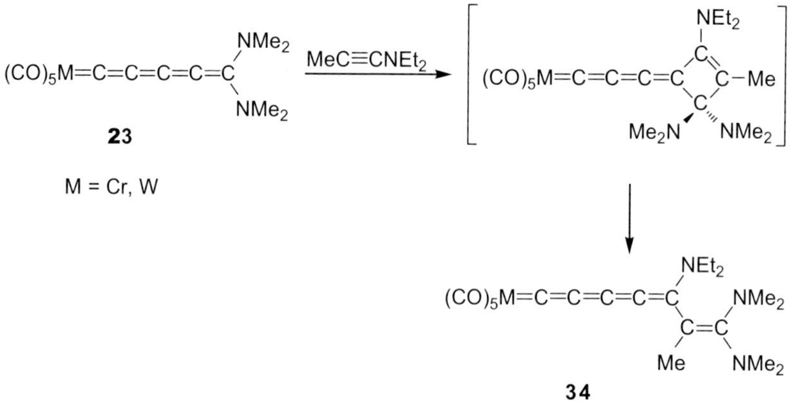

Scheme 3.34 Insertion of diethylaminopropyne into the C4–C5 bond of pentatetraenylidene complexes **23**.

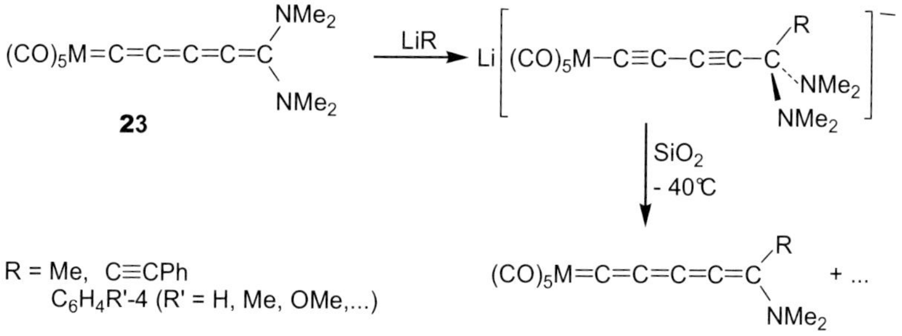

R = Me, C≡CPh
C$_6$H$_4$R'-4 (R' = H, Me, OMe,...)

Scheme 3.35 Addition of C-nucleophiles to the terminal carbon atom of the chain in **23** and subsequent SiO$_2$-induced amide elimination.

The addition of a nucleophile to C1 as well as to C3 was observed in the reaction of the pentatetraenylidene complex [(C$_6$Me$_6$)Cl(PMe$_3$)Ru=C=C=C=CPh$_2$]$^+$ generated *in situ* with MeOH. A mixture of two complexes derived from addition of methanol across the C1=C2 and the C3=C4 bond was obtained (Scheme 3.37) [29]. The 1,2 adduct, a butatrienyl(methoxy)carbene complex, rapidly reacted with water affording

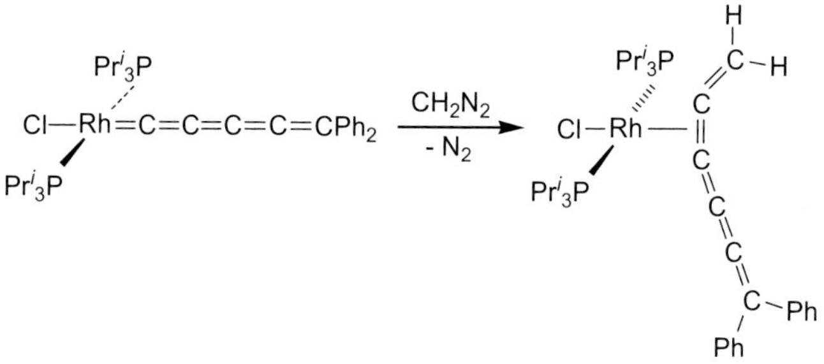

Scheme 3.36 Formation of a hexapentaene complex in the reaction of diazomethane with a pentatetraenylidene rhodium complex.

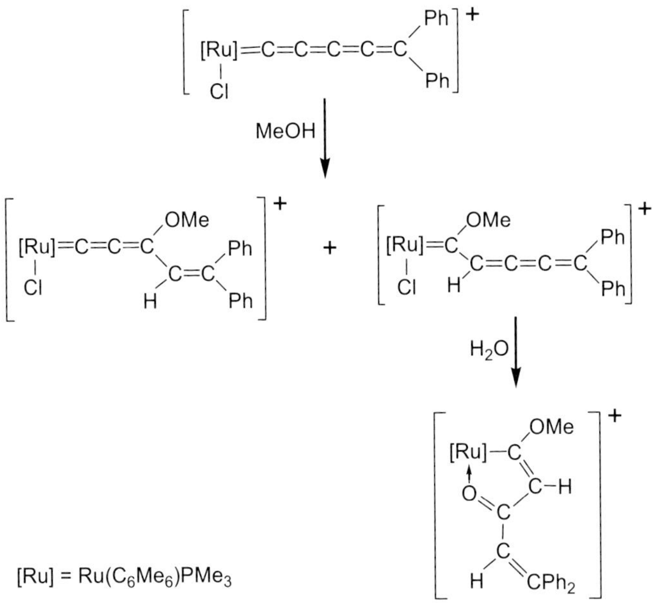

Scheme 3.37 Reaction of $[Cl(C_6Me_6)(PMe_3)Ru=(C)_5Ph_2]^+$ with methanol/water.

a chelated 3-oxapentadienyl ruthenium complex. Very likely, although not detected, pentatetraenylidene and butatrienyl(methoxy)carbene complexes are also intermediates in the reaction of $[RuCl_2(PR_3)(C_6Me_6)]$ ($PR_3 = PMe_3$, PMe_2Ph) with $HC\equiv CC\equiv C\text{-}CR_2OH$ (R = Me, Ph) in MeOH in the presence of $NaPF_6$ to form analogous chelated 3-oxapentadienyl ruthenium complexes [26].

However, alcohols or secondary amines usually add to the C3=C4 bond of pentatetraenylidene complexes of ruthenium generated *in situ* and the finally isolated products are the corresponding alkenylallenylidene complexes.

3.4.3
Hexapentaenylidene Complexes

Since hexapentaenylidene complexes have until now neither been isolated nor spectroscopically detected nothing is known about their reactivity. However, a C-nucleophile was suggested to add to C5 of a hexapentaenylidene complex generated *in situ* by protonation of a triynyl ruthenium complex (see Scheme 3.20) [35].

3.4.4
Heptahexaenylidene Complexes

The reactivity of heptahexaenylidene complexes is reminiscent of that of the corresponding pentacarbonyl pentatetraenylidene complexes. Like **23**, complex **26**

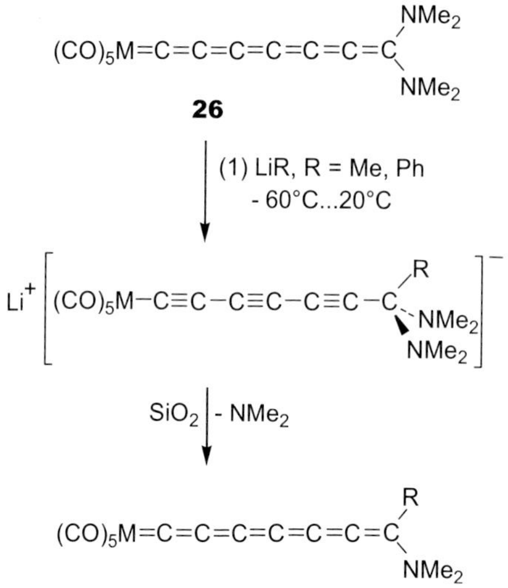

$$(CO)_5M=C=C=C=C=C=C=C\begin{matrix}NMe_2\\NMe_2\end{matrix}$$

26

↓ (1) LiR, R = Me, Ph
    −60 °C...20 °C

$$Li^+\left[(CO)_5M-C\equiv C-C\equiv C-C\equiv C-C\begin{matrix}R\\NMe_2\\NMe_2\end{matrix}\right]^-$$

↓ SiO$_2$ | − NMe$_2$

$$(CO)_5M=C=C=C=C=C=C=C\begin{matrix}R\\NMe_2\end{matrix}$$

Scheme 3.38 Addition of C-nucleophiles to the terminal carbon atom of the chain in complexes **26** and SiO$_2$-induced amide elimination.

is inert toward methanol, but adds dimethylamine across the second-last C=C bond of the chain to alkenyl(amino)pentatetraenylidene complexes (see Scheme 3.21). Carbon-nucleophiles add to the last carbon atom of the chain (C7) to form hexatriynyl complexes. Subsequent SiO$_2$-induced elimination of dimethylamide then affords new heptahexaenylidene complexes (Scheme 3.38) [38].

3.5
Summary and Conclusion

Metallacumulenes containing more than three carbon atoms in the cumulene chain are still rather rare. Until now, only a few syntheses have been developed and only a few complexes have been fully characterized. Most of these complexes have a d^6 electron configuration. The investigation of their reactivity towards nucleophiles has revealed some trends. Nucleophiles usually add to the third carbon atom of the chain (fifth carbon atom in heptahexaenylidene complexes) although there are a few exceptions known. For instance, C-nucleophiles will add to the last atom of the chain in strongly dipolar metallacumulenes, [(CO)$_5$M=(C)$_x$(NMe$_2$)$_2$] (M = Cr, W; x = 3, 5, 7). The reactivity of d^8 metallacumulenes deviates from that of the d^6 complexes.

A comparison of the stability and reactivity of analogous metallacumulenes differing by the length of the carbon chain reveals that on lengthening the chain the thermal stability decreases only slightly whereas the electrophilicity increases strongly. Thus, complexes with an even longer chain of carbon atoms should be isolable species provided suitable syntheses can be developed and the problems connected with the increasing electrophilicity can be overcome. Thus, there are no obvious reasons for the C5 diyne and homologous C7 triyne substrates not to be involved in catalytic transformations via metallacumulene activated species.

References

1 Some reviews on metallacumulenes contain also sections on the chemistry of butatrienylidene and pentatetraenylidene complexes: (a) Le Bozec, H. and Dixneuf, P.H. (1995) *Russian Chemical Bulletin*, **44**, 801–812. (b) Bruce, M.I. (1998) *Chemical Reviews*, **98**, 2797–2858. (c) Cadierno, V., Gamasa, M.P. and Gimeno, J. (2001) *European Journal of Inorganic Chemistry*, 571–591. (d) Winter, R.F. and Záliš, S. (2004) *Coordination Chemistry Reviews*, **248**, 1565–1583. (e) Bruce, M.I. (2004) *Coordination Chemistry Reviews*, **248**, 1603–1625. (f) Fischer, H. and Szesni, N. (2004) *Coordination Chemistry Reviews*, **248**, 1659–1677.
2 Ilg, K. and Werner, H. (2000) *Angewandte Chemie*, **112**, 1691–1693; (2000) *Angewandte Chemie-International Edition*, **39**, 1632–1634.
3 Ilg, K. and Werner, H. (2002) *Chemistry – A European Journal*, **8**, 2812–2820.
4 Venkatesan, K., Fernández, F.J., Blacque, O., Fox, T., Alfonso, M., Schmalle, H.W. and Berke, H. (2003) *Chemical Communications*, 2006–2008.
5 Venkatesan, K., Blacque, O., Fox, T., Alfonso, M., Schmalle, H.W. and Berke, H. (2004) *Organometallics*, **23**, 4661–4671.
6 Lass, R.W., Steinert, P., Wolf, J. and Werner, H. (1996) *Chemistry – A European Journal*, **2**, 19–23.
7 Touchard, D., Haquette, P., Daridor, A., Toupet, L. and Dixneuf, P.H. (1994) *Journal of the American Chemical Society*, **116**, 11157–11158.
8 Roth, G. and Fischer, H. (1996) *Organometallics*, **15**, 1139–1145.
9 Roth, G., Fischer, H., Meyer-Friedrichsen, T., Heck, J., Houbrechts, S. and Persoons, A. (1998) *Organometallics*, **17**, 1511–1516.
10 Re, N., Sgamellotti, A. and Floriani, C. (2000) *Organometallics*, **19**, 1115–1122.
11 Marrone, A. and Re, N. (2002) *Organometallics*, **21**, 3562–3571.
12 Auger, N., Touchard, D., Rigaut, S., Halet, J.-F. and Saillard, J.-Y. (2003) *Organometallics*, **22**, 1638–1644.
13 Marrone, A., Coletti, C. and Re, N. (2004) *Organometallics*, **23**, 4952–4963.
14 Bruce, M.I., Hinterding, P., Tiekink, E.R.T., Skelton, B.W. and White, A.H. (1993) *Journal of Organometallic Chemistry*, **450**, 209–218.
15 Lomprey, J.R. and Selegue, J.P. (1993) *Organometallics*, **12**, 616–617.
16 Selegue, J.P. (2004) *Coordination Chemistry Reviews*, **248**, 1543–1563.
17 Haquette, P., Touchard, D., Toupet, L. and Dixneuf, P. (1998) *Journal of Organometallic Chemistry*, **565**, 63–73.
18 Coat, F., Guillemot, M., Paul, F. and Lapinte, C. (1999) *Journal of Organometallic Chemistry*, **578**, 76–84.
19 Bruce, M.I., Hinterding, P., Low, P.J., Skelton, B.W. and White, A.H. (1996) *Chemical Communications*, 1009–1110.
20 Bruce, M.I., Hinterding, P., Low, P.J., Skelton, B.W. and White, A.H. (1998) *Journal of the Chemical Society-Dalton Transactions*, 467–473.
21 Winter, R.F. and Hornung, F.M. (1997) *Organometallics*, **16**, 4248–4250.
22 Winter, R.F. and Hornung, F.M. (1999) *Organometallics*, **18**, 4005–4014.
23 Guillaume, V., Thominot, P., Coat, F., Mari, A. and Lapinte, C. (1998) *Journal of Organometallic Chemistry*, **565**, 75–80.

24 Coat, F., Thominot, P. and Lapinte, C. (2001) *Journal of Organometallic Chemistry*, **629**, 39–43.

25 Romero, A., Peron, D. and Dixneuf, P.H. (1990) *Journal of the Chemical Society: Chemical Communications*, 1410–1412.

26 Romero, A., Vegas, A. and Dixneuf, P.H. (1990) *Angewandte Chemie*, **102**, 210–211; (1990) *Angewandte Chemie-International Edition in English*, **29**, 215.

27 Wolinska, A., Touchard, D., Dixneuf, P.H. and Romero, A. (1991) *Journal of Organometallic Chemistry*, **420**, 217–226.

28 Pirio, N., Touchard, D., Dixneuf, P.H., Fettouhi, M. and Ouahab, L. (1992) *Angewandte Chemie*, **104**, 664–666; (1992) *Angewandte Chemie-International Edition in English*, **31**, 651–653.

29 Peron, D., Romero, A. and Dixneuf, P.H. (1995) *Organometallics*, **14**, 3319–3326.

30 Touchard, D., Pirio, N., Toupet, L., Fettouhi, M., Ouahab, L. and Dixneuf, P.H. (1995) *Organometallics*, **14**, 5263–5272.

31 Pirio, N., Touchard, D., Toupet, L. and Dixneuf, P.H. (1991) *Journal of the Chemical Society: Chemical Communications*, 980–982.

32 Peron, D., Romero, A. and Dixneuf, P.H. (1994) *Gazzetta Chimica Italiana*, **124**, 497–502.

33 Kovacik, I., Laubender, M. and Werner, H. (1997) *Organometallics*, **16**, 5607–5609.

34 Szafert, S., Haquette, P., Falloon, S.B. and Gladysz, J.A. (2000) *Journal of Organometallic Chemistry*, **604**, 52–58.

35 Rigaut, S., Olivier, C., Costuas, K., Choua, S., Fadhel, O., Massue, J., Turek, P., Saillard, J.-Y., Dixneuf, P.H. and Touchard, D. (2006) *Journal of the American Chemical Society*, **128**, 5859–5876.

36 Dembinski, R., Lis, T., Szafert, S., Mayne, C.L., Bartik, T. and Gladysz, J.A. (1999) *Journal of Organometallic Chemistry*, **578**, 229.

37 Roth, G. and Fischer, H. (1996) *Organometallics*, **15**, 5766–5768.

38 Dede, M., Drexler, M. and Fischer, H. (2007) *Organometallics*, **26**, 4294–4299.

39 Werner, H., Ilg, K., Lass, R. and Wolf, J. (2002) *Journal of Organometallic Chemistry*, **661**, 137–147.

40 Ilg, K. and Werner, H. (2002) *Chemistry – A European Journal*, **8**, 2812–2820.

41 Winter, R.F., Hartmann, S., Záliš, S. and Klinkhammer, K.W. (2003) *Dalton Transactions*, 2342–2352.

42 Winter, R.F. (1999) *European Journal of Inorganic Chemistry*, 2121–2126.

43 Harbort, R.-C., Hartmann, S., Winter, R.F. and Klinkhammer, K.W. (2003) *Organometallics*, **22**, 3171–3174.

44 Winter, R.F., Klinkhammer, K.-W. and Záliš, S. (2001) *Organometallics*, **20**, 1317–1333.

45 Hartmann, S., Winter, R.F., Scheiring, T. and Wanner, M. (2001) *Journal of Organometallic Chemistry*, **637–639**, 240–250.

46 Winter, R.F. (1998) *Chemical Communications*, 2209–2210.

47 Hartmann, S., Winter, R.F., Sarkar, B. and Lissner, F. (2004) *Dalton Transactions*, **20**, 3273–3282.

48 Rigaut, S., Perruchon, J., Le Pichon, L., Touchard, D. and Dixneuf, P.H. (2003) *Journal of Organometallic Chemistry*, **670**, 37–44.

49 Rigaut, S., Massue, J., Touchard, D., Fillaut, J.-L., Golhen, S. and Dixneuf, P.H. (2002) *Angewandte Chemie*, **114**, 4695–4699; (2002) *Angewandte Chemie-International Edition*, **41**, 4513–4517.

50 Rigaut, S., Touchard, D. and Dixneuf, P.H. (2003) *Journal of Organometallic Chemistry*, **684**, 68–76.

51 Bruce, M.I., Ellis, B.G., Skelton, B.W. and White, A.H. (2005) *Journal of Organometallic Chemistry*, **690**, 1772–1783.

52 Bruce, M.I., Hinterding, P., Ke, M., Low, P.J., Skelton, B.W. and White, A.H. (1997) *Journal of the Chemical Society: Chemical Communications*, 715–716.

53 Bruce, M.I., Ke, M., Kelly, B.D., Low, P.J., Smith, M.E., Skelton, B.W. and White,

A.H. (1999) *Journal of Organometallic Chemistry*, **590**, 184–201.
54 Rigaut, S., Le Pichon, L., Daran, J.-C., Touchard, D. and Dixneuf, P.H. (2001) *Chemical Communications*, 1206–1207.
55 Rigaut, S., Maury, O., Touchard, D. and Dixneuf, P.H. (2001) *Chemical Communications*, 373–374.
56 Dede, M. and Fischer, H. unpublished results.

4
Theoretical Aspects of Metal Vinylidene and Allenylidene Complexes
Jun Zhu and Zhenyang Lin

4.1
Introduction

Species (**A**) and (**B**) constitute the main class of "unsaturated carbenes" and play important roles as reactive intermediates due to the very electron-deficient carbon C1 [1]. Once they are coordinated with an electron-rich transition metal, metal vinylidene (**C**) and allenylidene (**D**) complexes are formed (Scheme 4.1). Since the first example of mononuclear vinylidene complexes was reported by King and Saran in 1972 [2] and isolated and structurally characterized by Ibers and Kirchner in 1974 [3], transition metal vinylidene and allenylidene complexes have attracted considerable interest because of their role in carbon–heteroatom and carbon–carbon bond-forming reactions as well as alkene and enyne metathesis [4]. Over the last three decades, many reviews [4–18] have been contributed on various aspects of the chemistry of metal vinylidene and allenylidene complexes. A number of theoretical studies have also been carried out [19–43]. However, a review of the theoretical aspects of the metal vinylidene and allenylidene complexes is very limited [44]. This chapter will cover theoretical aspects of metal vinylidene and allenylidene complexes. The following aspects will be reviewed:

- Electronic structures of metal vinylidene and allenylidene complexes
- Barrier of rotation of vinylidene ligands
- Tautomerization between η^2-acetylene and vinylidene on transition metal centers
- Metal vinylidene mediated reactions
- Heavier group 14 analogs of metal vinylidene complexes
- Allenylidene complexes.

Metal Vinylidenes and Allenylidenes in Catalysis: From Reactivity to Applications in Synthesis
Edited by Christian Bruneau and Pierre Dixneuf
Copyright © 2008 WILEY-VCH Verlag GmbH & Co. KGaA, Weinheim
ISBN: 978-3-527-31892-6

Scheme 4.1

$R_2R_1C=C:$ — **A**

$R_2R_1C=C=C:$ — **B**

$R_2R_1C=C:ML_n$ — **C**

$R_2R_1C=C=C:ML_n$ — **D**

4.2
Electronic Structures of Metal Vinylidene and Allenylidene Complexes

4.2.1
Metal Vinylidene Complexes

It is well known that metal carbenes can be classified as Fisher and Schrock carbenes. The classification is mainly based on the π electron density distribution on the M=C moiety (Scheme 4.2). On the basis of the π electron density distribution, carbene complexes of the Fisher-type (**E**) are normally electrophilic at the carbene carbon while carbene complexes of the Schrock-type (**F**) are nucleophilic at the carbene carbon. Similarly, metal vinylidenes could also be classified into the two types: Fisher-type (**G**) and Schrock-type (**H**). The majority of isolated metal vinylidenes belong to the Fisher-type. On the basis of the π electron density distribution shown in

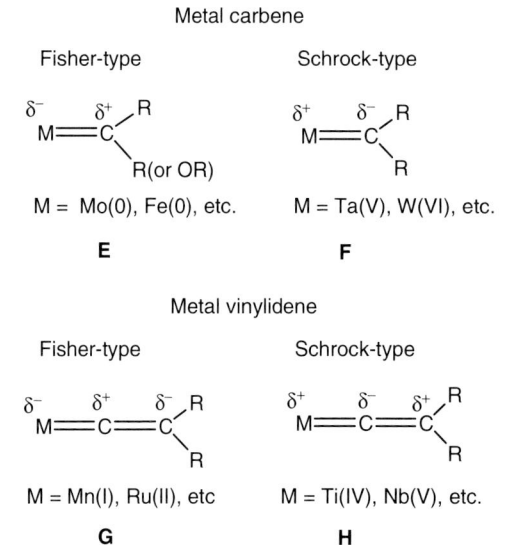

Metal carbene

Fisher-type: $M^{\delta-}=C^{\delta+}(R)(R \text{ or } OR)$, M = Mo(0), Fe(0), etc. — **E**

Schrock-type: $M^{\delta+}=C^{\delta-}(R)(R)$, M = Ta(V), W(VI), etc. — **F**

Metal vinylidene

Fisher-type: $M^{\delta-}=C^{\delta+}=C^{\delta-}(R)(R)$, M = Mn(I), Ru(II), etc. — **G**

Schrock-type: $M^{\delta+}=C^{\delta-}=C^{\delta+}(R)(R)$, M = Ti(IV), Nb(V), etc. — **H**

Scheme 4.2

Scheme 4.2, we can see that for metal vinylidenes of the Fisher-type nucleophilic attack usually occurs at the α-carbon and electrophilic attack at the β-carbon [45–52].

A simplified π-orbital interaction diagram, shown in Figure 4.1, can be used to understand the π electron density distribution given in Scheme 4.2 for complexes of the Fisher-type. In the lowest unoccupied molecular orbital (LUMO) (π4), the α-carbon p_π orbital contributes most, explaining the general observation that nucleophilic attack usually occurs at the α-carbon. The highest occupied molecular orbital (HOMO) (π3) corresponds to an antibonding interaction between the C=C π bonding orbital and the symmetry-adapted metal d_π orbital. The orbital characters shown in the HOMO indicate that electrophilic attack can occur at the β-carbon and/or the metal center [19, 53].

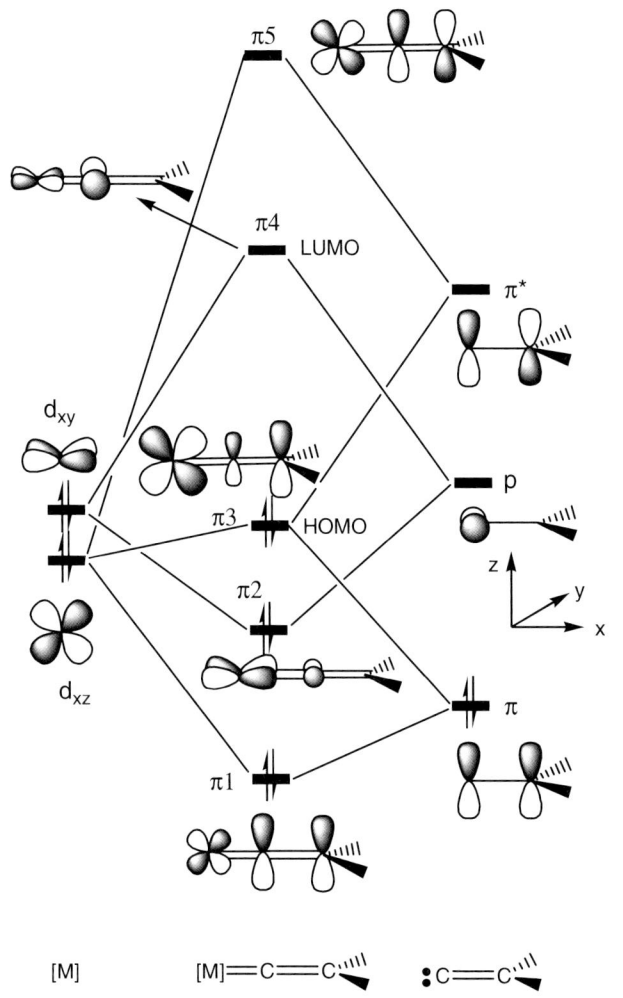

Figure 4.1 Simplified π-orbital interaction diagram for metal vinylidene complexes of the Fisher-type.

Examples showing electrophilic attack at the metal center can be found in the literature. For the square-planar complexes IrCl(PR$_3$)$_2$=C=CHR' [54] and ReCl(dppe)$_2$=C=CHPh [55] electrophiles were found to attack the metal center. An exception has been found for the metal complexes CpRh(PR$_3$)=C=CHR', where electrophilic attack occurs at the α-carbon [56–58]. Delbecq performed extended Hückel theory calculations and provided an explanation for the exception [20]. The strong electron-donating Cp and phosphine ligands in the complexes make the metal center electron-rich, leading to a strong metal(d)-to-vinylidene(p) backdonation. In other words, the contribution of the p orbital on the vinylidene α-carbon in the π2 orbital (Figure 4.1), which represents the backdonation interaction, increases with the electron-richness of the metal center. A strong metal(d)-to-vinylidene(p) backdonation increases the π electron population on the vinylidene α-carbon, leading to the observation that electrophilic attack occurs at the α-carbon.

4.2.2
Metal Allenylidene Complexes

Similarly, a simplified π-orbital interaction for metal allenylidene complexes of the Fisher-type can be constructed, as shown in Figure 4.2. The LUMO is mainly localized at C$_α$ and C$_γ$. As a result, nucleophilic attack favors the C$_α$ and C$_γ$ positions [59–61]. It has been found that the contributions of C$_α$ and C$_γ$ to the LUMO are similar. Therefore, a clear preference for a nucleophilic attack at either C$_α$ or C$_γ$ could not be deduced [59]. On the basis of the orbital characters in the HOMO, we can again deduce that electrophilic attack occurs at the β-carbon and/or the metal center.

4.3
Barrier of Rotation of Vinylidene Ligands

A vinylidene ligand has two substituents. When the two substituents are different, rotational isomers are possible. A recent theoretical study was carried out on the rotational barriers for a variety of 5-coordinate metal vinylidene complexes shown in Scheme 4.3 [42]. The effects of the substituent R, the metal center M and the monovalent ligand X on the rotational barriers were examined. The theoretical study was used to explain experimental observations that for some complexes, such as OsHCl(=C=CHPh)(P^{i}Pr$_3$)$_2$ and Os(C(C≡C^{t}Bu)=CH^{t}Bu)Cl(=C=CH^{t}Bu)(PPh$_3$)$_2$ [62, 63], rotational isomers were detected at −60 °C while for some other complexes, such as RuHCl(=C=CHSiMe$_3$)(P^{t}Bu$_2$Me)$_2$, RuHCl(=C=CHPh)(P^{i}Pr$_3$)$_2$, RuHCl(=C=CHPh)(P^{t}Bu$_2$Me)$_2$ and OsHCl(=C=CHSiMe$_3$)P^{i}Pr$_3$)$_2$, there was no evidence for coexistence of two isomers, even down to −90 °C [64]. It was considered that when the rotational barriers are large, the rotational isomers are interconverted slowly and therefore can be detected. When the rotational barriers are small, interconversion occurs rapidly, making the detection of the rotational isomers difficult on the NMR time scale.

The theoretical work showed that the rotational barriers of the vinylidene ligand increased with X from π acceptor, σ donor to π donor properties. Ligands (X) with

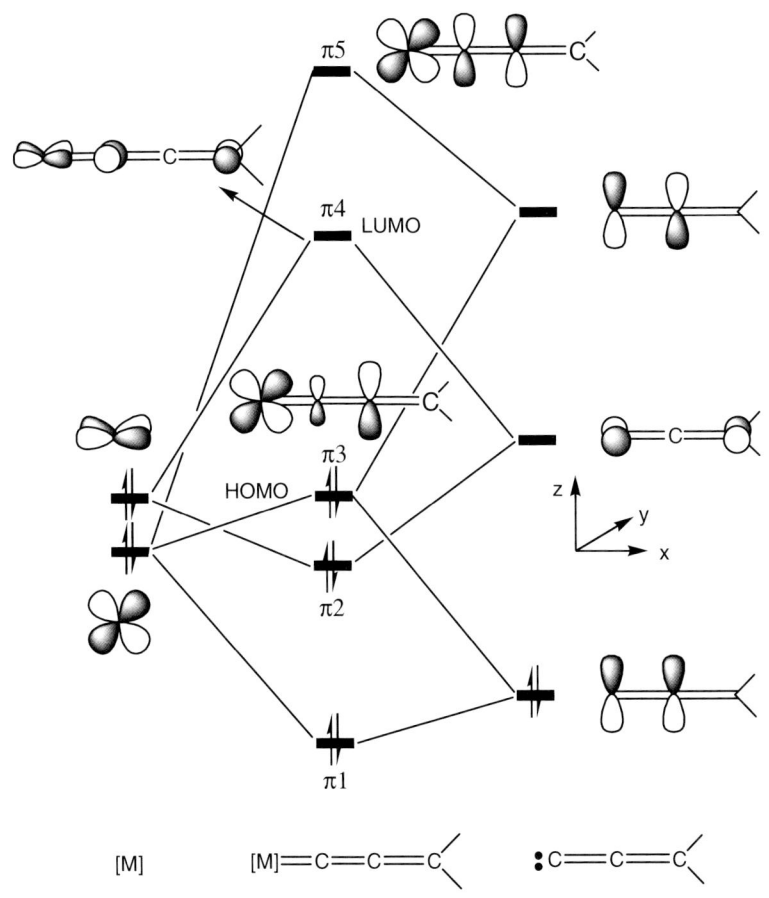

Figure 4.2 Simplified π-orbital interaction diagram for metal allenylidene complexes of the Fisher-type.

π-acceptor properties stabilize the transition structures through interactions with the d orbital in the HOMO, which is used for the metal–vinylidene π bonding (see π2 in Figure 4.1) in the most stable conformations. It was also found that complexes with silyl substituents on the vinylidene ligand had smaller rotational barriers. The

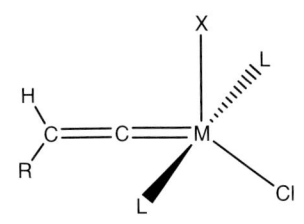

M = Ru or Os; L = phosphine;
X = H, CH_3, SiH_3, SiF_3 or Cl;
R = H, Ph, SiH_3 or SiF_3

Scheme 4.3

π-acceptor properties of the silyl substituents are capable of stabilizing the HOMO in the transition structures. The calculated rotational barriers for Os complexes were found to be generally higher in comparison with the Ru analogs. The strong metal–ligand interactions for osmium, due to its more diffuse d orbitals, are responsible for the higher rotational barriers.

In a subsequent study [65], Ariafard and his coworker studied theoretically the effect of phosphine ligands on the rotational barriers. They found that the rotational barrier increases with the increasing π-accepting ability of phosphine. When the Os–PX$_3$ π-backbonding is strong, the metal–vinylidene π bonding interaction in the rotational transition state is significantly weakened, leading to higher barriers.

4.4
Tautomerization Between η^2-Acetylene and Vinylidene on Transition Metal Centers

4.4.1
η^2-Acetylene to Vinylidene

Conversion of the unsubstituted acetylene to vinylidene has been widely investigated both experimentally [66–68] and theoretically [69–71]. These studies showed that when metals were not involved, the formation of vinylidene from free acetylene (Scheme 4.4) was very endothermic (44–47 kcal mol^{-1}). Since most transition metal fragments can stabilize a vinylidene ligand, the tautomerization of η^2-acetylene to vinylidene on transition metal centers becomes feasible. Recently, Clot and Eisenstein have thoroughly reviewed theoretical studies on various tautomerization pathways [44]. For the completeness of this chapter, we here briefly summarize the relevant theoretical findings.

Scheme 4.5 shows several possible pathways from η^2-acetylene metal complexes **RE** to metal vinylidenes **PR**. In the first pathway (a1 + a2), metal vinylidenes **PR** can be obtained from an intermediate (**IN1**) with a 1,2 hydrogen shift from C$_\alpha$ to C$_\beta$. The second pathway (b1 + b2) is through an intermediate (**IN2**) with an η^2 agostic interaction between the metal center and one C–H bond, which undergoes a 1,2 hydrogen shift to **PR**. The third pathway (b1 + b3 + b4) also starts from **IN2** but then goes into another intermediate, the hydrido-alkynyl **IN3**, which leads to **PR** with a 1,3 hydrogen shift from the metal center to C$_\beta$.

A few theoretical studies have been carried out to clarify the mechanism of the metal-assisted acetylene to vinylidene tautomerization. The first work done by Silvestre and Hoffmann using extended Hückel calculations on the complex

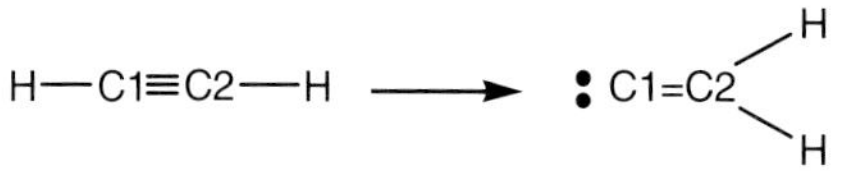

Scheme 4.4

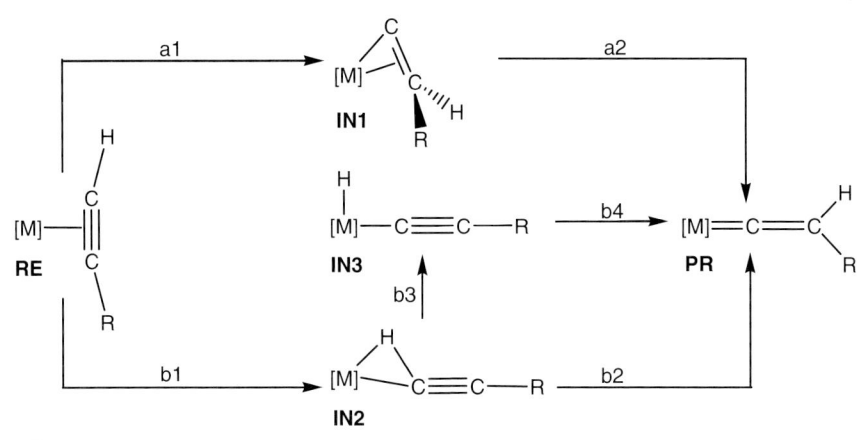

4.4 Tautomerization Between η^2-Acetylene and Vinylidene on Transition Metal Centers

Scheme 4.5

$(\eta^5\text{-}C_5H_5)(CO)_2Mn(HC{\equiv}CH)$ indicated that the second pathway (b1 + b2) with a direct 1,2 hydrogen shift was favorable and the alternative route (b1 + b3 + b4) involving a hydrido-alkynyl intermediate with a 1,3 hydrogen shift has a higher activation barrier [53]. A later density functional theory (DFT) study by Angelis and Sgamellotti on the same model complex also showed that the favorable pathway was the direct 1,2 hydrogen shift (b1 + b2) from the intermediate **IN2**, having a free energy barrier of 27.3 kcal mol^{-1}, and the barrier for the oxidative addition pathway (b1 + b3 + b4) was 45.5 kcal mol^{-1} [35]. The pathway (a1 + a2) through the intermediate **IN1** was also investigated and a free energy barrier of 45.3 kcal mol^{-1} was found.

A related theoretical work by Wakatsuki and coworkers on the acetylene–vinylidene rearrangement with a different metal center also showed the same preference for the 1,2 hydrogen shift [29]. The calculations by Wakatsuki and coworkers were done on the model complex $Cl_2Ru(PH_3)_2(HC{\equiv}CH)$. The results of the calculations suggested that the 1,2 hydrogen shift (b1 + b2) through the intermediate **IN2** having an agostic structure was kinetically favored and the hydrido-alkynyl intermediate **IN3**, $Cl_2Ru(PH_3)_2(H)(C{\equiv}CH)$, was found to be very unstable. In other words, oxidative addition leading to the formation of **IN3** was thermodynamically very unfavorable. Calculations on other model complexes with the same metal center Ru(II) but different ligands, $[(Cp)(PMe_3)_2Ru(HC{\equiv}CH)]^+$ and $[(Cp)(PMe_3)_2Ru(HC{\equiv}CMe)]^+$, were also carried out, leading to similar conclusions.

A more systematic theoretical study [31] on five different model complexes, $[(\eta^5\text{-}C_5Me_5)Ru(dippe)]^+$, $[(\eta^5\text{-}C_5Me_5)Ru(dmpe)]^+$, $[(\eta^5\text{-}C_5H_5)Ru(PMe_3)_2]^+$, $[(\eta^6\text{-}C_6Me_6)(PMe_3)ClRu]^+$, $[(\eta^5\text{-}C_5H_5)Ru(CO)(PPh_3)]^+$ and $[(\eta^6\text{-}C_6H_6)(PMe_3)ClRu]^+$ with a wide range of electron-richness at the metal center indicated that for the most electron-rich among these considered fragments, $[(Cp^*)(dippe)Ru(HC{\equiv}CH)]^+$, the hydrido-alkynyl intermediate was found essentially isoenergetic with the alkyne complex, which was only 1.9 kcal mol^{-1} higher in energy than the η^2-alkyne complex. The results are in agreement with the experimental evidence showing that for this system an

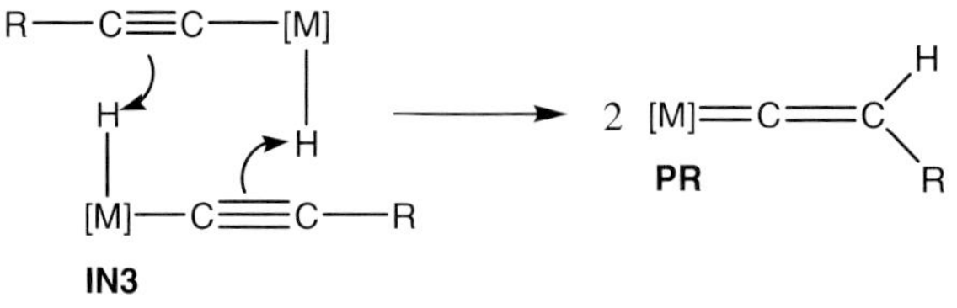

```
R—C≡C—[M]
   ↙
  H
  |            →    2 [M]=C=C(H)(R)
[M]—C≡C—R              PR

IN3
```

Scheme 4.6

equilibrium between the two forms [72] exists and that the 1,3 hydrogen shift (b1 + b3 + b4) is kinetically favorable with respect to the 1,2 hydrogen shift (b1 + b2).

Another theoretical study also showed that the third pathway (b1 + b3 + b4), 1,3 hydrogen shift, through a hydrido-alkynyl intermediate could compete with the 1,2 hydrogen shift pathway (b1 + b2) when the metal center is electron-rich enough [29, 30]. Indeed several hydrido-alkynyl intermediates have been detected or even isolated during the η^2-1-alkyne-to-vinylidene rearrangement on electron-rich metal centers, such as Co(I), Rh(I) and Ir(I) [73–78]. The *ab initio* MP2 calculations by Wakatsuki, Koga and their coworkers on the transformation of the model complex $RhCl(PH_3)_2(HC\equiv CH)$ to the vinylidene form $RhCl(PH_3)_2(C=CH_2)$ indicated that the transformation proceeded via the oxidative addition intermediate $RhCl(PH_3)_2(H)(C\equiv CH)$ [30].

A bimolecular hydrogen shift, shown in Scheme 4.6, was also studied theoretically [30]. Based on a localized molecular orbital (LMO) analysis, it was also concluded that the migrating hydrogen was regarded as a proton rather than a hydride in the bimolecular mechanism and that the bimolecular hydrogen shift was more favorable than the intramolecular 1,3 hydrogen shift (b1 + b2). In 2006, Grotjahn and coworkers studied the alkyne-to-vinylidene transformation on the square-planar complex *trans*-(Cl)Rh(phosphine)$_2$. With isotope labeling crossover experiments (Scheme 4.7), they, however, concluded that the bimolecular mechanism was unlikely [79].

In 1998, Caulton, Eisenstein and coworkers carried out labeling experiments to investigate the formation mechanism of a hydrido vinylidene complex from reactions of terminal alkynes with ruthenium and osmium hydrides [33]. Treatment of $RuHX(H_2)L_2$ (L = P^iPr_3) with $PhC\equiv CD$ gives $RuDX(=C=CHPh)L_2$ as the only isotopomer (Scheme 4.8). The results of these experiments indicated that the hydride ligand was not a spectator ligand but an active participant in the tautomerization process. On the basis of their theoretical calculations [33], a facile route (Scheme 4.9a) was suggested for the ruthenium-mediated tautomerization process. With a small barrier (6.6 kcal mol^{-1}), insertion of the η^2-coordinated alkyne into the Ru–H bond gives a vinyl complex. The vinyl ligand in the vinyl complex then undergoes a simple rotation along the Ru–C(vinyl) bond to give a new conformer of the vinyl complex. Finally, a hydride migration from the vinyl ligand to the metal center leads to the formation of the hydrido vinylidene complex. Interestingly, the tautomerization mediated by the osmium hydride proceeds through a slightly different pathway (Scheme 4.9b). The alkyne insertion gives an η^2-vinyl intermediate, instead of an η^1-vinyl intermediate. The intermediate then undergoes a hydride migration to give the hydrido vinylidene

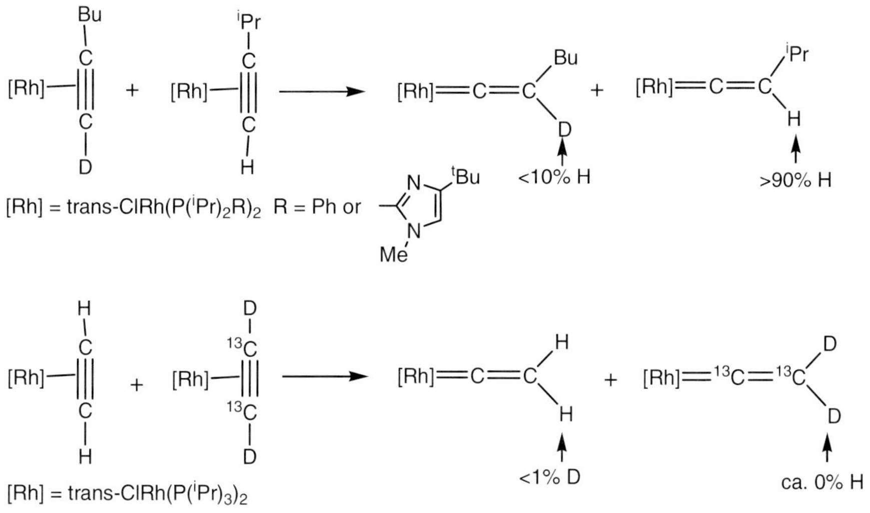

4.4 Tautomerization Between η²-Acetylene and Vinylidene on Transition Metal Centers

Scheme 4.7

product. The difference between the ruthenium- and osmium-mediated processes can be easily understood. Compared with ruthenium, osmium has more diffuse d orbitals and is capable of maximizing the number of metal–ligand bonds. In the literature, a number of osmium η²-alkenyl complexes have been reported [80, 81] and the relevant metal-η²-alkenyl bonding interactions have also been discussed [82].

Since electron-rich metal centers can easily stabilize a vinylidene ligand, it is no wonder that, in most cases, the metal centers in the tautomerization from η²-acetylene to vinylidene complexes are late transition metals. It is worth noting that the tautomerism involving high-valent transition metals has also been investigated theoretically [24, 34]. In 1993, Frenking and coworkers studied the structures and energetics of acetylene and difluoroacetylene complexes of W(VI) and Mo(VI) together with their vinylidene isomers [24]. In 1998, the same group carried out quantum mechanical calculations of the complexes at the CCSD(T) level of theory using BP86-optimized geometries and found that the high-valent d⁰ tungsten vinylidene complex [F₄W(=C=CH₂)] was 10.4 kcal mol⁻¹ higher in energy than the isomeric acetylene complex [F₄W(HC≡CH)] [34]. The pathway via a direct 1,2-H migration (a1 + a2 in Scheme 4.5) has a barrier of 84.8 kcal mol⁻¹. A hydrido-alkynyl intermediate was located at high energy (50.5 kcal mol⁻¹ higher than the acetylene complex) and the corresponding 1,3-H shift (b1 + b3 + b4 in Scheme 4.5) was found to have a very high

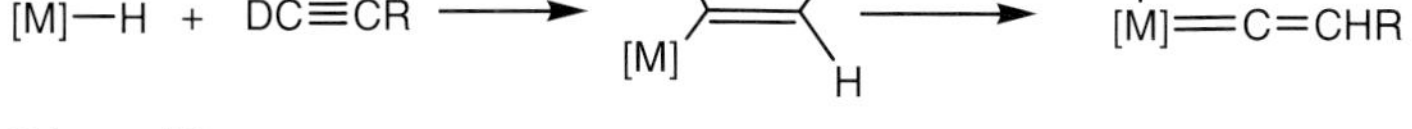

Scheme 4.8

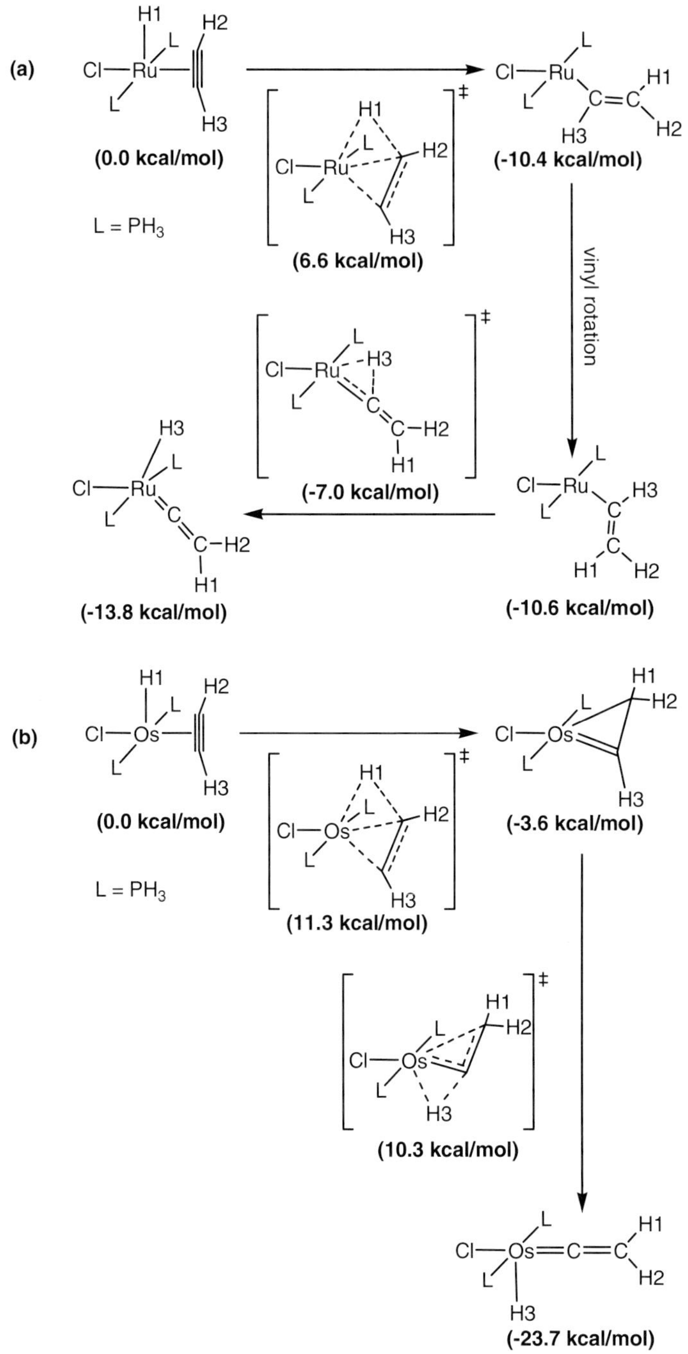
(a)
H1
H2
L
Cl—Ru—
L
H3
(0.0 kcal/mol)
L = PH3
H1
L
H2
Cl—Ru
L
H3
(6.6 kcal/mol)
L
Cl—Ru
L
C=C
H1
H3
H2
(-10.4 kcal/mol)
vinyl rotation
L
Cl—Ru
L
H3
C
C
H1
H2
(-10.6 kcal/mol)
L
Cl—Ru
H3
L
C
C—H2
H1
(-7.0 kcal/mol)
H3
L
Cl—Ru
L
C
C—H2
H1
(-13.8 kcal/mol)
(b)
H1
H2
L
Cl—Os—
L
H3
(0.0 kcal/mol)
L = PH3
H1
L
H2
Cl—Os
L
H3
(11.3 kcal/mol)
H1
H2
L
Cl—Os
L
H3
(-3.6 kcal/mol)
H1
H2
L
Cl—Os
L
H3
(10.3 kcal/mol)
L
H1
Cl—Os=C=C
L
H2
H3
(-23.7 kcal/mol)

Scheme 4.9

barrier of 85.5 kcal mol^{-1}. It was concluded that an intramolecular interconversion between the η^2-acetylene and vinylidene forms was unlikely for complexes containing high-valent metal centers.

4.4.2
Vinylidene to η^2-Acetylene

Vinylidene complexes of electron-rich transition metals are thermodynamically more stable than the isomeric η^2-alkyne complexes. Because of the better π-accepting property of a vinylidene ligand versus an η^2-alkyne ligand, electron-rich transition metal fragments are normally better stabilizers for vinylidene complexes. For η^2-alkyne complexes having electron-rich metal fragments, there is repulsive interaction between metal d electrons and the alkyne π$_\perp$ electrons. Therefore, the conversion of metal-η^2-1-alkyne to metal-vinylidene has a thermodynamic driving force [29, 37, 38, 72, 83–86]. However, for complexes in which the metal centers are not electron-rich enough, it is possible that the relative stability of the two isomers is switched. Indeed, reverse vinylidene-to-η^2-alkyne tautomerization processes have been reported experimentally [87–95].

Scheme 4.10 shows a few examples of the reverse vinylidene-to-η^2-alkyne tautomerization. Ipaktschi and coworkers investigated the conversion of vinylidene to η^2-alkyne on the transition metal fragment CpW(CO)(NO) at 130–150 °C in aromatic solvents (Scheme 4.10a) [95]. Based on their kinetic deuterium isotope experiments, they found that for R = silyl the migrating group was the silyl substituent, rather than the hydrogen substituent at the terminal position, in the rearrangement process of CpW(CO)(NO)[C=C(H)(SiMe$_2^i$Pr)] to CpW(CO)(NO)(^{i}PrMe$_2$Si)C≡CH]. Further experiments showed that the rate of the silyl 1,2-shift in the reverse vinylidene-to-η^2-alkyne tautomerization process followed the order SiMe$_2^i$Pr > Si(Ph)$_2$Me [96–98]. Another interesting example (Scheme 4.10b) is the interconversion process of chromium alkyne and vinylidene complexes [90]. Scheme 4.10b shows that a normal η^2-alkyne-to-vinylidene tautomerization is observed when the complex is neutral. However, when the complex is cationic, a reverse vinylidene-to-η^2-alkyne tautomerization occurs. Scheme 4.10c shows another reverse vinylidene-to-η^2-alkyne tautomerization process for the niobocene complexes [Cp$_2$Nb(PH$_3$)(HC≡CMe)]$^+$/[Cp$_2$Nb(PH$_3$)(=C=CHMe)]$^+$ [87]. DFT calculations with the B3LYP functional showed that a direct 1,2-shift was the most favorable pathway for the reverse vinylidene-to-η^2-alkyne tautomerization process and that the pathway through the hydrido alkynyl intermediate was unfavorable [87]. Scheme 4.10d shows a reverse tautomerization process on a molybdenum complex [95]. Experimentally, it was shown that the reverse vinylidene-to-η^2-alkyne tautomerization pathway was sensitive to the solvent used. In a nonpolar solvent, the molybdenum vinylidene complex undergoes a direct 1,2-H migration to give the η^2-alkyne isomer. In a polar solvent, such as ethanol, the reverse vinylidene-to-η^2-alkyne tautomerization proceeds by a multi-step process via deprotonation–protonation and subsequent reductive elimination, as illustrated in Scheme 4.11.

Although theoretical calculations on the reverse vinylidene-to-η^2-alkyne tautomerization are limited, we can see from the examples shown in Scheme 4.10 that in order

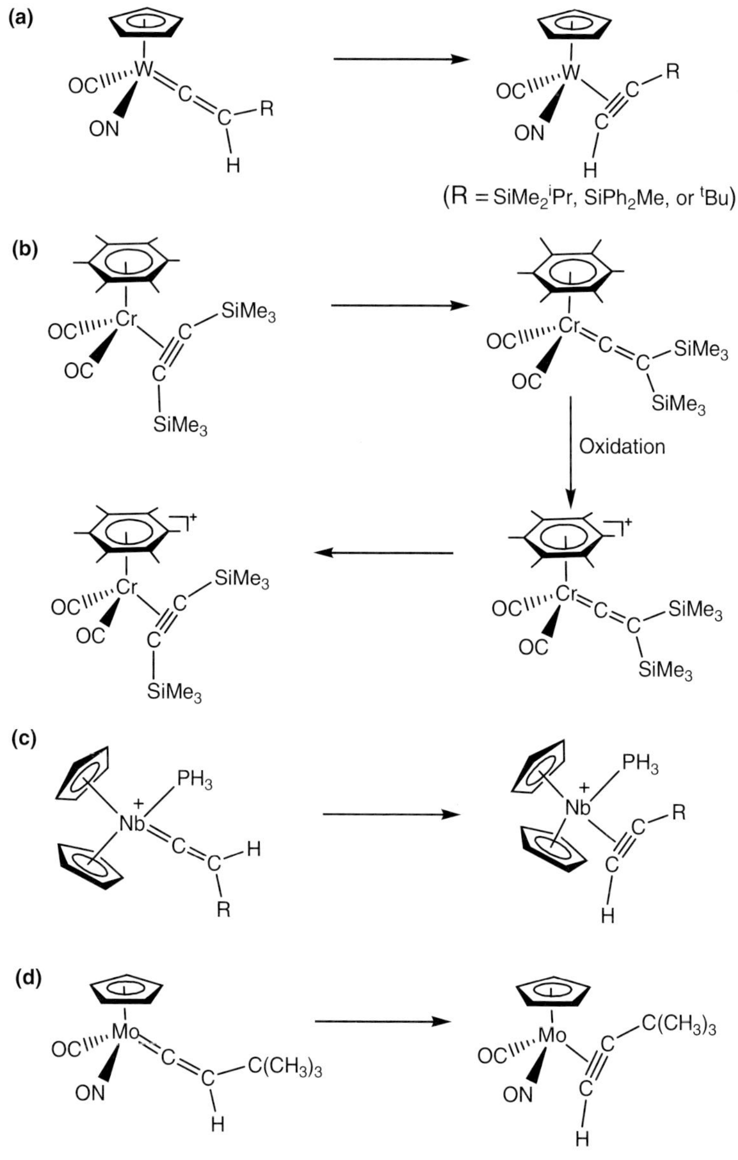

Scheme 4.10

to make the reverse vinylidene-to-η^2-alkyne tautomerization possible the metal centers of the complexes involved cannot be electron-rich. The examples discussed here have metal centers of early transition metals or middle/late transition metals containing strong π-acceptor ligands. More theoretical calculations are necessary in order to gain a deeper understanding.

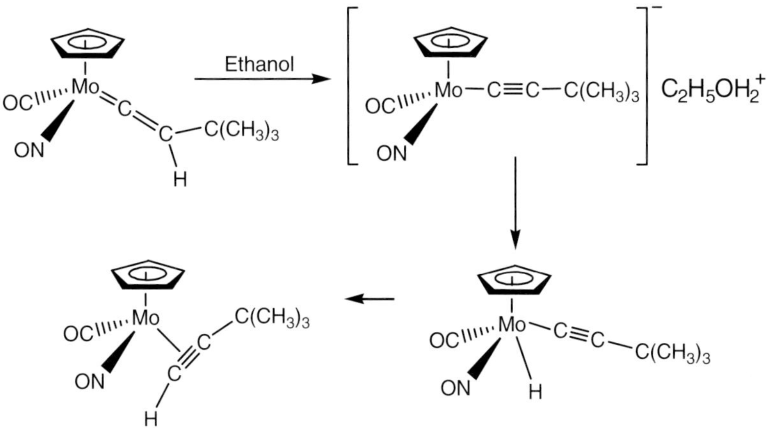

Scheme 4.11

4.5
Reversible C–C σ-bond Formation by Dimerization of Metal Vinylidene Complexes

In 2004, Berke and coworkers reported that treatment of the metal vinylidene complexes [(C_5H_4Me)(dmpe)Mn=C=CHR] (**Mn1**) with [Cp_2Fe](PF$_6$) in CH_2Cl_2 for 1 h yielded the dinuclear, dicationic biscarbyne complexes [(C_5H_4Me)(dmpe)Mn≡CCHRCHRC≡Mn(dmpe)(C_5H_4Me)]$^{2+}$ (**Mn2**) (Scheme 4.12). The dinuclear, dicationic biscarbyne complexes were characterized spectroscopically and crystallographically [99]. Density functional theory calculations on the model complexes [(C_5H_4Me)(dHpe)Mn=C=CHR]$^+$ (R = H, Me, Ph, C_6H_4Me, Si(^{t}Bu)(Me)$_2$) and the dinuclear, dicationic biscarbyne model complexes [(C_5H_4Me)(dHpe)Mn≡CCHRCHRC≡Mn(dHpe)(C_5H_4Me)]$^{2+}$ were carried out to investigate the role of the substituents on the C_β atom in the coupling process leading to dinuclear, dicationic biscarbyne manganese complexes [100, 101]. The optimized mononuclear cationic radical complexes showed spin densities localized at the manganese metal center and the C_β atom. The probability of finding the unpaired electron on the terminal carbon atom C_β decreases in the order R = H (+0.367α) > Me (+0.366α) > Ph (+0.358α) > C_6H_4Me (+0.344α) > Si(^{t}Bu)(Me)$_2$ (+0.332α). Furthermore, the energy calculations taking into account solvation effects on the model

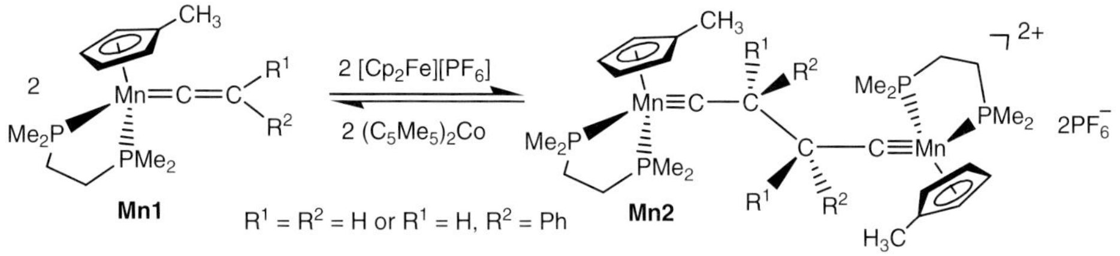

Scheme 4.12

complexes showed that the dimerization energies ΔE were of the order of -24.2 (R = H), -12.3 (Me), -3.9 (Ph), -2.1 (C_6H_4Me), and $+12.1$ kcal mol^{-1} (Si(^{t}Bu)(Me)$_2$). Greater dimerization energies were calculated for those **Mn1** having higher spin electron densities on their C_β atoms. The results showed that dimerization was easiest for [(C_5H_4Me)(dHpe)Mn=C=CH$_2$]$^+$ and was not possible for [(C_5H_4Me)(dHpe)Mn=C=CH(Si(^{t}Bu)(Me)$_2$)]$^+$. Scheme 4.12 also shows the reversibility of the dimerization process in which the dinuclear, dicationic biscarbyne complexes can be converted back to the mononuclear neutral complexes when (C_5Me_5)$_2$Co is added. The results of the DFT calculations showed that the LUMO of a given dinuclear, dicationic biscarbyne model complex consisted of π^*-antibonding character between Mn and C_α and of σ^*-antibonding character between C_β and C_β'. Therefore, the reduction of a given dinuclear, dicationic biscarbyne complex resulting formally from an addition of two electrons into the LUMO breaks one Mn≡C_α π bond and the C_β–C_β' σ bond, leading to the formation of the corresponding mononuclear neutral vinylidene complex **Mn1**. The reversibility shows the potential of utilizing the ubiquitous C–C bonds in organometallic systems as electron reservoirs.

4.6
Metal Vinylidene Mediated Reactions

4.6.1
Alkynol Cycloisomerization Promoted by Group 6 Metal Complexes

One of the important challenges in modern synthetic and pharmaceutical chemistry is to design and develop efficient, clean, and fast routes to obtain architectural molecules from simple starting materials [102, 103]. The *endo*-selective cycloisomerization (Scheme 4.13a) of terminal alkynes tethered to alcohols, nitrogen, and sulfur nucleophiles is one of the interesting transformations widely utilized in efficient syntheses of antiviral nucleosides [104, 105], polycyclic ethers [106], and oligosac-

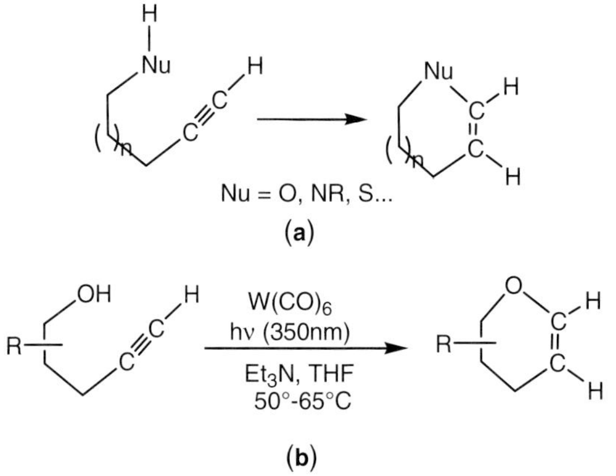

Scheme 4.13

charides [107, 108]. Various transition metals, such as platinum [109], ruthenium [110], rhenium [111] chromium [112] molybdenum [113–116], and tungsten [117, 118] complexes have been developed as catalysts for such cycloisomerizations. For instance, *endo*-cycloisomerization of substituted 4-alkyn-1-ols can be catalyzed by $W(CO)_6$ when the reactants are photolyzed at 350 nm at or near the reflux point of THF in the presence of triethylamine (Scheme 4.13b). The hypothesized mechanism of the cycloisomerization of terminal alkynyl alcohols involves an initial tautomerization of an η^2 metal alkyne complex to a vinylidene complex [117–120].

In 2002, Musaev and coworkers performed the first theoretical investigation of the mechanism with the aid of density functional theory calculations [26]. They first studied the mechanism of cycloisomerization in the absence of a tungsten catalyst, as shown in Scheme 4.14. The DFT calculations showed that the *exo*-cycloisomerization of 4-pentyn-1-ol via a concerted transition state leading to a five-membered-ring exo product had a high barrier (52.0 kcal mol^{-1}) (path a of Scheme 4.14). The pathways leading to a six-membered-ring *endo* product have also been calculated (paths b and c

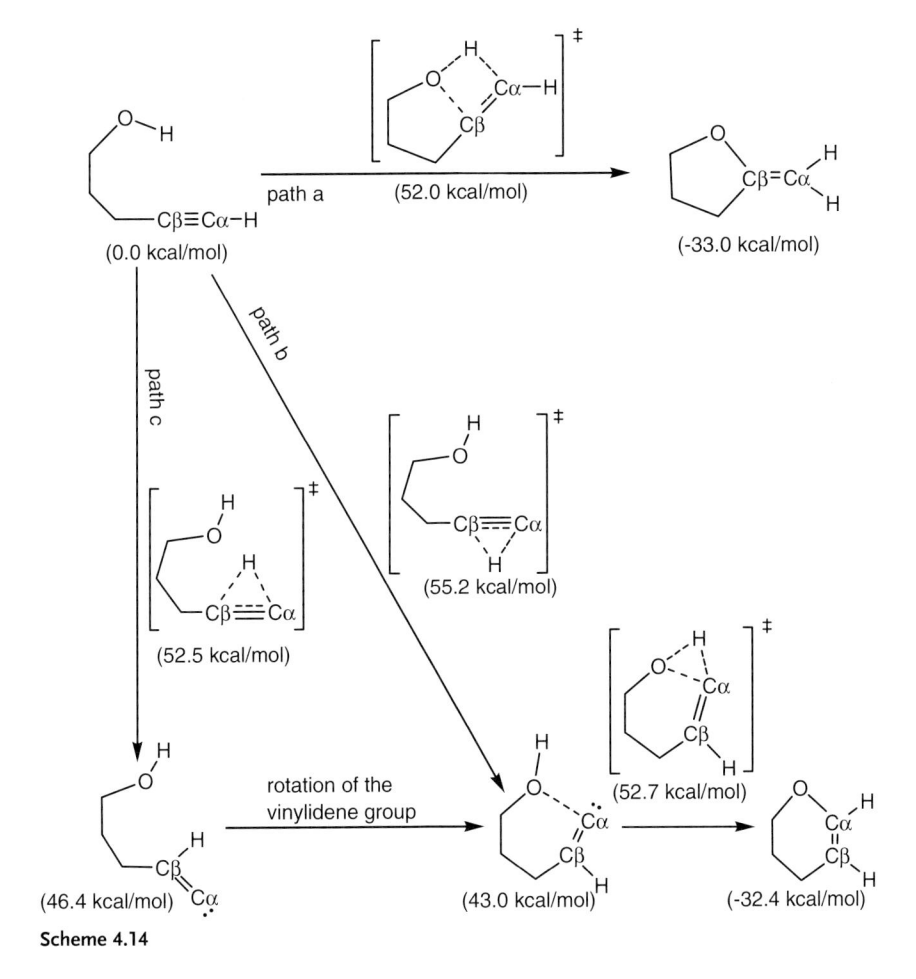

Scheme 4.14

of Scheme 4.14). In path b, an alkyne-to-vinylidene tautomerization gives a cyclic vinylidene intermediate with a barrier of 55.2 kcal mol^{-1}. The cyclic vinylidene intermediate then undergoes a proton migration to give the final six-membered-ring *endo* product. In path c, an alternative alkyne-to-vinylidene tautomerization gives a conformational isomer of the cyclic vinylidene intermediate from which a simple rotation of the vinylidene group gives the cyclic vinylidene intermediate. The high barriers calculated for the uncatalyzed cycloisomerization pathways suggest that, without a catalyst, neither *endo*- nor *exo*-cycloisomerization can occur under mild conditions.

With a tungsten pentacarbonyl catalyst, the calculated mechanisms are summarized in Scheme 4.15 [26]. Coordination of the 4-pentyn-1-ol substrate to the pentacarbonyl tungsten leads to the formation of the π-alkyne-W(CO)$_5$ adduct **W1**. This coordination process was calculated to be exothermic by 24.3 kcal mol^{-1}. The cycloisomerization leading to a five-membered-ring *exo* product starts with the π-complex **W1** via a one-step process with a barrier of 46.5 kcal mol^{-1} (path a of Scheme 4.15). The barrier calculated here is comparable with that calculated for the catalyst-free process. From **W1** to **W3**, the tungsten metal center does not play a significant role in the isomerization process.

The cycloisomerization leading to a six-membered-ring endo product also starts with the π-complex **W1** (paths b1 + b2 of Scheme 4.15). An alkyne-to-vinylidene tautomerization takes place with a barrier of 26.4 kcal mol^{-1}, leading to the formation of the tungsten vinylidene intermediate **W5**. In **W5**, the hydroxy group nucleophilically attacks the α-carbon of the vinylidene ligand to give a six-membered-ring tungsten alkenyl species **W7**. From **W7**, two pathways exist for the formation of the six-membered-ring endo product. In the one-step pathway (path b1), a direct proton migration to the α-carbon occurs. In the two-step pathway (path b2), the proton of the hydroxy group migrates to the metal center followed by a reductive elimination. The two-step pathway (path b2) is much more favorable than the one-step pathway (path b1), suggesting that the metal-assisted proton migration facilitates the formation of the the six-membered-ring *endo* product.

In path a (Scheme 4.15), the barrier for the rate-determining step is 46.5 kcal mol^{-1}. In path b2 (Scheme 4.15), the barrier for the rate-determining step (the alkyne-to-vinylidene tautomerization) is only 26.4 kcal mol^{-1}. The significant barrier difference (about 20 kcal mol^{-1}) between paths a and b2 explains the *endo*-regioselectivity observed experimentally in the cycloisomerization of 4-pentyn-1-ol [117, 118].

In 2005, Sordo and coworkers observed both the *endo*- and *exo*-cyclization products (**P1/P2** and **P3** in Scheme 4.16) in their experiments [27]. The *endo/exo* mixtures were also observed by other research groups with related systems [117, 120, 121]. These experimental results suggest that the pathways leading to both the *endo*- and *exo*-products should be comparable in the barriers of their rate-determining steps. The barrier difference should not be as large as the 20 kcal mol^{-1} mentioned above.

To understand the interesting experimental results, Sordo and coworkers studied theoretically the mechanism by carrying out PCM-UAHF calculations in which the solvent effect was modeled as a bulk [27]. They found that the reaction barriers as well

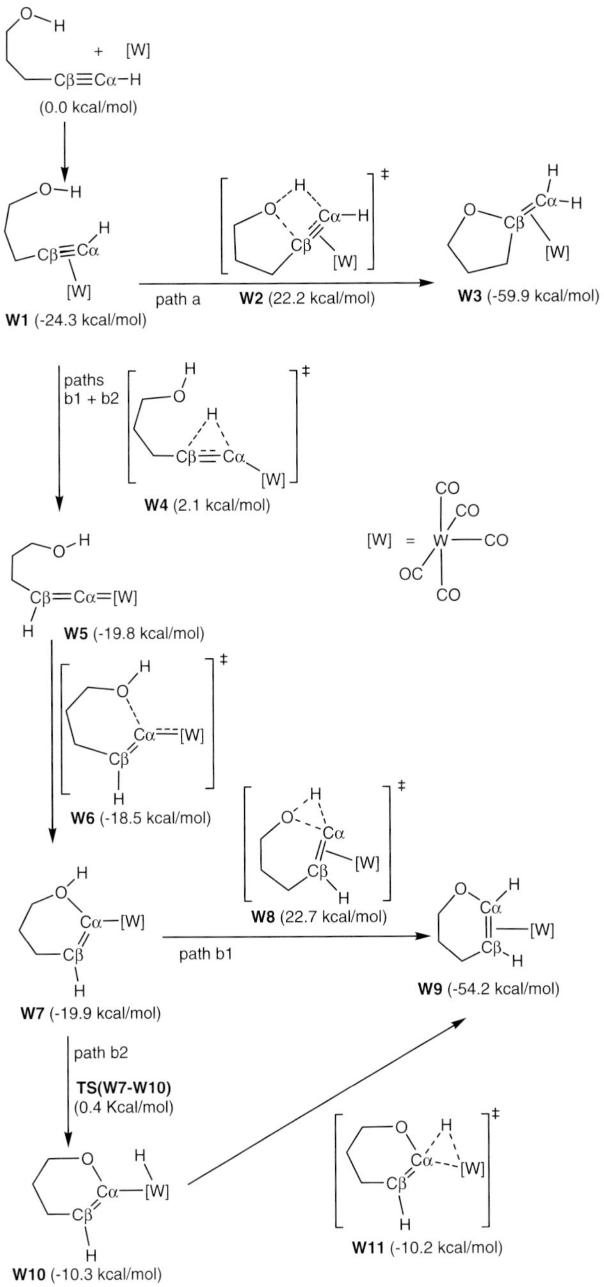

4.6 Metal Vinylidene Mediated Reactions | 145

Scheme 4.15

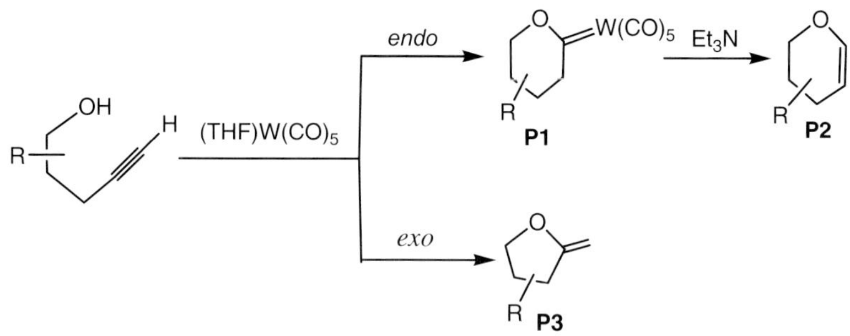

Scheme 4.16

as the reaction pathways were affected significantly when the solvent effect was considered. Scheme 4.17 summarizes the PCM-UAHF calculation results. The pathway leading to the five-membered-ring exo product (path a of Scheme 4.17) is very different from that obtained from the gas-phase calculations (path a of Scheme 4.15). When the solvent effect is considered, the nucleophilic attack of the hydroxy group on the β-carbon of the starting π-complex **M1** gives an intermediate (**M2** in Scheme 4.17). From **M2**, a proton migration and then a reductive elimination occur to give the five-membered ring *exo* product. The overall barrier for this pathway (path a of Scheme 4.17) was calculated to be 18.3 kcal mol^{-1}.

For the pathway that leads to the six-membered-ring *endo* product (paths b1 + b2 of Scheme 4.17), the first step is again the alkyne-to-vinylidene tautomerization. However, this step is no longer the rate-determining step when the solvent effect is taken into account. In the vinylidene intermediate **M5**, nucleophilic attack of the hydroxy group at the α-carbon of the vinylidene ligand gives a six-membered ring tungsten alkenyl species **M6**. From **M6**, two pathways are possible. In path b1, a 1,3-proton migration occurs to give a metal-carbene complex (**M7**), a model for the product **P1** shown in Scheme 4.16. In path b2, a proton migration to the metal center followed by a reductive elimination occurs to give a six-membered ring *endo* product. The overall barrier for path b1 is 45.4 kcal mol^{-1} while the overall barrier for path b2 is 25.1 kcal mol^{-1}.

Considering the solvent effect, one comes to the conclusion that the pathway leading to the five-membered ring exo product is more favorable than those leading to the six-membered ring *exo* product. The results again cannot well explain the experimental observation that both the *endo*- and *exo*-cyclizations are possible. Although the results obtained from the PCM-UAHF calculations are still inconsistent with the experimental observation, one can see that the solvent effect is important. Therefore, Sordo and coworkers further investigated the mechanism by including a THF molecule explicitly in the PCM-UAHF calculations. The results show that the overall barriers for paths a and b2 are 19.8 and 28.6 kcal mol^{-1}, respectively, 1.5 and 3.5 kcal mol^{-1} higher than those calculated without the explicit inclusion of a THF molecule. Significant barrier reduction was found for path b1 leading to the formation of the metal-carbene complex **M7**, a model for the product **P1**. The overall barrier for path b1 is only 18.7 kcal mol^{-1}. The 1,3-proton migration

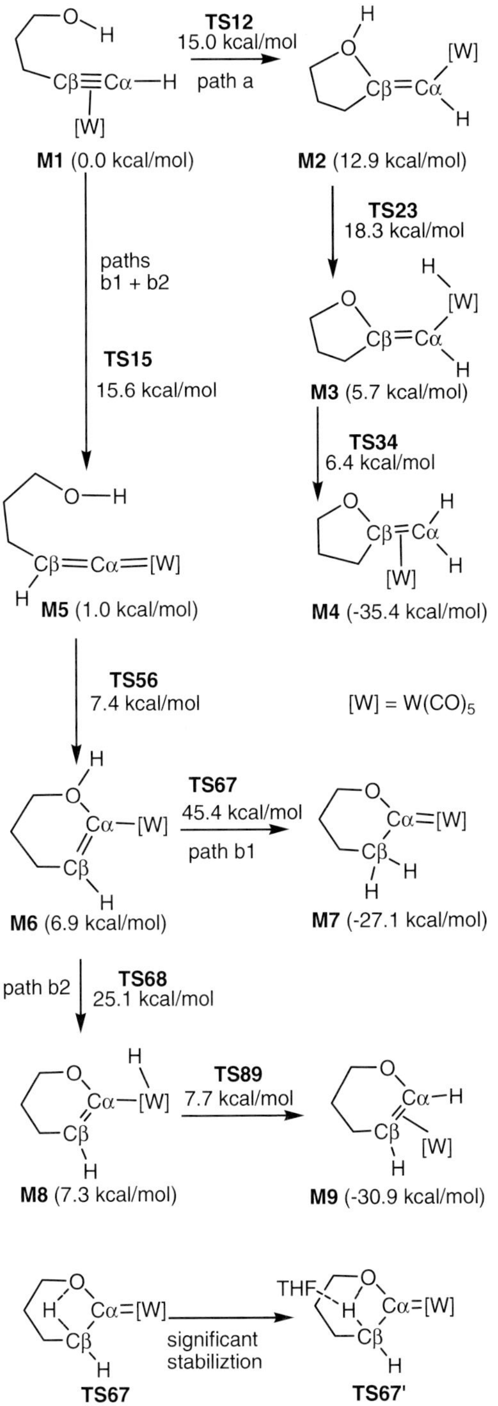

4.6 Metal Vinylidene Mediated Reactions | 147

Scheme 4.17

transition state **TS67** (Scheme 4.17) is significantly stabilized by a THF molecule. Clearly, the computational results rendered by this mixed model are able to rationalize the experimental findings.

It is worth noting that a subsequent theoretical work by Musaev, Morokuma and coworkers also demonstrated the stabilization effect of the THF solvent molecule [28]. In the subsequent theoretical work, the corresponding Mo-catalyzed *endo*-cycloisomerizations were also studied. The overall energy barriers calculated for the *endo*- and *exo*-cycloisomerizations are within a range of 25–30 kcal mol^{-1}, suggesting that both W(CO)$_5$ and Mo(CO)$_5$ are capable of catalyzing the cycloisomerization reactions.

In 2005, Barluenga and coworkers investigated theoretically the role of amine in the [W(CO)$_5$]-catalyzed *endo*- or *exo*-cylcoisomerization reactions [122]. The barriers calculated for the rate determining steps of both the *endo*- and *exo*-mechanisms are comparable (the difference is only 1.3 kcal mol^{-1}), demonstrating that small changes in the reaction conditions or the structures of the starting alkynols may affect the distribution of the *endo*- and *exo*-cycloisomerization products.

4.6.2
Unusual Intramolecular [2 + 2] Cycloaddition of a Vinyl Group with a Vinylidene C=C Bond

[2 + 2] cycloaddition is the most versatile and efficient method for forming four-membered rings containing both heterocyclic and isocyclic skeletons [123]. According to the Woodward–Hoffmann rules [124–126], the [2 + 2] thermal cycloaddition reactions are symmetry-forbidden when proceeding through a concerted path. Only allenes or activated alkenes (bearing electron-donating or electron-withdrawing substituents) are able to undergo [2 + 2] cycloadditions under thermal conditions [127].

In 2003, Gimeno, Bassetti and coworkers reported an unusual diastereoselective [2 + 2] cycloaddition of two C=C bonds under mild thermal conditions (Scheme 4.18) [128]. Heating the vinylidene complexes **Ru1** leads to the bicyclic alkylidene complexes **Ru2**. In 2004, Sordo and coworkers investigated the mechanism of this [2 + 2] cycloaddition theoretically [25]. With model complexes in which the indenyl ligand was modeled with a Cp ligand, two different pathways (paths a and b) were studied, shown in Scheme 4.19. Path a considers a concerted process. In the stepwise pathway (path b), the vinylidene to-alkyne tautomerization of **R1** followed by

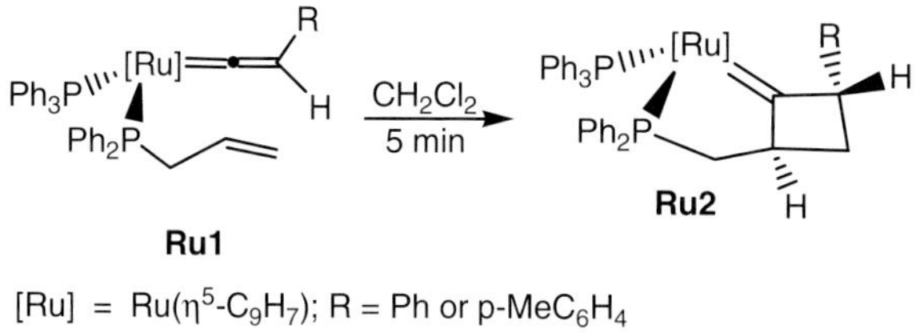

[Ru] = Ru(η^5-C$_9$H$_7$); R = Ph or p-MeC$_6$H$_4$

Scheme 4.18

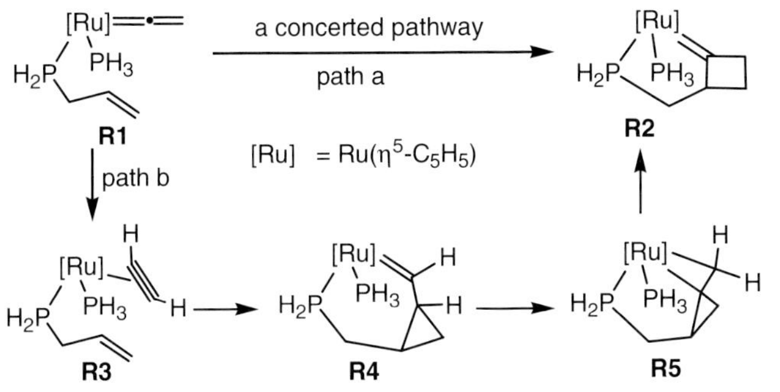

Scheme 4.19

a [1 + 2] cycloaddition between one of the coordinated alkyne carbons and the terminal carbon of the allyl substituent at P affords a three-membered ring intermediate **R4**. The three-membered ring intermediate then undergoes 1,2-hydrogen migration to give another three-membered ring intermediate **R5**. A rearrangement of the new three-membered ring intermediate gives the final four-membered ring product **R2**. The calculated barriers for the two studied pathways are comparable but substantially high (43–49 kcal mol^{-1}). Using the indenyl ligand instead of η^5-C$_5$H$_5$ in the model calculations, Sordo and coworkers found that the activation free energy barrier for path a is reduced to a value between 19.2 and 26.3 kcal mol^{-1}, in good agreement with the experimental value (23.8 ± 3.2 kcal mol^{-1} at 298.15 K). The significantly reduced barrier was considered to result from the stabilization of bonding interactions between the metal center and the internal carbon of the allyl group in the transition state [129, 130]. In the Cp model complexes, such bonding interactions are significantly weakened because of the much stronger metal–Cp bonding interactions compared with the metal–indenyl bonding interactions.

4.6.3
Intramolecular Methathesis of a Vinyl Group with a Vinylidene C=C Double Bond

In the subsection described above, we see that the C=C bond of a vinylidene ligand is able to undergo [2 + 2] cycloaddition with another C=C bond. It is interesting to think that an olefin metathesis would become possible if the [2 + 2] cycloaddition product could further undergo a ring-opening process from which a new vinylidene complex could be formed. In 2005, Lin and coworkers reported a novel transformation involving metathesis between the C=C double bond of the vinylidene ligand and a terminal vinyl group tethered on the ligand in a cationic ruthenium vinylidene complex [131]. As shown in Scheme 4.20, the cationic complex [Ru]=C=CHC(Ph)$_2$CH$_2$CH=CH$_2$$^+$ (**S1**) undergoes a regiospecific [2 + 2] cycloaddition of the two C=C double bonds, leading to the formation of a four-membered ring intermediate **S2**. A ring-opening process then gives a metathesis product, the ruthenium vinylidene complex **S3**.

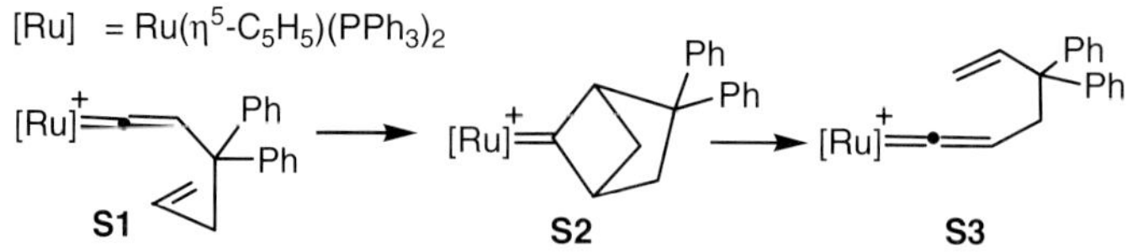

[Ru] = Ru(η⁵-C₅H₅)(PPh₃)₂

Scheme 4.20

4.6.4
[2 + 2] Cycloaddition of Titanocene Vinylidene Complexes with Unsaturated Molecules

[2 + 2] cycloaddition of titanocene vinylidene complexes (Schrock-type) with unsaturated molecules was studied theoretically by Böhme's group recently [32]. Different from metal vinylidene complexes of the Fischer-type, the titanocene vinylidene fragment $Cp_2Ti=C=CH_2$ undergoes [2 + 2] cycloaddition reactions with unsaturated molecules via the Ti=C double bond rather than the C=C bond. As shown in Scheme 4.21a, alkenes or alkynes react with titanocene-vinylidene via a [2 + 2] cycloaddition to give titancyclobutanes or titancyclobutenes.

The theoretical studies showed that the [2 + 2] cycloaddition reactions with alkenes or alkynes were almost barrierless and very exothermic. The exothermicity for alkynes ($-42.8\,\text{kcal mol}^{-1}$) was found to be greater than that for alkenes ($-21.3\,\text{kcal mol}^{-1}$). Scheme 4.21b shows the calculated [2 + 2] cycloaddition with polar unsaturated molecules, formaldehyde and HCN. The barriers ($3.6\,\text{kcal mol}^{-1}$ for formaldehyde and $9.0\,\text{kcal mol}^{-1}$ for HCN) are slightly higher than those in the reactions with alkenes and alkynes. The exothermicity is also significant (-26.5 for formaldehyde and -37.4 for HCN).

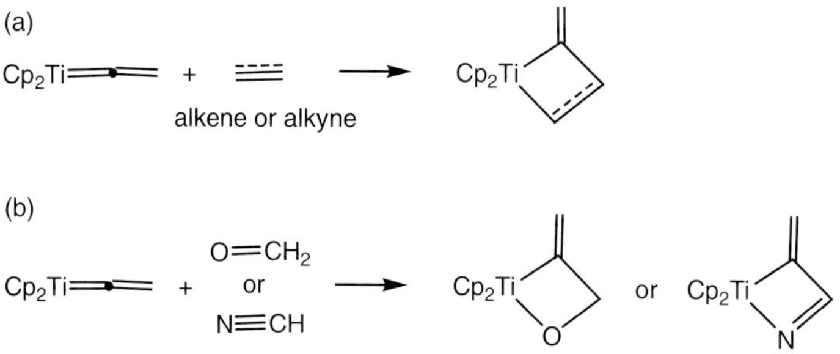

Scheme 4.21

4.7
Heavier Group 14 Analogs of Metal Vinylidene Complexes

Although metal vinylidene complexes are quite abundant [4–18], heavier group 14 analogs of metal vinylidene complexes containing M=C=E or M=E=C moiety (E = heavier group 14 elements) have not yet been found. A recent theoretical study of the stability of the heavier Group 14 analogs of vinylidene complexes

$M(Cl)_2(=C=EH_2)(PH_3)_2$, $M(Cl)_2(=E=CH_2)(PH_3)_2$ (M = Ru, Os), $Cp_2M(=C=EH_2)$(Cl), and $Cp_2M(=E=CH_2)(Cl)$ (M = Nb, Ta), where E = C, Si, Ge, Sn, was carried out by evaluating the reaction energies of various theoretically-designed isodesmic reactions [41]. The results of the theoretical work showed that the heavier Group 14 analogs of vinylidene complexes containing $M=E=CH_2$ were relatively stable, regardless of the type of metal center. It was also predicted that for analogs containing $M=C=EH_2$, the d^6 osmium complexes, $Os(Cl)_2(=C=EH_2)(PR_3)_2$, were the most promising targets for synthesis.

4.8
Allenylidene Complexes

Theoretical work on metal allenylidene complexes is very limited compared with that on metal vinylidene complexes. In 2002, Gimeno, Rodríguez and coworkers reported that the indenyl-ruthenium(II) complex [RuCl(η^5-C_9H_7)(PPh_3)$_2$] reacted with ethisterone, 17α-ethynylestradiol, and mestranol, in methanol and in the presence of NaPF$_6$, to afford equilibrium mixtures containing the corresponding allenylidene and vinylvinylidene tautomers [132]. They performed molecular orbital calculations to study the interconversion between allenylidene and vinylvinylidene on the model complexes [Ru{=C=C=C(H)CH$_3$}(η^5-C$_5$H$_5$)(PH$_3$)$_2$]$^+$ and [Ru{=C=C(H)CH=CH$_2$}(η^5-C$_5$H$_5$)(PH$_3$)$_2$]$^+$ (Scheme 4.22). The vinylvinylidene tautomer is only 2.1 kcal mol^{-1} more stable than the allenylidene one. The calculated barrier for the [1,3]-hydrogen sigmatropic rearrangement in the spontaneous tautomerization process is 66.5 kcal mol^{-1} which seems too high for the process observed in the experiments. The authors suspected that the level of theory used might have overestimated the activation barrier [132].

In 2002, Winter and coworkers reported that aminoallenylidene complexes trans-[Cl(dppm)$_2$Ru=C=C=C(NRR')(CH$_3$)]$^+$ were obtained from the regioselective addition of secondary amines to trans-[Cl(dppm)$_2$Ru=C=C=C=CH$_2$]$^+$ [133]. They also found that unsymmetrically substituted amines gave rise to Z/E isomeric mixtures. To study the Z/E isomeric interconversion, they calculated the rotational barrier around the C–N bond of the model complex shown in Scheme 4.23. The orthogonal

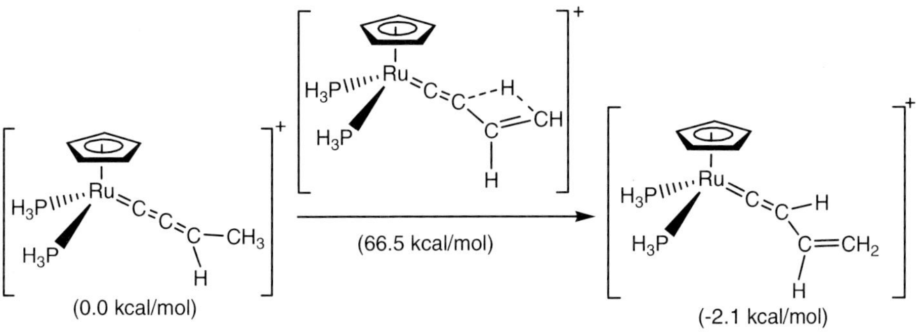

Scheme 4.22

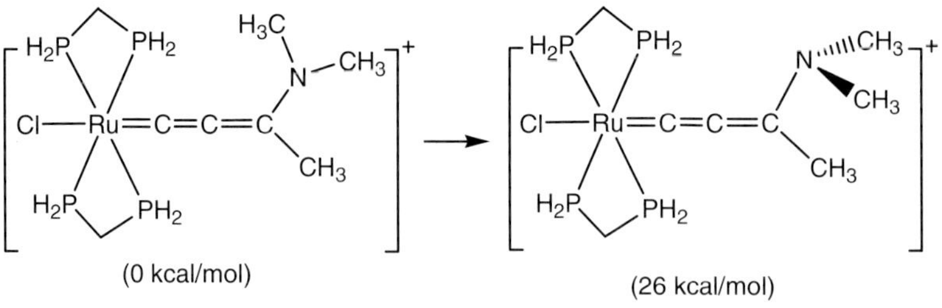

Scheme 4.23

(0 kcal/mol) → (26 kcal/mol)

form in which the $N(CH_3)_2$ moiety is perpendicular to the $Ru=C=C=C(CH_3)N$ plane is 26 kcal mol^{-1} higher in energy. The calculation results indicate that rotation around the iminium type CN bond decouples the nitrogen lone pair and the π-system of the allenylidene ligand, giving a high energy structure and resulting in a significantly longer CN bond and a tetrahedrally coordinated nitrogen atom.

4.9
Summary

In this chapter, we first analyzed the electronic structures of metal vinylidene and allenylidene complexes. The electronic structures allow us to understand the reactivities of these complexes. For metal vinylidene complexes of the Fischer-type, nucleophilic attack usually occurs at the α-carbon and electrophilic attack at the β-carbon. For the corresponding metal allenylidenes, electrophilic attack occurs at the β-carbon and/or the metal center. Then we briefly reviewed the theoretical study of the barriers of rotation of vinylidene ligands in various five-coordinate complexes $M(X)Cl(=C=CHR)L_2$ ($M = Os$, Ru; $L =$ phosphine). The study showed that π-acceptor ligands (X), electron-withdrawing substituents and lighter metals gave smaller barriers.

The alkyne-to-vinylidene tautomerization processes on various transition metal centers have also been discussed. Three different pathways for the formation of vinylidene from η^2-acetylene on electron-rich transition metals were the most theoretically studied. Most studies suggested that the favorable pathway proceeded via an intermediate with an η^2 agostic interaction between the metal center and one C–H bond followed by a 1,2 hydrogen shift (the b1 + b2 pathway shown in Scheme 4.5). The reverse process, the vinylidene-to-η^2-acetylene tautomerization, was also discussed. It was found that complexes with electron-poor metal centers were able to mediate the reverse process.

The dimerization process of $[(C_5H_4Me)(dmpe)Mn=C=CHR]^+$ to $[(C_5H_4Me)(dmpe)Mn\equiv CCHRCHRC\equiv Mn(dmpe)(C_5H_4Me)]^{2+}$, studied both experimentally and theoretically, was briefly discussed. Density functional theory calculations showed that the dimerization process was favorable when R = H, Me, Ph and C_6H_4Me but unfavorable when R = silyl. The calculations also showed that the LUMO of a given dimer consisted of π^*-antibonding character between Mn and C_α

and of σ^*-antibonding character between C_β and C_β', explaining the experimental observation that reducing the dicationic dimmers easily gives the corresponding neutral monomeric complexes [$(C_5H_4Me)(dmpe)Mn=C=CHR$].

Another focus of this chapter is the alkynol cycloisomerization mediated by Group 6 metal complexes. Experimental and theoretical studies showed that both *exo*- and *endo*- cycloisomerization are feasible. The cycloisomerization involves not only alkyne-to-vinylidene tautomerization but alo proton transfer steps. Therefore, the theoretical studies demonstrated that the solvent effect played a crucial role in determining the regioselectivity of cycloisomerization products. [2 + 2] cycloaddition of the metal vinylidene C=C bond in a ruthenium complex with the C=C bond of a vinyl group, together with the implication in metathesis reactions, was discussed. In addition, [2 + 2] cycloaddition of titanocene vinylidene with different unsaturated molecules was also briefly discussed.

We also discussed a recent theoretical study of the stability of the heavier group 14 analogs of vinylidene complexes $M(Cl)_2(=C=EH_2)(PH_3)_2$, $M(Cl)_2(=E=CH_2)(PH_3)_2$ (M = Ru, Os), $Cp_2M(=C=EH_2)(Cl)$, and $Cp_2M(=E=CH_2)(Cl)$ (M = Nb, Ta), where E = C, Si, Ge, Sn. The work predicted that the d^6 osmium complexes, $Os(Cl)_2(=C=EH_2)(PR_3)_2$, would be the most promising targets for synthesis.

Compared with vinylidene complexes, allenylidne complexes are not well studied theoretically. In this chapter, theoretical calculations of an allenylidene–vinylvinylidene equilibrium and the Z/E isomeric interconversion of aminoallenylidene ruthenium complexes were summarized.

In this chapter, we summarized the theoretical studies carried out on metal vinylidene complexes. Special emphasis was placed on aspects of their electronic structures, reactivities and their roles in organic reactions. Theoretical studies on the related metal allenylidene complexes have been quite limited. More theoretical studies on various aspects of these complexes, particularly on their metathesis reactivities, are clearly necessary.

Acknowledgement

This work was supported by the Research Grant Council of Hong Kong (HKUST 602304 and DAG05/06.SC19).

References

1 Stang, P.J. (1978) *Chemical Reviews*, 78, 383.
2 King, R.B. and Saran, M.S. (1972) *Journal of the Chemical Society. Chemical Communications*, 1053.
3 Kirchner, R.M. and Ibers, J.A. (1974) *Inorganic Chemistry*, 13, 1667.
4 Bruneau, C. and Dixneuf, P.H. (2006) *Angewandte Chemie-International Edition*, 45, 2176, and references therein.
5 Bruce, M.I. and Swincer, A.G. (1983) *Advances in Organometallic Chemistry*, 22, 59.
6 Antonova, A.B. and Ioganson, A.A. (1989) *Russian Chemical Reviews*, 58, 1197.

7 Bruce, M.I. (1991) *Chemical Reviews*, **91**, 197.
8 Werner, H. (1994) *Journal of Organometallic Chemistry*, **475**, 45.
9 Werner, H. (1997) *Chemical Communications*, 903.
10 Bruce, M.I. (1998) *Chemical Reviews*, **98**, 2797.
11 Puerta, M.C. and Valerga, P. (1999) *Coordination Chemical Reviews*, **193–195**, 977.
12 Rigaut, S., Touchard, D. and Dixneuf, P.H. (2004) *Coordination Chemical Reviews*, **248**, 1585.
13 Cadierno, V., Gamasa, M.P. and Gimeno, J. (2004) *Coordination Chemical Reviews*, **248**, 1627.
14 Werner, H. (2004) *Coordination Chemical Reviews*, **248**, 1693.
15 Valyaev, D.A., Semeikin, O.V. and Ustynyuk, N.A. (2004) *Coordination Chemical Reviews*, **248**, 1679.
16 Selegue, J.P. (2004) *Coordination Chemical Reviews*, **248**, 1543.
17 Bruce, M.I. (2004) *Coordination Chemical Reviews*, **248**, 1603.
18 Herndon, J.W. (2004) *Coordination Chemical Reviews*, **248**, 3.
19 Kostic, N.M. and Fenske, R.F. (1982) *Organometallics*, **1**, 974.
20 Delbecq, F. (1991) *Journal of Organometallic Chemistry*, **406**, 171.
21 Slugovc, C., Sapunov, V.N., Wiede, P., Merceiter, K., Schmid, R. and Kirchner, K. (1997) *Journal of the Chemical Society, Dalton Transactions*, 4209.
22 Moigno, D., Kiefer, W., Callejas-Gasper, B., Gil-Rubio, J. and Werner, H. (2001) *New Journal of Chemistry*, **25**, 1389.
23 Moigno, D., Callejas-Gasper, B., Gil-Rubio, J., Werner, H. and Kiefer, W. (2002) *Journal of Organometallic Chemistry*, **661**, 181.
24 Stegmann, R., Neuhaus, A. and Frenking, G. (1993) *Journal of the American Chemical Society*, **115**, 11930.
25 Brana, P., Gimeno, J. and Sordo, J.A. (2004) *Journal of Organic Chemistry*, **69**, 2544.
26 Sheng, Y.H., Musaev, D.G., Reddy, K.S., McDonald, F.E. and Morokuma, K. (2002) *Journal of the American Chemical Society*, **124**, 4149.
27 Sordo, T., Campomanes, P., Dieguez, A., Rodriguez, R. and Fananas, F.J. (2005) *Journal of the American Chemical Society*, **127**, 944.
28 Nowroozi-Isfahani, T., Musaev, D.G., McDonald, F.E. and Morokuma, K. (2005) *Organometallics*, **24**, 2921.
29 Wakatsuki, Y., Koga, N., Yamazaki, H. and Morokuma, K. (1994) *Journal of the American Chemical Society*, **116**, 8105.
30 Wakatsuki, Y., Koga, N., Werner, H. and Morokuma, K. (1997) *Journal of the American Chemical Society*, **119**, 360.
31 De Angelis, F., Sgamellotti, A. and Re, N. (2004) *Dalton Transactions*, 3225.
32 Böhme, U. (2003) *Journal of Organometallic Chemistry*, **671**, 75.
33 Oliván, M., Clot, E., Eisenstein, O. and Caulton, K.G. (1998) *Organometallics*, **17**, 3091.
34 Stegmann, R. and Frenking, G. (1998) *Organometallics*, **17**, 2089.
35 De Angelis, F. and Sgamellotti, A. (2002) *Organometallics*, **21**, 2715.
36 De Angelis, F., Sgamellotti, A. and Re, N. (2002) *Organometallics*, **21**, 5944.
37 Cadierno, V., Gamasa, M.P., Gimeno, J., Perez-Carreno, E. and Garcia-Granda, S. (1998) *Organometallics*, **18**, 2821.
38 Peréz-Carreño, E., Paoli, P., Ienco, A. and Mealli, C. (1999) *European Journal of Inorganic Chemistry*, 1315.
39 Bustelo, E., Carbó, J.J., Lledós, A., Mereiter, K., Puerta, M.C. and Valerga, P. (2003) *Journal of the American Chemical Society*, **125**, 3311.
40 Cadierno, V., Gamasa, M.P., Gimeno, J., Gonzalez-Bernardo, C., Perez-Carreno, E. and Garcia-Granda, S. (2001) *Organometallics*, **20**, 5177.
41 Ariafard, A. and Lin, Z. (2005) *Organometallics*, **24**, 6283.
42 Yang, S., Wen, T., Jia, G. and Lin, Z. (2000) *Organometallics*, **19**, 5477.

43 Wakatsuki, Y. (2004) *Journal of Organometallic Chemistry*, **689**, 4092.

44 Clot, E. and Eisenstein, O. (2002) *Computational Modeling of Homogeneous Catalysis*, (eds F. Maseras and A. Lledós), Kluwer, Boston, p. 137.

45 Birdwhistell, K.R., Tanker, T.L. and Templeton, J.L. (1985) *Journal of the American Chemical Society*, **107**, 4474.

46 Mayr, A., Schaefer, K.C. and Huang, E.Y. (1984) *Journal of the American Chemical Society*, **106**, 1517.

47 Beevor, R.G., Green, M., Orpen, A.G. and Williams, I.D. (1983) *Journal of the Chemical Society. Chemical Communications*, 673.

48 Lewis, L.N., Huffman, J.C. and Caulton, K.G. (1980) *Journal of the American Chemical Society*, **102**, 403.

49 Dawkins, G.M., Green, M., Jeffery, J.C., Sambale, C. and Stone, F.G.A. (1983) *Journal of the Chemical Society Dalton Transactions*, 499.

50 Colbom, R.E., Davies, D.L., Dyke, A.F., Endesfelder, A., Knox, S.A.R., Orpen, A.G. and Plaas, D. (1983) *Journal of the Chemical Society Dalton Transactions*, 2661.

51 Kolobova, N.E., Ivanov, L.L., Zhvanko, O.S., Khitrova, O.M., Batsavov, A.S. and Struchkov, Yu., T. (1984) *Journal of Organometallic Chemistry*, **262**, 39.

52 Boland-Lussier, B.E. and Hughes, R.P. (1982) *Organometallics*, **1**, 635.

53 Silvestre, J. and Hoffmann, R. (1985) *Helvetica Chimica Acta*, **68**, 1461.

54 Hohn, A. and Werner, H. (1986) *Angewandte Chemie (International Edition in English)*, **25**, 737.

55 Carvalho, M.F.N.N., Henderson, R.A., Pombeiro, A.J.L. and Richards, R.L. (1989) *Journal of the Chemical Society. Chemical Communications*, 1796.

56 Wolf, J. and Werner, H. (1987) *Journal of Organometallic Chemistry*, **336**, 413.

57 Wolf, J., ZoJk, R., Schubert, U. and Werner, H. (1988) *Journal of Organometallic Chemistry*, **340**, 161.

58 Werner, H., Wolf, J., Muller, G. and Kruger, C. (1988) *Journal of Organometallic Chemistry*, **342**, 381.

59 Drexler, M., Haas, T., Yu, S.M., Beckmann, H.S.G., Weibert, B. and Fischer, H. (2005) *Journal of Organometallic Chemistry*, **690**, 3700.

60 Marrone, A. and Re, N. (2002) *Organometallics*, **21**, 3562.

61 Auger, N., Touchard, D., Rigaut, S., Halet, J.F. and Saillard, J.Y. (2003) *Organometallics*, **22**, 1638.

62 Bourgault, M., Castillo, A., Esteruelas, M.A., Oñate, E. and Ruiz, N. (1997) *Organometallics*, **16**, 636.

63 Wen, T.B., Yang, S.Y., Zhou, Z.Y., Lin, Z., Lau, C.P. and Jia, G. (2000) *Organometallics*, **19**, 3757.

64 Oliván, M., Clot, E., Eisenstein, O. and Caulton, K.G. (1998) *Organometallics*, **17**, 897.

65 Ariafard, A. and Zare, K. (2004) *Inorganic Chemistry Communications*, **7**, 999.

66 Chen, Y., Jonas, D.M., Kinsey, J.L. and Field, R.W. (1989) *Journal of Chemical Physics*, **91**, 3976.

67 Ervin, K.M., Ho, J. and Lineberger, W.C. (1989) *Journal of Chemical Physics*, **91**, 5974.

68 Erwin, K.M., Gronert, S., Barlow, S.E., Gilles, M.K., Harrison, A.G., Bierbaum, V.M., DePuy, C.H., Lineberger, W.C. and Ellison, G.B. (1990) *Journal of the American Chemical Society*, **112**, 5750.

69 Gallo, M.M., Hamilton, T.P. and Schaefer, H.F. (1990) *Journal of the American Chemical Society*, **112**, 8714.

70 Jensen, J.H., Morokuma, K. and Gordon, M.S. (1994) *Journal of Chemical Physics*, **100**, 1981.

71 Chen, W.-C. and Yu, C.-H. (1997) *Chemical Physics Letters*, **277**, 245, and references therein.

72 de Los Ríos, I., Tenorio, M.J., Puerta, M.C. and Valerga, P. (1997) *Journal of the American Chemical Society*, **119**, 6529.

73 Wolf, J., Werner, H., Serhadli, O. and Ziegler, M.L. (1983) *Angewandte Chemie (International Edition in English)*, **22**, 414.

74 Werner, H. and Höhn, A. (1984) *Journal of Organometallic Chemistry*, **272**, 105.
75 García-Alonso, F.J., Höhn, A., Wolf, J., Otto, H. and Werner, H. (1985) *Angewandte Chemie (International Edition in English)*, **24**, 406.
76 Dziallas, M. and Werner, H. (1987) *Journal of the Chemical Society. Chemical Communications*, 852.
77 Bianchini, C., Peruzzini, M., Vacca, A. and Zanobini, F. (1991) *Organometallics*, **10**, 3697.
78 Bianchini, C., Masi, D., Meli, A., Peruzzini, M., Ramírez, J.A., Vacca, A. and Zanobini, F. (1989) *Organometallics*, **8**, 2179.
79 Grotjahn, D.B., Zeng, X. and Cooksy, A.L. (2006) *Journal of the American Chemical Society*, **128**, 2798.
80 Buil, M.L., Eisenstein, O., Esteruelas, M.A., García-Yebra, C., Gutíerrez-Puebla, E., Oliván, M., Oñate, E., Ruiz, N. and Tajada, M.A. (1999) *Organometallics*, **18**, 4949.
81 Barrio, P., Esteruelas, M.A. and Onate, E. (2003) *Organometallics*, **22**, 2472.
82 Choi, S.H. and Lin, Z. (1999) *Organometallics*, **18**, 2473.
83 Bustelo, E., Jiménez-Tenorio, M., Puerta, M.C. and Valerga, P. (2001) *European Journal of Inorganic Chemistry*, 2391.
84 Tokunaga, M., Suzuki, T., Koga, N., Fukushima, T., Horiuchi, A. and Wakatsuki, Y. (2001) *Journal of the American Chemical Society*, **123**, 11917.
85 Baya, M., Crochet, P., Esteruelas, M.A., López, A.M., Modrego, J. and Oñate, E. (2001) *Organometallics*, **20**, 4291.
86 Bustelo, E., De los Ríos, I., Jiménez-Tenorio, M., Puerta, M.C. and Valerga, P. (2000) *Monatshefte fur Chemie*, **131**, 1311–1320.
87 García-Yebra, C., López-Mardomingo, C., Fajardo, M., Antiñolo, A., Otero, A., Rodríguez, A., Vallat, A., Lucas, D., Mugnier, Y., Carbó, J.J., Lledós, A. and Bo, C. (2000) *Organometallics*, **19**, 1749.
88 Bartlett, I.M., Connelly, N.G., Martín, A.J., Orpen, A.G., Paget, T.J., Rieger, A.L. and Rieger, P.H. (1999) *Journal of the Chemical Society, Dalton Transactions*, 691.
89 Katayama, H., Onitsuka, K. and Ozawa, F. (1996) *Organometallics*, **15**, 4642.
90 Connelly, N.G., Geiger, W.E., Lagunas, C., Metz, B., Rieger, A.L., Rieger, P.H. and Shaw, M.J. (1995) *Journal of the American Chemical Society*, **117**, 12202.
91 Nombel, P., Lugan, N. and Mathieu, R. (1995) *Journal of Organometallic Chemistry*, **503**, C22.
92 Bly, R.S., Zhong, Z., Kane, C. and Bly, R.K. (1994) *Organometallics*, **13**, 899.
93 Bly, R.S., Raja, M. and Bly, R.K. (1992) *Organometallics*, **11**, 1220.
94 Connelly, N.G., Orpen, A.G., Rieger, A.L., Rieger, P.H., Scott, C.J. and Rosair, G.M. (1992) *Journal of the Chemical Society. Chemical Communications*, 1293.
95 Ipaktschi, J., Mohesseni-Ala, J. and Ulig, S. (2003) *European Journal of Inorganic Chemistry*, 4313.
96 Rigby, S.S., Gupta, H.K., Werstiuk, N.K., Bain, A.D. and McGlinchey, M.J. (1995) *Polyhedron*, **14**, 2787.
97 Bonny, A., Holmes-Smith, R.D., Hunter, G. and Stobart, S.R. (1982) *Journal of the American Chemical Society*, **104**, 1855.
98 Ashe, A.J. III (1970) *Journal of the American Chemical Society*, **92**, 1233.
99 Venkatesan, K., Blacque, O., Fox, T., Alfonso, M., Schmalle, H.W. and Berke, H. (2004) *Organometallics*, **23**, 1183.
100 Venkatesan, K., Blacque, O., Fox, T., Alfonso, M., Schmalle, H.W., Kheradmandan, S. and Berke, H. (2005) *Organometallics*, **24**, 920.
101 Venkatesan, K., Blacque, O. and Berke, H. (2006) *Organometallics*, **25**, 5190.
102 Hall, N. (ed.) (2001) *The New Chemistry*, Cambridge University Press, Cambridge, U.K.
103 Vögtle, F., Stoddart J.F. and Shibasaki, M. (eds) (2000) *Stimulating Concepts in Chemistry*, Wiley-VCH, Weinheim, Germany.
104 McDonald, F.E. and Gleason, M.M. (1995) *Angewandte Chemie (International Edition in English)*, **34**, 350.

105 McDonald, F.E. and Gleason, M.M. (1996) *Journal of the American Chemical Society*, **118**, 6648.

106 Bowman, J.L. and McDonald, F.E. (1998) *Journal of Organic Chemistry*, **63**, 3680.

107 McDonald, F.E. and Zhu, H.Y.H. (1998) *Journal of the American Chemical Society*, **120**, 4246.

108 McDonald, F.E. and Reddy, K.S. (2001) *Angewandte Chemie-International Edition*, **40**, 3653.

109 Chisholm, M.H. and Clark, H.C. (1972) *Journal of the American Chemical Society*, **94**, 1532.

110 Bruce, M.I., Swincer, A.G., Thomson, B.J. and Wallis, R.C. (1980) *Australian Journal of Chemistry*, **33**, 2605.

111 Bianchini, C., Marchi, A., Mantovani, N., Marvelli, L., Masi, D., Peruzzini, M. and Rossi, R. (1998) *European Journal of Inorganic Chemistry*, 211.

112 McDonald, F.E., Burova, S.A. and Huffman, L.G. (2000) *Synthesis*, **7**, 970.

113 McDonald, F.E., Connolly, C.B., Gleason, M.M., Towne, T.B. and Treiber, K.D. (1993) *Journal of Organic Chemistry*, **58**, 6952.

114 McDonald, F.E. and Schultz, C.C. (1994) *Journal of the American Chemical Society*, **116**, 9363.

115 McDonald, F.E. and Chatterjee, A.K. (1997) *Tetrahedron Letters*, **38**, 7687.

116 McDonald, F.E. and Olson, T.C. (1997) *Tetrahedron Letters*, **38**, 7691.

117 McDonald, F.E., Reddy, K.S. and Díaz, Y. (2000) *Journal of the American Chemical Society*, **122**, 4304.

118 McDonald, F.E. and Reddy, K.S. (2001) *Journal of Organometallic Chemistry*, **617**, 444.

119 Weyershausen, B. and Dötz, K.H. (1999) *European Journal of Inorganic Chemistry*, 1057.

120 Wipf, P. and Graham, T.H. (2003) *Journal of Organic Chemistry*, **68**, 8798.

121 McDonald, E. (1999) *Chemistry–A European Journal*, **5**, 3103.

122 Barluenga, J., Dieguez, A., Rodriguez, F., Fananas, F.J., Sordo, T. and Campomanes, P. (2005) *Chemistry–A European Journal*, **11**, 5735.

123 Lautens, M., Klute, W. and Tam, W. (1996) *Chemical Reviews*, **96**, 49.

124 Woodward, R.B. and Hoffmann, R. (1965) *Journal of the American Chemical Society*, **87**, 395.

125 Woodward, R.B. and Hoffmann, R. (1969) *Angewandte Chemie (International Edition in English)*, **8**, 781.

126 Woodward, R.B. and Hoffmann, R. (1970) *The Conservation of Orbital Symmetry*, Verlag Chemie; Academic Press, New York.

127 Smith, M.B. and March, J. (2001) *March's Advanced Organic Chemistry: Reaction, Mechanisms and Structure*, 5th edn, John Wiley & Sons, New York, p. 1077.

128 Alvarez, P., Lastra, E., Gimeno, J., Bassetti, M. and Falvello, L.R. (2003) *Journal of the American Chemical Society*, **125**, 2386.

129 Suárez, D., Sordo, T.L. and Sordo, J.A. (1994) *Journal of the American Chemical Society*, **116**, 763.

130 Suárez, D., González, J., Sordo, T.L. and Sordo, J.A. (1994) *Journal of Organic Chemistry*, **59**, 8059.

131 Yen, Y.S., Lin, Y.C., Huang, S.L., Liu, Y.H., Sung, H.L. and Wang, Y. (2005) *Journal of the American Chemical Society*, **127**, 18037.

132 Cadierno, V., Conejero, S., Gamasa, M.P., Gimeno, J. and Rodriguez, M.A. (2002) *Organometallics*, **21**, 203.

133 Winter, R.F., Hartmann, S., Zalis, S. and Klinkhammer, K.W. (2003) *Dalton Transactions*, 2342.

5
Group 6 Metal Vinylidenes in Catalysis (Cr, Mo, W)
Nobuharu Iwasawa

5.1
Introduction

This chapter deals mainly with the synthetic aspects of the vinylidene complexes of Group 6 metals. Although there have been reports of many kinds of vinylidene complexes of Group 6 metals, most of them deal with the preparative methods and their characterization and have been summarized previously in several excellent reviews [1]. Thus, the emphasis of this chapter is placed on the chemistry of zero-valent pentacarbonyl complexes as almost all the synthetic reactions utilizing vinylidene complexes of Group 6 metals employ such pentacarbonyl complexes. Some historical background for the preparation of Fischer-type carbene complexes through vinylidene pentacarbonyl complexes is described as an introduction to the main subject, as these examples represent typical methods for the preparation of vinylidene complexes of Group 6 metals. The utilization in synthetic organic reactions of these methods of preparation of vinylidene complexes, in particular those starting from terminal alkynes, will be described, with the emphasis being on catalytic reactions [2].

5.2
Preparation of Fischer-type Carbene Complexes through the Generation of the Vinylidene Complexes

Historically, vinylidene complexes of zero-valent pentacarbonyl Group 6 metals appeared as a fleeting intermediate for the preparation of Fischer-type carbene complexes. Probably the first example of the formation of such a pentacarbonyl vinylidene complex of a Group 6 metal was proposed in 1974 by Fischer *et al.*, who examined the reaction of pentacarbonyl[hydroxy(methyl)carbene] chromium **1** with dicyclohexylcarbodiimide(DCC) [3]. Thus, treatment of **1** with DCC in CH_2Cl_2 at $-20\,°C \sim rt$ gave a novel azetidinylidene complex **2** in 47% yield. As a possible

Metal Vinylidenes and Allenylidenes in Catalysis: From Reactivity to Applications in Synthesis
Edited by Christian Bruneau and Pierre Dixneuf
Copyright © 2008 WILEY-VCH Verlag GmbH & Co. KGaA, Weinheim
ISBN: 978-3-527-31892-6

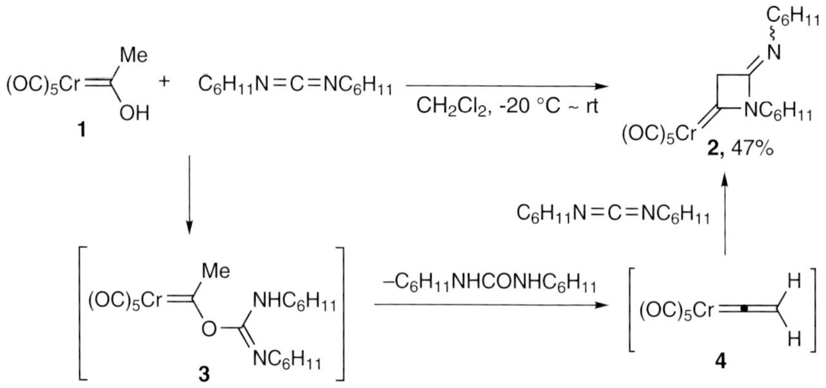

Scheme 5.1 The reaction of pentacarbonyl[hydroxy(methyl)carbene] chromium with DCC.

mechanism, [2 + 2] cycloaddition of the carbodiimide with the vinylidene complex **4**, generated by elimination of an urea derivative from the intermediate **3**, was proposed (Scheme 5.1).

In 1979, Rudler *et al.* reported another example of the presence of a vinylidene complex during the reaction of pentacarbonyl[methoxy(methyl)carbene] tungsten **5** with MeLi followed by acidification with TFA [4]. It was proposed that the vinylidene complex **7** was generated by deprotonation of the α-proton of the carbene complex followed by elimination of methoxide and reaction with the dimethylcarbene complex **8**, the addition–elimination product of MeLi with the starting carbene complex, to give the dinuclear complex **6** (Scheme 5.2).

In 1991 Fischer *et al.* observed the interesting phenomenon that treatment of benzylidene tungsten carbene complex **9** with triphenylketeneimine at −70 °C in CH_2Cl_2 gave zwitterionic intermediate **11**, which was derived from the rearrangement of the initially formed zwitterioinic intermediate **10** [5]. Careful analysis of

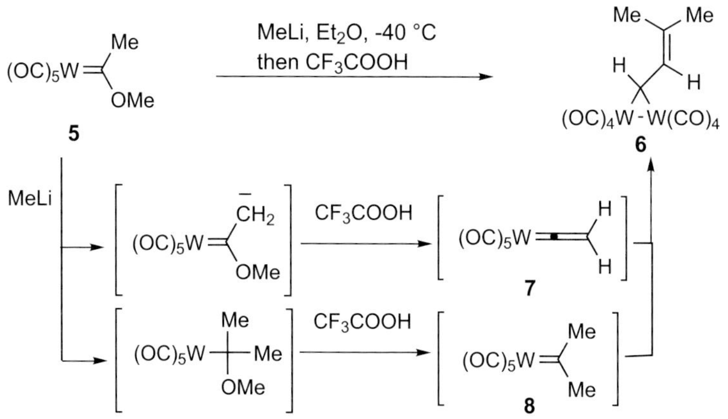

Scheme 5.2 The reaction of pentacarbonyl[methoxy(methyl)] carbene]tungsten with MeLi followed by TFA.

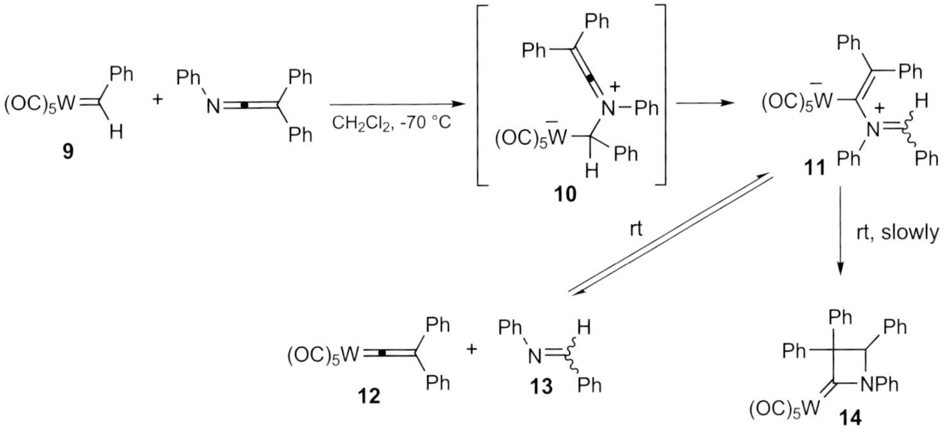

5.2 Preparation of Fischer-type Carbene Complexes through the Generation of the Vinylidene Complexes

Scheme 5.3 The reaction of a benzylidene tungsten complex with keteneimine.

the reaction revealed that the latter was found to be in equilibrium with the vinylidene complex **12** and imine **13** at rt and slowly undergoes ring closure to give the azetidinylidene complex **14** irreversibly. Furthermore, the vinylidene complex **12** could be isolated by carrying out the reaction in the presence of 1 equiv of $HBF_4 \cdot OEt_2$ to suppress the re-addition of **13** to **12** (Scheme 5.3).

In fact, the first isolation of the vinylidene pentacarbonyltungsten complexes was reported by Mayr et al. in 1984 [6]. The vinylidene complexes **16** were obtained by the alkylation of anionic pentacarbonyltungsten t-butylacetylide complex **15**, obtained by the reaction of $[Et_4N^+][W(CO)_5Cl^-]$ with lithium acetylide, with FSO_3Me or $[Et_3O^+][BF_4^-]$. Further protonation with CF_3SO_3H in the presence of $Me_4N^+I^-$ afforded a unique method for the preparation of carbyne complexes **17** (Scheme 5.4).

In situ preparation of the vinylidene pentacarbonylchromium complexes was also reported by Fischer et al. [7]. Treatment of the acylchromate complexes **18**, prepared by acylation of pentacarbonylchromium dianion, with trifluoroacetic anhydride in the presence of DBU afforded very labile pentacarbonylchromium vinylidene complexes **19**, which were, without isolation, reacted with several electron-rich alkynes such as ynamines, alkynyl ethers, and alkynyl thioethers to give the corresponding cyclobutenylidene chromium complexes in reasonable yield [8] (Scheme 5.5).

Furthermore, Fischer rendered this chemistry more practical by generating vinylidene complexes of pentacarbonylchromium and tungsten directly *in situ* from terminal alkynes [9]. For example, treatment of terminal alkynes with $W(CO)_5(CH_2Cl_2)$, generated by photolysis of $W(CO)_6$ in CH_2Cl_2, gave thermo-labile π-alkyne $W(CO)_5$

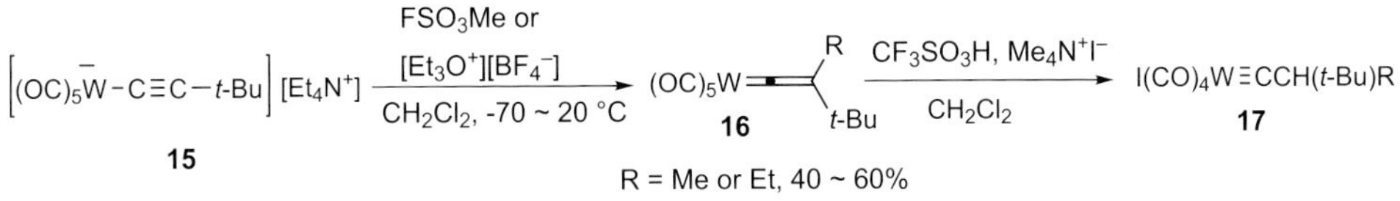

Scheme 5.4 Isolation of vinylidene complexes.

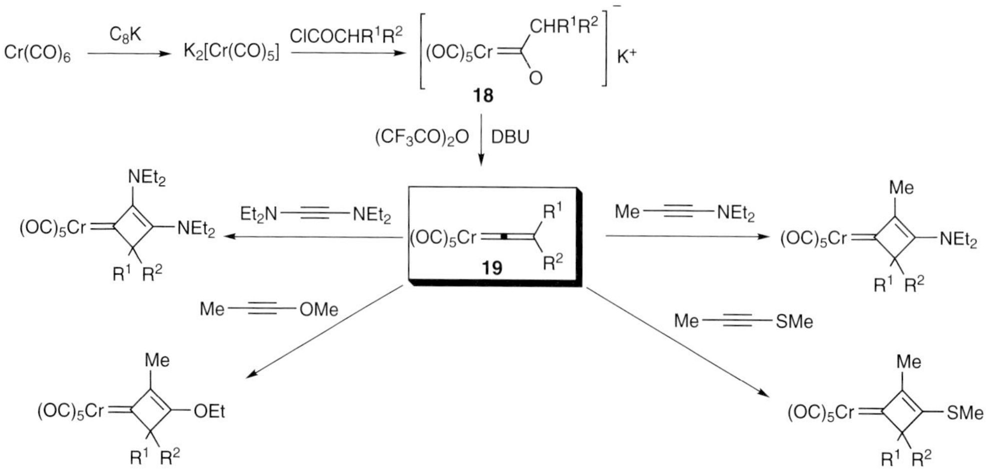

Scheme 5.5 Reaction of vinylidene complexes with electron-rich alkynes.

complexes **20**, which reacted through the vinylidene complex **21** with an equimolar amount of ynamines at $-40 \sim -20\,°C$ to give 3-aminocyclobutenylidene complexes **21** in moderate yield. In a similar manner, azetidinylidene complexes **22** were prepared directly from terminal alkynes and imines (Scheme 5.6).

An important contribution that developed into the catalytic use of the vinylidene complexes for the construction of carbon frameworks was reported by two research groups independently for the preparation of Fischer-type carbene complexes by the reaction of terminal alkynes with pentacarbonylchromium or tungsten species in the presence of oxygen nucleophiles.

In 1985, Dötz et al. reported during a study on the reaction of Fischer-type carbene complexes with alkynes [10] that 2-oxacyclopentylidene chromium complex **24** was obtained as a side product. Thus, treatment of methyl(methoxy)carbene complex with 3-butynol at 70 °C in dibutyl ether gave the cyclic carbene complex **24** in 23% yield along with the desired metathesis product **23**. The authors briefly commented that the cyclic carbene complex **24** might be obtained through the vinylidene complex **25**, generated by the reaction of the alkyne with the liberated pentacarbonylchromium species (Scheme 5.7).

In a similar manner, Rudler et al. reported in 1986 that treatment of the 4-penten-1-ylidene complex **25** with trimethylsilylacetylene gave the methyl carbene complex **27**

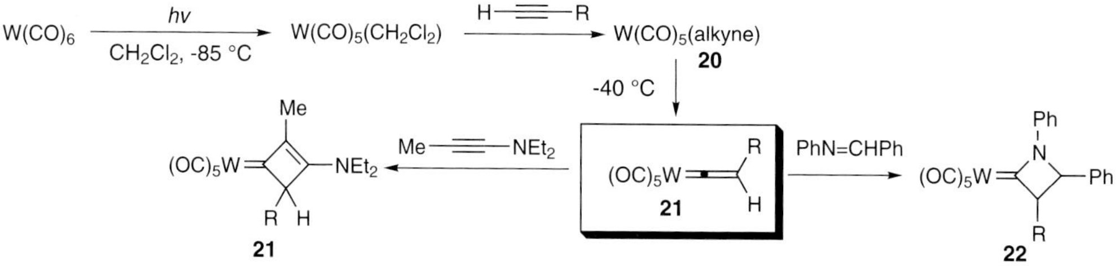

Scheme 5.6 Generation of Generation of vinylidene complexes directly from alkynes.

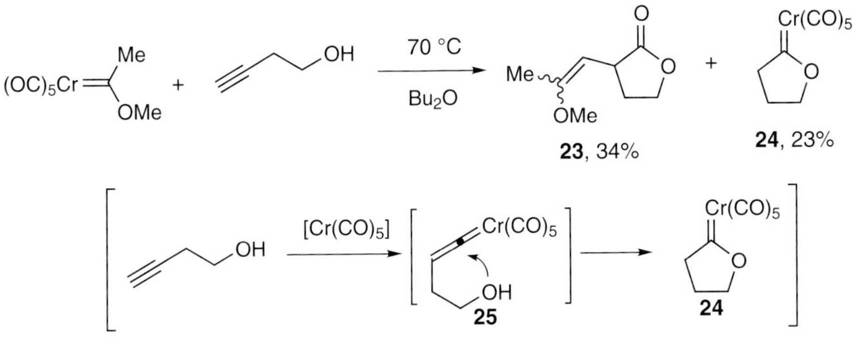

5.2 Preparation of Fischer-type Carbene Complexes through the Generation of the Vinylidene Complexes

Scheme 5.7 Formation of a cyclic Fischer-carbene complex via vinylidene by Dötz.

in 12% yield along with the desired cyclopropane derivative **26** [11]. Based on this result they optimized the reaction conditions and found that irradiation of $W(CO)_6$ and excess trimethylsilylacetylene in CH_2Cl_2 or hexane for 4 h, followed by treatment with MeOH in the presence of SiO_2 led to formation of methyl carbene complex **27** in 32% yield. The reaction was proposed to proceed through the vinylidene complex **28** generated from $W(CO)_5(L)$ and the terminal alkyne, and methanol attacked the vinylidene carbon to give the carbene complex **27** (Scheme 5.8).

Then, in 1987, Dötz reported an improved procedure for this transformation, for which the use of Et_2O as solvent improved the yield of the cyclic carbene complexes considerably [12]. For example, the five-membered Fischer-type carbene complex **29** ($n = 1$) was prepared in 58~66% yield by the reaction of preformed $M(CO)_5(L)$ ($M = Cr, W, L = Et_2O$) and 3-butynol in Et_2O at room temperature. The six-membered cyclic carbene complex could also be prepared by this method. This method has been applied to the preparation of functionalized cyclic Fischer-type carbene complexes from the corresponding alkynols. For example, Dötz et al. reported the preparation of various carbohydrate-functionalized cyclic Fischer-type carbene complexes, one of which is shown in Scheme 5.9.

Preparation of alkenylcarbene metal complexes was reported by Le Bozec and Dixneuf et al. in 1991 by activation of propargylic alcohols in the presence of methanol [13]. Thus, photolysis of $M(CO)_6$ ($M = Cr, W$) in the presence of 2-propyn-1-ol derivatives **30** in the presence of MeOH gave the corresponding

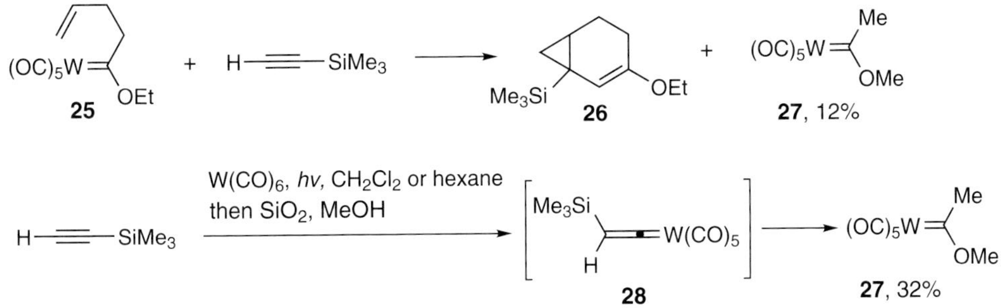

Scheme 5.8 Formation of a Fischer-carbene complex from vinylidene by Rudler.

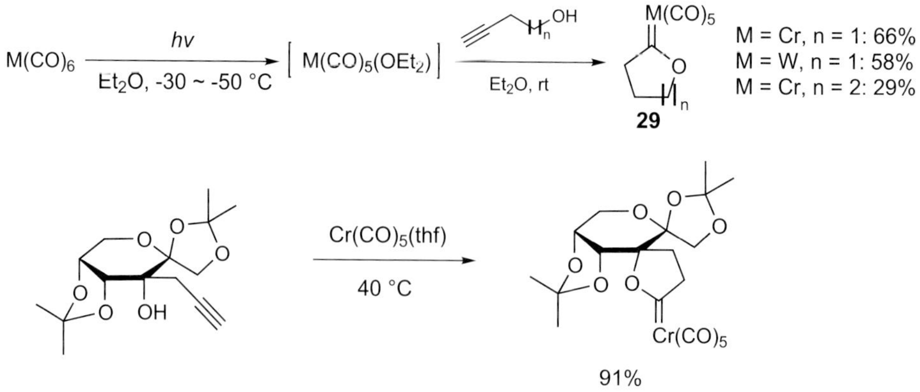

Scheme 5.9 Improved procedure for the formation of a cyclic Fischer-carbene complex via vinylidene by Dötz.

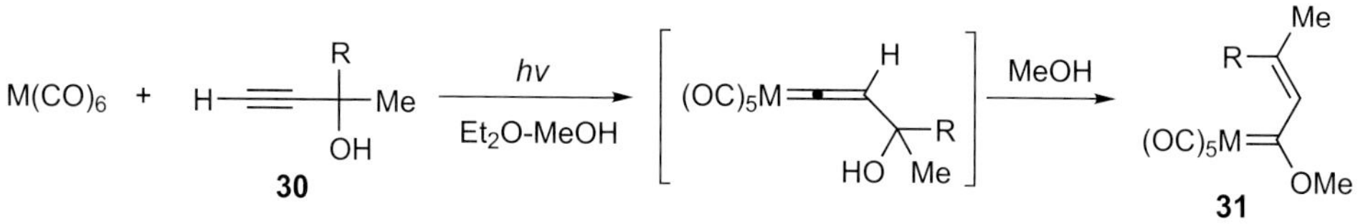

Scheme 5.10 Preparation of alkenylcarbene complexes.

alkenylcarbene complexes **31** in reasonable yield. Even dienyl and trienylcarbene complexes were prepared by this method (Scheme 5.10).

5.3
Utilization of Pentacarbonyl Vinylidene Complexes of Group 6 Metals for Synthetic Reactions

Before going into the details of the synthetic use of the vinylidene complexes of Group 6 metals, the work by Geoffroy on the polymerization of phenylacetylene should be noted, as this is probably the first example of the use of the vinylidene intermediate of Group 6 metals initiating polymerization [14]. Thus, photolysis of phenylacetylene in the presence of 3 mol% of W(CO)$_6$ in hexane was found to promote the polymerization of phenylacetylene. The reaction was thought to be initiated by the formation and metathesis reaction of the vinylidene complex **32**. Thus, the vinylidene complex **32**, generated from phenylacetylene and W(CO)$_6$, undergoes [2 + 2] cycloaddition with phenylacetylene, followed by ring-opening of the metallacyclobutene **33** to give the vinyl carbene complex **34**, which repeatedly undergoes a similar metathesis reaction to give polymerized product **35**. Generation of the vinylidene complex was indispensable for the initiation of this polymerization because polymerization of an internal alkyne did not occur under similar reaction conditions but could be triggered by the addition of a trace amount of phenylacetylene (Scheme 5.11).

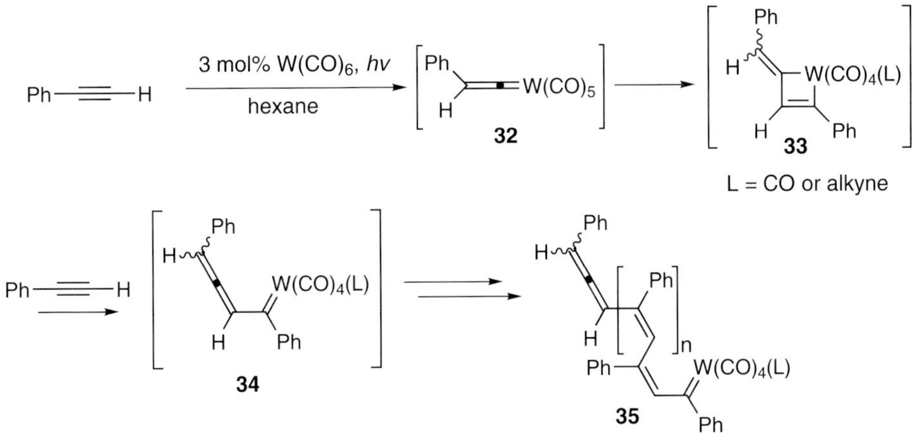

Scheme 5.11 Polymerization of phenylacetylene initiated by the vinylidene intermediate.

5.3.1
Catalytic Addition of Hetero-Nucleophiles

Utilization of vinylidene complexes of Group 6 metals for catalytic synthetic reactions was initiated by McDonald et al. based on the cyclic Fischer-type carbene complex synthesis. Thus, in 1993, they reported that this ω-alkynol cyclization could be applied to the preparation of various kinds of oxygen- and nitrogen-containing heterocycles [15]. In their first report, the single-step preparation of 2,3-dihydrofurans **38** was achieved by the reaction of 3-butyn-1-ol derivatives **36** with Mo(CO)$_6$ and trimethylamine N-oxide(TMNO) in the presence of Et$_3$N. In this reaction, it was thought that the vinylidene intermediate **40** was generated from the 3-butynol derivative and Mo(CO)$_5$(Et$_3$N), and nucleophilic attack of the hydroxy group on the central carbon of the vinylidene complex gave the alkenylmolybdenum intermediate **41**. Although protonation of the intermediate at the β-carbon of the molybdenum would provide the cyclic Fischer-type carbene complex, protonation in fact occurred directly at the α-carbon of the metal to give the endocyclic enol ether **38** along with regeneration of Mo(CO)$_5$(Et$_3$N). It should be noted that similar reactions employing Cr(CO)$_5$(Et$_3$N) or W(CO)$_6$–TMNO gave the corresponding cyclic Fischer-type carbene complexes **33** instead of 2,3-dihydrofurans. The most important aspect of this reaction is that the reaction could proceed catalytically although 0.5 equiv of Mo(CO)$_6$ was in fact employed in this initial stage [15a]. The efficiency of the reaction was partially improved by using Mo(CO)$_5$(Et$_3$N) which was prepared by irradiation of Mo(CO)$_6$ in the presence of Et$_3$N in Et$_2$O [15b]. Reaction of 3-butynylamine derivatives also proceeded to give the corresponding N-containing five-membered heterocyclic compounds where the amount of the complex employed was dependent on the nature of the substrate [15c]. Furthermore, the alkenylmolybdenum intermediate could be trapped by Bu$_3$SnOTf to give alkenylstannane derivatives **39**, which are useful intermediates for further coupling reactions [15d] (Scheme 5.12 and Table 5.1).

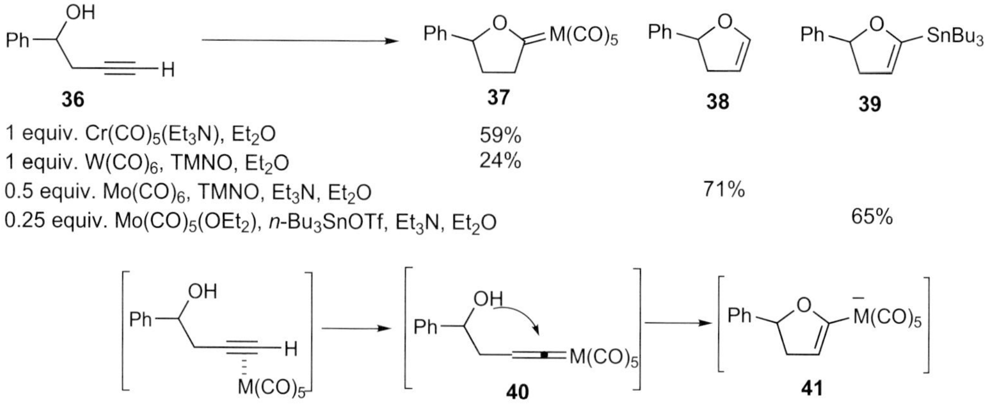

Scheme 5.12 3-Butynol cyclization leading to dihydrofuran derivatives.

Conditions for 36 → products:
- 1 equiv. Cr(CO)$_5$(Et$_3$N), Et$_2$O → 37, 59%
- 1 equiv. W(CO)$_6$, TMNO, Et$_2$O → 37, 24%
- 0.5 equiv. Mo(CO)$_6$, TMNO, Et$_3$N, Et$_2$O → 38, 71%
- 0.25 equiv. Mo(CO)$_5$(OEt$_2$), n-Bu$_3$SnOTf, Et$_3$N, Et$_2$O → 39, 65%

Table 5.1 3-Alkynol cyclization.

Alkynol	Products
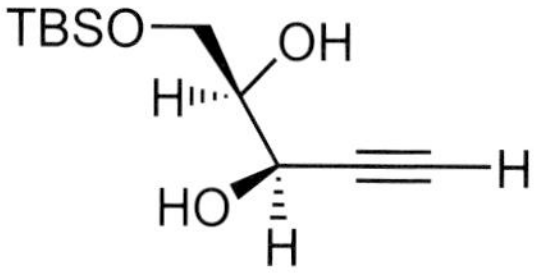 TBSO-CH$_2$-C(H,OH)-C(OH,H)-C≡C-H	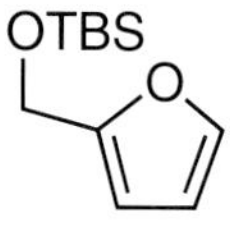 2-(OTBS-methyl)furan, 58%[a]
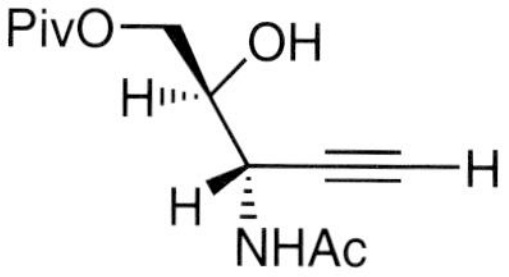 PivO-CH$_2$-C(H,OH)-C(H,NHAc)-C≡C-H	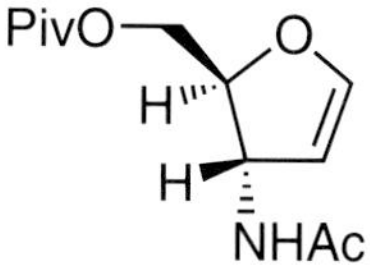 PivO-CH$_2$-(2,3-dihydrofuran with NHAc), 89%[a]
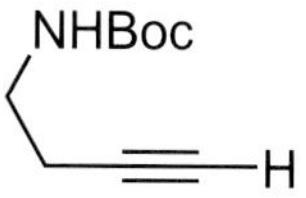 NHBoc-CH$_2$CH$_2$-C≡C-H	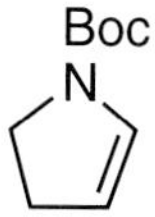 N-Boc-2,3-dihydropyrrole, 68%[a]
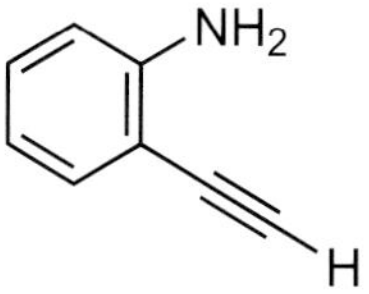 2-aminophenyl-C≡C-H	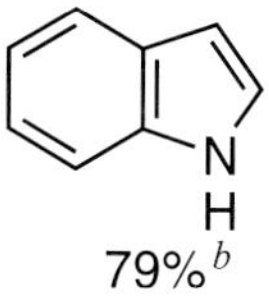 indole, 79%[b]

[a] 1 equiv Mo(CO)$_5$(NEt$_3$), Et$_3$N, Et$_2$O.
[b] 0.1 equiv Mo(CO)$_5$(NEt$_3$), Et$_3$N, Et$_2$O.

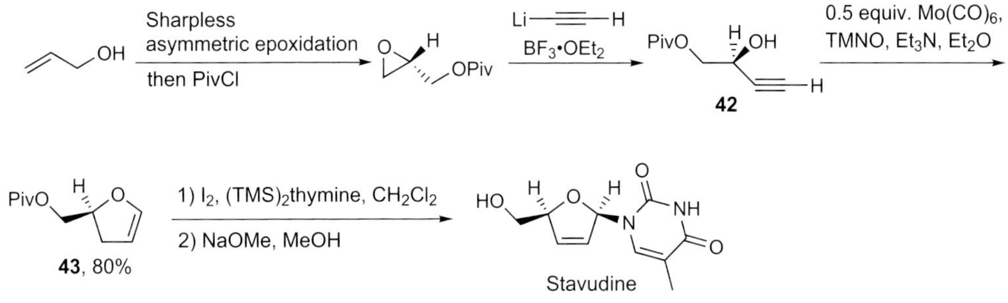

5.3 Utilization of Pentacarbonyl Vinylidene Complexes of Group 6 Metals for Synthetic Reactions

Scheme 5.13 Synthesis of stavudine.

Asymmetric synthesis of stavudine and cordycepin, anti-HIV agents, and several 3′-amino-3′-deoxy-β-nucleosides was achieved utilizing this cycloisomerization of 3-butynols to dihydrofuran derivatives [16]. For example, $Mo(CO)_6$-TMNO-promoted cyclization of the optically active alkynyl alcohol **42**, prepared utilizing Sharpless asymmetric epoxidation, afforded dihydrofuran **43** in good yield. Iodine-mediated introduction of a thymine moiety followed by dehydroiodination and hydrolysis of the pivaloate gave stavudine in only six steps starting from allyl alcohol (Scheme 5.13).

In 1994, Quayle et al. reported the application of this cyclic Fischer-carbene synthesis from 3-butynols to spirolactone synthesis, although the process was stepwise and a stoichiometric amount of the complex was employed [17]. The key transformation was the chromium or tungsten carbene complex formation followed by the CAN oxidation of the complex to give γ-lactone. The reaction was further applied to the synthesis of andirolactone and muricatacin, the former being shown in Scheme 5.14.

Schmidt et al. reported similar reactions of 3-butynols with $Cr(CO)_5(L)$ and $Mo(CO)_5(L)$ [18]. In most cases, a similar tendency to that reported by McDonald was observed, that is, five-membered cyclic carbene complexes were obtained when $Cr(CO)_5(OEt_2)$ was employed, while dihydrofurans were obtained when $Mo(CO)_5(NEt_3)$ was employed, however, in one specific case, a unique difference of the reaction pathway was observed. Thus, when hemiacetal **44** was treated with $Cr(CO)_5(OEt_2)$, the corresponding carbene complex **45** was obtained, which was further converted to dihydrofuran by treatment with DMAP. On the other hand, when **44** was

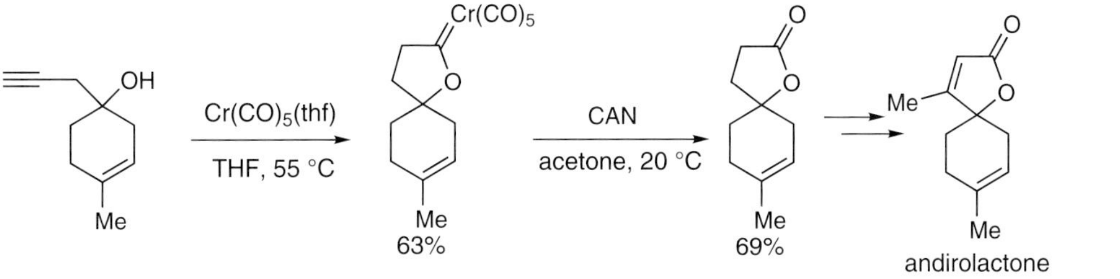

Scheme 5.14 Synthesis of andirolactone.

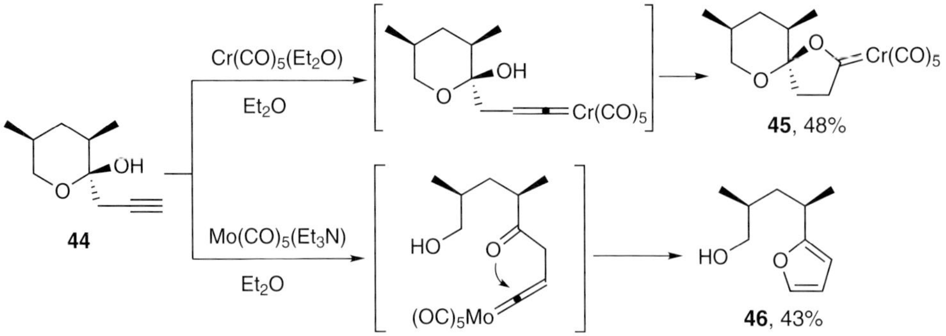

Scheme 5.15 Reaction of hemiacetal **44** with Cr(CO)₅(L) and Mo(CO)₅(L).

treated with Mo(CO)₅(NEt₃) in Et₂O at rt, furan **46** was obtained. It was proposed that under the reaction conditions, ring-opening of hemiacetal occurred and carbonyl oxygen attacked the vinylidene carbon, as shown in Scheme 5.15.

The preparation of dihydropyrans is somewhat different from that of dihydrofurans [19]. Mo(CO)₅(Et₃N) was not so effective for this class of substrates, while W (CO)₅(thf) is sufficiently reactive to give tetrahydropyranylidene complexes as product, albeit in moderate yield. The complexes were converted to dihydropyrans in good yield by further treatment with Et₃N [19a–d]. In the initial stage, the reaction was stepwise and thus stoichiometric, however, the reaction was made catalytic and more efficient by carrying out the reaction under direct photoirradiation in the presence of Et₃N or DABCO as base [19e,f]. Thus, irradiation of 4-pentynols **47** with 10 ∼ 25 mol% of W(CO)₆ in the presence of Et₃N or DABCO in THF at 50 ∼ 65°C gave the endo cyclized dihydropyran derivatives **48** in high yield (Schemes 5.16 and 5.17).

McDonald *et al.* applied their dihydropyran formation to the oligosaccharide synthesis featuring reiterative alkynol cycloisomerization [19c–f]. Dihydropyran **49** was prepared by the method described above and then NIS-promoted glycosylation with the acyclic alkynol **50** followed by Ph₃SnH-promoted dehalogenation and

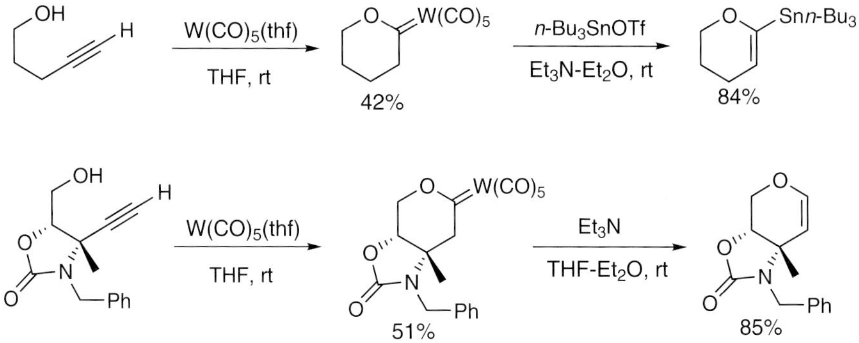

Scheme 5.16 4-Pentynol cyclization leading to dihydropyran derivatives: Stepwise.

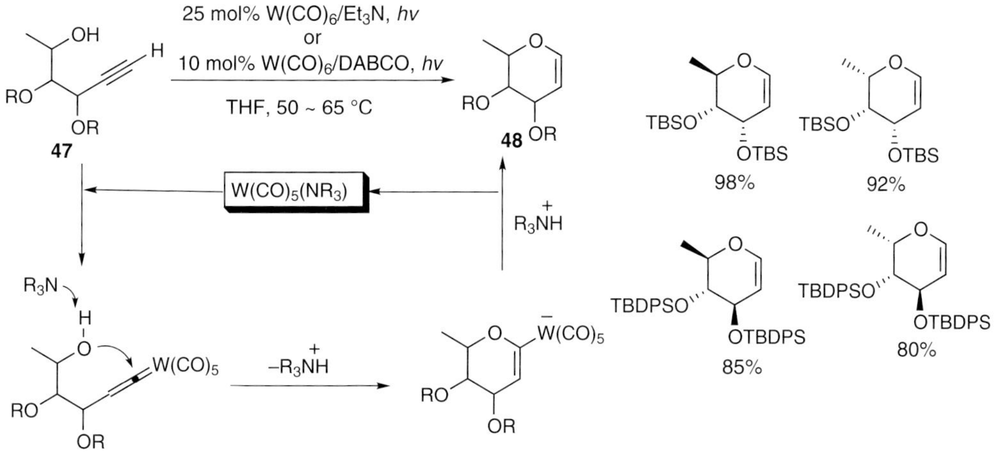

5.3 Utilization of Pentacarbonyl Vinylidene Complexes of Group 6 Metals for Synthetic Reactions

Scheme 5.17 4-Pentynol cyclization leading to dihydropyran derivatives: Catalytic.

then desilylation afforded the alkynyl glycoside **51**, the precursor for the next tungsten-promoted cycloisomerization. The acid-catalyzed glycosylation could be carried out with high β-selectivity, as shown in the second glycosylation. By repetition of this strategy, trisaccharide glycal **52** could be obtained in a concise manner. In this example, a stoichiometric amount of W(CO)$_5$(thf) was employed but, later, the catalytic version was employed for the preparation of several kinds of biologically active glycosides, one example being shown here. Stereoselective synthesis of C-3 branched 3-aminoglycals of vancosamine, saccharosamine, and desosamine, is also achieved utilizing this method [19g,h] (Schemes 5.18 and 5.19).

Further study of this type of alkynol cyclization by Wipf et al. showed that the 5-exo vs 6-endo cyclization pathway is competitive and highly dependent on the structure of the substrate [20]. For example, syn-diol **53** gave the 5-exo cyclized furan derivative **54**, obtained by the addition of the hydroxy group to the π-alkyne complex, while the corresponding anti-diol **55** gave 6-endo cyclized dihydropyran derivative **56** selectively. Based on the effect of the protective group of the propargylic hydroxy group on the exo versus endo selectivity, they proposed that an oxygen substituent at the propargylic position which can allow chelation with the metal would slow the rate of the vinylidene formation allowing exo-cyclization, while a bulky protective group at this hydroxy group would diminish such interaction allowing endo-cyclization. The steric compression could destabilize the π-complex, thus favoring the vinylidene complex formation and the conformation of the molecule might be another important factor for the selection of the reaction pathway (Scheme 5.20).

More recently McDonald et al. reported a rare example of 7-membered oxacycle formation based on the same concept [21]. Thus, polyoxygenated terminal alkynes **60** containing the dioxolane structure in the tether gave 7-endo cyclized glycals **61** in good yield. It is necessary to have the dioxolane ring in the tether for efficient reaction, probably favoring the suitable conformation for cyclization (Scheme 5.21).

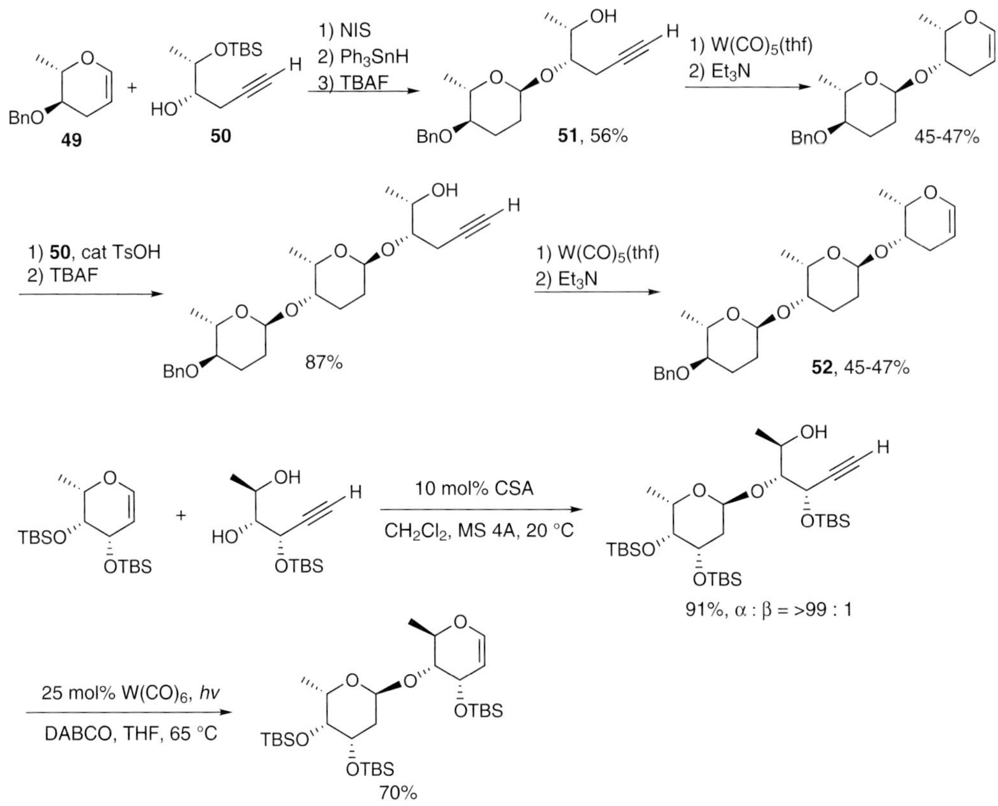

Scheme 5.18 Reiterative alkynol cycloisomerization.

Another related reaction developed by McDonald et al. is the synthesis of substituted furans from epoxyalkynes [15b]. Treatment of epoxyalkynes **62** with 15 mol% Mo(CO)$_5$(NEt$_3$) in Et$_2$O–Et$_3$N at rt gave furan derivatives **65** in good yield. The reaction was proposed to proceed through the epoxyvinylidene complex **63**, which underwent concerted rearrangement to the cyclic α,β-unsaturated cyclic carbene complex **64**. Finally, deprotonation of the α-proton of the carbene moiety, followed by protonation of the alkenylmolybdenum anion, gave the product. Interestingly, the related epoxyalkyne **66** containing a 3-butynol unit underwent nucleo-

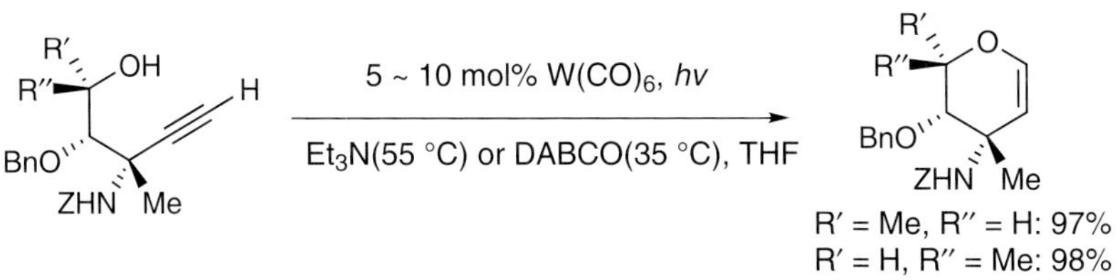

Scheme 5.19 Stereoselective synthesis of C-3 branched 3-aminoglycals.

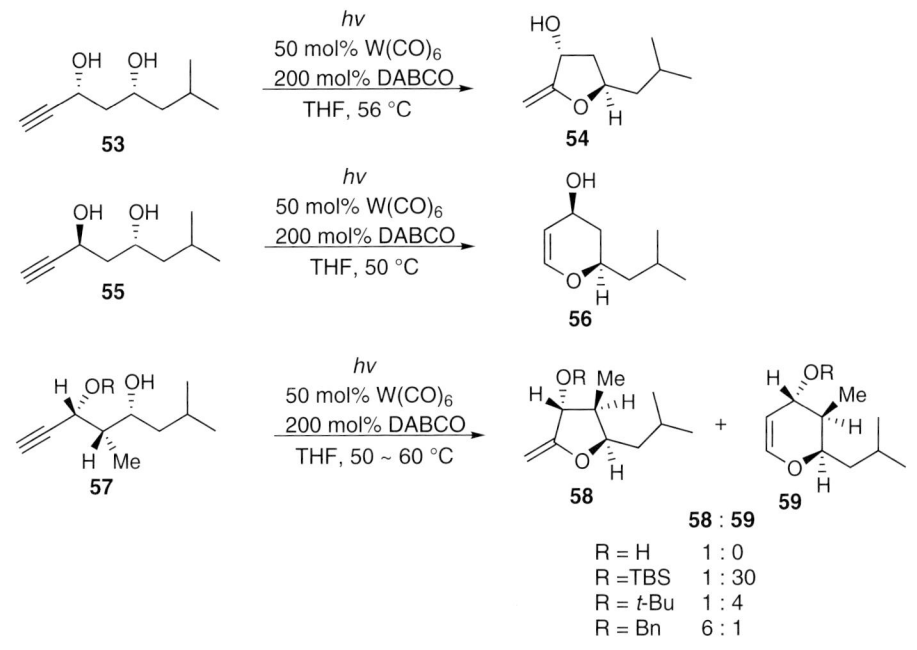

5.3 Utilization of Pentacarbonyl Vinylidene Complexes of Group 6 Metals for Synthetic Reactions | 171

Scheme 5.20 *Exo* vs *endo* cyclization.

philic addition of the hydroxy group onto the vinylidene carbon to give the alkenylmolybdenum intermediate **67**, which further underwent isomerization to give another furan derivative **68** in good yield (Scheme 5.22).

Use of the produced carbene complex intermediate for further carbon–carbon bond formation has been achieved by Barluenga and coworkers [22]. Treatment of the alkynol derivatives **69** with 25 mol% W(CO)₅(thf) in THF at rt gave tricyclic com-

Scheme 5.21 6-Alkynol cyclization leading to seven-membered ring formation.

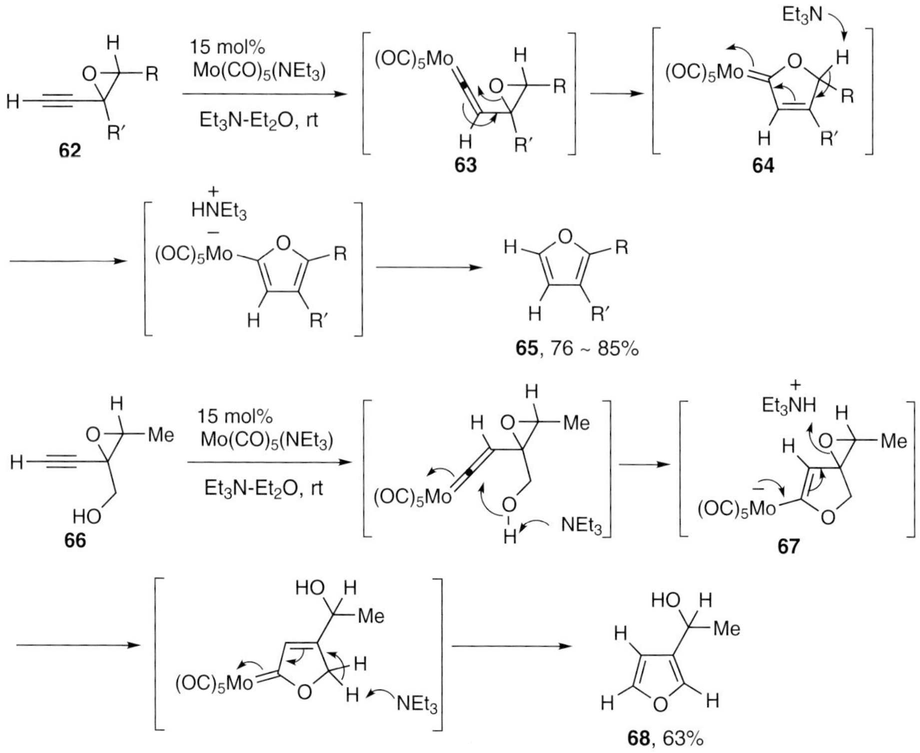

Scheme 5.22 Cycloisomerization of epoxyalkynes leading to furan derivatives.

pounds **70** diastereoselectively in high yield. The products were obtained by intramolecular cyclopropanation of the carbene complex intermediates **71** with regeneration of the catalyst. Furthermore, by carrying out the reaction under constant irradiation, higher efficiency of the reaction was achieved, giving similar results with only 10 mol% of the catalyst. This reaction was applied to variously substituted substrates to give the cyclopropanation product with complete diastereoselectivity. Further treatment of the products with hydrochloric acid in acetone gave the corresponding 8-membered carbocycles **72** in high yield (Scheme 5.23).

Computational analyses of these reactions have been carried out by several groups. As the *exo* versus *endo* selectivity of the reaction is sometimes dependent on the structure of the substrate, it is not necessarily easy to analyze the reaction in a general manner [22b, 23].

5.3.2
Catalytic Addition of Carbo-Nucleophiles

All the examples described in the previous section involve the addition of heteronucleophiles, such as alcohols and amines, to the vinylidene carbon. Addition of

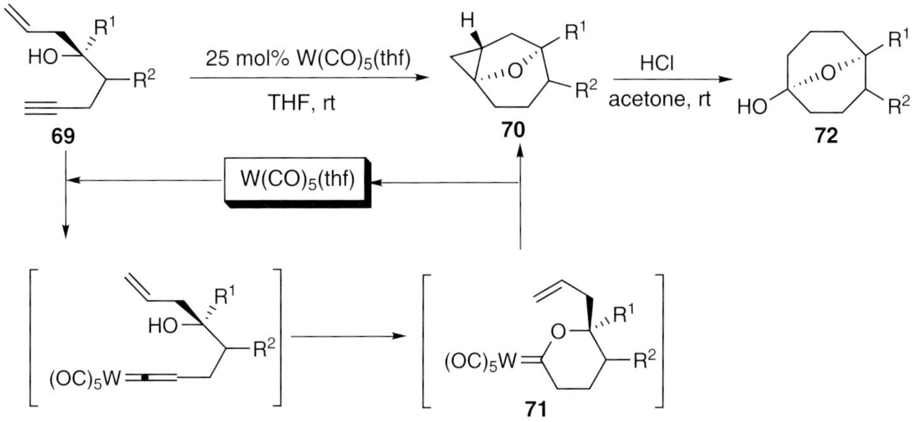

5.3 Utilization of Pentacarbonyl Vinylidene Complexes of Group 6 Metals for Synthetic Reactions

Scheme 5.23 Catalytic carbene formation-cyclopropanation.

carbon nucleophiles to the vinylidene carbon for the construction of carbocycles had not been reported until McDonald realized the first examples of this type of reaction using malonate anion as a carbon nucleophile [24]. Treatment of the sodium salt of active methylene compounds **73** with 50 mol% Mo(CO)$_5$(NEt$_3$) was found to give the *endo*-cyclized product **74** in reasonable yield. The homologous substrate **75** gave the *exo*-cyclized product **76**, indicating that here again π-alkyne and vinylidene complex pathways are competitive, depending on the substrate (Scheme 5.24).

In 1998, we reported an efficient method for the *endo*-selective cyclization of ω-acetylenic silyl enol ethers using a catalytic amount of W(CO)$_5$(L) [25a]. The most important feature of this reaction is the use of a silyl enol ether as a neutral, isolable carbon nucleophile. Treatment of 6-siloxy-5-en-1-yne **77** at rt with 1.5 equiv W(CO)$_5$(thf), prepared by irradiating W(CO)$_6$ in THF just before use, gave a cyclopentene derivative **78**, produced by intramolecular attack of the silyl enol ether on the terminal alkyne in an *endo* manner, in moderate yield. Other Group 6 metal complexes such as Cr(CO)$_5$(thf) and Mo(CO)$_5$(thf) failed to give better results, but

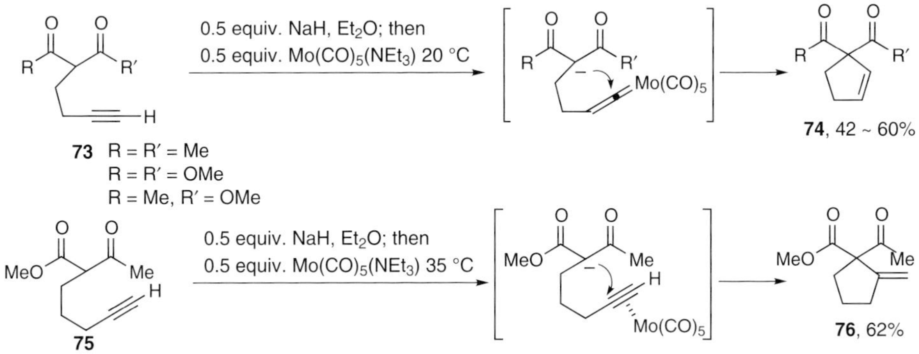

Scheme 5.24 Addition of carbo-nucleophile: active methylene compound.

the reaction did proceed in better yield (80%) in the presence of a proton source such as methanol or H_2O. The reaction is assumed to proceed as follows; treatment of the silyl enol ether **77** with $W(CO)_5(thf)$ generates a small amount of π–alkyne complexes **79** and/or a vinylidene complex **80**. On forming this complex, the alkyne part becomes electron deficient due to the electron-withdrawing nature of $W(CO)_5$, and intramolecular attack of the silyl enol ether occurs to give a vinyl metallic intermediate **81** and/or **82**, which is finally protonated to give the cyclopentene **78**. The reaction is thought to proceed via both pathways, based on the deuteration experiments in which both of the olefinic protons of **78** were partially deuterated (Ha, 33% D; Hb, 33% D; total D 66%) when the reaction of **77** was run in the presence of 10 equiv of D_2O. The most important aspect of this reaction is that $W(CO)_5(thf)$ is regenerated on protonation of the vinyl metallic intermediate **81** and/or **82**, and thus the reaction proceeds even with a catalytic amount of $W(CO)_5(thf)$. (30 mol%, 73% yield; 10 mol %, 56% yield) (Scheme 5.25).

Reactions of 5-siloxy-5-en-1-ynes are summarized in Table 5.2. In every case the reaction proceeds at rt in the presence of H_2O to give cyclized β,γ–unsaturated ketones in good yield without isomerization of the double bond, even using only 10 mol% $W(CO)_5(thf)$. The most characteristic feature of this reaction is that the *endo* mode of cyclization occurs more readily than in other related cyclization reactions such as $HgCl_2$-mediated reactions.

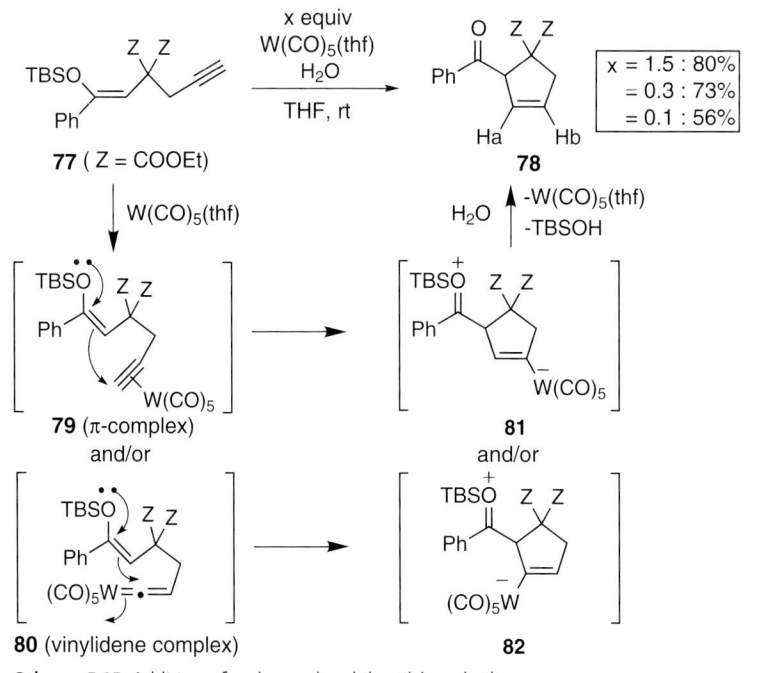

Scheme 5.25 Addition of carbo-nucleophile: silyl enol ether.

5.3 Utilization of Pentacarbonyl Vinylidene Complexes of Group 6 Metals for Synthetic Reactions

Table 5.2 Stoichiometric and catalytic cyclization of ω-acetylenic silyl enol ethers.[a]

Starting material	Product	Yield (%), Amount of W(CO)$_5$(thf)		
		1.2–1.5 eq	0.3 eq	0.1 eq
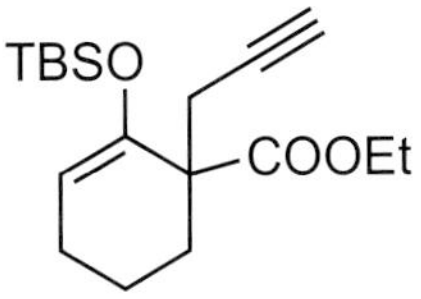 TBSO-cyclohexene-COOEt alkyne	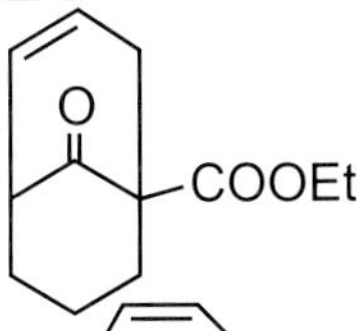 bicyclic ketone-COOEt	88	90	90
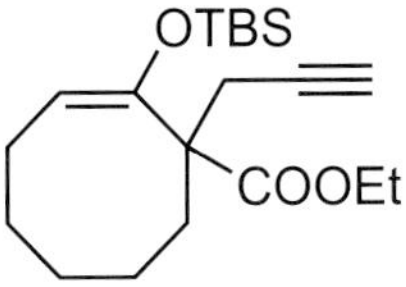 OTBS-cyclooctene-COOEt alkyne	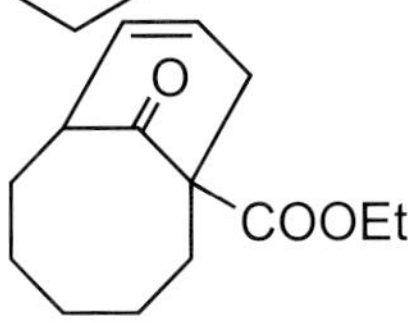 bicyclic ketone-COOEt	90	93	97
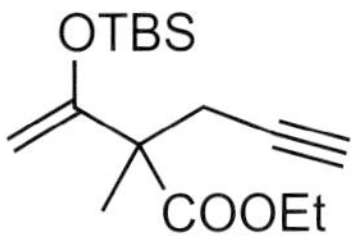 OTBS acyclic-COOEt alkyne	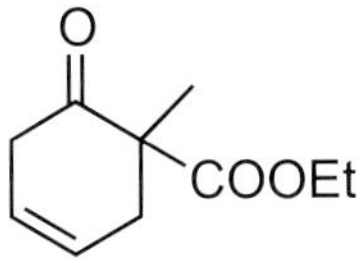 cyclohexenone-COOEt	74	77	64
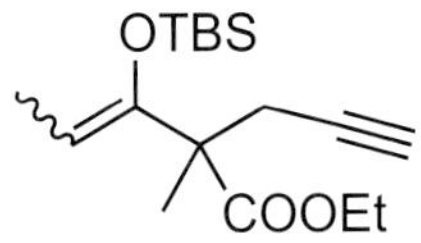 OTBS acyclic-COOEt alkyne	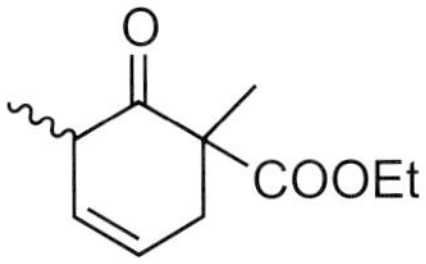 cyclohexenone-COOEt	80	82	83

[a] The reactions were performed in the presence of H$_2$O (2 equiv) in THF at rt.

The reaction of another type of substrate, 7-siloxy-6-en-1-yne **83** revealed the unique character of these W(CO)$_5$(L)-catalyzed reactions. Thus, the same reaction carried out under the same conditions gave the *exo*-cyclized product **84**, thought to be obtained via the π–alkyne complex, in high yield with a catalytic amount of W(CO)$_5$(thf). On the other hand, the reaction in a less polar solvent like CH$_2$Cl$_2$ gave the products with good *endo*-selectivity (**84** : **85** = 1 : 5). Thus, either *exo*- or *endo*-cyclized products could be obtained selectively by choosing appropriate reaction conditions (Scheme 5.26).

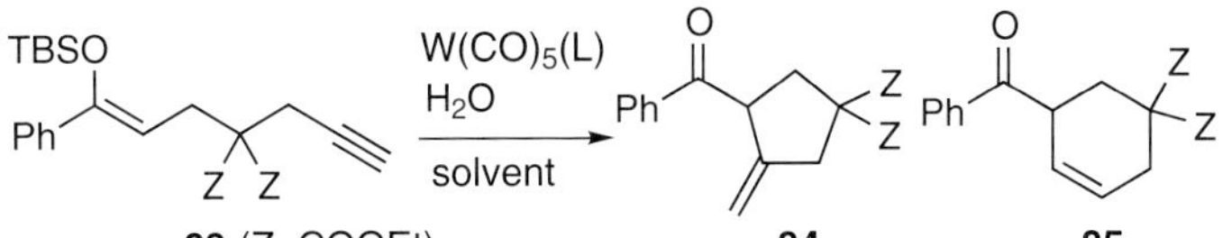

83 (Z=COOEt) → **84** + **85**

Conditions:
- 0.1 equiv of preformed W(CO)$_5$(thf)/THF: 94% **83** only
- 1.3 equiv of W(CO)$_6$/CH$_2$Cl$_2$, hv: 60% **84** : **85** = 1 : 5

Reagents: W(CO)$_5$(L), H$_2$O, solvent

Scheme 5.26 Reaction of 7-siloxy-6-en-1-yne: solvent dependent *exo* vs *endo* cyclization.

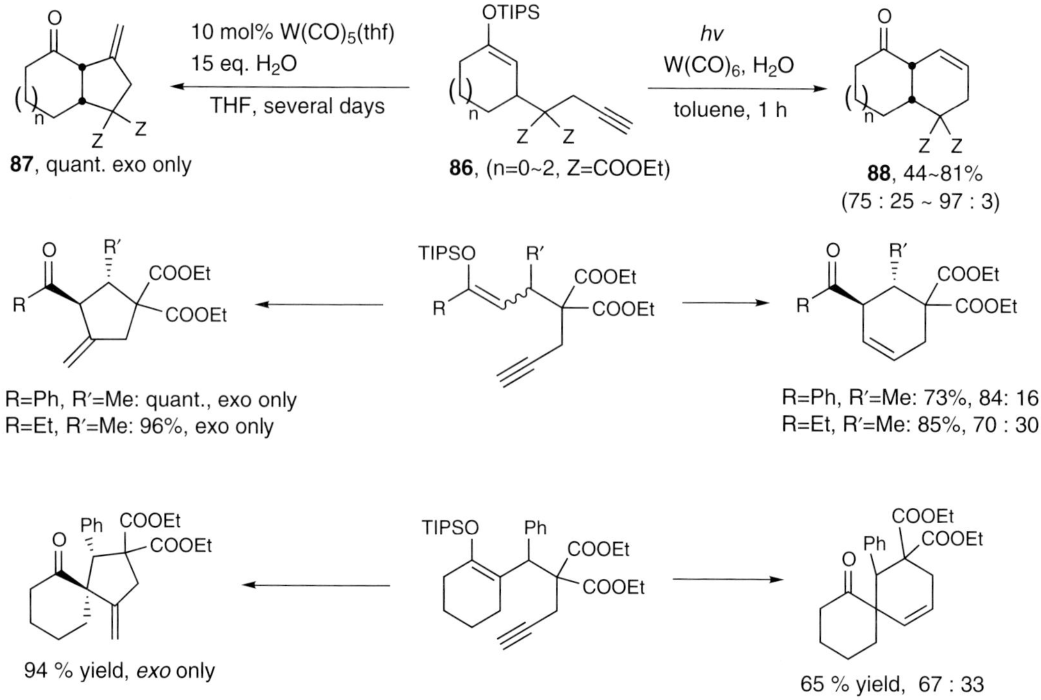

Scheme 5.27 Solvent controlled *exo* and *endo* selective cyclization of various 7-siloxy-6-en-1-ynes.

This reaction was further developed into 5- and 6-membered ring annulation onto α,β-unsaturated ketones through the following strategy [25b]. Thus, alkenones were converted to the corresponding 7-siloxy-6-en-1-ynes **86** in moderate to good yield by the reaction of the Na salt of propargylmalonate with cycloalkenone, followed by addition of TIPSOTf. Then, the *exo*-cyclizations proceeded smoothly to give the corresponding bicyclic methylenecyclopentane derivatives **87** in high yields by the use of 10 mol% of preformed W(CO)$_5$(thf) in the presence of H$_2$O in THF. For the *endo*-selective cyclization, a modest to good level of selectivity for formation of cyclohexene derivatives **88** was realized in most cases by directly irradiating a mixture of W(CO)$_6$ and a substrate in toluene in the presence of H$_2$O. Thus, synthetically useful bicyclo[m.3.0] and bicyclo[m.4.0] systems can be prepared from the same starting materials via the same intermediates simply by changing the reaction conditions in the final step. This reaction could also be applicable to acyclic α,β-unsaturated ketones (Scheme 5.27).

Furthermore, by carrying out the reaction in the presence of DABCO as base, the *exo*-cyclized silyl enol ether **89** was obtained in high yield by deprotonation of the α-proton of silyloxonium moiety of the zwitterionic intermediate **91** followed by protonation of the tungsten–carbon bond [25c]. *Endo*-cyclized silyl enol ether **90** could be obtained in good selectivity by carrying out the reaction in toluene by using bulky tributylamine. Thus again, both the *exo*- and *endo*-cyclized silyl enol ethers could be obtained in good yield by the appropriate choice of amine (Scheme 5.28).

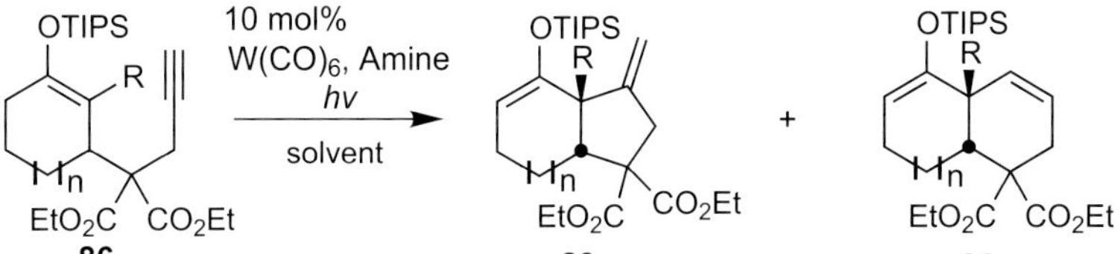

5.3 Utilization of Pentacarbonyl Vinylidene Complexes of Group 6 Metals for Synthetic Reactions

Scheme 5.28 The reaction of 7-siloxy-6-en-1-ynes in the presence of amine.

n	R	DABCO, THF	n-Bu₃N, toluene
0	H	88%, 90 : 10	97%, 16 : 84
0	Me	quant., >99 : 1	90%, .10 : 90
1	H	92%, 94 : 6	92%, 17 : 83
1	Me	quant., >99 : 1	73%, 11 : 89

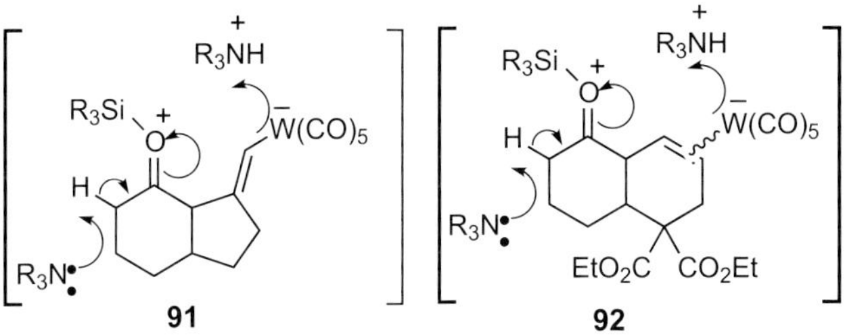

This reaction in the presence of base was applied to a tandem cyclization. When bis-alkynyl silyl enol ether **93a** was irradiated in toluene in the presence of 10 mol % W(CO)$_6$ and DABCO with 1 equiv of H$_2$O, the expected tricyclic ketone **94a** was obtained in 80% yield. The five-membered substrate **93b** also gave the corresponding tricyclic ketone **94b** having the basic carbon skeleton of the cedranes. Thus we can prepare synthetically useful tricyclic compounds utilizing this W(CO)$_5$(L)-catalyzed tandem cyclization in the presence of DABCO [25c] (Scheme 5.29).

Although the exact reaction mechanisms and origin of *endo* versus *exo* selectivity remain to be clarified, these reactions should find wide use in the

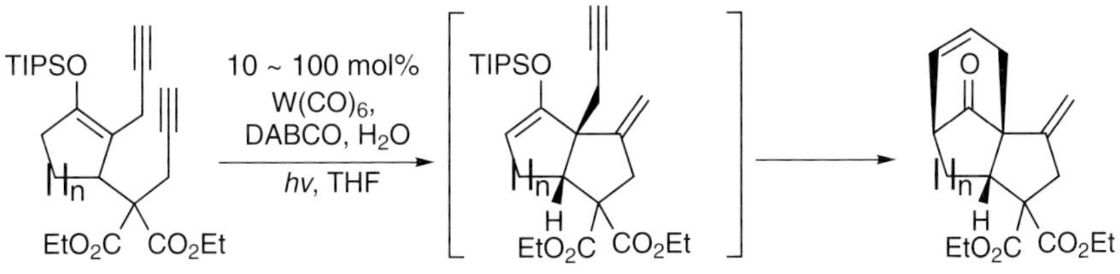

Scheme 5.29 Tandem cyclization of bis-alkynyl silyl enol ethers.

synthesis of such cyclic molecules due to their simple reaction procedures and wide generality.

5.3.3
Electrocyclization and Related Reactions

Electrocyclization is another major mode of reaction utilizing the vinylidene complexes of Group 6 metals. It should be noted that the first example of this type of reaction was reported by Merlic et al. using a ruthenium complex employing nonaromatic diene-yne substrates [26].

We examined $W(CO)_5(L)$ for the same type of reaction employing o-ethynylstyrene derivatives **95** as substrate [27]. Treatment of **95** with 5 mol% preformed $W(CO)_5(thf)$ at rt in THF afforded the substituted naphthalene derivatives **96** in high yield. The mechanism of the reaction was thought to be as follows: treatment of **95** with $W(CO)_5(thf)$ gives an alkyne–$W(CO)_5$ π-complex, which gradually isomerizes to the vinylidene intermediate **97**. Then 6π-electrocyclization proceeds to give a carbene intermediate **98**, which gives **96** by 1,2-hydrogen migration with regeneration of $W(CO)_5(thf)$. This reaction showed wide generality, and o-ethynylstyrenes with an electron-donating or electron-withdrawing substituent on the alkene moiety can be employed for this reaction without problem. Even benzene ortho-substituted with an ethynyl and a heteroaromatic group such as furan, thiophene and pyrrole, can be employed as substrate to give hetero-polyaromatics in high yield with 5 mol% $W(CO)_5(thf)$ (Scheme 5.30).

Akiyama et al. extended this reaction to alkynylimines for the preparation of quinoline derivatives [28]. Treatment of N-aryl(alkynyl)imines **99** with 20 mol% $W(CO)_5(thf)$ in THF at reflux followed by oxidative work-up using NMO gave 2-arylquinolines **100** in reasonable yield through electrocyclization of the vinylidene intermediate (Scheme 5.31).

This electrocyclization was also applied to iodoalkyne derivatives, where generation of novel iodinated vinylidene complexes was realized [29]. Thus, treatment of iodoalkynes **101** with 10~100 mol% $W(CO)_5(thf)$ gave the corresponding 1-iodinated naphthalene derivatives **103** in good yield. It should be noted that generation of iodinated vinylidene complexes **102** from iodoalkynes has rarely been achieved and the obtained iodides would be employable for further carbon–carbon bond formation. This iodovinylidene complex formation was also applicable to the silyl enol ether addition reaction, although a stoichiometric amount of $W(CO)_5(thf)$ was necessary for good conversion (Scheme 5.32).

Such electrocyclization through the vinylidene complex was further extended into the o-ethynylphenylcarbonyl or β-ethynyl α,β-unsaturated carbonyl compounds by Ohe et al. and our group, independently. In this case, dienone-pyran-type electrocyclization of the yne-ene-carbonyl moiety occurs through the vinylidene intermediate to give novel pyranylidene-type complexes. For example, we reported that treatment of o-ethynylphenylketone derivatives **104** with 3 equiv of preformed $W(CO)_5(thf)$ gave benzopyranylidene complexes **106** in good yield when carrying out the reaction at rt for 1 day [30] (Scheme 5.33).

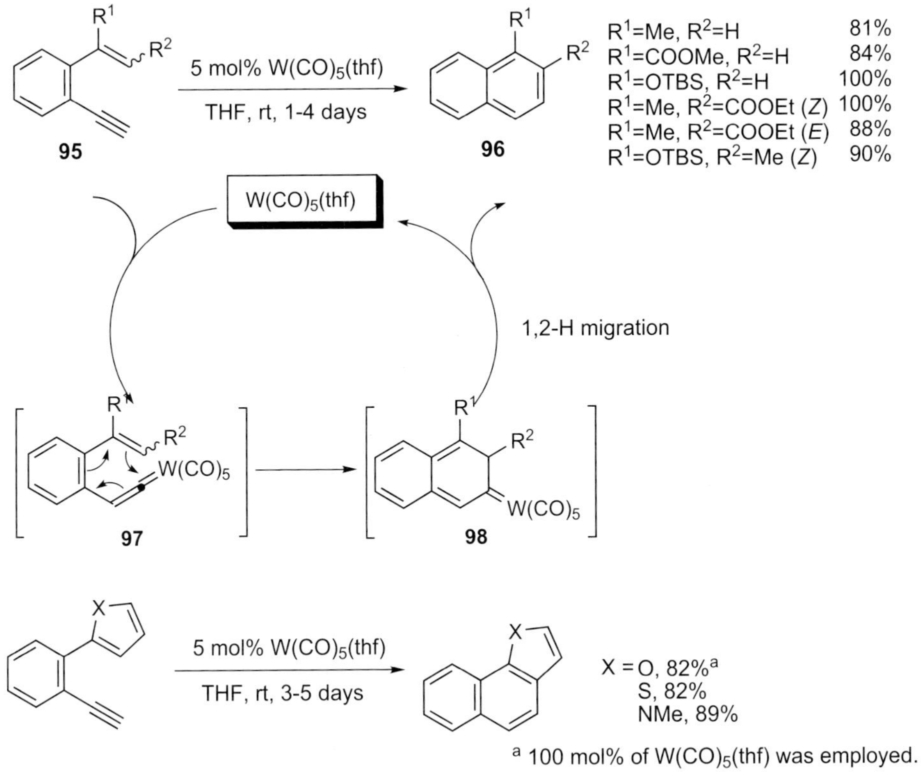

5.3 Utilization of Pentacarbonyl Vinylidene Complexes of Group 6 Metals for Synthetic Reactions | 179

R^1=Me, R^2=H	81%
R^1=COOMe, R^2=H	84%
R^1=OTBS, R^2=H	100%
R^1=Me, R^2=COOEt (Z)	100%
R^1=Me, R^2=COOEt (E)	88%
R^1=OTBS, R^2=Me (Z)	90%

X = O, 82%[a]
S, 82%
NMe, 89%

[a] 100 mol% of W(CO)$_5$(thf) was employed.

Scheme 5.30 Tungsten-catalyzed electrocyclization of dienynes.

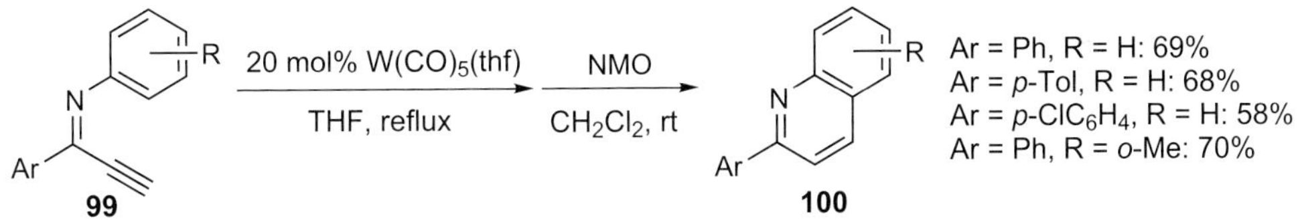

Ar = Ph, R = H: 69%
Ar = p-Tol, R = H: 68%
Ar = p-ClC$_6$H$_4$, R = H: 58%
Ar = Ph, R = o-Me: 70%

Scheme 5.31 Electrocyclization of N-aryl(alkynyl)imines leading to quinoline derivatives.

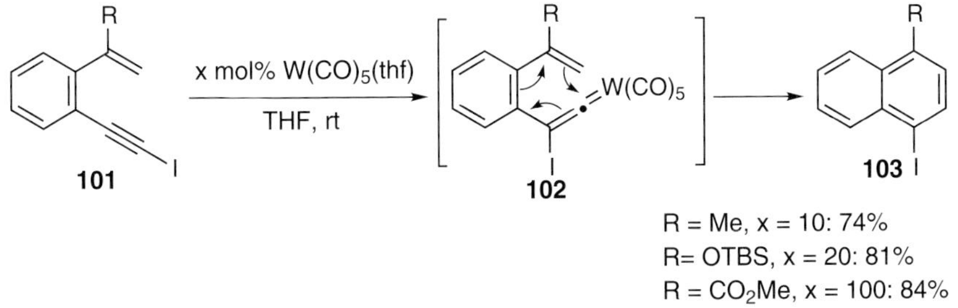

R = Me, x = 10: 74%
R = OTBS, x = 20: 81%
R = CO$_2$Me, x = 100: 84%

Scheme 5.32 Generation of iodovinylidene complexes for electrocyclization.

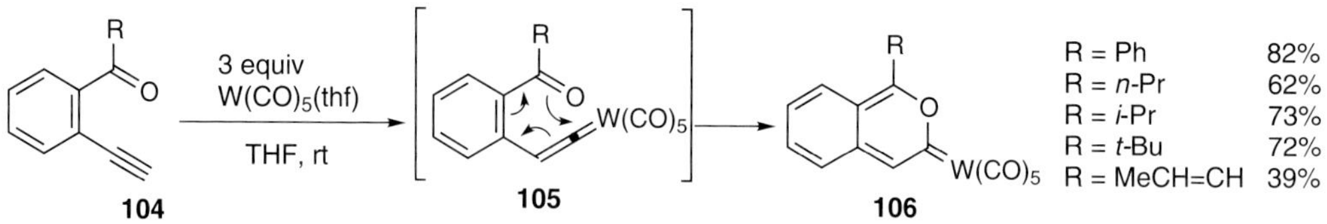

Scheme 5.33 Electrocyclization of o-ethynylphenylketone derivatives leading to benzopyranylidene complexes.

R = Ph	82%
R = n-Pr	62%
R = i-Pr	73%
R = t-Bu	72%
R = MeCH=CH	39%

Ohe, Uemura et al. reported similar electrocyclization of non-benzannulated 5-yn-3-en-1-one derivatives [31]. Thus, treatment of β-ethynyl α,β-unsaturated carboxylic esters and amides **107** with 3 equiv of preformed M(CO)$_5$(L) such as Mo(CO)$_5$(Et$_3$N) and W(CO)$_5$(thf) gave the corresponding pyranylidene complexes **109** in good yield. In the case of the Cr complex, it is necessary to carry out the reaction with Cr(CO)$_5$(thf) in the presence of Et$_3$N. Interestingly, the reaction of the corresponding phenyl ketone derivative **110** gave rather unstable (2-furyl) carbene complexes **111** in reasonable yield. They could be converted to the corresponding aldehydes by exposure to air. In this instance, the reaction likely proceeds through the 5-exo nucleophilic attack of the carbonyl oxygen on the alkyne electrophilically activated by π-complex formation with M(CO)$_5$. The produced zwitterioinic intermediate is the resonance isomer of the furyl carbene complex. The reaction of the same phenylketone derivative **110** in the presence of Et$_3$N gave the corresponding pyranylidene complex **113**, albeit in low yield (8%). It is likely that Et$_3$N accelerates the formation of the vinylidene complex **112**. This reaction of phenylketone derivatives was developed into the catalytic method for generation and utilization of furylcarbene complexes [31c,d]. Treatment of a mixture of **110** and alkenes with 5 mol% Cr(CO)$_5$(thf) afforded cyclopropane derivatives **114** in good yield through the furylcarbene complex intermediate **111** (Scheme 5.34).

The reactivity of the produced complexes was also examined [30a,b]. Since the benzopyranylidene complex **106** has an electron-deficient diene moiety due to the strong electron-withdrawing nature of W(CO)$_5$ group, **106** is expected to undergo inverse electron-demand Diels–Alder reaction with electron-rich alkenes. In fact, naphthalenes **116** variously substituted at the 1-, 2-, and 3-positions were prepared by the reaction of benzopyranylidene complexes **106** and typical electron-rich alkenes such as vinyl ethers, ketene acetals, and enamines through the Diels–Alder adducts **115**, which simultaneously eliminated W(CO)$_6$ and an alcohol or an amine at rt (Scheme 5.35).

In order to carry out this reaction by a one-pot procedure, we examined the same complexation reaction *in the presence of* a ketene acetal, which led us to the discovery of the novel reaction pathways described below [32]. Thus, treatment of o-ethynylphenyl ketone **104** with a catalytic amount of W(CO)$_5$(thf) *in the presence of* 4 equiv of 1,1-diethoxyethylene at rt gave a novel polycyclic compound **120** in good yield as a single stereoisomer. Both vinyl ethers and ketene acetals can be employed as the electron-

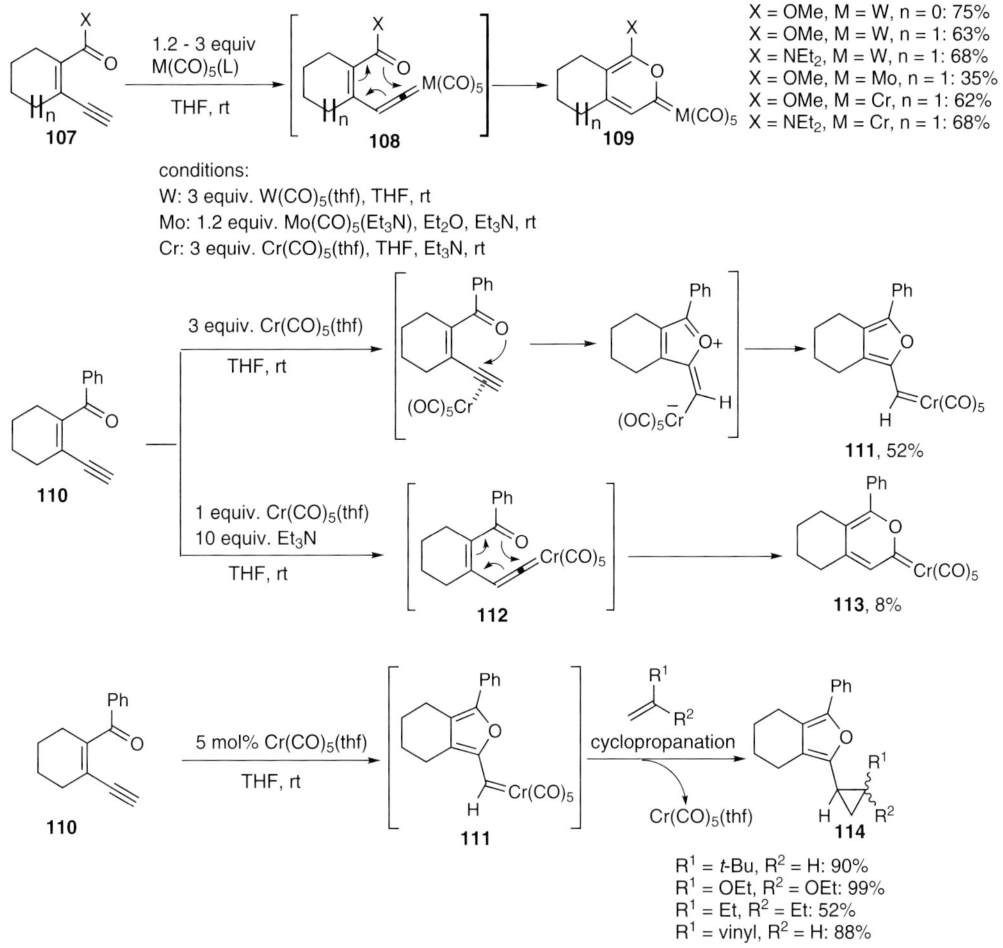

5.3 Utilization of Pentacarbonyl Vinylidene Complexes of Group 6 Metals for Synthetic Reactions | 181

X = OMe, M = W, n = 0: 75%
X = OMe, M = W, n = 1: 63%
X = NEt$_2$, M = W, n = 1: 68%
X = OMe, M = Mo, n = 1: 35%
X = OMe, M = Cr, n = 1: 62%
X = NEt$_2$, M = Cr, n = 1: 68%

conditions:
W: 3 equiv. W(CO)$_5$(thf), THF, rt
Mo: 1.2 equiv. Mo(CO)$_5$(Et$_3$N), Et$_2$O, Et$_3$N, rt
Cr: 3 equiv. Cr(CO)$_5$(thf), THF, Et$_3$N, rt

111, 52%

113, 8%

R^1 = t-Bu, R^2 = H: 90%
R^1 = OEt, R^2 = OEt: 99%
R^1 = Et, R^2 = Et: 52%
R^1 = vinyl, R^2 = H: 88%

Scheme 5.34 Electrocyclization of non-aromatic 5-yn-3-en-1-ones.

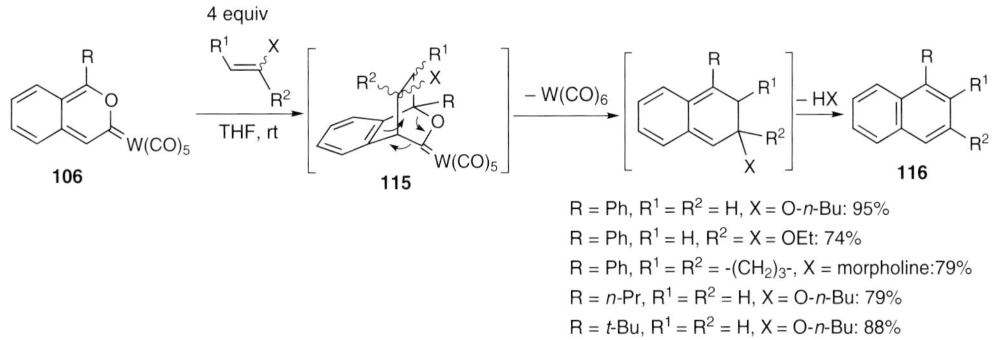

R = Ph, R^1 = R^2 = H, X = O-n-Bu: 95%
R = Ph, R^1 = H, R^2 = X = OEt: 74%
R = Ph, R^1 = R^2 = -(CH$_2$)$_3$-, X = morpholine: 79%
R = n-Pr, R^1 = R^2 = H, X = O-n-Bu: 79%
R = t-Bu, R^1 = R^2 = H, X = O-n-Bu: 88%

Scheme 5.35 Inverse electron demands Diels–Alder reaction of benzopyranylidene complexes with electron-rich olefins.

rich alkene component to give the corresponding polycyclic products in good yield with 10 or 20 mol% W(CO)$_5$(thf).

The mechanism of formation of this compound is thought to be as follows. Treatment of W(CO)$_5$(thf) with o-ethynylphenyl ketone **104** gives a metal-containing carbonyl ylide intermediate **118** through the *endo*-mode of attack of the carbonyl oxygen on the W(CO)$_5$–π-complexed alkynyl part of the molecule. Then, [3 + 2]-cycloaddition of **118** with the ketene acetal proceeds to give an unstable carbene complex **119**. Finally, the W(CO)$_5$–carbene moiety thus generated inserts into a C–H bond of the neighboring ethoxy group to give the product **120** while regenerating W(CO)$_5$(thf).

Furthermore, when o-ethynylphenyl methyl ketone **104** (R = Me) was treated with W(CO)$_5$(thf) in the presence of 5 equiv of water, o-acetylacetophenone **123** (R = Me) was obtained in about 50% yield. Direct observation of the reaction mixture in THF-d$_8$ clearly showed the formation of a methyleneisobenzofuran derivative **122** (R' = H) at rt. This compound is produced by the exo-mode of attack of the carbonyl oxygen on the W(CO)$_5$–π-complexed alkyne to give **121** (R = Me), followed by protonation of the C–W bond and deprotonation from the methyl group (Scheme 5.36).

All these phenomena can be explained as follows. Thus, the reaction of o-ethynylphenyl and W(CO)$_5$(thf) can proceed through the following three pathways; (i) *exo*-attack of the carbonyl oxygen on the π-complexed alkyne to give **121**, which is a resonance isomer of benzofuranylidene complex **124**; (ii) *endo*-attack of the carbonyl oxygen on the π-complexed alkyne to give carbonyl ylide **118**; and (iii) 1,2-H shift to give vinylidene complex **105**, which then undergoes irreversible electrocyclization to give the pyranylidene complex **106**. We currently suppose that the reactions (i) and (ii) are faster than (iii), but that pathways (i) and (ii) are under rapid equilibrium. Thus, in the presence of a reagent capable of trapping intermediate **121** or **118**, such as H$_2$O or electron-rich alkenes, the reaction proceeds through either pathway (i) or (ii) to give the corresponding hydrolysis product **123** (path (i)) or [3 + 2]-cycloaddition product **120** (path (ii)). However, in the absence of such trapping reagents, an equilibrium exists between **121**, **118** and the π-complex **117**, and formation of the vinylidene complex **105** occurs as a relatively slower process, which gives the benzopyranylidene complex **106** through an irreversible unimolecular electrocyclization. Thus, this reaction involves dynamic equilibria of three reaction pathways, which can be partially controlled by the order of addition of the reagents.

Ohe, Uemura et al. further developed this electrocyclization into the [3,3]-sigmatropy of a cyclopropane system substituted with ethynyl and an acyl or an alkenyl group [33]. Thus, treatment of *cis*-1-acyl-2-ethynylcyclopropanes **125** with 5 mol% Cr CO)$_5$(thf) in the presence of Et$_3$N at rt induced isomerization to phenol derivatives through the [3,3]-sigmatropy of the vinylidene intermediate **126** to give 1-oxa-2,5-cycloheptadiene-7-ylidene complex **127**, which undergoes 1,2-hydrogen shift with elimination of Cr(CO)$_5$ to give oxepin intermediates **128**. The oxepins were not isolated but were converted into phenol derivatives **130** via arene oxides **129** under the reaction conditions. Furthermore, the same kind of reaction

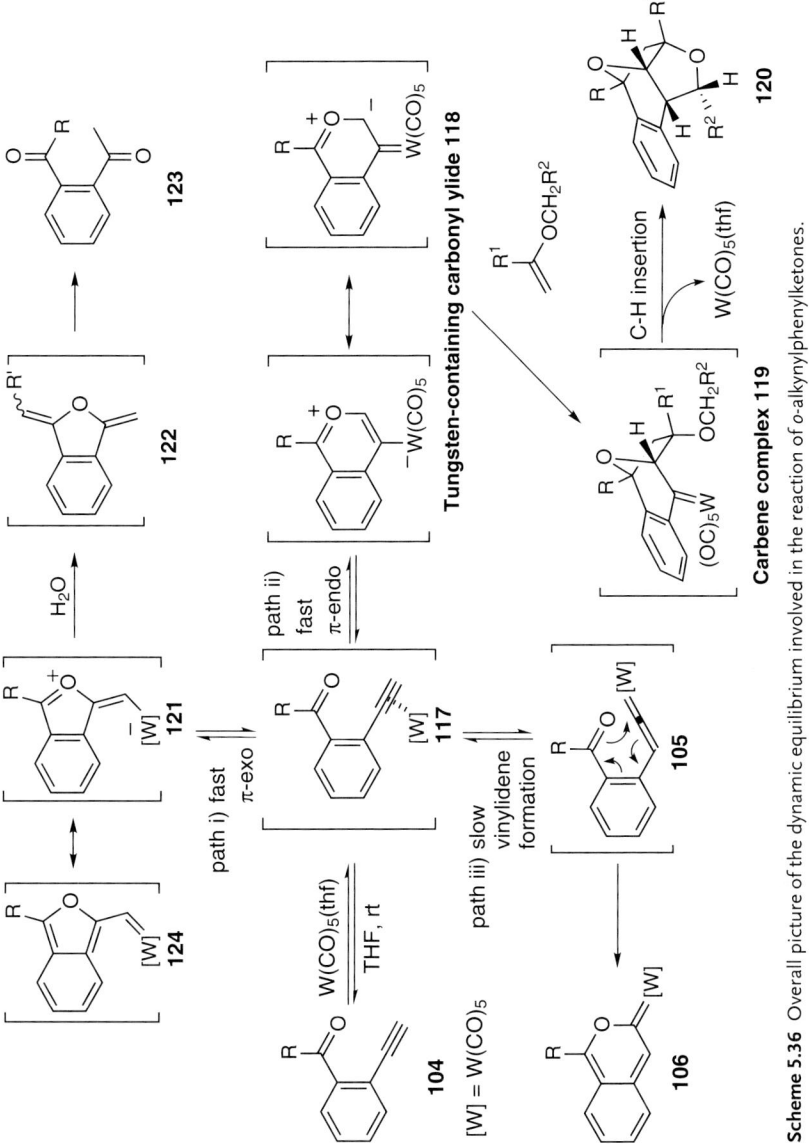

Scheme 5.36 Overall picture of the dynamic equilibrium involved in the reaction of o-alkynylphenylketones.

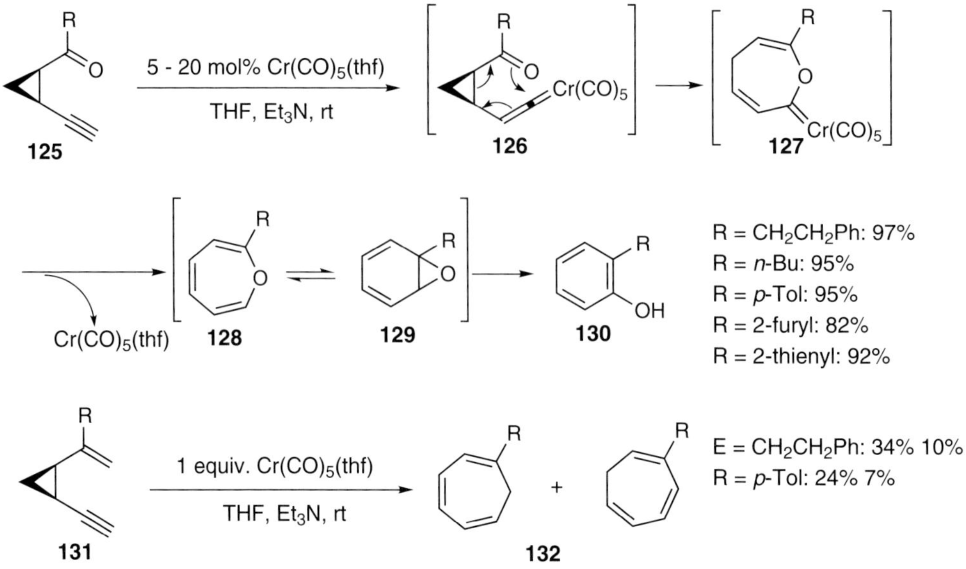

Scheme 5.37 Electrocyclization of cis-1-acyl-2-ethynylcyclopropanes.

employing cis-1-alkenyl-2-ethynylcyclopropanes **131** gave cycloheptatrienes **132** in a similar sequence although a stoichiometric amount of Cr(CO)$_5$(thf) was employed (Scheme 5.37).

5.4
Utilization of Vinylidene to Alkyne Conversion

In this short section are described two unique examples utilizing the inverse transformation of vinylidene complexes to alkynes.

Söderberg et al. reported that treatment of α,β-unsaturated acyl chromate complexes **133** with 4-methoxybenzoyl chloride gave terminal alkynes **134** in moderate yield together with the γ-deprotonation product **135** [34]. They proposed the [3,3]-sigmatropic rearrangement of in situ formed α,β-unsaturated Fischer acyloxy carbene complexes **136**, which gave vinylidene complex intermediates **137**. Then these complexes underwent inverse hydrogen shift to give the terminal alkynes **134** (Scheme 5.38).

Wulff et al. recently reported another unique example of this inverse transformation [35]. Thus, treatment of α,β-unsaturated Fischer carbene complexes **138** with an isopropoxy group on the carbene carbon with ketene acetal **139** at 80 °C in THF under CO pressure gave 4-pentynoate derivatives **140** in good yield. The reaction was proposed to proceed through 1,4-addition of ketene acetal to the carbene complex to give a zwitterionic intermediate **141**. This underwent internal

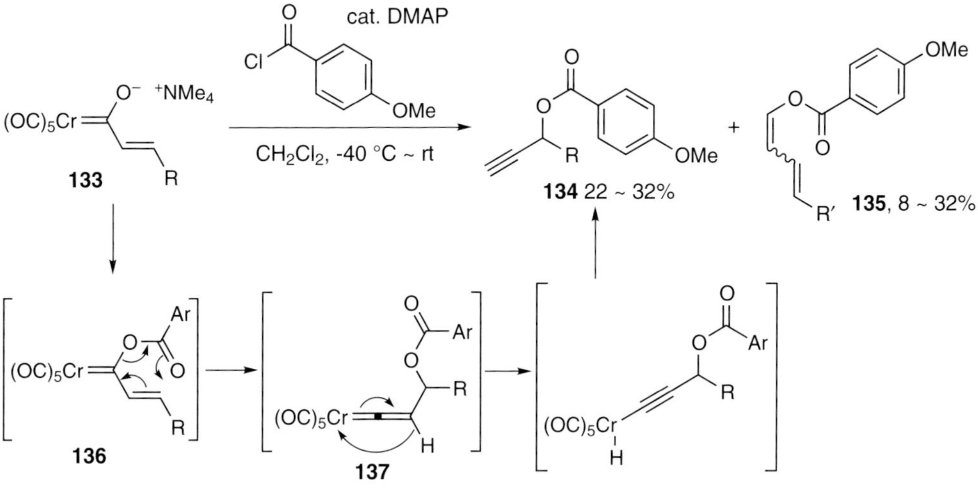

Scheme 5.38 Vinylidene to alkyne conversion-1.

isopropoxide transfer to generate a vinylidene complex **142**, from which a hydrogen, or more notably, a phenyl group, on the vinylidene carbon underwent 1,2-migration to give 4-pentynoate esters after hydrolysis of the orthoester (Scheme 5.39).

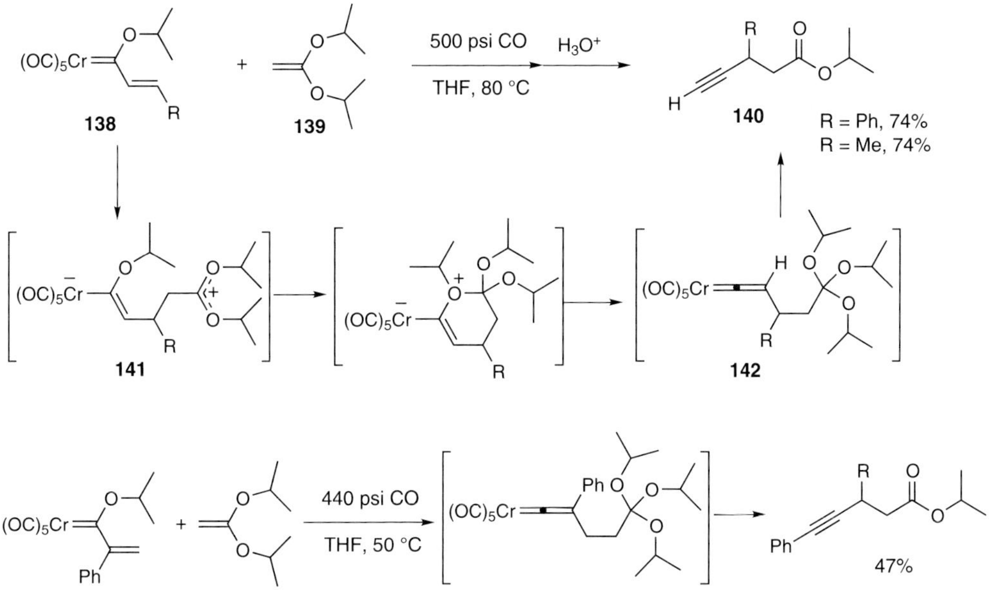

Scheme 5.39 Vinylidene to alkyne conversion-2.

5.5
Synthetic Reactions Utilizing Other Kinds of Vinylidene Complexes of Group 6 Metals

Liu et al. reported a novel cycloalkenylation through vinylidene intermediates starting from alkynyltungsten complexes [36]. Treatment of tungsten-η^1-alkynol compounds **143**, prepared by the reaction of CpW(CO)$_3$Cl and alkynols in Et$_2$NH in the presenceof CuI, with PhCHO and BF$_3$·OEt$_2$ in Et$_2$O at $-40\,°C$ gave air-sensitive cyclized furanylidene or pyranylidene tungsten complexes **145** in nearly quantitative yield. The reaction was thought to proceed through the vinylidene complex intermediates **144** generated by the nucleophilic attack of alkynyltungsten on activated PhCHO at the β-position of the tungsten. Treatment of the obtained complexes **145** with water under air in CH$_2$Cl$_2$ gave oxidized unsaturated γ- and δ-lactones **146** in high yields. Further examination of the reaction of the complexes revealed that double nucleophilic addition occurs at the carbene moiety to give various useful compounds [36a] (Scheme 5.40).

This reaction was further developed into the intramolecular [3 + 2]-cycloaddition of epoxides to alkynes to give bicyclic lactones [36b]. Treatment of alkynyltungstens containing an epoxide moiety **147** with BF$_3$·OEt$_2$ in CH$_2$Cl$_2$ at $-40\,°C$ gave, after aqueous work-up, cis-γ-lactones **150** fused with a five- or six-membered ring in good yield. In this case again, alkynyltungsten attacks the epoxide activated by BF$_3$·OEt$_2$ to give vinylidene intermediates **148**, the vinylidene carbon of which is attacked by the generated oxy group to give formal [3 + 2]-cycloaddition products **149** between epoxides and alkynes. The carbene moiety was hydrolyzed during the work-up procedure. When optically active epoxides were employed as

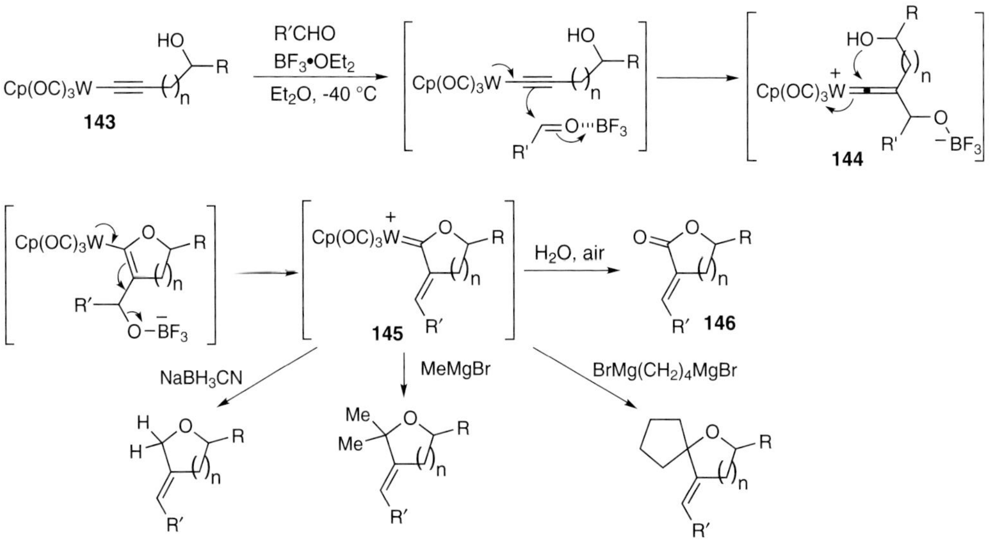

Scheme 5.40 The reaction of alkynyltungsten with aldehydes: synthesis of cyclic ethers.

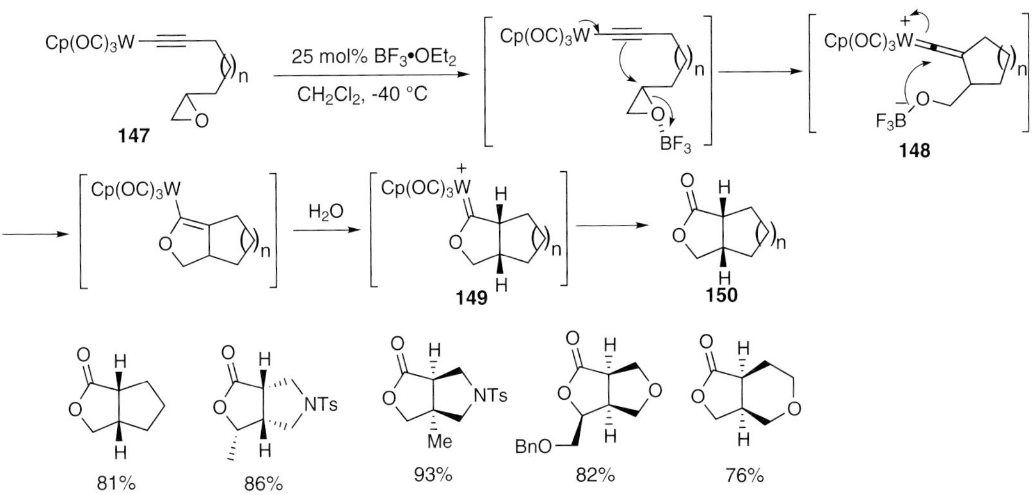

Scheme 5.41 The reaction of alkynyltungsten with epoxides; synthesis of bicyclic lactones.

substrate, the reaction mostly proceeded with inversion at the epoxide carbon to give synthetically useful, optically active bicyclic lactones stereospecifically (Scheme 5.41).

Murakami et al. reported a ring-closing metathesis reaction of allenynes using Schrock's molybdenum alkylidene complex [37]. Treatment of allenynes **151** with a catalytic amount of the complex **152** in toluene at rt gave cyclopentene derivatives **153** in good yield. Two possible reaction mechanisms were proposed, one through a vinylidene complex **154** and the other through a carbene complex, but based on several mechanistic studies, they favored the vinylidene complex pathway, which is shown here (Scheme 5.42).

5.6 Conclusion

As described in this chapter, vinylidene complexes of Group 6 metals have been utilized for the preparation of various synthetically useful compounds through electrophilic activation or electrocyclization of terminal alkyne derivatives. These intermediates are quite easily generated from terminal alkynes and $M(CO)_6$, mostly by photo-irradiation and will have abundant possibilities for the catalytic activation of terminal alkynes. Furthermore, it should be emphasized that one of the most notable characteristic features of the vinylidene complexes of Group 6 metals is their dynamic equilibrium with the π-alkyne complex. Control of such an equilibrium would bring about new possibilities for unique metal catalysis in synthetic reactions.

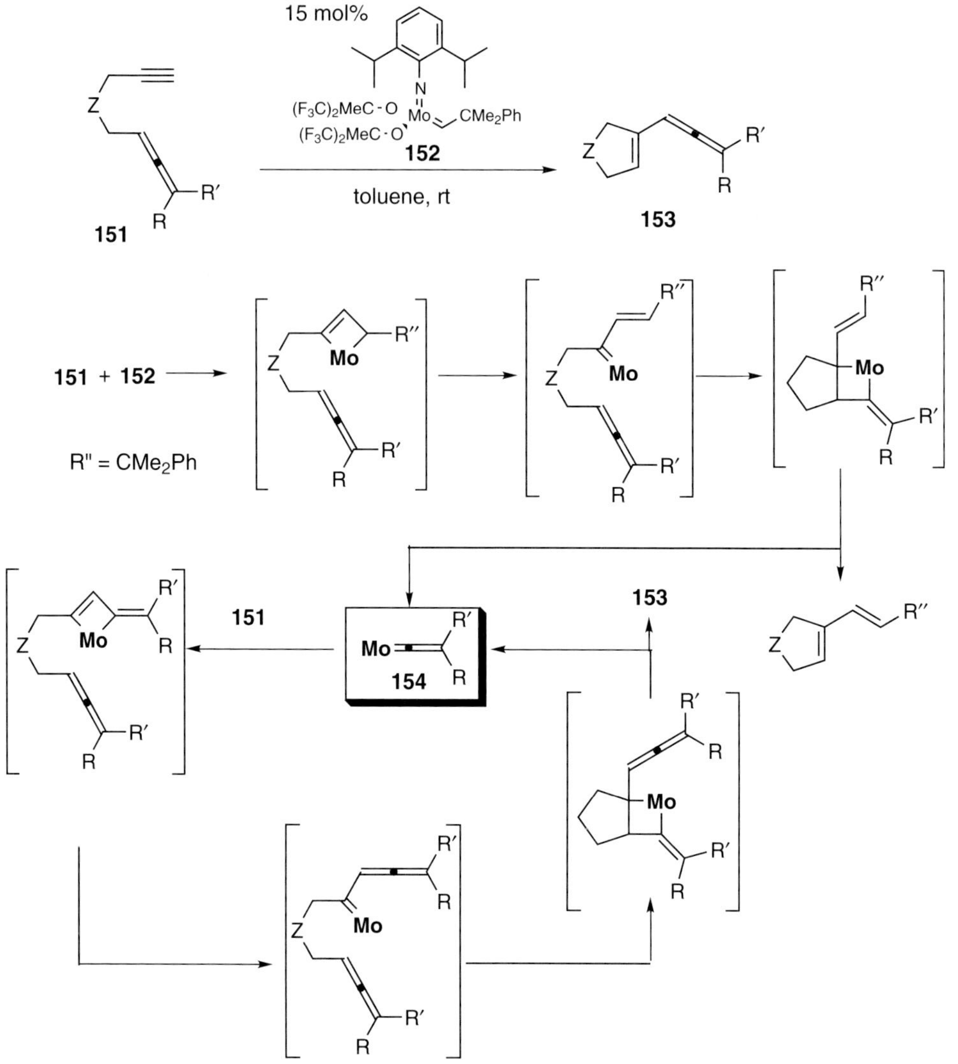

Scheme 5.42 Ring-closing metathesis reaction of allenynes via vinylidene complex.

References

1 (a) Bruce, M.I. and Swincer, A.G. (1983) *Advances in Organometallic Chemistry*, **22**, 59–128. (b) Bruce, M.I. (1991) *Chemical Reviews*, **91**, 197–257. (c) Bruce, M.I. (1998) *Chemical Reviews*, **98**, 2797–2858. (d) Selegue, J.P. (2004) *Coordination Chemistry Reviews*, **248**, 1543–1563.

2 (a) McDonald, F.E. (1999) *Chemistry – A European Journal*, **5**, 3103–3106. (b) Bruneau, C. and Dixneuf, P.H. (1999) *Accounts of Chemical Research*, **32**, 311–323.

(c) Bruneau, C. and Dixneuf, P.H. (2006) *Angewandte Chemie-International Edition*, **45**, 2176–2203.

3 (a) Weiss, K., Fischer, E.O. and Müller, J. (1974) *Chemische Berichte*, **107**, 3548–3553. See also. (b) Barrett, A.G.M., Brock, C.P. and Sturgess, M.A., (1985) *Organometallics*, **4**, 1903–1905. (c) Barrett, A.G.M., Mortier, J., Sabat, M. and Sturgess, M.A. (1988) *Organometallics*, **7**, 2553–2561.

4 Levisalles, J., Rudler, H., Jeannin, Y. and Dahan, F. (1979) *Journal of Organometallic Chemistry*, **178**, C8–C12.

5 Fischer, H., Schlageter, A., Bidell, W. and Früh, A. (1991) *Organometallics*, **10**, 389–391.

6 Mayr, A., Schaefer, K.C. and Huang, E.Y. (1984) *Journal of the American Chemical Society*, **106**, 1517–1518.

7 (a) Fischer, H., Karl, C.C. and Roth, G. (1996) *Chemische Berichte*, **129**, 615–622. (b) Fischer, H., Podschadly, O., Roth, G., Herminghaus, S., Klewitz, S., Heck, J., Houbrechts, S. and Meyer, T. (1997) *Journal of Organometallic Chemistry*, **541**, 321–332. (c) Fischer, H., Kirchbauer, F., Früh, A., Abd-Elzaher, M.M., Roth, G., Karl, C.C. and Dede, M. (2001) *Journal of Organometallic Chemistry*, **620**, 165–173.

8 Fischer, H., Podschadly, O., Früh, A., Troll, C., Stumpf, R. and Schlageter, A. (1992) *Chemische Berichte*, **125**, 2667–2673.

9 (a) Fischer, H., Volkland, H.-P., Früh, A. and Stumpf, R. (1995) *Journal of Organometallic Chemistry*, **491**, 267–273. (b) Abd-Elzaher, M.M. and Fischer, H. (1999) *Journal of Organometallic Chemistry*, **588**, 235–241. (c) Abd-Elzaher, M.M., Froneck, T., Roth, G., Gvozdev, V. and Fischer, H. (2000) *Journal of Organometallic Chemistry*, **599**, 288–297.

10 Dötz, K.H. and Sturm, W. (1985) *Journal of Organometallic Chemistry*, **285**, 205–211.

11 Parlier, A. and Rudler, H. (1986) *Journal of the Chemical Society: Chemical Communications*, 514–515.

12 (a) Dötz, K.H., Sturm, W. and Alt, H.G. (1987) *Organometallics*, **6**, 1424–1427.

(b) Weyershausen, B. and Dötz, K.H. (1999) *European Journal of Inorganic Chemistry*, 1057–1066.

13 (a) Le Bozec, H., Cosset, C. and Dixneuf, P.H. (1991) *Journal of the Chemical Society: Chemical Communications*, 881–882. (b) Cosset, C., Rio, I.D. and Le Bozec, H. (1995) *Organometallics*, **14**, 1938–1944. (c) Dötz, K.H., Paetsch, D. and Bozec, H.L. (1999) *Journal of Organometallic Chemistry*, **589**, 11–20.

14 Landon, S.J., Shulman, P.M. and Geoffroy, G.L. (1985) *Journal of the American Chemical Society*, **107**, 6739–6740.

15 (a) McDonald, F.K., Connolly, C.B., Gleason, M.M., Towne, T.B. and Treiber, K.D. (1993) *The Journal of Organic Chemistry*, **58**, 6952–6953. (b) McDonald, F.K. and Schultz, C.C. (1994) *Journal of the American Chemical Society*, **116**, 9363–9364. (c) McDonald, F.K. and Chatterjee, A.K. (1997) *Tetrahedron Letters*, **38**, 7687–7690. (d) McDonald, F.E., Schultz, C.C. and Chatterjee, A.K. (1995) *Organometallics*, **14**, 3628–3629. (e) McDonald, F.E., Burova, S.A. and Huffman, L.G. Jr. (2000) *Synthesis*, 970–974.

16 (a) McDonald, F.K. and Gleason, M.M. (1995) *Angewandte Chemie-International Edition in English*, **34**, 350–352. (b) McDonald, F.K. and Gleason, M.M. (1996) *Journal of the American Chemical Society*, **118**, 6648–6659.

17 (a) Quayle, P., Rahman, S., Ward, E.L.M. and Herbert, J. (1994) *Tetrahedron Letters*, **35**, 3801–3804. (b) Quayle, P., Ward, E.L.M. and Taylor, P. (1994) *Tetrahedron Letters*, **35**, 8883–8884. (c) Quayle, P., Rahman, S. and Herbert, J. (1995) *Tetrahedron Letters*, **36**, 8087–8088.

18 Schmidt, B., Kocienski, P. and Reid, G. (1996) *Tetrahedron*, **52**, 1617–1630.

19 (a) McDonald, F.E. and Bowman, J.L. (1996) *Tetrahedron Letters*, **37**, 4675–4678. (b) McDonald, F.E. and Zhu, H.Y.H. (1997) *Tetrahedron*, **53**, 11061–11068. (c) McDonald, F.E. and Zhu, H.Y.H. (1998) *Journal of the*

American Chemical Society, **120**, 4246–4247. (d) Bowman, J.L. and McDonald, F.E. (1998) *The Journal of Organic Chemistry*, **63**, 3680–3682. (e) McDonald, F.E., Reddy, K.S. and Díaz, Y. (2000) *Journal of the American Chemical Society*, **122**, 4304–4309. (f) McDonald, F.E. and Reddy, K.S. (2001) *Journal of Organometallic Chemistry*, **617–618**, 444–452. (g) Cutchins, W.W. and McDonald, F.E. (2002) *Organic Letters*, **4**, 749–752. (h) Davidson, M.H. and McDonald, F.E. (2004) *Organic Letters*, **6**, 1601–1603. (i) Koo, B. and McDonald, F.E. (2007) *Organic Letters*, **9**, 1737–1740. See also, (j) Parker, K.A. and Chang, W., (2003) *Organic Letters*, **5**, 3891–3893.

20 (a) Wipf, P. and Graham, T.H. (2003) *The Journal of Organic Chemistry*, **68**, 8798–8807. (b) Moilanen, S.B. and Tan, D.S. (2005) *Organic Biomolecular Chemistry*, **3**, 798–803.

21 Alcázar, E., Pletcher, J.M. and McDonald, F.E. (2004) *Organic Letters*, **6**, 3877–3880.

22 (a) Barluenga, J., Diéguez, A., Rodríguez, F. and Fañanás, F.J. (2005) *Angewandte Chemie-International Edition*, **44**, 126–128. (b) Barluenga, J., Diéguez, A., Rodríguez, F., Fañanás, F.J., Sordo, T. and Campomanes, P. (2005) *Chemistry – A European Journal*, **11**, 5735–5741.

23 (a) Sheng, Y., Musaev, D.G., Reddy, K.S., McDonald, F.E. and Morokuma, K. (2002) *Journal of the American Chemical Society*, **124**, 4149–4157. (b) Nowroozi-Isfahani, T., Musaev, D.G., McDonald, F.E. and Morokuma, K. (2005) *Organometallics*, **24**, 2921–2929. (c) Campomanes, P., Menéndez, M.I. and Sordo, T.L. (2006) *Chemistry – A European Journal*, **12**, 7929–7934.

24 McDonald, F.K. and Olson, T.C. (1997) *Tetrahedron Letters*, **38**, 7691–7692.

25 (a) Maeyama, K. and Iwasawa, N. (1998) *Journal of the American Chemical Society*, **120**, 1928–1929. (b) Iwasawa, N., Maeyama, K. and Kusama, H. (2001) *Organic Letters*, **3**, 3871–3873. (c) Kusama, H., Yamabe, H. and Iwasawa, N. (2002) *Organic Letters*, **4**, 2569–2571. (d) Iwasawa, N., Miura, T., Kiyota, K., Kusama, H., Lee, K. and Lee, P.H. (2002) *Organic Letters*, **4**, 4463–4466. See also, (e) Kusama, H., Yamabe, H., Onizawa, Y., Hoshino, T. and Iwasawa, N., (2005) *Angewandte Chemie-International Edition*, **44**, 468–470. (f) Kusama, H., Onizawa, Y. and Iwasawa, N. (2006) *Journal of the American Chemical Society*, **128**, 16500–16501.

26 Merlic, C.A. and Pauly, M.E. (1996) *Journal of the American Chemical Society*, **118**, 11319–11320.

27 Maeyama, K. and Iwasawa, N. (1999) *The Journal of Organic Chemistry*, **64**, 1344–1346.

28 Sangu, K., Fuchibe, K. and Akiyama, T. (2004) *Organic Letters*, **6**, 353–355.

29 (a) Miura, T. and Iwasawa, N. (2002) *Journal of the American Chemical Society*, **124**, 518–519. (b) Miura, T., Murata, H., Kiyota, K., Kusama, H. and Iwasawa, N. (2004) *Journal of Molecular Catalysis A-Chemical*, **213**, 59–71.

30 (a) Iwasawa, N., Shido, M., Maeyama, K. and Kusama, H. (2000) *Journal of the American Chemical Society*, **122**, 10226–10227. (b) Kusama, H., Shiozawa, F., Shido, M. and Iwasasa, N. (2002) *Chemistry Letters*, **31**, 124–125.

31 (a) Ohe, K., Miki, K., Yokoi, T., Nishino, F. and Uemura, S. (2000) *Organometallics*, **19**, 5525–5528. (d) Miki, K., Yokoi, T., Nishino, F., Ohe, K. and Uemura, S. (2002) *Journal of Organometallic Chemistry*, **645**, 228–234. (c) Miki, K., Nishino, F., Ohe, K. and Uemura, S. (2002) *Journal of the American Chemical Society*, **124**, 5260–5261. (d) Miki, K., Yokoi, T., Nishino, F., Kato, Y., Washitake, Y., Ohe, K. and Uemura, S. (2004) *The Journal of Organic Chemistry*, **69**, 1557–1564. (e) Miki, K., Uemura, S. and Ohe, K. (2005) *Chemistry Letters*, **34**, 1068–1073.

32 (a) Iwasawa, N., Shido, M. and Kusama, H. (2001) *Journal of the American Chemical Society*, **123**, 5814–5815. (b) Kusama, H., Funami, H., Shido, M., Hara, Y., Takaya, J. and Iwasawa, N. (2005) *Journal of the American Chemical Society*, **127**, 2709–2716. (c) Kusama, H. and Iwasawa, N. (2006) *Chemistry Letters*, **35**, 1082–1087. See also: (a) Kusama, H., Takaya, J. and Iwasawa, N., (2002) *Journal of the American Chemical Society*, **124**, 11592–11593. (b) Takaya, J., Kusama, H. and Iwasawa, N. (2004) *Chemistry Letters*, **33**, 16–17.

33 Ohe, K., Yokoi, T., Miki, K., Nishino, F. and Uemura, S. (2002) *Journal of the American Chemical Society*, **124**, 526–527.

34 Söderberg, B.C., O' Neil, S.N., Chisnell, A.C. and Liu, J. (2000) *Tetrahedron*, **56**, 5037–5044.

35 Wang, S.L.B., Liu, X., Ruiz, M.C., Gopalsamuthiram, V. and Wulff, W.D. (2006) *European Journal of Organic Chemistry*, 5219–5224.

36 (a) Liang, K.-W., Li, W.-T., Peng, S.-M., Wang, S.-L. and Liu, R.-S. (1997) *Journal of the American Chemical Society*, **119**, 4404–4412. (b) Madhushaw, R.J., Li, C.-L., Shen, K.-H., Hu, C.-C. and Liu, R.-S. (2001) *Journal of the American Chemical Society*, **123**, 7427–7428.

37 Murakami, M., Kadowaki, S. and Matsuda, T. (2005) *Organic Letters*, **7**, 3953–3956.

6
Ruthenium Vinylidenes in the Catalysis of Carbocyclization
Arjan Odedra and Rai-Shung Liu

6.1
Introduction

Highly reactive organic vinylidene and allenylidene species can be stabilized upon coordination to a metal center [1]. In 1979, Bruce *et al.* [2] reported the first ruthenium vinylidene complex from phenylacetylene and [RuCpCl(PPh$_3$)$_2$] in the presence of NH$_4$PF$_6$. Following this report, various ruthenium vinylidene complexes have been isolated and their physical and chemical properties have been extensively elucidated [3]. As the α-carbon of ruthenium vinylidenes and the α and γ-carbon of ruthenium allenylidenes are electrophilic in nature [4], the direct formation of ruthenium vinylidene and ruthenium allenylidene species, respectively, from terminal alkynes and propargylic alcohols provides easy access to numerous catalytic reactions since nucleophilic addition at these carbons is a viable route for new catalysis (Scheme 6.1).

In this chapter, we focus on catalytic carbocyclization involving ruthenium vinylidene as a reaction intermediate. The carbocyclization reactions involving allenylidene intermediates are described in Chapter 7. The reactions are categorized according to the types of substrates. Several stoichiometric carbocyclizations are also included in this chapter because of their mechanistic significance.

6.2
Stoichiometric Carbocyclization via Ruthenium Vinylidene

Ruthenium vinylidene species can be transformed into small carbocyclic rings via carbocyclization reactions. Ruthenium vinylidene complex **2**, generated from the electrophilic reaction of alkyne complex **1** with haloalkanes, was deprotonated with ^{n}Bu$_4$NOH to give the unprecedented neutral cyclopropenyl complex **3** (Scheme 6.2) [5]. Gimeno and Bassetti prepared ruthenium vinylidene species **4a** and **4b** bearing a pendent vinyl group; when these complexes were heated in chloroform for a brief period, cyclobutylidene products **5a** and **5b** formed via a [2 + 2] cycloaddition between the vinylidene C$_\alpha$=C$_\beta$ bond and olefin (Scheme 6.3) [6].

Metal Vinylidenes and Allenylidenes in Catalysis: From Reactivity to Applications in Synthesis
Edited by Christian Bruneau and Pierre Dixneuf
Copyright © 2008 WILEY-VCH Verlag GmbH & Co. KGaA, Weinheim
ISBN: 978-3-527-31892-6

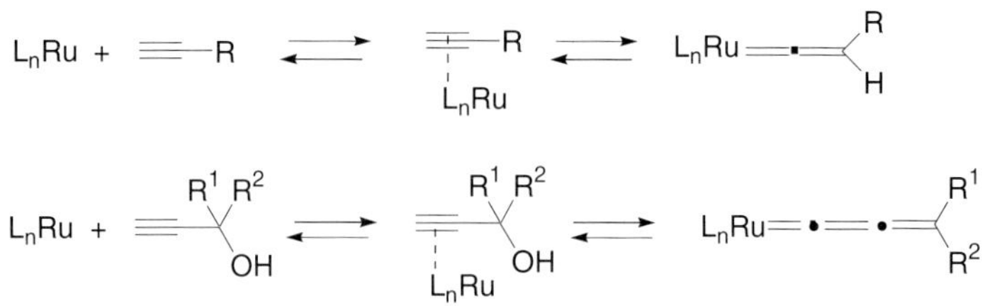

Scheme 6.1

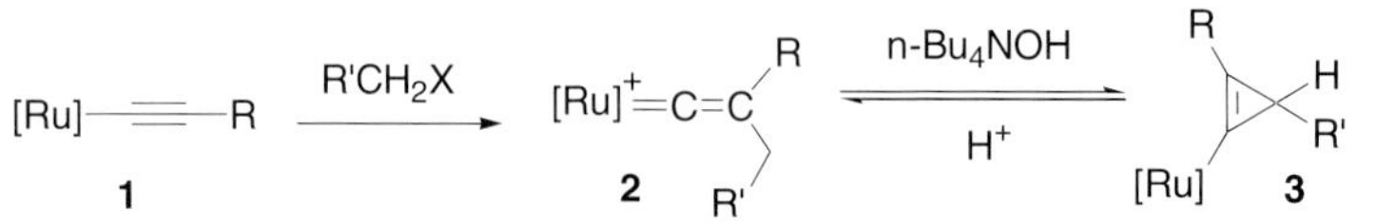

Scheme 6.2

[Ru] = RuCp(PPh$_3$)$_2$

R = C$_6$H$_5$, R' = CN, C$_6$H$_5$, CH=CH$_2$, CH=CMe$_2$

Scheme 6.3

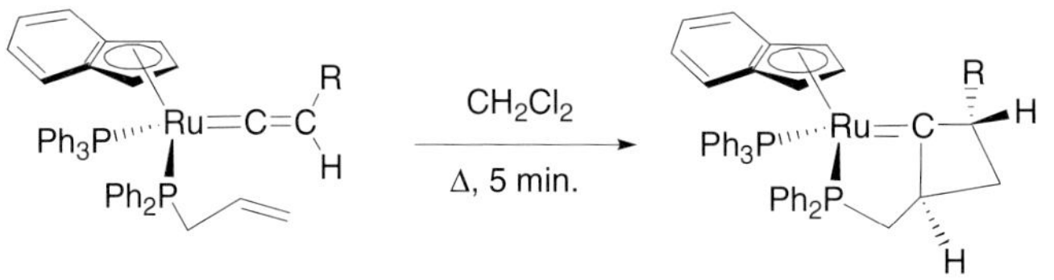

R = Ph (**4a**), p-MeC$_6$H$_4$ (**4b**)   R = Ph (**5a**), p-MeC$_6$H$_4$ (**5b**)

Finn et al. reported the first instance of a metal-catalyzed aromatization of enediynes via vinylidene intermediates [7]. Aromatization of unstrained enediynes is known as Bergman cyclization and occurs at 200–250 °C via diradical intermediates [8]. Ruthenium-vinylidene complex **7** was formed when 1,2-benzodiyne **6** was treated with RuCp(PMe$_3$)$_2$Cl and NH$_4$PF$_6$ at 100 °C, ultimately giving good naphthalene product **8** in good yields (Scheme 6.4). This process mimics Myers-Saito cyclization of 5-allene-3-

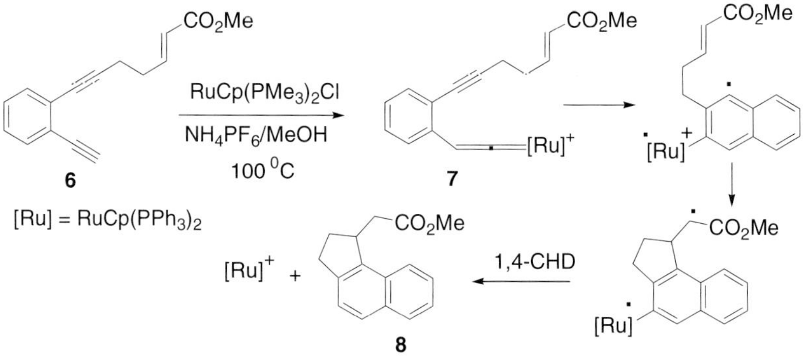

[Ru] = RuCp(PPh$_3$)$_2$

Scheme 6.4

en-1-ynes occurring at 50–100 °C [9]. Despite its stoichiometric nature, the concept behind this cyclization has stimulated intensive investigations including a rhodium-catalyzed aromatization of enediynes by Ohe and Uemura [10].

6.3
Catalytic Carbocyclization via Electrocyclization of Ruthenium-Vinylidene Intermediates

6.3.1
Cyclization of cis-3-En-1-Ynes

The first catalytic cyclization of cis-3-en-1-ynes was implemented by [Ru(Tp)(PPh$_3$)(CH$_3$CN)$_2$]PF$_6$ (Tp = tris(1-pyrazolyl)borate) catalyst to give cyclopentadiene derivatives. This process mimics the thermal rearrangement of cis-1-allen-4-enes via the generation of ruthenium-vinylidene intermediates (10) as depicted in Scheme 6.5 [11]. A crucial ^{2}H-labeling experiment provides direct support for a 1,5-sigmatropic hydrogen shift, whereas an oxidative C–H bond insertion by a RuTp(II) species can be completely excluded.

This ruthenium catalyst efficiently transforms cis-3-en-1-yne **d$_3$-9** into cyclopentadienyl species **d$_3$-11**; selected examples are depicted in Table 6.1. The cyclization works not only for cis-3-en-1-ynes bearing a benzylic hydrogen but also for those bearing aliphatic C–H bonds. Table 6.1 also manifests additional use of this cyclization that 5-siloxyl-3-en-1-ynes were transformed into cyclopentenone derivatives using the same ruthenium catalyst.

This hydrogen shift is extensible to the cyclization of 2-substituted-1-ethynylbenzene **12** to give 1-substituted 1H-indene derivative **14** selectively using Ru(Tp)(PPh$_3$)(CH$_3$CN)$_2$SbF$_6$ [12]. The corresponding PF$_6$ salt was unsuitable for this catalytic reaction because of its facile hydrolysis under more severe conditions. This indene synthesis is more difficult to achieve because the preceding [1, 5]-hydrogen shift leads to dearomatization of a benzene ring. The success of the

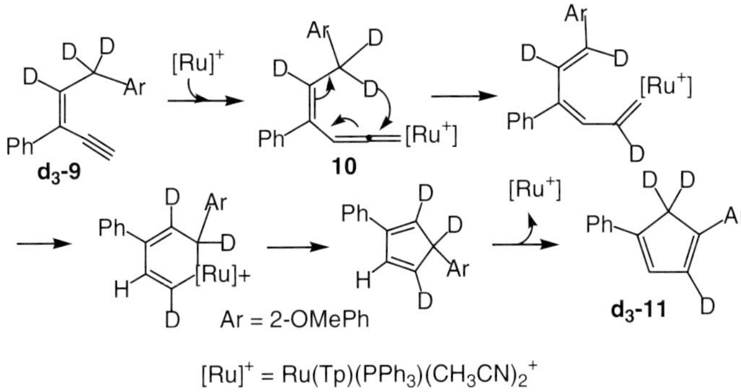

Scheme 6.5

Table 6.1 Ruthenium-catalyzed cyclization of 3-en-1-ynes.

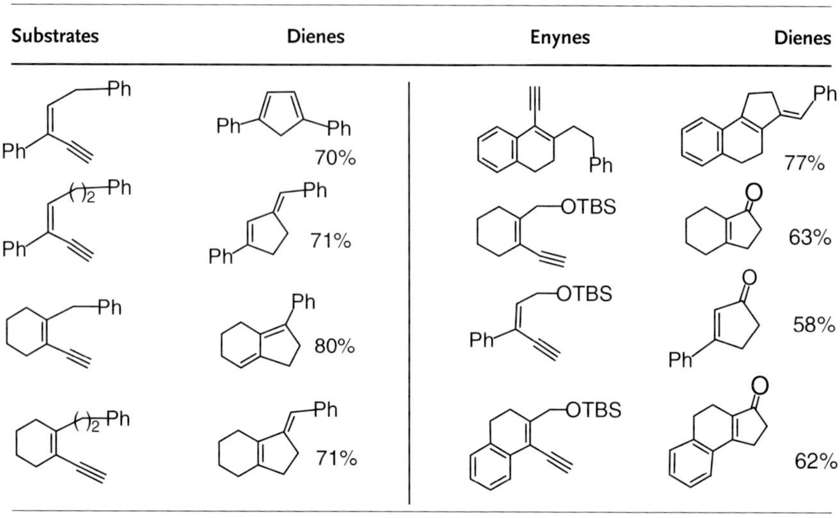

Substrates	Dienes	Enynes	Dienes
Ph-CH=C(Ph)-C≡CH with CH2Ph	Ph-cyclopentadiene-Ph 70%	naphthyl-CH2CH2-Ph enyne	tricyclic-Ph 77%
Ph-CH=C(Ph)-C≡CH with (CH2)2Ph	Ph-cyclopentene-Ph 71%	cyclohexenyl-CH2-OTBS enyne	bicyclic ketone 63%
cyclohexenyl-Ph enyne	indene-Ph 80%	Ph-CH2-CH2-C≡C with OTBS	Ph-cyclopentene ketone 58%
cyclohexenyl-(CH2)2Ph enyne	indene=CHPh 71%	naphthyl-CH2-C≡C with OTBS	tricyclic ketone 62%

cyclization relies on the stabilization of the transition state structure **13** having a pentadienyl cation character; the examples shown in Scheme 6.6 support this hypothesis as electron-rich ethynylbenzenes are more efficient for the cyclization.

6.3.2
Cycloaromatization of 3,5-Dien-1-Ynes

The cyclization of dienynes provides an interesting example of ruthenium vinylidene species in catalysis. This cyclization reaction is based on the nucleophilic

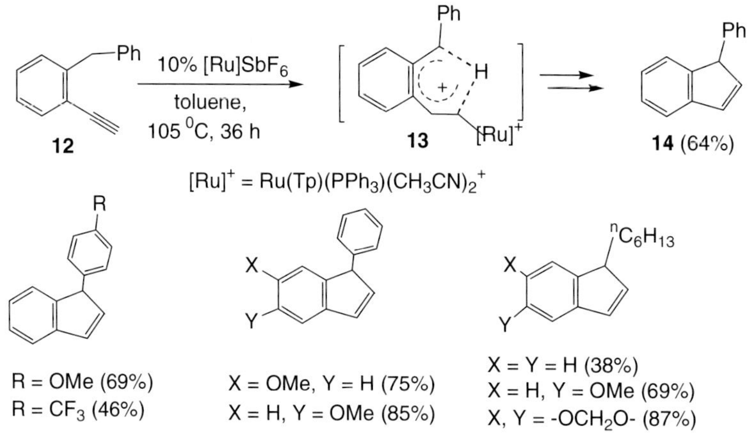

$$\text{12} \xrightarrow[\text{toluene, 105 °C, 36 h}]{10\% \text{ [Ru]SbF}_6} \left[\begin{array}{c} \text{Ph} \\ \text{H} \\ \text{[Ru]}^+ \end{array} \right] \longrightarrow \text{14 (64\%)}$$

$$[\text{Ru}]^+ = \text{Ru(Tp)(PPh}_3)(\text{CH}_3\text{CN})_2{}^+$$

R = OMe (69%)
R = CF$_3$ (46%)

X = OMe, Y = H (75%)
X = H, Y = OMe (85%)

X = Y = H (38%)
X = H, Y = OMe (69%)
X, Y = -OCH$_2$O- (87%)

Scheme 6.6

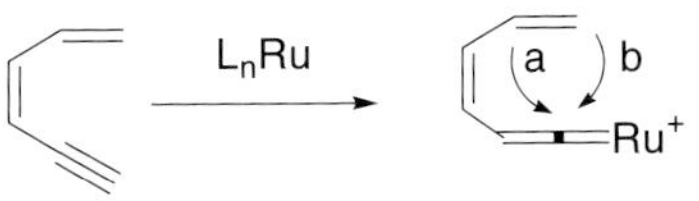

6.3 Catalytic Carbocyclization via Electrocyclization of Ruthenium-Vinylidene Intermediates

Scheme 6.7

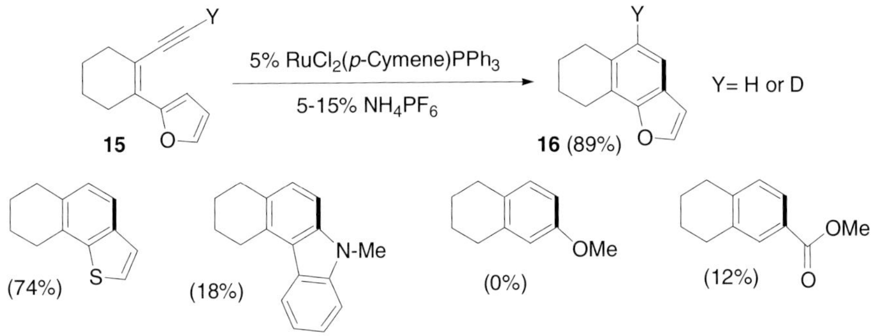

Scheme 6.8

attack of an electron-rich carbon center on the electrophilic vinylidene moiety. As shown in Scheme 6.7, there are two paths for the cyclization of ruthenium-vinylidene intermediates: 5-*endo-dig* cyclization (path a) and 6-*endo-dig* cyclization (path b).

Merlic's group was the first to realize a catalytic aromatization of dienynes via a 6-*endo-dig* cyclization of ruthenium vinylidene intermediates [13]. [RuCl$_2$(PPh$_3$) (p-cymene)]/NH$_4$PF$_6$ efficiently catalyzed the cycloaromatization of conjugated dienylalkynes to benzofurans, pyrroles and thiophenes in good yields (Scheme 6.8). A deuterium labeling experiment indicates a 1,2-migration of the terminal alkyne hydrogen, which strongly supports the intermediacy of ruthenium vinylidene.

The proposed mechanism involves either path a in which initially formed ruthenium vinylidene undergoes nonpolar pericyclic reaction or path b in which a polar transition state was formed (Scheme 6.9). According to Merlic's mechanism, the cyclization is followed by aromatization of the ruthenium cyclohexadienylidene intermediate, and reductive elimination of phenylruthenium hydride to form the arene derivatives (path c). A direct transformation of ruthenium cyclohexadienylidene into benzene product (path d) is more likely to occur through a 1,2-hydride shift of a ruthenium alkylidene intermediate. A similar catalytic transformation was later reported by Iwasawa using W(CO)$_5$THF catalyst [14].

Various cyclization products have been observed in the cycloisomerization of 3,5-dien-1-ynes using [Ru(Tp)(PPh$_3$)(CH$_3$CN)$_2$]PF$_6$ catalyst; the cyclization chemoselectivity is strongly dependent on the type of substrate structures, which alters the cyclization pathway according to its preferred carbocation intermediate. The reaction protocols are summarized below; ruthenium vinylidene intermediates are responsible for these cyclizations (Scheme 6.10).

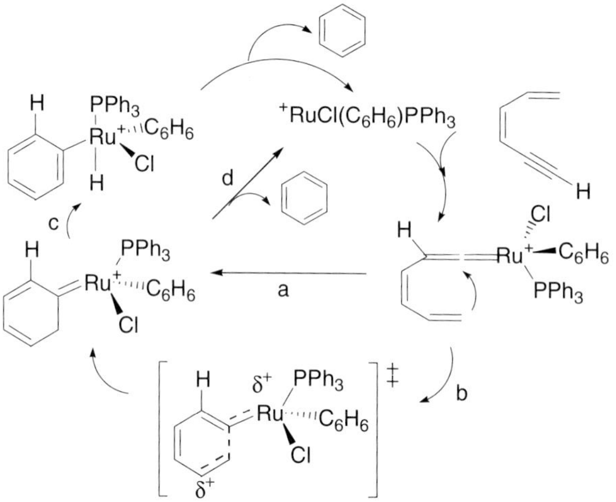

Scheme 6.9

A 1,2-halo migration was observed in the catalytic cyclization of 3,5-dien-1-ynes using [Ru(Tp)(PPh$_3$)(CH$_3$CN)$_2$]PF$_6$ catalyst, as depicted in Scheme 6.11 [15]. When a toluene solution of *o*-(ethynyl)styrene **17a** was heated in the presence of [Ru(Tp)(PPh$_3$)(CH$_3$CN)$_2$]PF$_6$ (10 mol%), naphthalene product **18a** was obtained in 95% yield. Under similar conditions, 2′-iodo derivatives **17b** gave naphthalene derivative **18b** via a 1,2-iodo migration. For 2′-bromo derivatives **d-17c**, the catalytic cyclization afforded bromonaphthalene **d-18c** via a 1,2-bromo migration but in low yield. Representative examples of this 1,2-halo migration are depicted in Scheme 6.11.

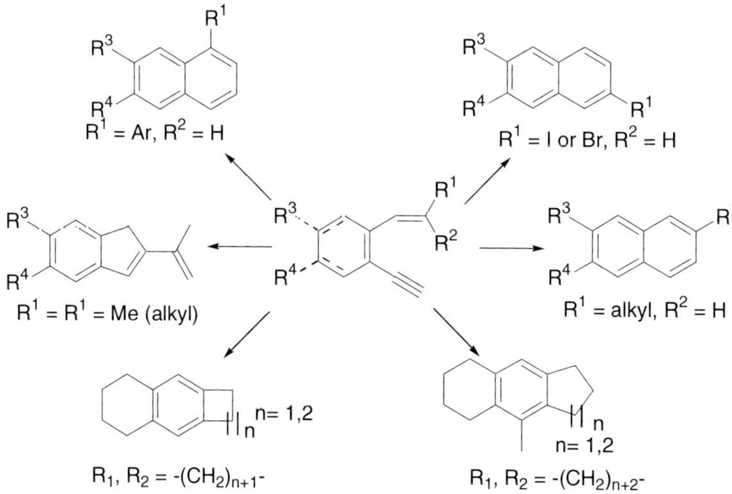

Scheme 6.10 Catalytic cyclization of dienynes using Ru(Tp)(PPh$_3$)(CH$_3$CN)$_2$PF$_6$.

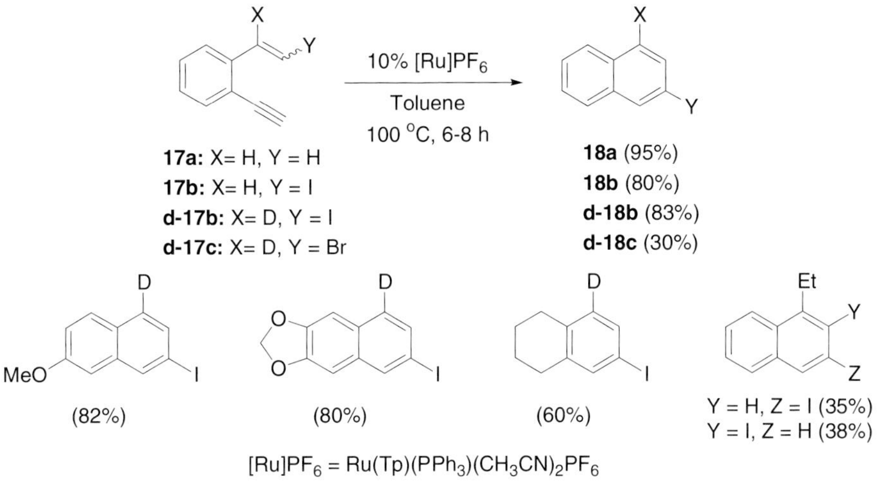

6.3 Catalytic Carbocyclization via Electrocyclization of Ruthenium-Vinylidene Intermediates | 199

17a: X= H, Y = H
17b: X= H, Y = I
d-17b: X= D, Y = I
d-17c: X= D, Y = Br

18a (95%)
18b (80%)
d-18b (83%)
d-18c (30%)

(82%) (80%) (60%)

Y = H, Z = I (35%)
Y = I, Z = H (38%)

[Ru]PF$_6$ = Ru(Tp)(PPh$_3$)(CH$_3$CN)$_2$PF$_6$

Scheme 6.11

The aromatization of 2-(2′-arylvinyl)ethynylbenzene substrates might be accompanied by a 1,2-aryl shift even though the parent compound **19a** failed to show such a phenomenon. As shown in Scheme 6.12, the 2′-aryl group preferentially underwent a 1,2-shift to the 1′-vinyl carbon rather than to the terminal alkyne carbon. This 1,2-aryl shift is favored by the electron-donating R groups on the migrating aryl, as reflected by the respective yields: **20c** (60%) > **20b** (35%) > **20a** (0%).

On the basis of ^{2}H- and ^{13}C-labeling experiments, a plausible mechanism is proposed in Scheme 6.13, which involves electrocyclization of ruthenium vinylidene **22** via a 6-*endo-dig* (path b) to give the ruthenium naphthylidene species **23**. The latter subsequently undergoes a 1,2-iodo shift to give ruthenium-η^2-naphthalene, ultimately giving the observed naphthalene product and an active ruthenium species. On the basis of ^{13}C-labeling experiments, a 1,2-aryl shift is proposed to arise from the 5-*endo-dig* electrocyclization (path a) of species **22** to give the ruthenium fluorenyl

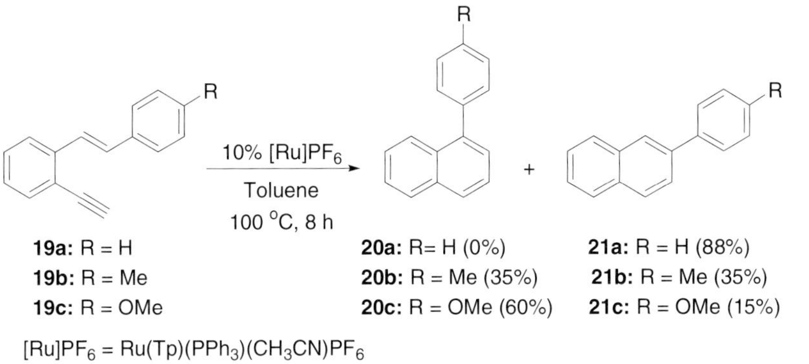

19a: R = H
19b: R = Me
19c: R = OMe

20a: R= H (0%)
20b: R = Me (35%)
20c: R = OMe (60%)

21a: R = H (88%)
21b: R = Me (35%)
21c: R = OMe (15%)

[Ru]PF$_6$ = Ru(Tp)(PPh$_3$)(CH$_3$CN)PF$_6$

Scheme 6.12

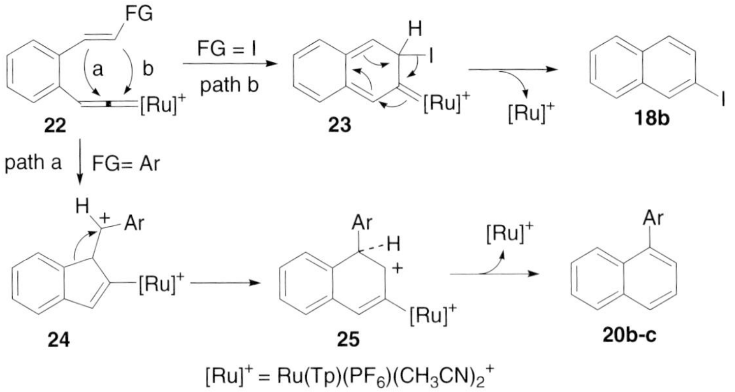

Scheme 6.13

species **24**, which bears a benzyl cation to induce a 1,2-shift of a carbon–carbon bond to generate intermediates **25**, ultimately to produce naphthalene derivatives **20b** and **20c** and active catalyst.

In the preceding cyclization, one concern is the behavior of intermediate **24** if its benzyl cation is replaced with a tertiary cation to avoid forming a tertiary carbon, the 1,2-aryl migration of this intermediate would be unlikely to occur. When this cyclization was extended to 2,2-dimethyl(o-ethynyl)styrene **26**, the 2-alkenyl-1H-indene product **28** [16] was obtained in 76% yield, as depicted in Scheme 6.14. In this transformation, the alkenyl double bond of the 3,5-dien-1-ynes is cleaved and a

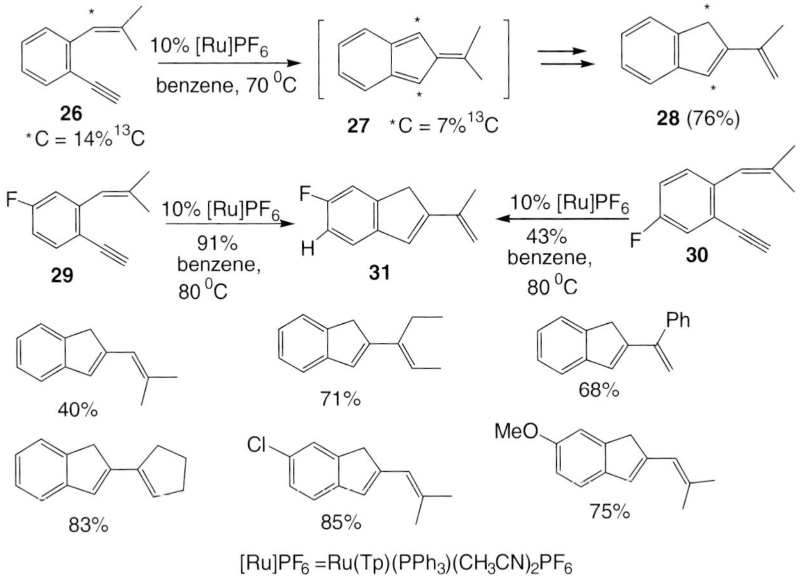

Scheme 6.14

6.3 Catalytic Carbocyclization via Electrocyclization of Ruthenium-Vinylidene Intermediates

terminal alkynyl C(1)-carbon is inserted, which represents a 1,2-alkylidene shift. The reaction mechanism was elucidated by a series of ^{2}H and ^{13}C labeling experiments, as well as by varying the substituents at the phenyl moieties. In the case of (o-ethynyl) styrene **26** bearing a 14% ^{13}C content at its ethenyl C(1′)-carbon, the resulting indene derivative **28** had equal ^{13}C content at its C(1) and C(3) carbons. As depicted in Scheme 6.14, 4-fluoro-2-ethynylstyrene **29** and its 5-fluoro analog **30** gave the same indene product **31**. These observations strongly indicate that the isobenzofulven species **27** is likely to be a reaction intermediate. This new catalytic rearrangement proceeds well with various 2,2-disubstituted (o-ethynyl)styrenes, as shown by selected examples in Scheme 6.14 [16].

For 6,6-disubstituted 3,5-dien-1-yne, the corresponding vinylidene intermediate is prone to 6-π-electrocyclization; one hence expects to observe a skeletal reorganization of this carbenium to form a stable benzene product, as depicted in Scheme 6.15. Heating 6,6-disubstituted 3,5-dien-1-yne **32** in hot toluene (100 °C, 10 h) with [Ru(Tp)(PPh$_3$)(CH$_3$CN)$_2$]PF$_6$ gave 1-(n-propyl)naphthalene **33** in 42% yield [17]. In contrast, the cyclopropylidenyl and cyclobutylidenyl derivatives **34a** and **34b** preferentially gave benzene products **35a** and **35b**, resulting from a regioselective 1,2-alkyl shift. The observation of a 1,2-deuterium shift in the case of **d-34b** indicates the formation of a vinylidene intermediate.

When the reaction was performed with cyclopentyl or cyclohexyl species **36a–d**, benzene products **37a–d** containing an unexpected methyl group were obtained (Scheme 6.16) in useful yields [17]. The benzene framework in this cyclization is constructed through a unique pattern, that is, forming the C(2)–C(7) bond in the starting dienynes. The mechanism of this cyclization has been elucidated by experiments including the extension of these cyclizations to internal alkynes [18]; the pathways are proposed to involve a 1,7-hydrogen shift of the dienylvinylidene intermediate **40** in Scheme 6.17, which ultimately gives the desired benzene products via the intermediates **41** and **42**.

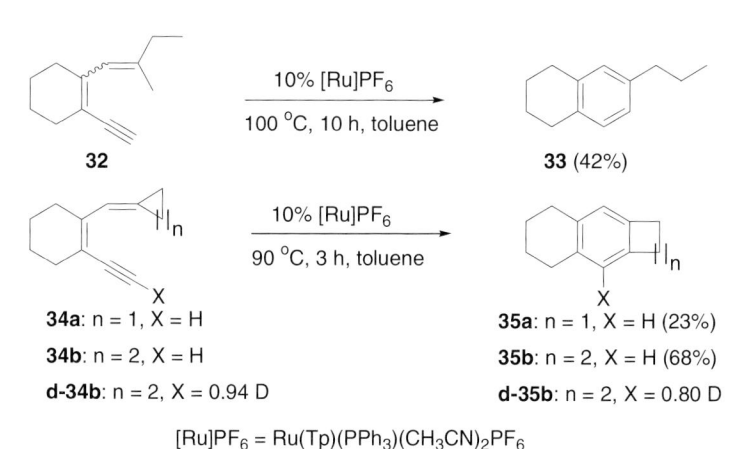

32 → 10% [Ru]PF$_6$, 100 °C, 10 h, toluene → **33** (42%)

34a: n = 1, X = H
34b: n = 2, X = H
d-34b: n = 2, X = 0.94 D

10% [Ru]PF$_6$, 90 °C, 3 h, toluene →

35a: n = 1, X = H (23%)
35b: n = 2, X = H (68%)
d-35b: n = 2, X = 0.80 D

[Ru]PF$_6$ = Ru(Tp)(PPh$_3$)(CH$_3$CN)$_2$PF$_6$

Scheme 6.15

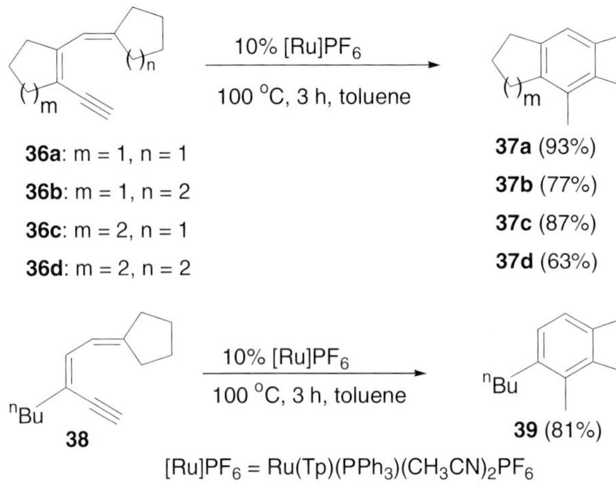
10% [Ru]PF6
100 °C, 3 h, toluene
36a: m = 1, n = 1
36b: m = 1, n = 2
36c: m = 2, n = 1
36d: m = 2, n = 2
37a (93%)
37b (77%)
37c (87%)
37d (63%)
nBu
38
39 (81%)
[Ru]PF6 = Ru(Tp)(PPh3)(CH3CN)2PF6

Scheme 6.16

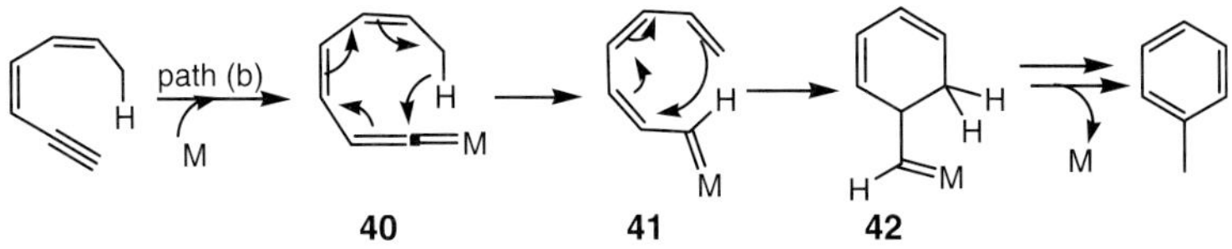
path (b)
H
M
40
41
42

Scheme 6.17

One use in the cyclization of 3,dien-ynes is demonstrated by a facile synthesis of coronene derivatives via fourfold cyclization of ethynylbenzenes, as depicted in Scheme 6.18. This approach was first reported by Scott [19] who obtained coronenes in 12% yield using $[RuCl_2(m\text{-cymene})]_2$ (10 mol%) and $AgPF_6$ (40 mol%). $[Ru(Tp)(PPh_3)(CH_3CN)_2]PF_6$ catalyst (10 mol%) provided coronene **44** in 67% yield through the cyclization of substrate **43** [20] whereas $PtCl_2$, $AuCl_3$ and $W(CO)_5(THF)$ failed to produce the desired coronene **44** in a significant amount (yields <2%). $[Ru(Tp)(PPh_3)(CH_3CN)_2]PF_6$ catalyst is applicable also to the synthesis of dibenzo[*a,h*]anthracene, chrysene, benzo[*c*]phenanthrene and benzo[*ghi*]perylene via twofold cyclization of ethynylbenzenes [0.05 *M*]. ^{2}H-labeling experiments reveal that ruthenium vinylidene intermediates were responsible for production of these aromatic products.

6.3.3 Ruthenium-Catalyzed Cyclization of 3-Azadienynes

Movassaghi *et al.* [21] reported the synthesis of substituted pyridine derivatives via ruthenium-catalyzed cycloisomerization of 3-azadienynes. To avoid the isolation of the chemically active alkynyl imines, trimethysilyl alkynyl amines served as initial substrates, as shown in Scheme 6.19. The formation of ruthenium vinylidene intermediates is accompanied by a 1,2-silyl migration according to controlled

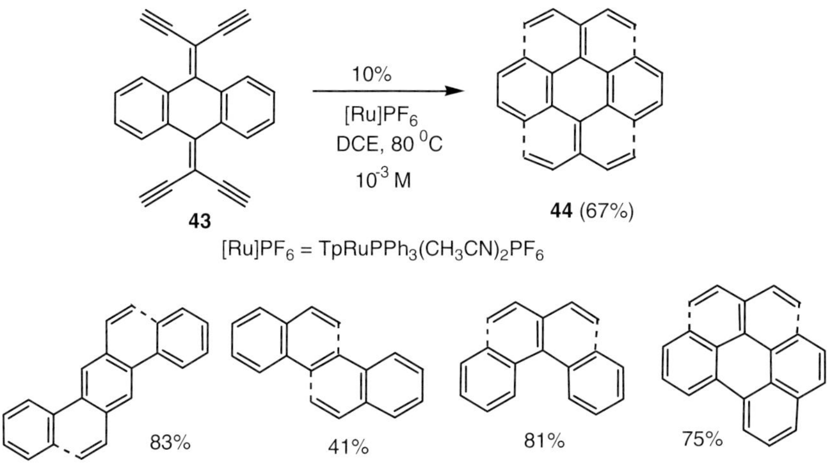

6.3 Catalytic Carbocyclization via Electrocyclization of Ruthenium-Vinylidene Intermediates

Scheme 6.18

experiments. Diversely substituted pyridines have been synthesized using a combination of RuCpCl(PPh$_3$)$_2$ CpRu(PPh$_3$)$_2$Cl, (10 mol%), 2-dicyclohexyl-phosphino-2′,6′-dimethoxy-1,1′-biphenyl (SPhos, 10 mol%) and NH$_4$PF$_6$ (1 equiv) in toluene.

6.3.4
Cycloisomerization of *cis*-1-Ethynyl-2-Vinyloxiranes

Catalytic cycloisomerization of *cis*-1-ethynyl-2-vinyloxiranes was implemented by [Ru(Tp)(PPh$_3$)(CH$_3$CN)$_2$]PF$_6$, which afforded substituted phenols, as shown in

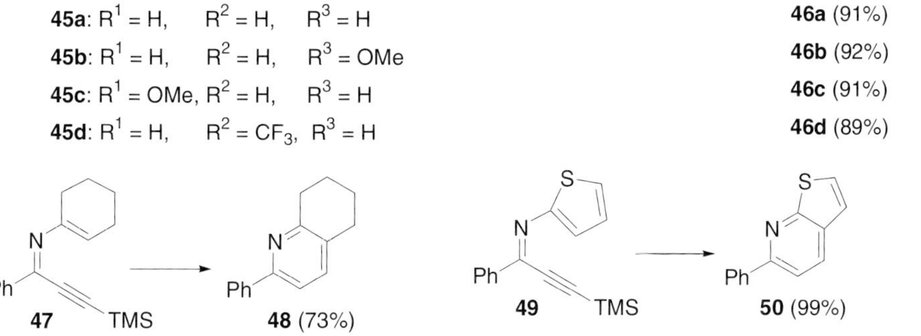

45a: R^1 = H,   R^2 = H,   R^3 = H
45b: R^1 = H,   R^2 = H,   R^3 = OMe
45c: R^1 = OMe, R^2 = H,   R^3 = H
45d: R^1 = H,   R^2 = CF$_3$, R^3 = H

46a (91%)
46b (92%)
46c (91%)
46d (89%)

Scheme 6.19

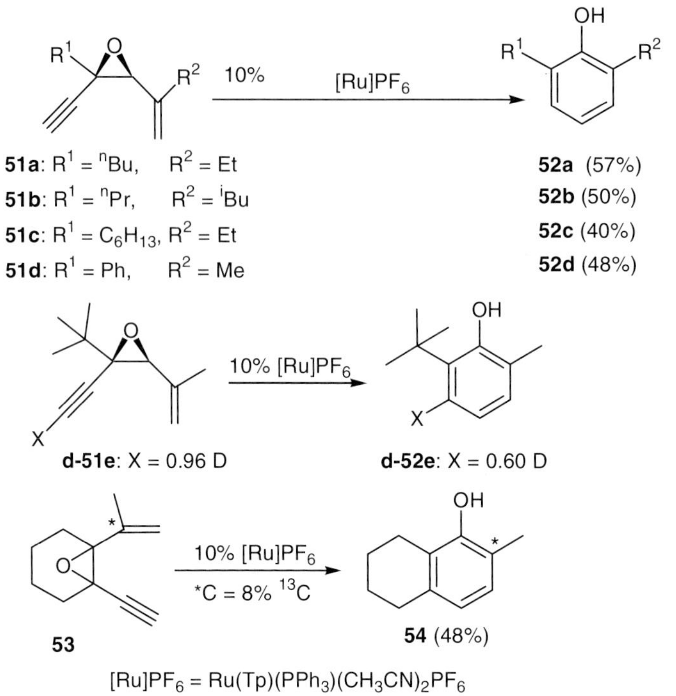

Scheme 6.20

Scheme 6.20. This ruthenium catalyst (10 mol%) was active for the cyclization of cis-1-ethynyl-2-vinyloxiranes to afford various 2,6-disubstituted phenols in reasonable yields. Under similar conditions, 1,1,2,2,-tetrasubstituted oxiranes gave the 2,3,6-trisubstituted phenols with a skeleton reorganization [22]. The 1,2-deuterium shift of the alkynyl deuterium of **d-51e** was indicative of ruthenium vinylidene intermediates (Scheme 6.20).

Based on the labeling experiments, a plausible mechanism involving ruthenium vinylidene intermediates **55** is proposed in Scheme 6.21. Cyclization of this vinylidene intermediate leads to the formation of the epoxy carbenium **56**, which then undergoes an epoxide opening to form 1,4-dien-3-ol **57**. A subsequent pinacol rearrangement of this alcohol furnishes ketone **58**, providing the required skeleton for the observed phenol product **54**.

6.3.5
Catalytic Cyclization of Enynyl Epoxides

$[Ru(Tp)(PPh_3)(CH_3CN)_2]PF_6$-catalyzed cyclization of enynyl epoxides is shown to give various carbocyclic compounds depending on the types of epoxides; a summary of the reaction protocol is provided in Scheme 6.22. Notably, these cyclized products are generated from dienyl ketene intermediate **59**, which was trapped efficiently by alcohols to form the ester product.

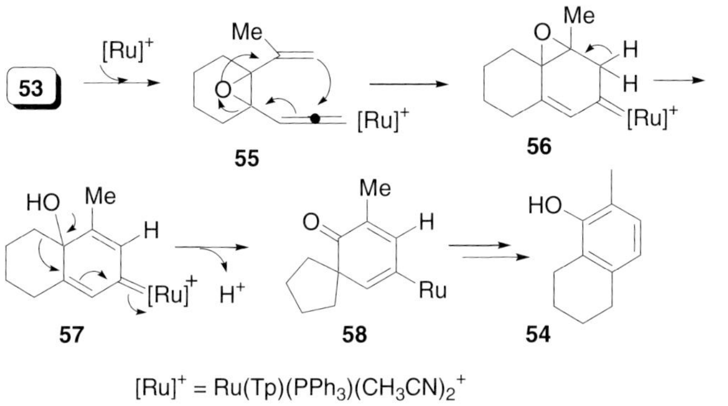

6.3 Catalytic Carbocyclization via Electrocyclization of Ruthenium-Vinylidene Intermediates

Scheme 6.21

Treatment of various substituted (o-ethynyl)phenyl epoxides with [Ru(Tp)(PPh$_3$)(MeCN)$_2$]PF$_6$ (10 mol%) in hot toluene (100 °C, 3–6 h) gave 2-naphthols or 1-alkylidene-2-indanones highly selectively, depending on the nature of the epoxide substituents [23]. As shown in Scheme 6.23, various 1′,2′-disubstituted epoxides gave 2-naphthol products in high yields. In contrast, 1′,2′,2′-trisubstituted epoxides gave 1-alkylidene-2-indenones efficiently under similar conditions. Some representative examples are shown in Scheme 6.24.

Scheme 6.25 shows a plausible mechanism involving ruthenium vinylidene and ruthenium-stabilized ketene intermediates. The ketene intermediate was verified through efficient trapping of this species with isobutanol to produce esters [23]. Nucleophilic attack by epoxide oxygen at the C$_\alpha$-carbon of ruthenium vinylidene produces the seven-membered ether species **64**, which ultimately forms ruthenium

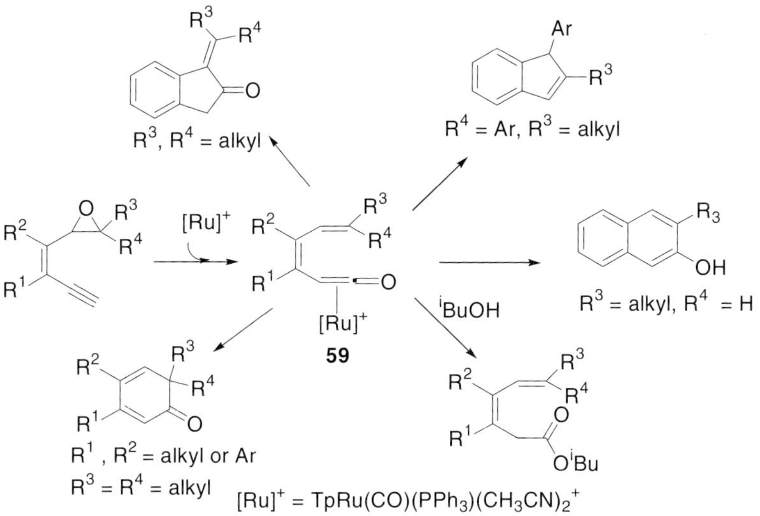

Scheme 6.22 Substrate-dependent cyclization of enynyl epoxides catalyzed by ruthenium.

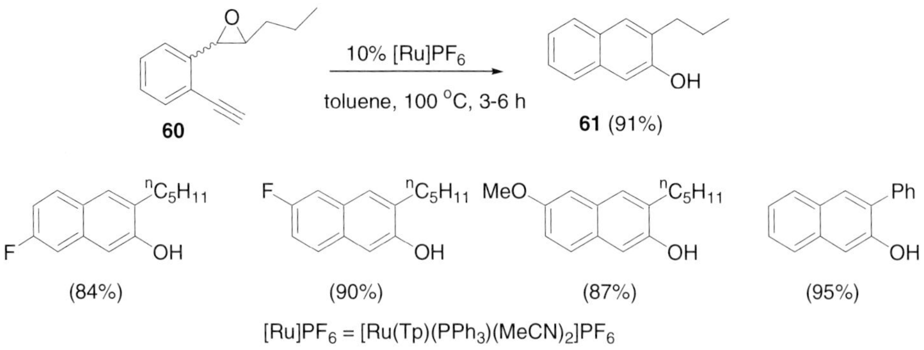
10% [Ru]PF6
toluene, 100 °C, 3-6 h
60
61 (91%)
OH
nC5H11
F
MeO
Ph
(84%)
(90%)
(87%)
(95%)
[Ru]PF6 = [Ru(Tp)(PPh3)(MeCN)2]PF6

Scheme 6.23

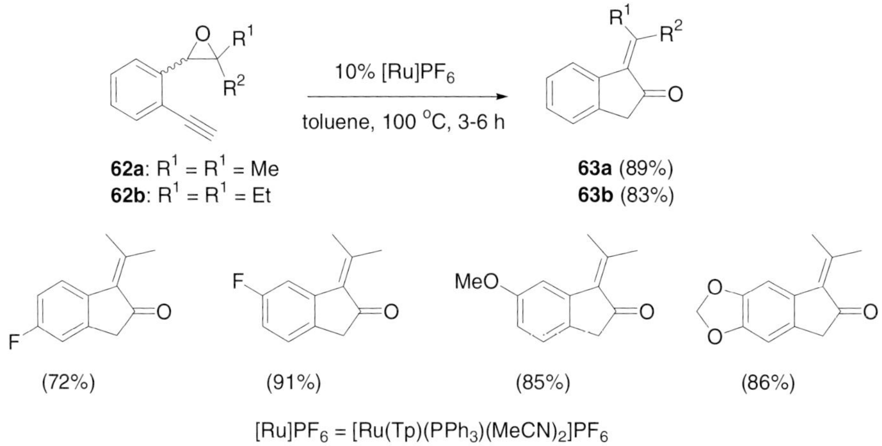
10% [Ru]PF6
toluene, 100 °C, 3-6 h
62a: R1 = R1 = Me
62b: R1 = R1 = Et
63a (89%)
63b (83%)
F
MeO
(72%)
(91%)
(85%)
(86%)
[Ru]PF6 = [Ru(Tp)(PPh3)(MeCN)2]PF6

Scheme 6.24

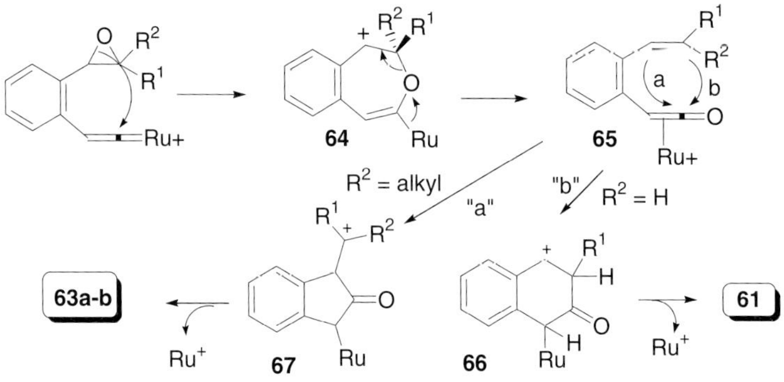
64
65
R2 = alkyl
"a"
"b"
R2 = H
63a-b
61
Ru+
67
66

Scheme 6.25

stabilized ketene **65**. For 1′,2′-disubstituted epoxide, species **65** undergoes 6-*endo-dig* electrocyclization (path b) [24] to form the six-membered ketone **66**, ultimately giving naphthol products. 1′,2′,2′-Trisubstituted epoxide species **65** undergoes 5-*endo-dig* cyclization (path a) to give the ketone species **67**, finally producing 1-alkylidene-2-indanones. The dialkyl substituent of the epoxide enhances the 5-*endo-dig* cyclization of species **65** via formation of a stable tertiary carbocation **67**. We observed similar behavior for the cyclization of (*o*-styryl)ethynylbenzenes [15, 16]. Formation of 2,4-cyclohexadien-1-one is explicable according to 6-*endo-dig* cyclization of a ruthenium-stabilized ketene, which ultimately afforded the observed products [25].

For 2,2-disubstituted epoxides **68a–d** bearing a 2-phenyl substituent, their corresponding catalytic cyclizations gave good yields of 1-phenyl-2-methyl-1H- indenes **69a–c** using the same ruthenium catalyst under similar conditions [25]. It is interesting to note that the same product **69c** was obtained for different epoxides **68c** and **68d**, bearing a fluoro substituent at their phenyl C4 and C5, respectively.

In this cyclodecarbonylation reaction, a ketene species is unlikely to be the reaction intermediate as added alcohols produce no esters. As shown in Scheme 6.26, the ruthenium acyl species **72** is likely to be the intermediate [25], which is prone to decarbonylation to give ruthena-cyclohexadiene **73**; this species undergoes subsequent reductive elimination to form 2H-indene. Addition of proton or Ru$^+$ to species **74** generated the benzylic cation **75**, which after a 1,2-aryl shift gave the observed products. The symmetric character of species **74** provides a rationale for both 4-fluoro **68c** and 5-fluoro **68d** substituted epoxides giving the same product (Table 6.2, entries 3–4).

We also studied the catalytic cyclization of such epoxide/alkyne functionalities via the selective formation of ruthenium-2-iodovinylidene **77** or π-iodoalkyne **78** intermediate in suitable solvents [26]. The former was preferentially generated in DMF whereas the latter was the dominant species in benzene, as depicted in Scheme 6.27. A polar solvent like DMF is known to accelerate the oxidative addition of a C−I bond by low-valent metals, which is thought to be the preceding step for the formation of 2-iodovinylidene intermediates **77**. Scheme 6.27 illustrates one application of this solvent-dependent chemoselectivity, that 1-iodo-naphthen-2-ols **79** and 6-iodo-7-*oxa*-benzocycloheptenes **80** can be selectively produced in DMF and benzene using the same substrate.

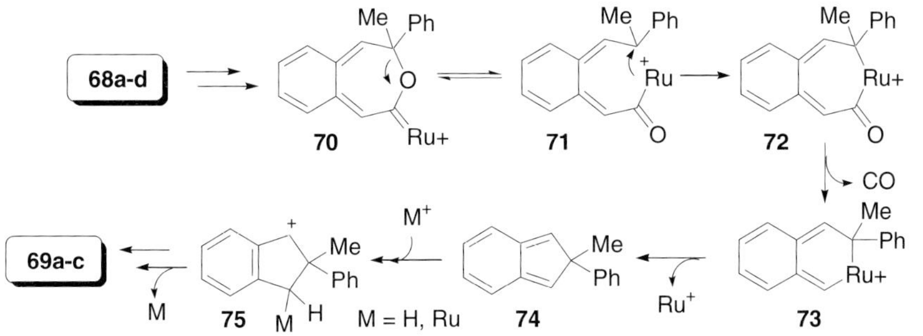

Scheme 6.26

6 Ruthenium Vinylidenes in the Catalysis of Carbocyclization

Table 6.2 Ruthenium catalyzed indene formation reaction.

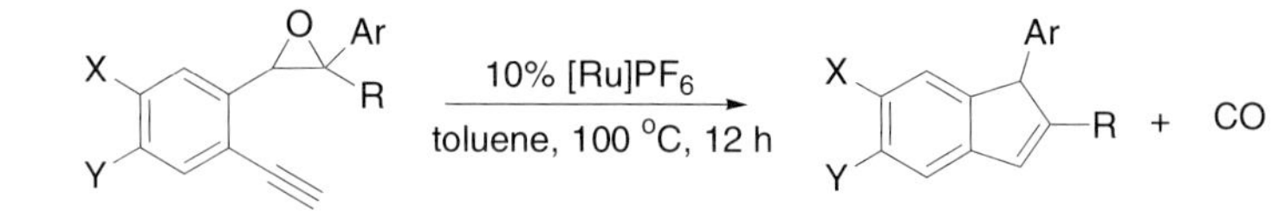

Epoxides	Indenes
(1) 68a (E/Z = 1.3):X = Y = H, Ar = Ph, R = Me	69a (78%)
(2) 68b (E/Z = 2.5):X = Y = H, Ar = p-MeC$_6$H$_4$, R = Me	69b (76%)
(3) 68c (E):X = H, Y = F, Ar = Ph, R = Me	69c (75%)
(4) 68d (E):X = F, Y = H, Ar = Ph, R = Me	69c: X = H, Y = F (80%)

[Ru]PF$_6$ = [Ru(Tp)(PPh$_3$)(MeCN)$_2$]PF$_6$.

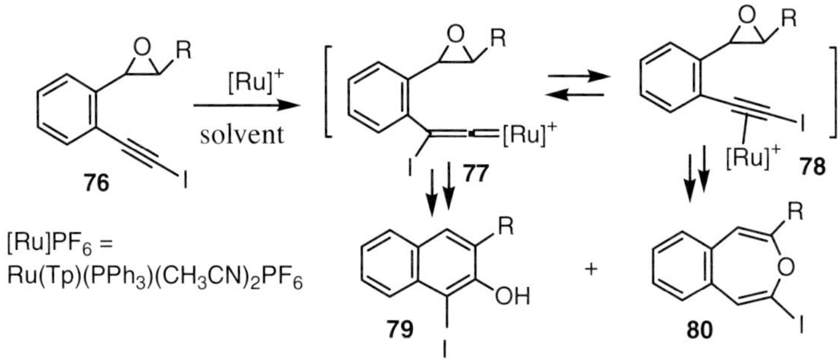

[Ru]PF$_6$ = Ru(Tp)(PPh$_3$)(CH$_3$CN)$_2$PF$_6$

R = ^{n}Pr	DMF (95 °C)	88%	1%
	benzene (80 °C)	12%	78%
R = ^{n}C$_5$H$_{11}$	DMF (95 °C)	87%	trace
	benzene (80 °C)	18%	62%

Scheme 6.27

6.4
Catalytic Carbocyclization via Cycloaddition of Ruthenium Vinylidene Intermediates

6.4.1
Cyclocarbonylation of 1,1′-Bis(silylethynyl)ferrocene

Onitsuka et al. [27] have developed a novel cyclocarbonylation of 1,1′-bis[(trimethylsilyl)ethynyl]ferrocene catalyzed by [Ru$_3$(CO)$_{12}$] under CO. Treatment of 1,1′-bis((trimethylsilyl)-ethynyl)ferrocene **81** with Ru$_3$(CO)$_{12}$ (5 mol%) under CO pressure (10 kg cm^{-2}) in toluene at 150 °C for 24 h gave the cyclocarbonylation product **82** in

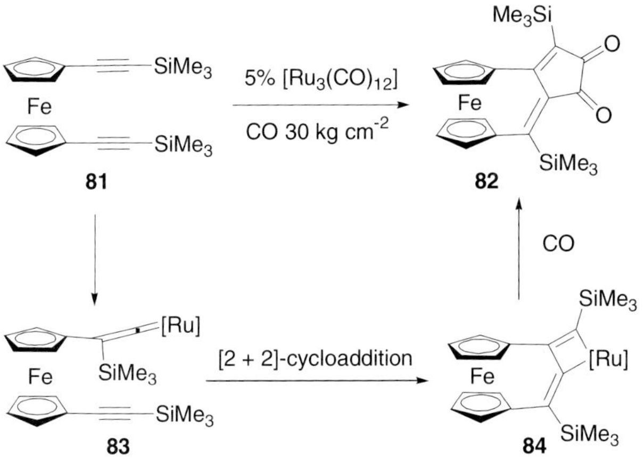

6.4 Catalytic Carbocyclization via Cycloaddition of Ruthenium Vinylidene Intermediates

Scheme 6.28

36% yield together with traces of a dinuclear ruthenacycle complex. A mechanism involving insertion of a triple bond into a vinylidene Ru=C double bond has been postulated for this catalysis according to Scheme 6.28. The catalytic cycle involves the 1,2-migration of a silyl group in the reaction with $Ru_3(CO)_{12}$ to form a vinylidene intermediate which interacts with the triple bond in [2 + 2] manner to produce a ruthenacyclobutene. A double insertion of carbon monoxide followed by reductive elimination gives the new ferrocene derivative.

6.4.2
Dimerization of 1-Arylethynes to 1-Aryl-Substituted Naphthalenes

Aryl acetylenes undergo dimerization to give 1-aryl naphthalenes at 180 °C in the presence of ruthenium and rhodium porphyrin complexes. The reaction proceeds via a metal vinylidene intermediate, which undergoes [4 + 2]-cycloaddition with the same terminal alkyne or another internal alkyne, and then H migration and aromatization furnish naphthalene products [28] (Scheme 6.29).

6.4.3
Ruthenium-Catalyzed Cycloaddition Reaction between Enyne and Alkene

Cyclohexadiene compounds can be prepared through ruthenium-catalyzed coupling between conjugated enynes and alkenes. Murakami et al. [29] have developed a catalytic coupling of olefins with unactivated alkynes. The coupling of enyne **87** with styrene was achieved smoothly by $RuCl(Cp)(PPh_3)_2$ (10%)/$NaPF_6$ (12%) to give initially the triene intermediate **89**, which subsequently formed bicyclic diene **90** via 6-π-electrocyclization. The same final products are notably obtainable from propargylic alcohol substrates such as 1-ethynyl cyclohexanol **88** (Scheme 6.30).

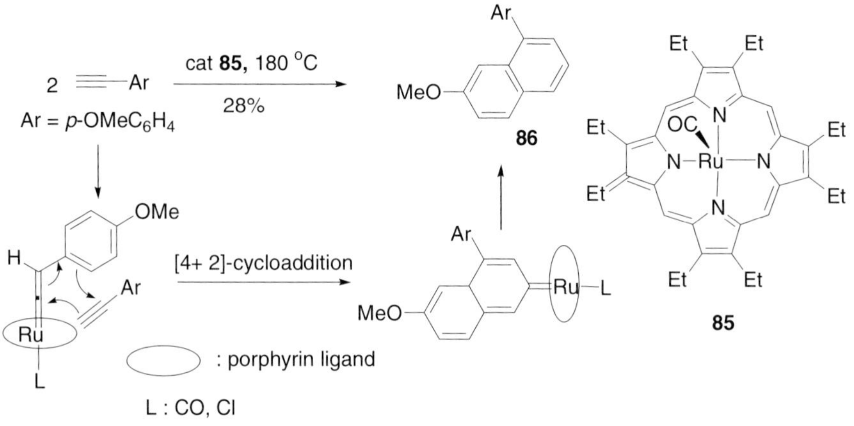

Scheme 6.29

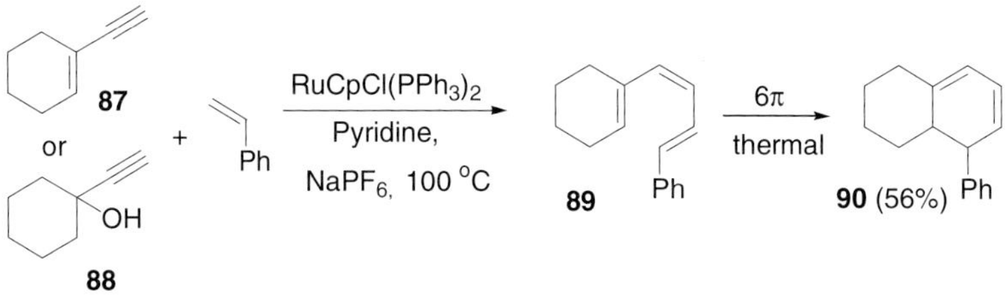

Scheme 6.30

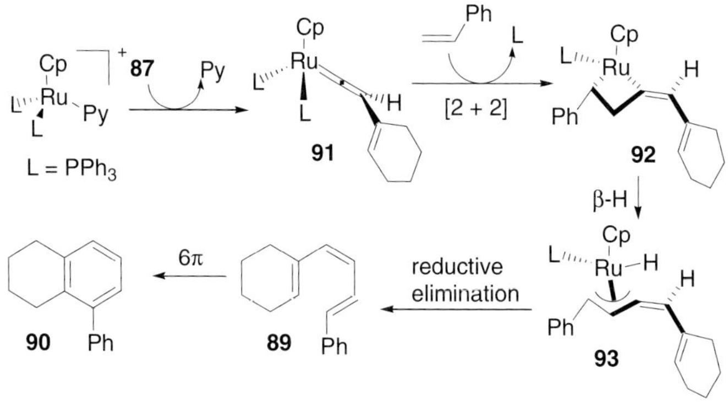

Scheme 6.31

The working mechanism involves a [2 + 2] cycloaddition between the Ru=C bond of ruthenium vinylidene and olefin to form the metallacyclobutane **92**, which subsequently undergoes β-hydride elimination leading to the π-allyl hydride complex **93** and reductive elimination to furnish the conjugated trienes **89** (Scheme 6.31), and eventually to give the observed aromatic product **90**.

6.5
Catalyzed Cyclization of Alkynals to Cycloalkenes

Saa and coworkers reported a remarkable cycloisomerization of alkynal to cycloalkene derivatives with loss of a CO molecule [30]; some examples are shown in Table 6.3. Heating 5-alkynal **94a** in AcOH (90 °C, 24 h) with [Ru(Cp)(CH$_3$CN)$_3$]PF$_6$ (5 mol%) afforded the cyclopentene derivative **95a** in 90% yield. Ketone **94e** afforded the cyclopentene **95e** in moderate yield, whereas the ester **94f** gave the noncyclized product **96** with loss of one carbon unit (Table 6.3) (Scheme 6.32).

The proposed mechanism involves the formation of ruthenium vinylidene **97** from an active ruthenium complex and alkyne, which upon nucleophilic attack of acetic acid at the ruthenium vinylidene carbon affords the vinylruthenium species **98**. A subsequent intramolecular aldol condensation gives acylruthenium hydride **99**, which is expected to give the observed cyclopentene products through a sequential decarbonylation and reductive elimination reactions.

6.6
Ruthenium-Catalyzed Hydrative Cyclization of 1,5-Enynes

Ruthenium-catalyzed 1,1-difunctionalization of alkynes can be achieved through ruthenium vinylidene intermediates. In this context, Lee's group reported the

Table 6.3 Ru-catalyzed cyclization of alkynals.

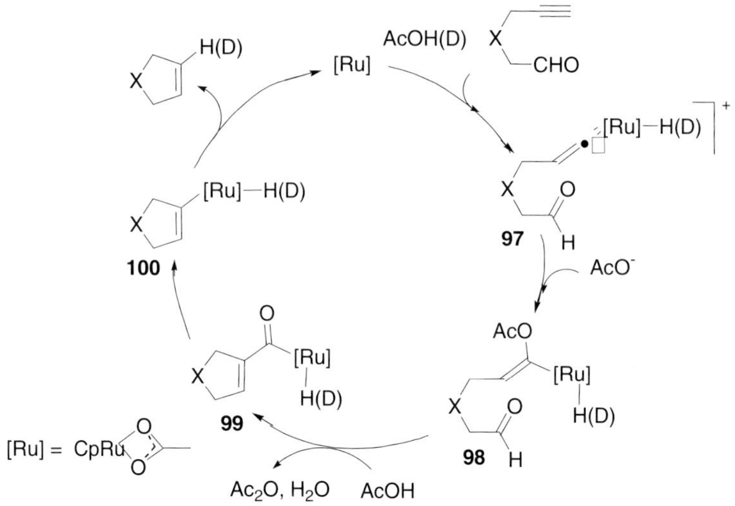

Scheme 6.32

hydrative cyclization of 1,5-enynes catalyzed by [Ru$_3$Cl$_3$(dppm)$_3$]PF$_6$ [31]. Representative examples of this novel hydrative cyclization are given in Scheme 6.33.

This reaction achieves an umpolung cyclization in which a terminal alkyne is hydrated and undergoes an intramolecular Michael addition according to the mechanism depicted in Scheme 6.34.

The hydrative cyclization involves the formation of a ruthenium vinylidene, an anti-Markovnikov addition of water, and cyclization of an acylmetal species onto the alkene. Although the cyclization may occur through a hydroacylation [32] (path A) or Michael addition [33] (path B), the requirement for an electron-withdrawing substituent on the alkene and lack of aldehyde formation indicate the latter pathway to be the more likely mechanism. Notably, acylruthenium complex underwent no decarbonylation in this instance.

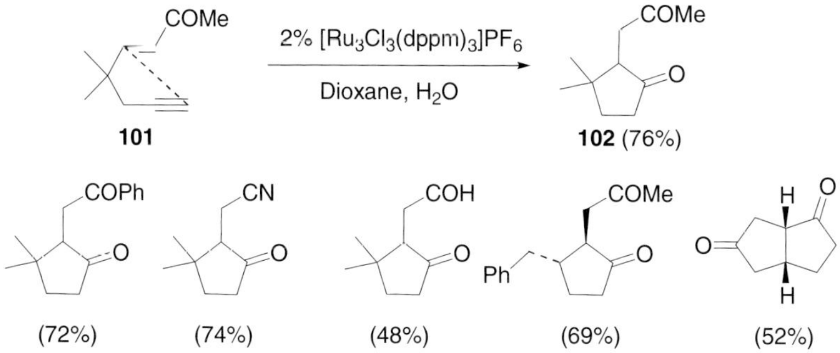

Scheme 6.33

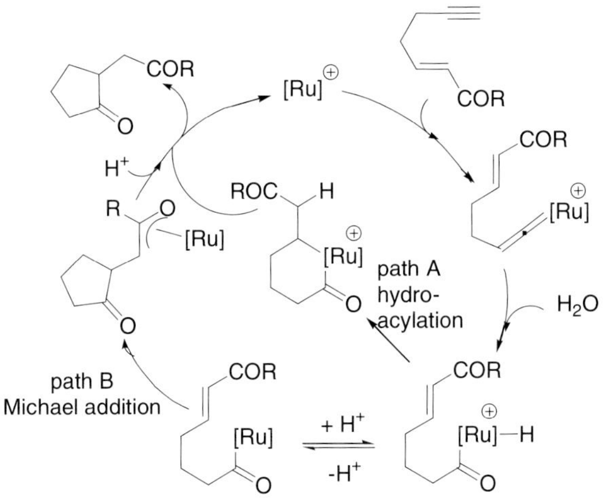

Scheme 6.34

6.7
Carbocyclization Initiated by Addition of C-Nucleophile to Ruthenium Vinylidene

Lee's group has also reported ruthenium-catalyzed carbonylative cyclization of 1,6-diynes. The noteworthy aspect of this cyclization is the unprecedented *anti* nucleophile attack on a π-alkyne complex bearing a ruthenium vinylidene functionality. A catalytic system based on [Ru(p-cymene)Cl$_2$]$_2$/P(4-F-C$_6$H$_4$)$_3$/DMAP was active for the cyclization of 1,6-diyne **103** and benzoic acid in dioxane at 65 °C to afford cyclohexenylidene enol ester **104a** in 74% yield after 24 h [34]. Additional examples are shown in Scheme 6.35.

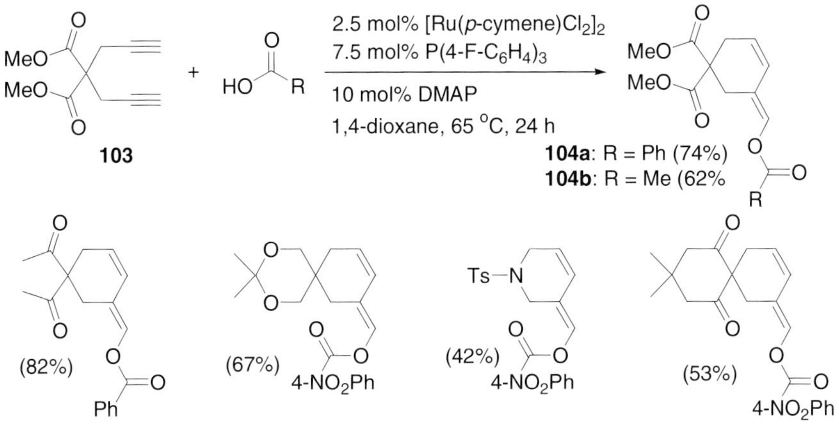

Scheme 6.35

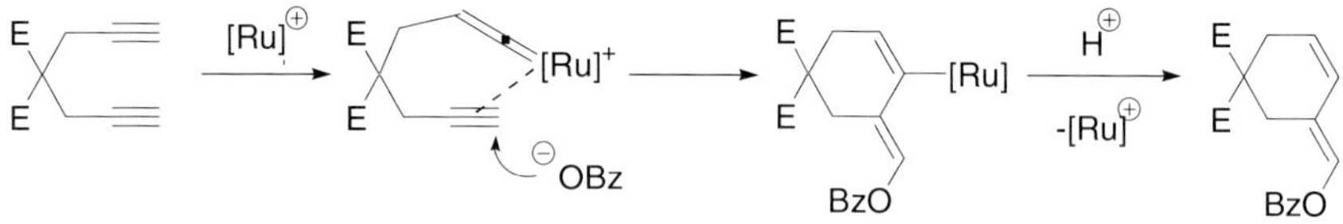

Scheme 6.36

103: E = CO₂Me — 105 — 106 — 104a

This cyclization represents a rare example in which ruthenium vinylidene is capable of activating a tethered π-alkyne toward nucleophilic addition, as shown in species **105**; this species ultimately formed the observed products via intermediate **106** (Scheme 6.36).

6.8
Conclusion

As terminal alkynes and ethynyl alcohols are the convenient sources to generate ruthenium vinylidene and allenylidene intermediates, many carbocyclizations have been achieved via nucleophilic addition and other activations at the two intermediates. Most reported carbocyclizations appear to be synthetically useful, not only because of their chemoselectivities but also because of their tolerance toward organic functional groups. Additional examples of catalytic carbocyclization based on ruthenium vinylidenes are still growing, and on the basis of the concepts developed here one can expect to see many new applications in the near future.

References

1. (a) Bruce, M.I. and Swincer, A.G. (1983) *Advances in Organometallic Chemistry*, **22**, 59. (b) Bruce, M.I. (1991) *Chemical Reviews*, **91**, 197. (c) Bruce, M.I. (1998) *Chemical Reviews*, **98**, 2797.
2. Bruce, M.I. and Wallis, R.C. (1979) *Australian Journal of Chemistry*, **32**, 1471.
3. Puerta, M.C. and Valerga, P. (1999) *Coordination Chemistry Reviews*, **193–195**, 977.
4. For recent reviews, see Bruneau, C. and Dixneuf, P.H. (2006) *Angewandte Chemie-International Edition*, **45**, 2176. (b) Bruneau, C. (2004) *Topics in Organometallic Chemistry*, Vol. 11, (eds C. Bruneau and P.H. Dixneuf), Springer, Berlin. (c) Varela, J.A. and Saa, C. (2006) *Chemistry – A European Journal*, **12**, 6450. (d) Bruneau, C. and Dixneuf, P.H. (1999) *Accounts of Chemical Research*, **32**, 3110. (e) Trost, B.M. (2002) *Accounts of Chemical Research*, **35**, 695.
5. (a) Ting, P.-C., Lin, Y.-C., Lee, G.-H., Cheng, M.-C. and Wang, Y. (1996) *Journal of the American Chemical Society*, **118**, 6433. (b) Lin, Y.-C. (2001) *Journal of Organometallic Chemistry*, **617–618**, 141.
6. (a) Alvarez, P., Lastra, E., Gimeno, J., Bassetti, M. and Falvello, L.R. (2003) *Journal of the American Chemical Society*,

125, 2386. (b) Bassetti, M., Alvarez, P., Gimeno, J. and Lastra, E. (2004) *Organometallics*, **23**, 5127. (c) Brana, P., Gimeno, J. and Sordo, J.A. (2004) *Journal of Organic Chemistry*, **69**, 2544.

7 Wang, Y. and Finn, M.G. (1995) *Journal of the American Chemical Society*, **117**, 8045.

8 (a) Bergman, R.G. (1973) *Accounts of Chemical Research*, **6**, 25. (b) Lockhart, T.P., Comita, P.B. and Bergman, R.G. (1981) *Journal of the American Chemical Society*, **103**, 4082.

9 (a) Myers, A.G., Kuo, E.Y. and Finney, N.S. (1989) *Journal of the American Chemical Society*, **111**, 8057. (b) Nagata, R., Yamanaka, H., Okazaki, E. and Saito, I. (1989) *Tetrahedron Letters*, **30**, 4995.

10 (a) Ohe, K., -a Kojima, M., Yonehara, K. and Uemura, S. (1996) *Angewandte Chemie-International Edition in English*, **35**, 1823. (b) Manabe, T., Yanagi, S., Ohe, K. and Uemura, S. (1998) *Organometallics*, **17**, 2942.

11 Datta, S., Odedra, A. and Liu, R.-S. (2005) *Journal of the American Chemical Society*, **127**, 11606.

12 Odedra, A., Datta, S. andLiu, R.-S. (2007) *Journal of Organic Chemistry*, **72**, 3289.

13 Merlic, C.A. and Pauly, M.E. (1996) *Journal of the American Chemical Society*, **118**, 11319.

14 (a) Maeyama, K. and Iwasawa, N. (1999) *Journal of Organic Chemistry*, **64**, 1344 (b) Miura, T. and Iwasawa, N. (2002) *Journal of the American Chemical Society*, **124**, 518.

15 Shen, H.-C., Pal, S., Lian, J.-J. and Liu, R.-S. (2003) *Journal of the American Chemical Society*, **125**, 15762.

16 Madhushaw, R.J., Lo, C.-Y., Hwang, C.-W., Su, M.-D., Shen, H.-C., Pal, S., Shaikh, I.R. and Liu, R.-S. (2004) *Journal of the American Chemical Society*, **126**, 15560.

17 Lian, J.-J., Odedra, A., Wu, C.-J. and Liu, R.-S. (2005) *Journal of the American Chemical Society*, **127**, 4186.

18 Lian, J.-J., Lin, C.-C., Chang, H.-K., Chen, P.-C. and Liu, R.-S. (2006) *Journal of the American Chemical Society*, **128**, 9661.

19 Donovan, P.M. and Scott, L.T. (2004) *Journal of the American Chemical Society*, **126**, 3108.

20 Shen, H.-C., Tang, J.-M., Chang, H.-K., Yang, C.-W. and Liu, R.-S. (2005) *Journal of Organic Chemistry*, **70**, 10113.

21 Movassaghi, M. and Hill, M.D. (2006) *Journal of the American Chemical Society*, **128**, 4592.

22 Maddirala, S.J., Odedra, A., Taduri, B.P. and Liu, R.-S. (2006) *Synlett*, 1173.

23 Madhushaw, R.J., Lin, M.-Y., Abu Sohel, S.M. and Liu, R.-S. (2004) *Journal of the American Chemical Society*, **126**, 6895.

24 (a) Danheiser, R.L., Brisbois, R.G., Kowalczyk, J.J. and Miller, R.T. (1990) *Journal of the American Chemical Society*, **112**, 3093. (b) Merlic, C.A. and Xu, D. (1991) *Journal of the American Chemical Society*, **113**, 7418 (c) Masamune, S. and Fukumoto, K. (1965) *Tetrahedron Letters*, **6**, 4647. (d) Mayr, H. and Huisgen, R. (1976) *Journal of the Chemical Society Chemical Communications*, 57.

25 Lin, M.-Y., Madhushaw, R.J. and Liu, R.-S. (2004) *Journal of Organic Chemistry*, **69**, 7700.

26 Lin, M.-Y., Maddirala, S.J. and Liu, R.-S. (2005) *Organic Letters*, **7**, 1745.

27 Onitsuka, K., Katayama, H., Sonogashira, K. and Ozawa, F. (1995) *Journal of the Chemical Society Chemical Communications*, 2267.

28 Elakkari, E., Floris, B., Galloni, P. and Tagliatesta, P. (2005) *European Journal of Organic Chemistry*, 889.

29 (a) Murakami, M., Ubukata, M. and Ito, Y. (1998) *Tetrahedron Letters*, **39**, 7361 (b) Murakami, M., Ubukata, M. and Ito, Y. (2002) *Chemistry Letters*, **31**, 294.

30 Varela, J.A., Gonzalez-Rodriguez, C., Rubin, S.G., Castedo, L. and Saa, C. (2006) *Journal of the American Chemical Society*, **128**, 9576.

31 Chen, Y., Ho, D.M. and Lee, C. (2005) *Journal of the American Chemical Society*, **127**, 12184.

32 Kondo, T., Tsuji, Y. and Watanabe, Y. (1987) *Tetrahedron Letters*, **28**, 6229.

33 (a) Yamashita, M., Tashika, H. and Suemitsu, R. (1989) *Chemistry Letters*, **18**, 691. (b) Seyferth, D. and Hui, R.C. (1985) *Journal of the American Chemical Society*, **107**, 4551.

34 Kim, H., Goble, S.D. and Lee, C. (2007) *Journal of the American Chemical Society*, **129**, 1030.

7
Allenylidene Complexes in Catalysis
Yoshiaki Nishibayashi and Sakae Uemura

7.1
Introduction

Metal allenylidene complexes (M=C=C=CR$_2$) are organometallic species having a double bond between a metal and a carbon, such as metal carbenes (M=CR$_2$), metal vinylidenes (M=C=CR$_2$), and other metal cumulenylidenes like M=C=C=C=CR$_2$ [1]. These metal–carbon double bonds are reactive enough to be employed for many organic transformations, both catalytically and stoichiometrically [1, 2]. Especially, the metathesis of alkenes via metal carbenes may be one of the most useful reactions in the field of recent organic synthesis [3], while metal vinylidenes are also revealed to be the important species in many organic syntheses such as alkyne polymerization and cycloaromatization [4, 5].

A variety of metal allenylidene complexes, first discovered in 1976 [6] independently by Fischer and Berke, are also known to be the reactive species, among which cationic ruthenium complexes (Ru$^+$=C=C=CR$_2$) have so far been studied the most and applied successfully to organic syntheses. These cationic species are readily available via dehydration of propargylic alcohols and can be regarded as stabilized propargylic cations because of the extensive contribution of the ruthenium-alkynyl resonance form (Ru–C≡C–C$^+$R$_2$), as shown in Scheme 7.1 [7]. In stoichiometric reactions of the species, a variety of nucleophiles attack either the allenylidene C$_\alpha$ or C$_\gamma$ atom to afford Fischer-type carbenes or alkynyl complexes, respectively (Scheme 7.1) [1]. It is known that nucleophilic addition occurs at the C$_\gamma$ position regioselectively when electron-rich nucleophiles and/or bulky ruthenium fragments are employed. In fact, Gimeno and coworkers have developed a novel synthetic procedure for the propargylic substitution reaction of 2-propyn-1-ols mediated by the bulky monoruthenium complex [(η^5-C$_9$H$_7$)Ru(PPh$_3$)$_2$]$^+$ where an allenylidene complex is formed in the first step and then transformed into the corresponding σ-alkynyl derivative which undergoes a selective protonation to afford a vinylidene complex [8]. The result agrees well with that of theoretical studies which indicate that the C$_\alpha$ and C$_\gamma$ atoms of allenylidene ligands can work as electrophilic centers, while the C$_\beta$ atom

Metal Vinylidenes and Allenylidenes in Catalysis: From Reactivity to Applications in Synthesis
Edited by Christian Bruneau and Pierre Dixneuf
Copyright © 2008 WILEY-VCH Verlag GmbH & Co. KGaA, Weinheim
ISBN: 978-3-527-31892-6

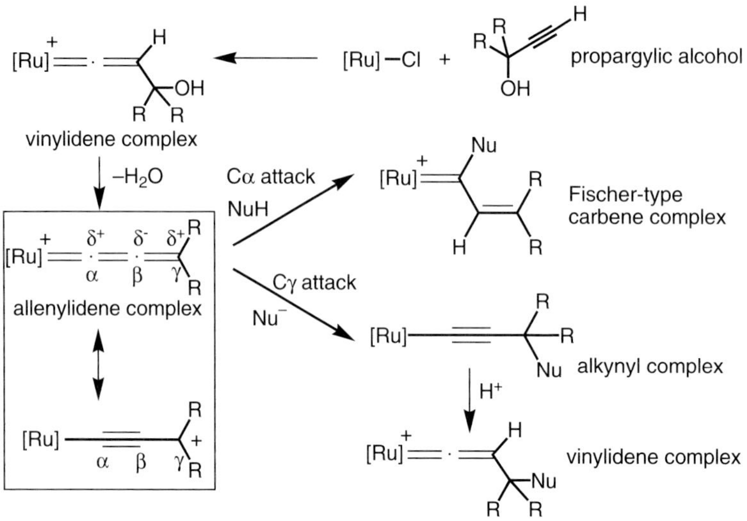

Scheme 7.1 Reactivity of allenylidene complex.

behaves as a nucleophilic site [1]. This synthetic methodology is considered to be an alternative to the well-known Nicholas reaction using $Co_2(CO)_8$, shown in Scheme 7.2 [9]. In addition to these stoichiometric reactions, some unprecedented reactivities of allenylidene complexes have been reported, but also in stoichiometric reactions [10].

All reactions described above need a stoichiometric amount of reagents such as ruthenium complex and cobalt carbonyl and also several steps are necessary to obtain the desired products. From the viewpoint of organic synthesis, catalytic and single reactions are preferable. The first example of a catalytic reaction via allenylidene complexes as key intermediates was reported in 1992 by Trost and a coworker (Scheme 7.3) [11]. In this reaction system, intramolecular nucleophilic attack of a hydroxy group on the C_γ atom of allenylidene intermediates gave the corresponding vinylidene species, which further reacted with allylic alcohols to afford β,γ-unsaturated ketones. During our study of the catalysis of thiolate-bridged diruthenium complexes such as $[Cp^*RuCl(\mu_2\text{-}SR)_2RuCp^*Cl]$, first prepared by Hidai and coworkers in 1989 (Scheme 7.4) [12], we eventually found that these complexes are good catalysts for the propargylic substitution reaction of propargylic alcohols bearing a terminal alkyne with a variety of nucleophiles. Not only heteroatom-centered but also carbon-centered nucleophiles can be employed for this catalytic reaction that clearly proceeds via ruthenium-allenylidene intermediates. In this chapter, we review the role of transition metal allenylidene complexes in catalytic

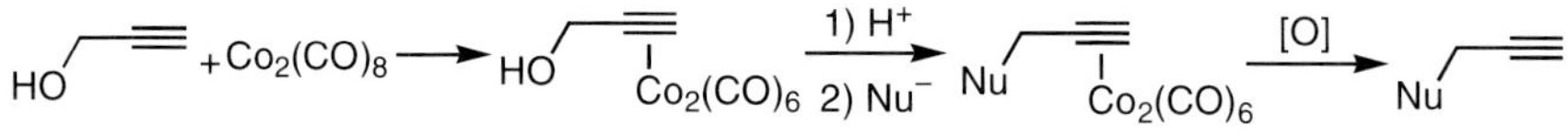

Scheme 7.2 Nicholas reaction.

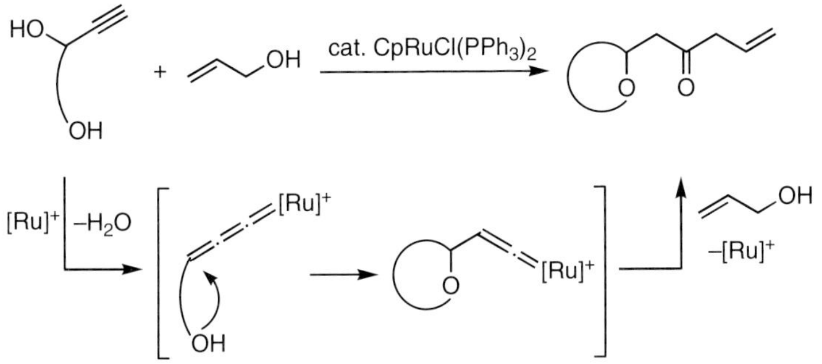

Scheme 7.3 The first catalytic reaction via allenylidene complexes as key intermediates.

[Cp*RuCl(μ$_2$-Cl)]$_2$ + 2 RSH or RSSiMe$_3$ ⟶ [Cp*RuCl(μ$_2$-SR)$_2$RuCp*Cl] (1)

Cp* = η^5-C$_5$Me$_5$ R = PhCH$_2$, Et, ^{i}Pr, ^{t}Bu

Scheme 7.4 The first thiolate-bridged diruthenium complexes.

organic transformations, except in alkene metathesis which will be studied in Chapter 8, focusing especially on our recent finding of thiolate-bridged diruthenium complex-catalyzed substitution reactions at the propargylic position of propargylic alcohols and their derivatives [13]. Some reactions, where allenylidene complexes work only as catalyst precursors, are not included in this chapter, they will be discussed later (see Chapter 8).

7.2
Propargylic Substitution Reactions

In sharp contrast to the well-studied transition metal-catalyzed allylic substitution reactions [14], much less attention has been paid to the propargylic substitution reactions of propargylic alcohol derivatives. Recently, thiolate-bridged diruthenium complexes were shown to be good catalysts for such transformations. A variety of heteroatom- and carbon-centered nucleophiles reacted with propargylic alcohols to give the corresponding propargylic substituted products in high yields with complete selectivity. An outline of catalytic propargylic substitution reactions of propargylic alcohols is given in this section.

7.2.1
Propargylic Substitution Reactions with Heteroatom-Centered Nucleophiles

In the presence of a catalytic amount of methanethiolate-bridged diruthenium complex (**1a**; abbreviated as met-DIRUX), reactions of propargylic alcohols (**2**) with a variety of heteroatom-centered nucleophiles such as alcohols, thiols, amines, amides, and diphenylphosphine oxide gave the corresponding propargylic substituted

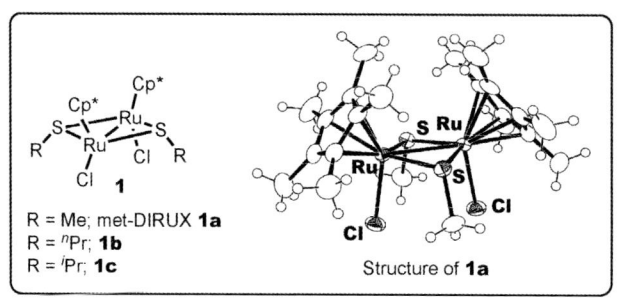

Scheme 7.5 Propargylic substitution reactions of propargylic alcohols with a variety of heteroatom-centered nucleophiles.

products in high to excellent yields with complete selectivity (Scheme 7.5) [15, 16]. Some propargylic esters such as acetates and carbonates can also be employed in place of propargylic alcohols under the same reaction conditions. The nature of the bridging thiolato ligands in the diruthenium complexes (**1b** and **1c**) did not have much effect on the catalytic activity. In contrast to the effective catalysis of thiolate-bridged diruthenium complexes, conventional monoruthenium complexes did not work as well as some diruthenium complexes having no ruthenium–ruthenium bond.

The scope and limitations of these ruthenium-catalyzed propargylic substitution reactions depend on the kind of nucleophiles employed [15]. With alcohols, amines or thiols, propargylic alcohols bearing not only aryl moieties but also alkyl moieties at the propargylic position undergo substitution reactions, substitution with alcohols being especially rapid. With amides or diphenylphosphine oxide, only propargylic alcohols bearing aryl moieties reacted to give the corresponding propargylic substituted products. With propargylic alcohols bearing two substituents at the propargylic position, the reactions were generally slow. In all cases, the reactions of 1-alkenyl-substituted propargylic alcohols proceeded smoothly to give the corresponding 1-alkenyl-substituted propargylic products in high yields without isomerization of the carbon–carbon double bond (Equation 7.1).

$$\text{Ph}\underset{\text{Ph OH}}{\diagup\!\!\!\diagdown}\diagup\!\!\!\equiv \ + \ \text{ROH} \ \xrightarrow[\text{60 °C, 1 h}]{\substack{\text{5 mol\% 1a} \\ \text{10 mol\% NH}_4\text{BF}_4}} \ \text{Ph}\underset{\text{Ph OR}}{\diagup\!\!\!\diagdown}\diagup\!\!\!\equiv \quad (7.1)$$

In addition to simple alcohols such as ethanol, methanol, and phenols, functionalized alcohols bearing a halogen, alkene or alkyne moiety are available as oxygen-centered nucleophiles. This method is considered to be useful for a direct approach to highly functionalized propargylic ethers. When reactions with chiral alcohols were investigated, a mixture of two diastereomeric isomers of propargylic ethers was obtained in all cases. On the other hand, intramolecular cyclization of propargylic alcohols bearing a hydroxy group at a suitable position in the same molecule afforded the corresponding cyclic ethers in moderate to high yields with complete selectivity (Equation 7.2).

$$(7.2)$$

Reactions with thiols proceeded quite smoothly to give the corresponding propargylic sulfides in high to excellent yields [17]. Although it is generally known that sulfur-containing compounds act as catalytic poisons because of their strong coordinating property to transition metals, a variety of thiols such as alkanethiols, thiophenol, and functionalized thiols bearing a halogen moiety were available as sulfur-centered nucleophiles in this catalytic reaction. The first synthetically useful method for propargylic sulfides is now provided because the preparative method for such sulfides by other methods is so far quite limited [18].

Anilines are used as typical nitrogen-centered nucleophiles for the propargylic substitution reactions. The presence of electron-releasing groups, such as alkyl and alkoxy moieties, in the aromatic ring of aniline decreased the reaction rate, while that of electron-withdrawing groups, such as nitro and ester moieties, increased it. Reactions with *secondary* alkylamine such as N-methylaniline were sluggish under identical conditions and a prolonged reaction time was required to improve the yield of propargylic amines. Unfortunately, the use of *primary* alkylamines for the propargylic amination was in vain, showing that amines with a high basicity are not applicable to this reaction. The scope of the propargylic amination extends beyond simple amine substrates. For example, selected amides, lactams, and sulfonamides produced useful aminated products directly from propargylic alcohols (Scheme 7.6).

Diphenylphosphine oxide can be used as a phosphorus-centered nucleophile for propargylic substitution reactions, where its tautomer (diphenylphosphinous acid) is

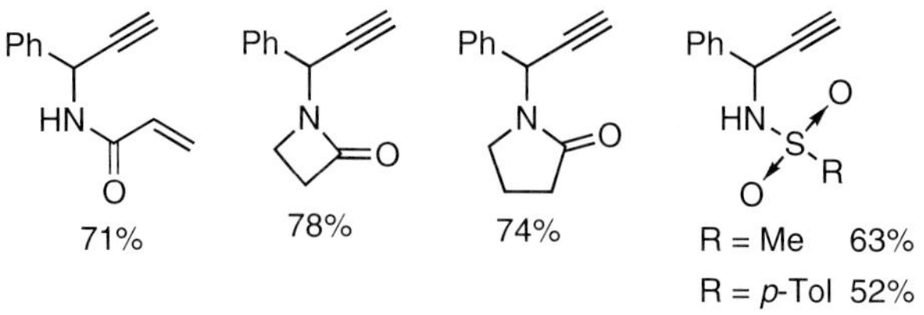

Scheme 7.6 Formation of various propargylic amides.

believed to work as a nucleophile. On the other hand, diphenylphosphine did not function well as a phosphorus-centered nucleophile for this catalytic reaction. The high basicity of diphenylphosphine, which might coordinate to ruthenium, may be one of the reasons for this unsuccessful substitution. When an excess of diphenylphosphine oxide was used, double phosphinylation of propargylic alcohols occurred, giving the corresponding 2,3-bis(diphenylphosphinyl)-1-propenes in high yields (Scheme 7.7) [19a]. Detailed investigation of the reaction pathway indicated that the double phosphinylation proceeded via an allenyldiphenylphosphine oxide as an intermediate. The double phosphinylated products are considered to be precursors of modified chiraphos (2,3-bis(diphenylphosphinyl)butane derivatives), which have the potential to work as a new type of chiral bidentate ligands for asymmetric synthesis. Furthermore, the unexpected and highly efficient formation of aryl(diphenyl)phosphine oxide was observed from the reactions of 1,1-diaryl-1-pentene-4-yn-3-ols with diphenylphosphine oxide (Scheme 7.8) [19b]. It is proposed that the reaction may involve the isomerization of an initially produced propargylic substituted compound to an allenyl intermediate, followed by cyclization and aromatization to afford the final product.

It is noteworthy that, in the absence of nucleophiles, the dehydration of propargylic alcohols occurred, giving the corresponding conjugated enynes in high yields (Equation 7.3) [15]. More recently, Gimeno and coworkers reported that a mononuclear ruthenium complex [Ru(η^3-2-C$_3$H$_4$Me)(CO)(dppf)]SbF$_6$ promotes a similar dehydration of propargylic alcohols to give the corresponding conjugated enynes in high yields [20].

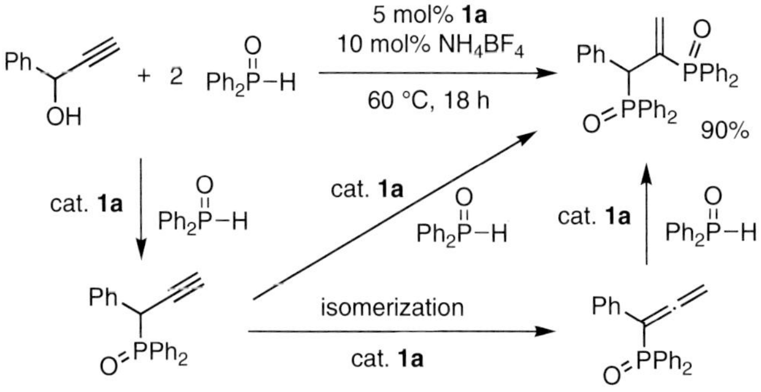

Scheme 7.7 Double phosphinylation of propargylic alcohols with diphenylphosphine oxide.

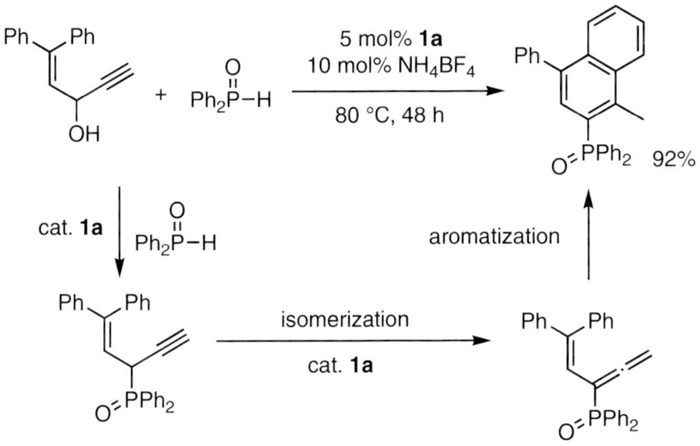
5 mol% 1a
10 mol% NH4BF4
80 °C, 48 h
92%
cat. 1a
aromatization
isomerization
cat. 1a

Scheme 7.8 Formation of aryl(diphenyl)phosphine oxide from propargylic alcohol and diphenylphosphine oxide.

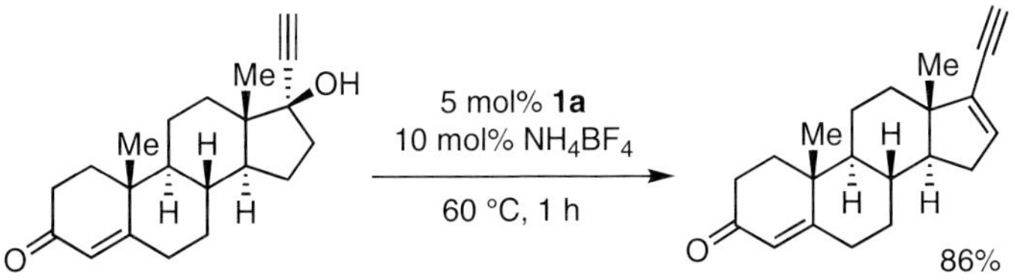
5 mol% 1a
10 mol% NH4BF4
60 °C, 1 h
86%

(7.3)

7.2.2
Propargylic Substitution Reactions with Carbon-Centered Nucleophiles

This propargylic substitution reaction also proceeds with carbon-centered nucleophiles such as simple dialkyl ketones and β-diketones to give the corresponding propargylic alkylated products in high yields with complete regioselectivity [21]. Thus, reactions of propargylic alcohols (**2**) with acetone at reflux temperature in the presence of a catalytic amount of methanethiolate-bridged diruthenium complex (**1a**) and NH_4BF_4 proceeded quite smoothly to give the corresponding alkylated products in high yields with complete selectivity (Scheme 7.9). The alkylation proceeded

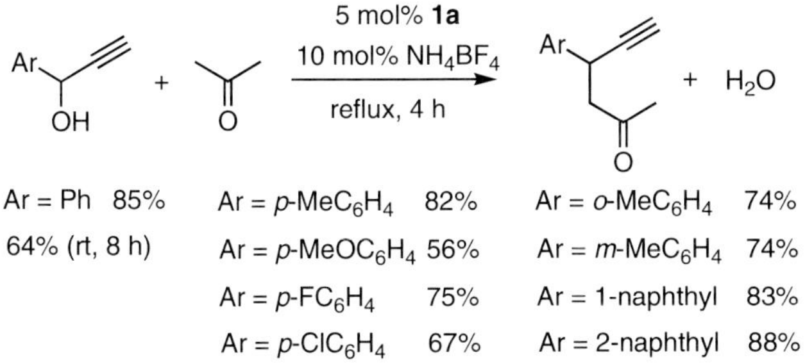
5 mol% 1a
10 mol% NH4BF4
reflux, 4 h
+ H2O
Ar = Ph 85%
64% (rt, 8 h)
Ar = p-MeC6H4 82%
Ar = p-MeOC6H4 56%
Ar = p-FC6H4 75%
Ar = p-ClC6H4 67%
Ar = o-MeC6H4 74%
Ar = m-MeC6H4 74%
Ar = 1-naphthyl 83%
Ar = 2-naphthyl 88%

Scheme 7.9 Propargylic substitution reaction with acetone as a carbon-centered nucleophile.

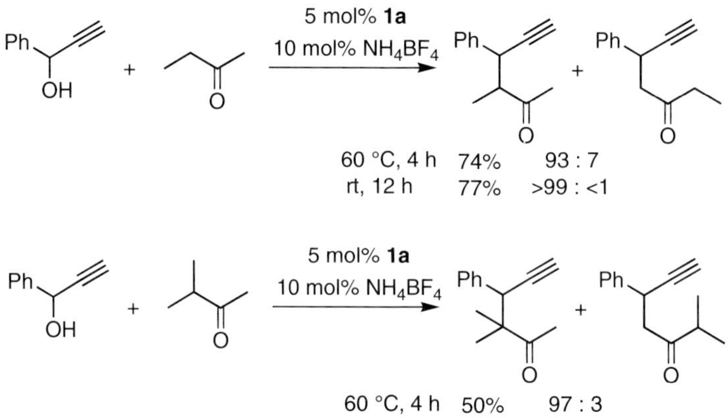

Scheme 7.10 Propargylic substitution reactions with unsymmetrical simple ketones.

smoothly even in the absence of NH_4BF_4, but its role is considered to afford the more reactive cationic diruthenium complex bearing a vacant site. It is noteworthy that the propargylic alkylation proceeds under extremely mild and neutral reaction conditions, in sharp contrast to the allylic alkylation catalyzed by a variety of transition metal complexes where a stoichiometric amount of base is required to activate carbon-centered nucleophiles. When reactions with unsymmetrical simple ketones such as ethyl methyl ketone and methyl iso-propyl ketone were carried out at room temperature to 60 °C, the more encumbered α-site of ketones was introduced into the propargylic position of propargylic alcohols (Scheme 7.10). Similar regioselectivity has also been observed in the stoichiometric Nicholas reaction with unsymmetrical simple ketones [9]. These results indicate that the enol tautomer of ketone works as a carbon-centered nucleophile [21]. In addition to simple ketones, reactions of propargylic alcohols with a silyl enol ether also afforded the corresponding alkylated products in high yields (Equation 7.4).

$$Ar = Ph; 50\%$$
$$Ar = Fc; 82\%$$

(7.4)

7.2.3
Reaction Pathway for Propargylic Substitution Reactions

Stoichiometric reactions of the isolated allenylidene complexes, which can be almost quantitatively prepared by treatment of **1a** with propargylic alcohols with various nucleophiles, except thiols, gave the corresponding propargylic substituted products [15, 21] (Scheme 7.11). Here, the addition of either NH_4Cl or another terminal

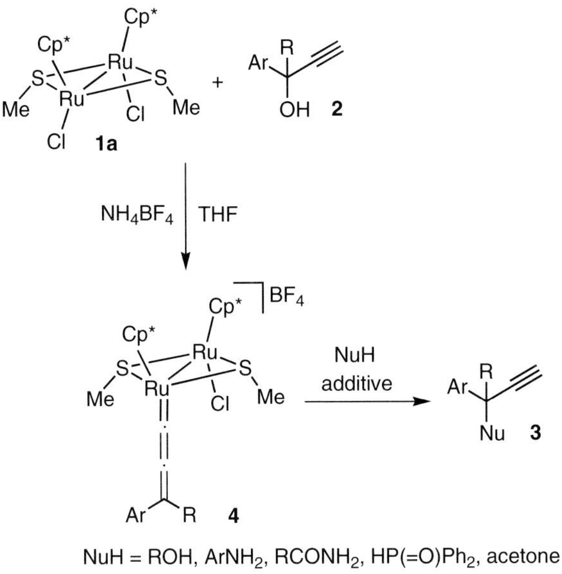

Scheme 7.11 Preparation and reactivity of allenylidene complexes.

alkyne much improved the yield of the substituted products. These results indicate that propargylic substitution reactions proceed via nucleophilic attack of nucleophiles on the electrophilic Cγ atom of allenylidene intermediates. The reason why only the diruthenium complexes are effective catalysts for this reaction is considered to be as follows. One Ru moiety, which is not involved in the allenylidene formation, is believed to work as an electron pool or a mobile ligand to the other Ru site, making the ligand exchange step easier (Scheme 7.12) [15, 21]. This proposition is based on the theoretical report of synergistic effects of two equal rhodium metals in the dirhodium-catalyzed reaction of a diazo compound with alkane [22]. To prove the possibility of the synergistic effects of two rutheniums as shown in Scheme 7.12, a series of chalcogenolate (S, Se, Te)-bridged diruthenium complexes were prepared and their catalytic activity toward propargylic substitution reactions as well as their electronic properties were investigated (Scheme 7.13) [16]. Both the results of the catalytic reactions and the electronic behavior of the complexes indicated that the ease

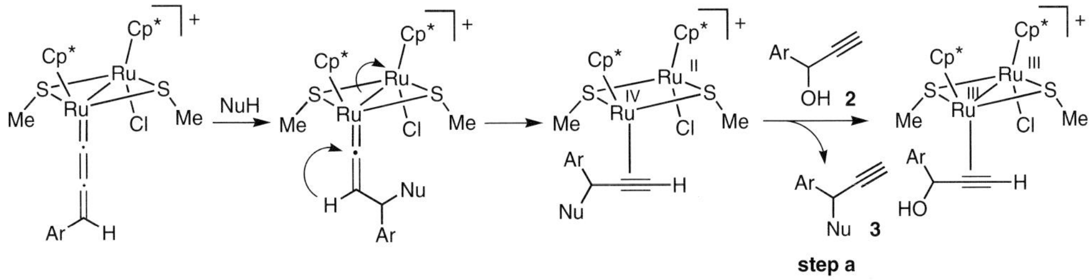

Scheme 7.12 Possibility of synergistic effect of two rutheniums in **1**.

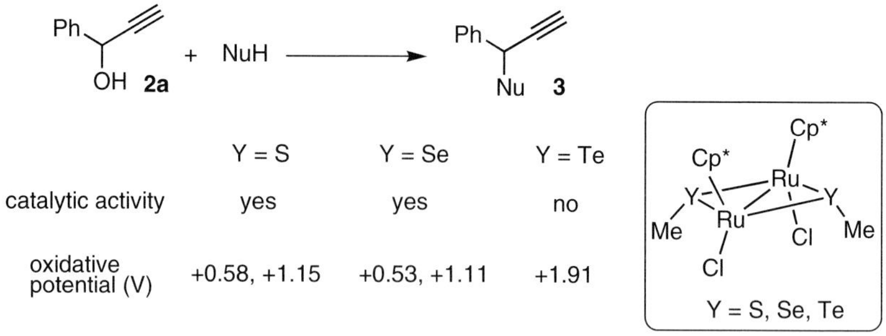

7 Allenylidene Complexes in Catalysis

Ph—C≡C—CH(OH) + NuH → Ph—C≡C—CH(Nu)
    2a   3

	Y = S	Y = Se	Y = Te
catalytic activity	yes	yes	no
oxidative potential (V)	+0.58, +1.15	+0.53, +1.11	+1.91

Y = S, Se, Te

Scheme 7.13 Catalytic activity toward propargylic substitution reactions and electronic properties of chalcogenolate-bridged diruthenium complexes.

of the charge transfer from one Ru atom to the other in the complexes (synergistic effect) may be one of the important factors in promoting the ligand exchange (step a in Scheme 7.12), which is a key step for the catalytic reaction [15, 21]. A proposed reaction pathway of this catalytic propargylic substitution reaction is summarized in Scheme 7.14. Theoretical studies using density functional calculations (B3LYP) on the detailed reaction mechanism also support this proposal [23]. Gimeno and coworkers have reported the propargylic etherification of propargylic alcohols with various alcohols catalyzed by a monoruthenium complex [Ru(η^3-2-C$_3$H$_4$Me)(CO)-(dppf)]SbF$_6$, where no allenylidene complexes were detected as key intermediates

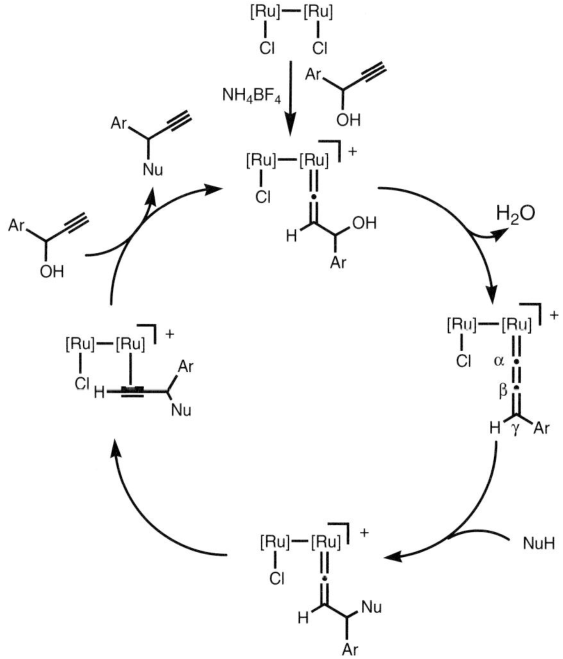

Scheme 7.14 Proposed reaction pathway for the catalytic propargylic substitution reaction.

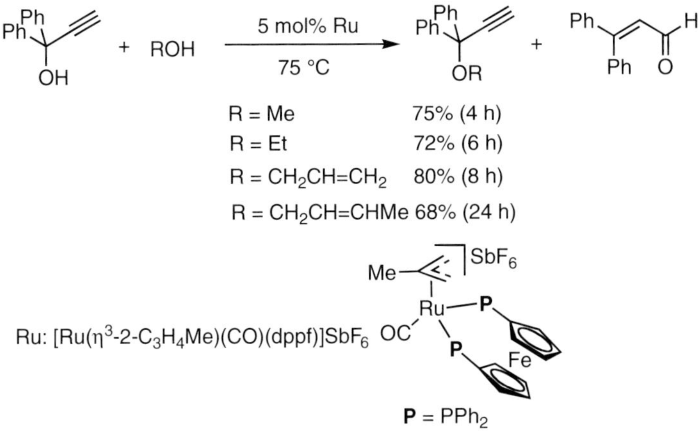

Scheme 7.15 Monoruthenium complex-catalyzed propargylic etherification of propargylic alcohols with various alcohols.

R = Me, 75% (4 h)
R = Et, 72% (6 h)
R = CH$_2$CH=CH$_2$, 80% (8 h)
R = CH$_2$CH=CHMe, 68% (24 h)

Ru: [Ru(η^3-2-C$_3$H$_4$Me)(CO)(dppf)]SbF$_6$

P = PPh$_2$

(Scheme 7.15) [24]. In all cases, the aldehydes corresponding to propargylic alcohol isomerization were formed as side-products.

7.2.4
Asymmetric Propargylic Alkylation with Acetone

In sharp contrast to the transition metal-catalyzed highly enantioselective allylic substitution reactions of allylic alcohol derivatives with nucleophiles [14], the propargylic substitution version has not yet been developed. So far, the only related work is that by Nicholas and a coworker who reported the stereospecific propargylic alkylation of chiral propargylic alcohols by using a stoichiometric amount of [Co$_2$(CO)$_5$L] (L = phosphite) for enantiomerically rich propargylic alkylated products, but several reaction steps as well as two separation procedures of the produced diastereoisomers were necessary to obtain the enantiomerically rich propargylic alkylated compounds (Scheme 7.16) [25]. Quite recently, enantioselective propargylic substitution reactions of racemic propargylic alcohols with lithium enolates have

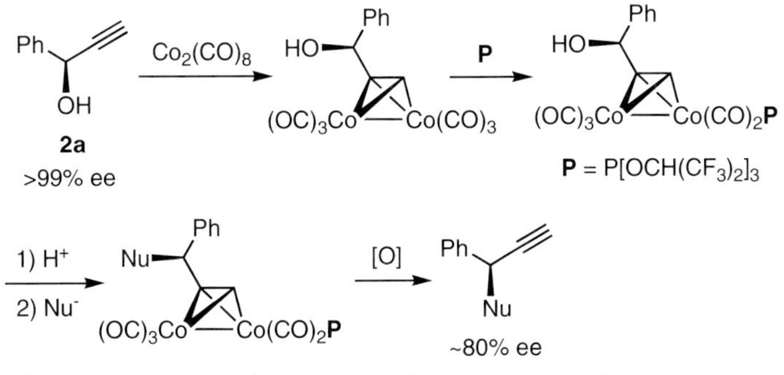

P = P[OCH(CF$_3$)$_2$]$_3$

Scheme 7.16 Diastereoselective propargylic substitution reactions.

Scheme 7.17 Enantioselective propargylic substitution reactions.

been achieved by using a stoichiometric amount of ruthenium complex bearing BINAP as a chiral ligand [26]. The column chromatographic separation of two diastereoisomers of σ-alkynyl complexes gives the enantiomerically pure propargylic substituted products bearing completely opposite configurations with almost 100% ee (Scheme 7.17). This means that a synthetic cycle for the formation of enantiomerically pure propargylic alkylated compounds from an achiral propargylic alcohol has been accomplished by starting from the ruthenium-BINAP complex (Scheme 7.18). This stepwise reaction provides the first synthetic approach to highly enantioselective propargylic substitution reactions.

As described in the previous section, the ruthenium-catalyzed propargylic alkylation of propargylic alcohols with acetone afforded the corresponding alkylated products in high yields with complete selectivity [27]. When an optically active 1-phenyl-2-propyn-1-ol was treated with acetone at room temperature in the presence of **1a** as catalyst, only a racemic alkylated product was obtained [27]. This result

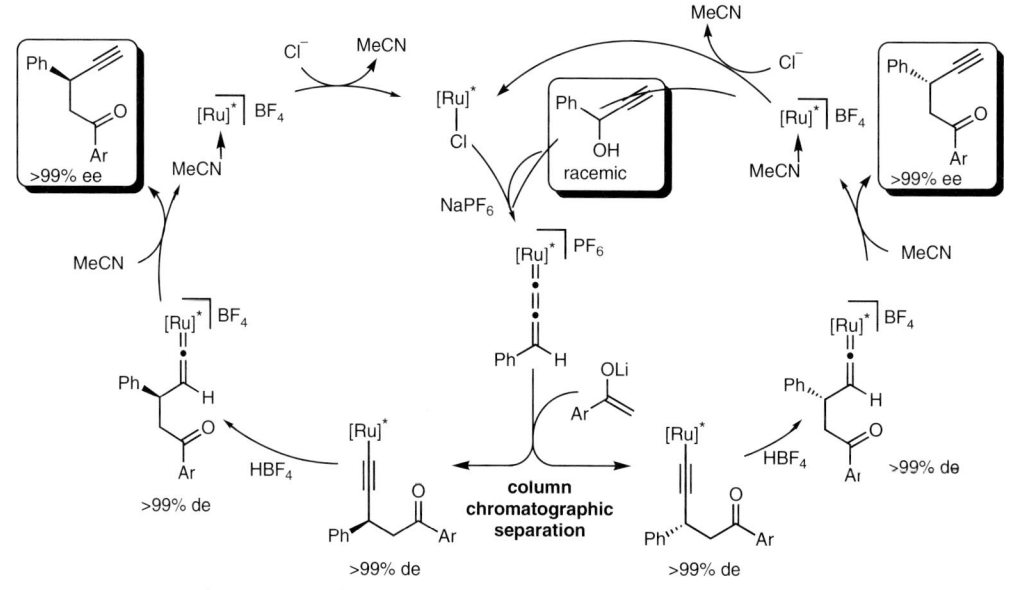

Scheme 7.18 Synthetic cycles for the formation of enantiomerically pure propargylic substituted products.

indicates that racemization occurred rapidly in a chiral allenylidene intermediate when an achiral thiolate-bridged diruthenium complex was used as a catalyst. This phenomenon prompted us to develop an enantioselective propargylic substitution reaction of propargylic alcohols with acetone in the presence of a chiral thiolate-bridged diruthenium complex.

The catalytic propargylic alkylation was investigated in the presence of thiolate-bridged diruthenium complexes as catalysts generated *in situ* from reactions of [Cp*RuCl(μ_2-Cl)]$_2$ with optically active thiols prepared from the corresponding optically active alcohols [27]. Typical results for the reaction of 1-phenyl-2-propyn-1-ol with acetone in the presence of a variety of catalysts are shown in Scheme 7.19. The best enantioselectivity (35% ee) was observed in the reaction of 1-(1-naphthyl)-2-propyn-1-ol with acetone in the presence of a complex bearing a 1-naphthylethylthiolato moiety as a chiral ligand. Although the enantioselectivity is not yet satisfactory, this was the first example of an enantioselective propargylic substitution reaction catalyzed by transition metal complexes [27]. It is noteworthy that the chiral thiolate-bridged ligands work to control the chiral environment around the diruthenium site.

More recently, a different concept other than steric repulsion between substrates and chiral ligands was introduced to improve the enantioselectivity: thus, a new type of chiral alkanethiolato ligand having a phenyl ring (second generation chiral ligands) has been prepared which might interact with a phenyl ring of ruthenium-allenylidene complexes by a π–π interaction (Scheme 7.20). In this system, nucleophilic attack of nucleophiles on the Cγ atom of the allenylidene ligand should occur from the side which is not blocked by a chiral ligand, as shown in Scheme 7.20. In fact, a quite high enantioselectivity (up to 82% ee) was observed in the catalytic propargylic alkylation (Scheme 7.21) [28]. The development of a more suitable catalytic system is still awaited.

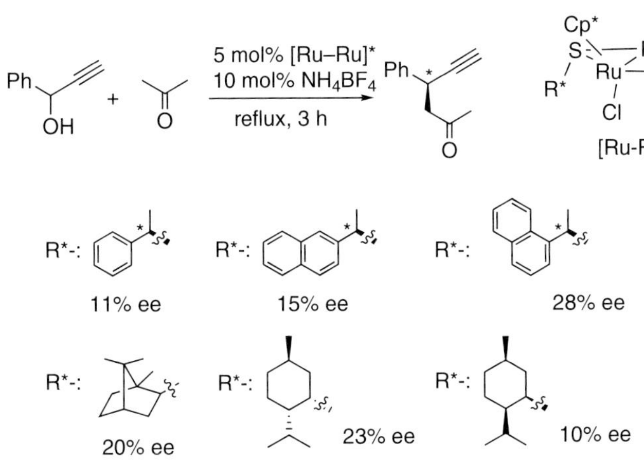

Scheme 7.19 The first example of enantioselective propargylic substitution reactions catalyzed by chiral thiolate-bridged diruthenium complexes.

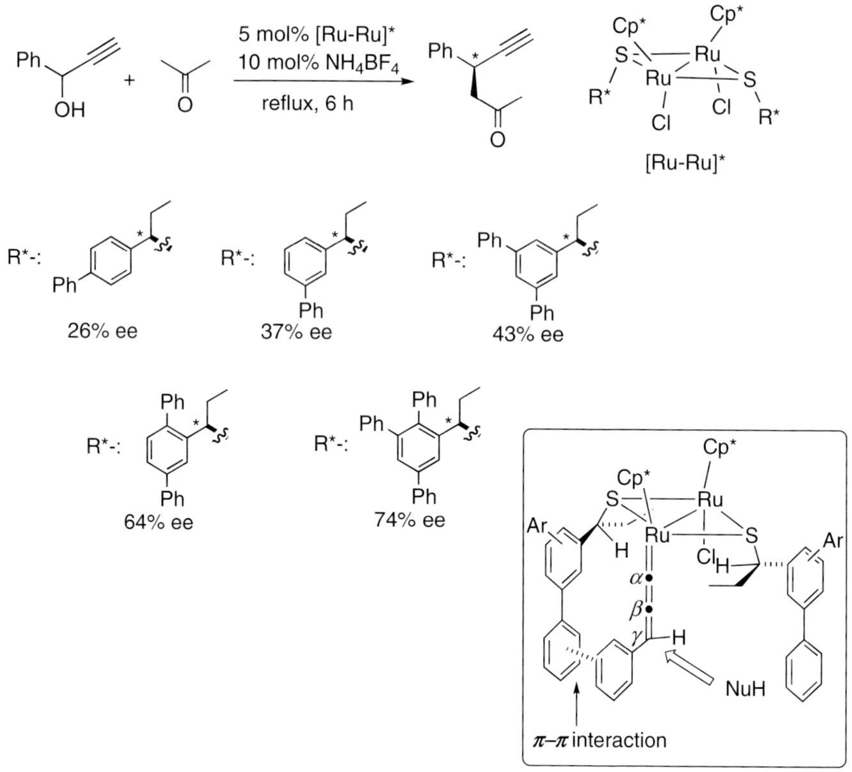

Scheme 7.20 Improved enantioselective propargylic substitution reactions catalyzed by chiral thiolate-bridged diruthenium complexes.

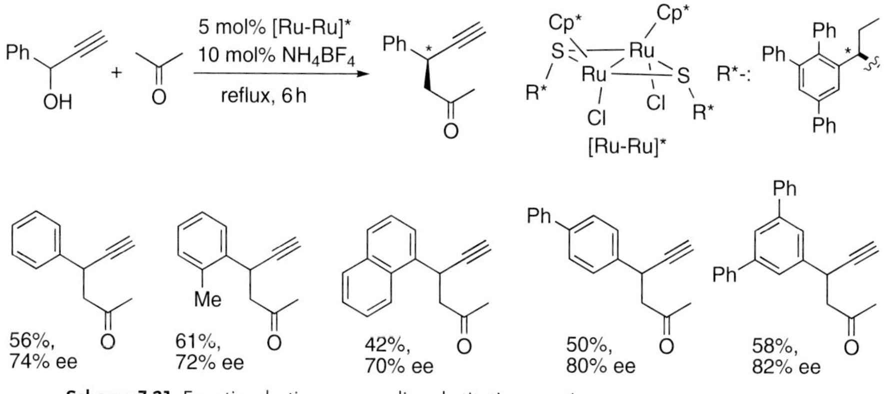

Scheme 7.21 Enantioselective propargylic substitution reactions of various propargylic alcohols catalyzed by a chiral thiolate-bridged diruthenium complex.

7.2.5
Cycloaddition between Propargylic Alcohols and Cyclic 1,3-Dicarbonyl Compounds

Reactions of propargylic alcohols with some *cyclic* 1,3-diketones did not give the corresponding propargylic alkylated products, but instead the corresponding cycloaddition products, chromenone derivatives, were obtained in high yields with complete regioselectivity [29]. This catalytic cycloaddition provides a simple and one-pot synthetic protocol for a variety of substituted chromenones, which are difficult to access by currently known synthetic methods.

Reactions of propargylic alcohols with 1,3-cyclohexanedione gave 4,6,7,8-tetrahydrochromen-5-ones in excellent yields with complete regioselectivity (Equation 7.5). Reactions with various six-membered or five-membered ring 1,3-diketones and β-keto esters also gave the corresponding 4,6,7,8-tetrahydrochromen-5-ones and pyranone derivatives, respectively (Scheme 7.22). In contrast, reactions with

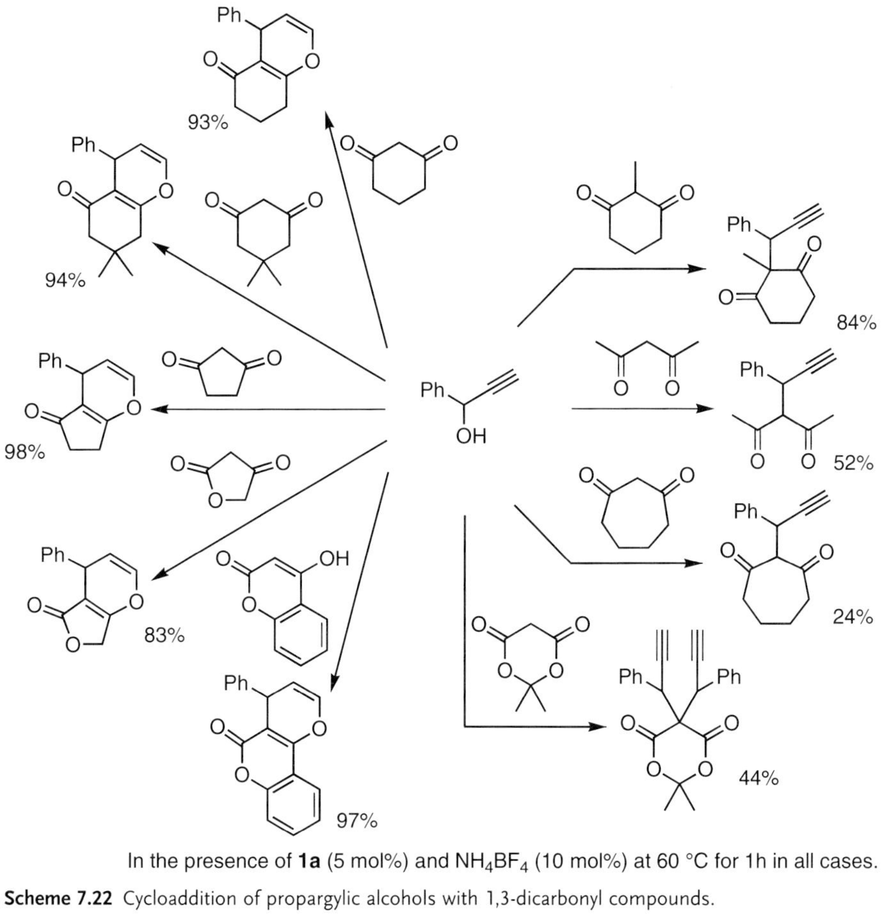

In the presence of **1a** (5 mol%) and NH_4BF_4 (10 mol%) at 60 °C for 1h in all cases.

Scheme 7.22 Cycloaddition of propargylic alcohols with 1,3-dicarbonyl compounds.

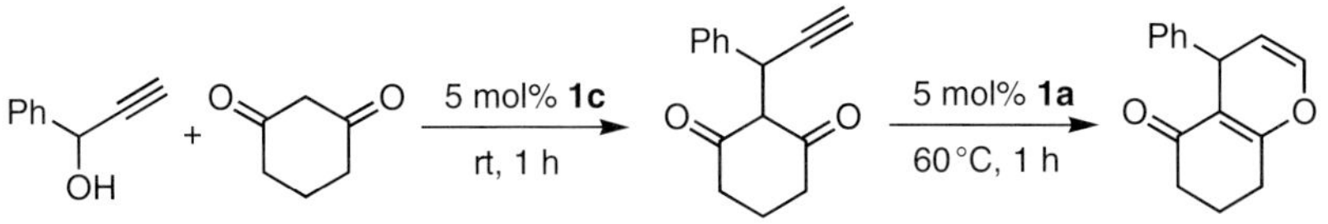

Scheme 7.23 Propargylic alkylated compound as an intermediate for cycloaddition.

7-membered ring 1,3-diketones, acyclic 1,3-diketones, acyclic β-keto esters and cyclic diesters gave only the propargylic alkylated products (Scheme 7.22), showing at first glance that cycloaddition proceeds only by the use of cyclic 1,3-diketones and β-keto esters. However, even for this apparent cycloaddition reaction, it was revealed that it actually proceeded via the propargylic alkylated compounds as intermediates. Thus, the reaction of propargylic alcohol with 1,3-cyclohexanedione in the presence of **1c** (5 mol%) at room temperature gave the corresponding propargylic alkylated product together with a small amount of the cycloaddition product. Treatment of this isolated propargylic alkylated product in the presence of **1a** (5 mol%) at 60 °C gave the corresponding cycloaddition product (Scheme 7.23).

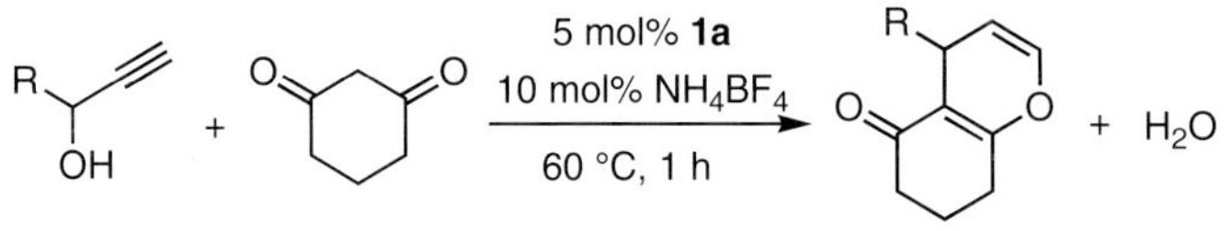

$$(7.5)$$

R = Ph 93%	R = p-MeC$_6$H$_4$ 91%	R = p-CF$_3$C$_6$H$_4$ 86%
	R = p-MeOC$_6$H$_4$ 94%	R = 2-naphthyl 98%
	R = p-ClC$_6$H$_4$ 99%	R = Ph$_2$C=CH 87%

On the basis of these findings, a pathway for this cycloaddition is proposed in Scheme 7.24. The first step is the nucleophilic attack of the carbon atom in the 2-position of 1,3-cyclohexanedione on the Cγ atom of the allenylidene complex to give a vinylidene complex, which is transformed into an alkenyl complex by intramolecular nucleophilic attack of the oxygen atom of a hydroxy group of an enol on the C$_α$ atom of the vinylidene complex. By the use of **1c** with its bulkier alkanethio moiety as a catalyst and at lower temperature, a subsequent intramolecular cyclization may be slow enough to make isolation of the alkylated product possible.

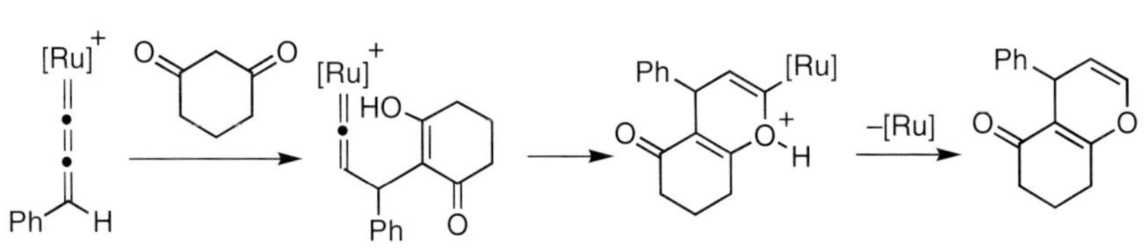

Scheme 7.24 Proposed reaction pathway for cycloaddition.

7.3
Propargylation of Aromatic Compounds with Propargylic Alcohols

7.3.1
Propargylation of Heteroaromatic and Aromatic Compounds with Propargylic Alcohols

As described in the previous sections, a variety of nucleophiles attack the Cγ atom of ruthenium-allenylidene intermediates. Aromatic compounds should also be suitable candidates and this was found to be the case [30]. Thus, reactions of propargylic alcohols with heteroaromatic compounds such as furans, thiophenes, pyrroles, and indoles in the presence of a diruthenium catalyst such as **1a** proceeded smoothly to afford the corresponding propargylated heteroaromatic compounds in high yields with complete regioselectivity (Scheme 7.25). The reaction is considered to be an electrophilic aromatic substitution if viewed from the side of aromatic compounds. In all cases, the propargylation occurred selectively at the α-position of the heterocyclic rings, while the β-propargylated indole was obtained in the reaction with indole. The regioselectivity observed here is exactly in accord with that of electrophilic substitution reactions of heteroaromatic compounds. Intramolecular propargylation of furan proceeded smoothly in the reactions of propargylic alcohols bearing a furan moiety (Equation 7.6) [30b]. Not only heteroaromatic compounds but also benzene derivatives, such as aniline derivatives, are available for this propargylation. In addition to various N-substituted anilines, electron-rich arenes such as 3,5-dimethoxyacetanilide, 1,3,5-trimethoxybenzene, and azulene can also be employed for this propargylation (Scheme 7.26). Unfortunately, however, an effective propargylation did not proceed when aromatic compounds such as acetanilide, 1,3-dimethoxybenzene,

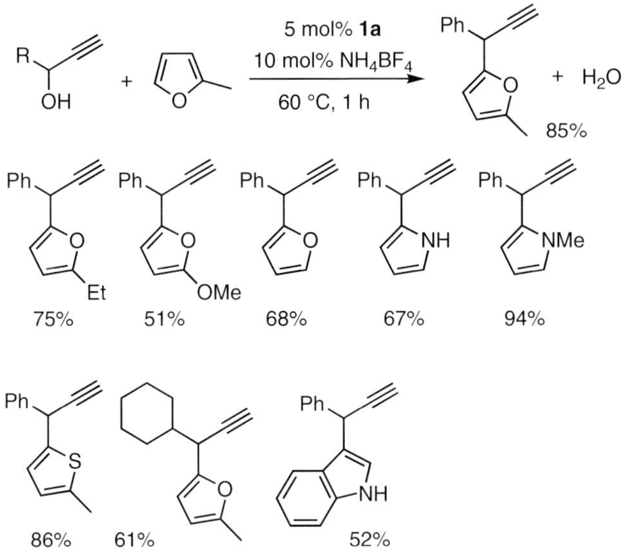

Scheme 7.25 Propargylation of heteroaromatic compounds with propargylic alcohols.

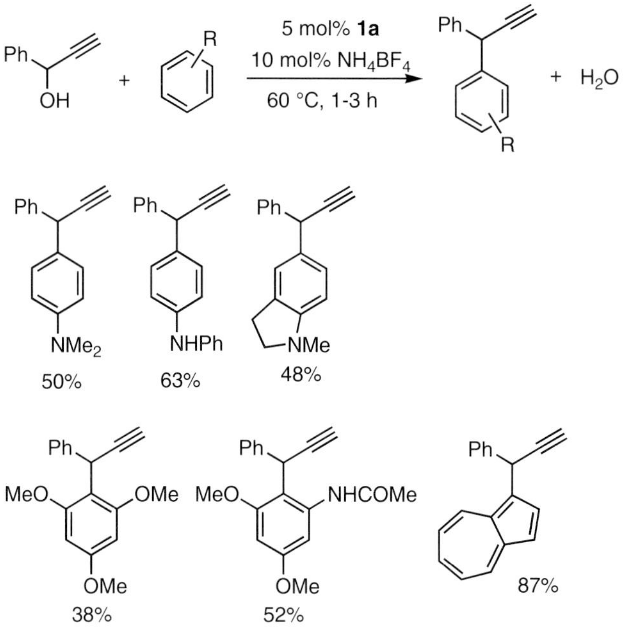

Scheme 7.26 Propargylation of aromatic compounds with propargylic alcohols.

1,3,5-trimethylbenzene, p-xylene, toluene, benzene, and ferrocene were used, indicating that this catalytic propargylation proceeded only when highly electron-rich arenes were used.

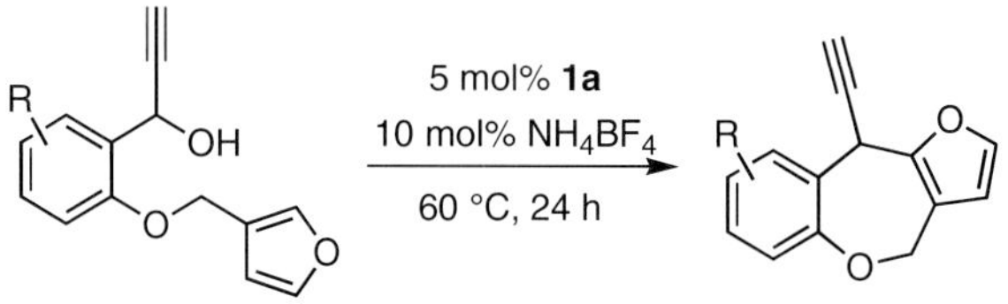

$$(7.6)$$

Quite recently, some mononuclear ruthenium complexes such as [(p-cymene)RuX-(CO)(PR$_3$)]OTf (X = Cl, OTf, R = Ph, Cy) have been found to work as catalysts for the propargylation of aromatic compounds such as furans, where some ruthenium complexes were isolated as catalytically active species from the stoichiometric reactions of propargylic alcohols (Scheme 7.27) [31]. The produced active species promoted the propargylation of furans with propargylic alcohols bearing not only a terminal alkyne moiety but also an internal alkyne moiety, indicating that this propargylation does not proceed via allenylidene complexes as key intermediates.

7.3.2
Cycloaddition between Propargylic Alcohols and Phenol and Naphthol Derivatives

Reactions of propargylic alcohols with 2-naphthol gave the cycloaddition products such as 1H-naphtho[2,1-b]pyrans in excellent yields with complete selectivity

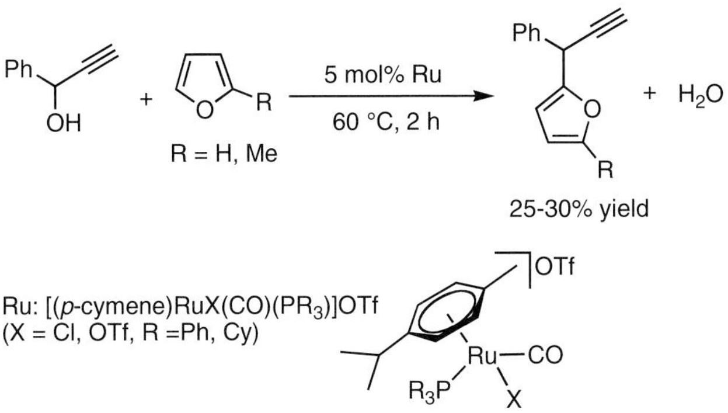

7.3 Propargylation of Aromatic Compounds with Propargylic Alcohols | 235

Scheme 7.27 Monoruthenium complex-catalyzed propargylation of aromatic compounds with propargylic alcohols.

Ru: [(p-cymene)RuX(CO)(PR$_3$)]OTf
(X = Cl, OTf, R = Ph, Cy)

(Scheme 7.28) [32]. Reactions with phenols bearing electron-releasing groups, such as 3,5-dimethoxyphenol and 3,4,5-trimethoxyphenol, proceeded smoothly to give the corresponding 4H-1-benzopyrans in excellent yields (Scheme 7.28). Unfortunately, when some 4-aminophenols were used, the corresponding pyrans were only obtained in moderate yields.

A proposed reaction pathway is shown in Scheme 7.29, where either the aromatic carbon or oxygen atom of naphthol may work as a nucleophile. Thus, the first step is the nucleophilic attack of the carbon atom of 1-position of 2-naphthol on the C$_\gamma$ atom of an allenylidene complex **A** to give a vinylidene complex **B**, which is then transformed into an alkenyl complex **C** by nucleophilic attack of the oxygen atom of a hydroxy group upon the C$_\alpha$ atom of **B**. Another possibility is the nucleophilic attack of the oxygen of 2-naphthol upon the C$_\alpha$ atom of the complex **A**. In this case, the initial attack of the naphthol oxygen results in the formation of a ruthenium-carbene complex, which subsequently leads to the complex **B** via the Claisen rearrangement of the carbene complex.

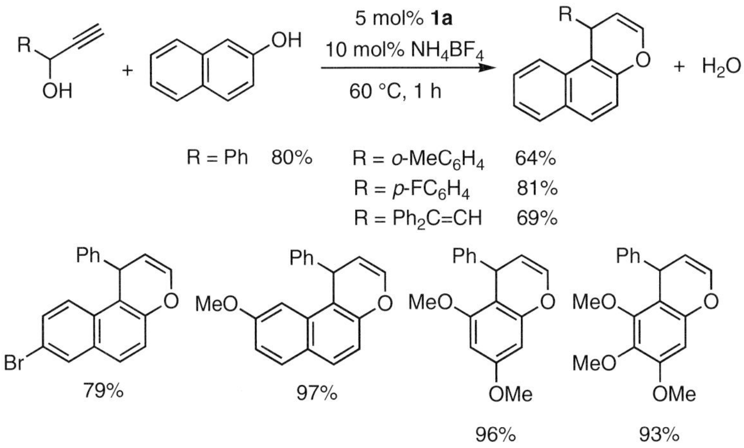

R = Ph   80%
R = o-MeC$_6$H$_4$ 64%
R = p-FC$_6$H$_4$ 81%
R = Ph$_2$C=CH   69%

79%   97%   96%   93%

Scheme 7.28 Cycloaddition of propargylic alcohols with phenol and naphthol derivatives.

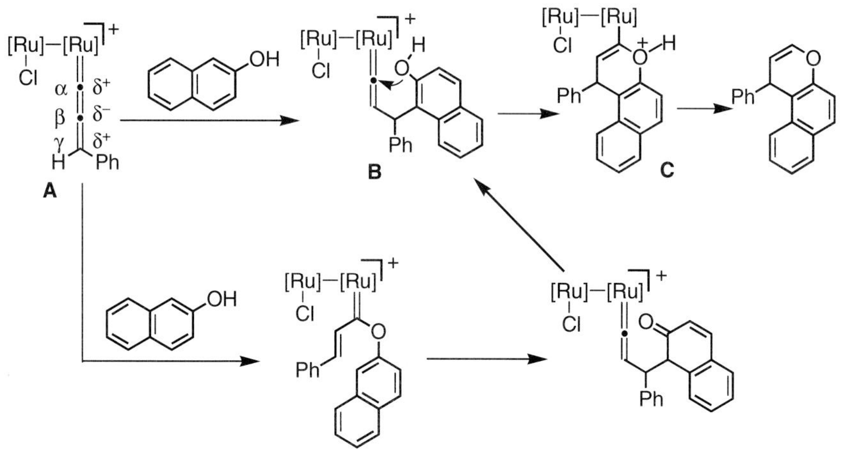

Scheme 7.29 Proposed reaction pathway for cycloaddition.

7.4
Carbon–Carbon Bond Formation via Allenylidene-Ene Reactions

Esteruelas and coworkers reported the stoichiometric Diels–Alder type addition of dienes to the C_β–C_γ double bond of allenylidene complexes to give the corresponding substituted vinylidene complexes (Equation 7.7) [33]. The results of this stoichiometric reaction prompted us to investigate the diruthenium complex-catalyzed allenylidene-ene reaction between alkenes and the C_β–C_γ double bond of an allenylidene moiety. Results of inter- and intramolecular allenylidene-ene reactions providing novel coupling products between alkynes and alkenes are described in this section [34].

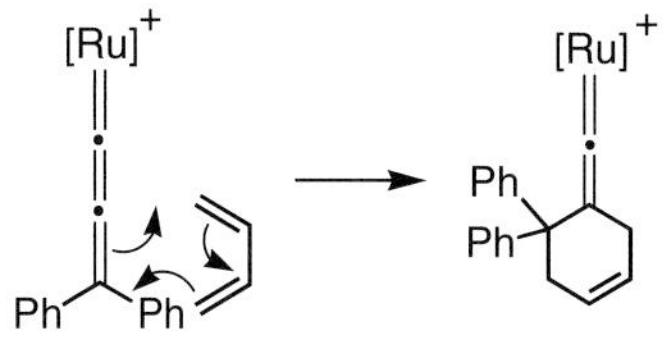

(7.7)

Intermolecular reactions of propargylic alcohols with α-methylstyrene gave the corresponding 1-hexene-5-ynes in moderate yields with complete regioselectivity (Scheme 7.30). The incorporation of a deuterium atom at the C-6 position (acetylenic terminal carbon) of the product and a substantial isotope effect ($k_H/k_D = 4$) were observed when α-methylstyrene-*methyl-d₃* was used in place of α-methylstyrene. It is considered that the C_β–C_γ double bond of an allenylidene complex reacts with α-methylstyrene, where the allenylidene complex works as an enophile, to afford the corresponding vinylidene complex via an allenylidene-ene reaction, as shown in Scheme 7.30.

Intramolecular reactions of propargylic alcohols bearing an alkene unit at a suitable position proceeded smoothly to give the corresponding substituted chro-

7.4 Carbon–Carbon Bond Formation via Allenylidene-Ene Reactions

Scheme 7.30 Intermolecular reactions of propargylic alcohols with α-methylstyrene.

manes in high yields as a mixture of two diastereoisomers, the *syn* isomer being the major product (Scheme 7.31). Use of the complexes bearing sterically more demanding groups, such as **1b** (R = ^{n}Pr) and **1c** (R = ^{i}Pr), increased the diastereoselectivity of the substituted chromanes dramatically, although a prolonged reaction time was required. The diastereoselectivity of the produced chromanes using the Nicholas

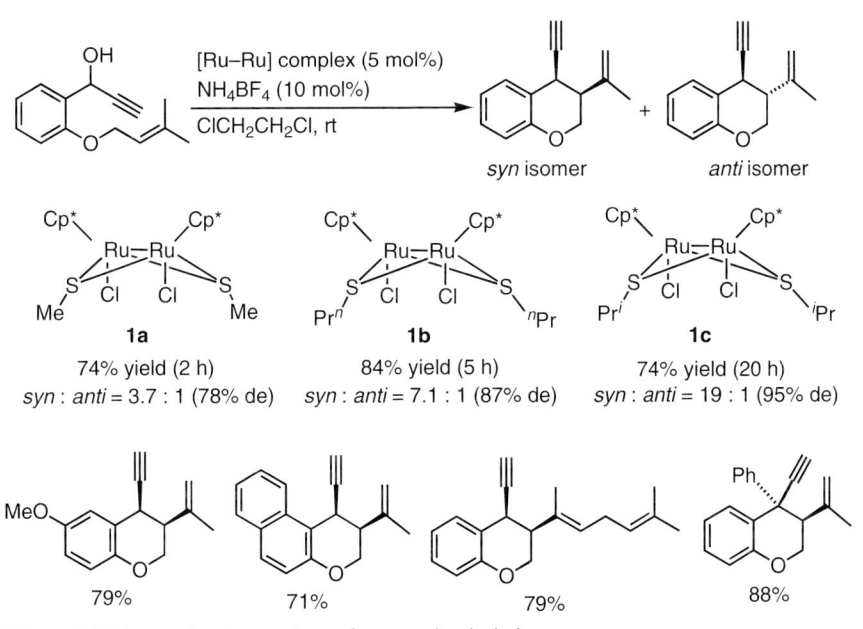

1a
74% yield (2 h)
syn : anti = 3.7 : 1 (78% de)

1b
84% yield (5 h)
syn : anti = 7.1 : 1 (87% de)

1c
74% yield (20 h)
syn : anti = 19 : 1 (95% de)

79%   71%   79%   88%

Scheme 7.31 Intramolecular reactions of propargylic alcohols bearing an alkene moiety at a suitable position.

reaction is completely the reverse of that of the chromanes obtained in diruthenium-catalyzed reactions [35].

7.5
Reductive Coupling Reaction via Hydroboration of Allenylidene Intermediates

Although hydroboration of carbon–carbon double and triple bonds is one of the most useful methods in organic synthesis [36], its application to an allenylidene moiety has not yet been reported [37]. When propargylic alcohols were treated with pinacolborane in the presence of a catalytic amount of **1a**, the reductive homocoupling products of the propargylic alcohols were obtained unexpectedly in place of the expected reduced (OH $\rightarrow$ H) compounds. The reaction seems to proceed via hydroboration of the allenylidene intermediates [38]. Reactions of propargylic alcohols with pinacolborane in the presence of a catalytic amount of **1a** afforded the propargylic group homocoupled products in moderate to good yields as a mixture of two diastereoisomers (*dl* and *meso* isomers) (Scheme 7.32). In contrast to the reaction with pinacolborane, the homocoupled compound was not produced in reactions with catecholborane and 9-borabicyclo[3.3.1]nonane (9-BBN). On the other hand, the stoichiometric reaction of the allenylidene complex with pinacolborane led to the formation of the homocoupled compound (Equation 7.8). These results support the proposed reaction pathway shown in Scheme 7.33. First, hydroboration with pinacolborane occurs at the C_β–C_γ double bond of an initially produced allenylidene

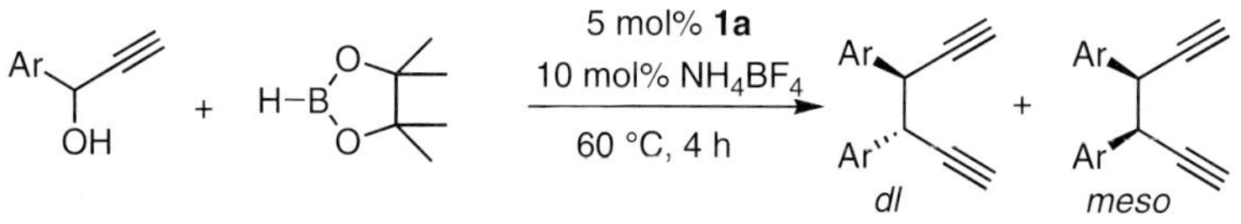

Ar = Ph 68% (93 : 7)

Ar = *p*-FC_6H_4 66% (96 : 4)
Ar = *p*-ClC_6H_4 60% (92 : 8)
Ar = *p*-$CH_3C_6H_4$ 30% (78 : 22)

Scheme 7.32 Reductive coupling reactions of propargylic alcohols.

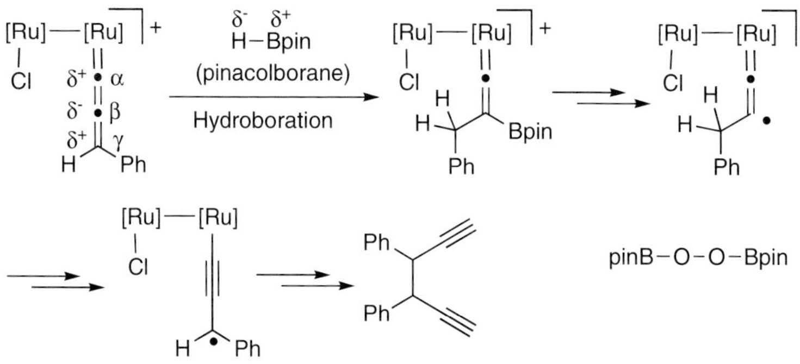

Scheme 7.33 Proposed reaction pathway for reductive coupling reactions.

complex to give the corresponding β-boravinylidene complex. Then, the vinylidene complex is converted into the cationic radical complex via radical fission assisted by adventitious molecular oxygen [39]. Because of the presence of a phenyl group, the radical at the C_γ atom of the complex is highly stabilized by spin delocalization over the phenyl ring, and hydrogen radical elimination forms the neutral radical complex preferentially [40]. Finally, intermolecular coupling between two radical species results in the formation of the final product. The fact that only propargylic alcohols bearing an aryl moiety at the propargylic position are available for this catalytic coupling reaction may support this proposed reaction pathway.

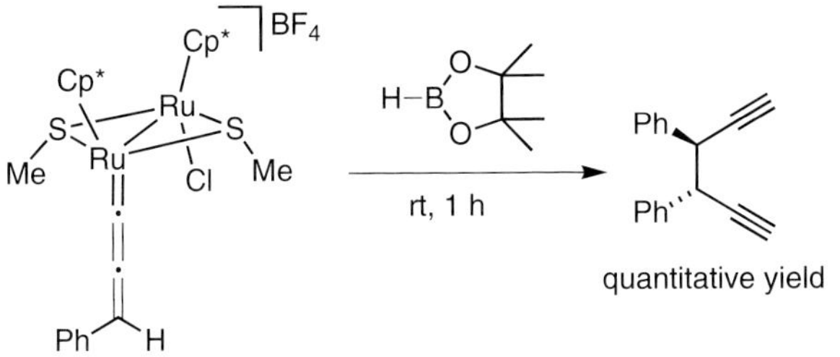
rt, 1 h
quantitative yield

(7.8)

7.6 Selective Preparation of Conjugated Enynes

Reactions of propargylic alcohols bearing a cyclopropyl group at the propargylic position with anilines in the presence of a catalytic amount of **1a** gave the conjugated enynes (*E* only) in high to excellent yields with complete selectivity (Scheme 7.34) [41]. The reaction did not proceed at all in the presence of a catalytic amount of other transition metal salts and a Brønsted acid such as $AuCl_3$, $FeCl_3$, and *p*-toluenesulfonic acid [42]. Reactions of some propargylic alcohols with water also proceeded smoothly to give the corresponding conjugated (*E*)-enynes in good to high yields (Equation 7.9). This is the first successful example of catalytic reactions with water, where allenylidene complexes are involved as important intermediates.

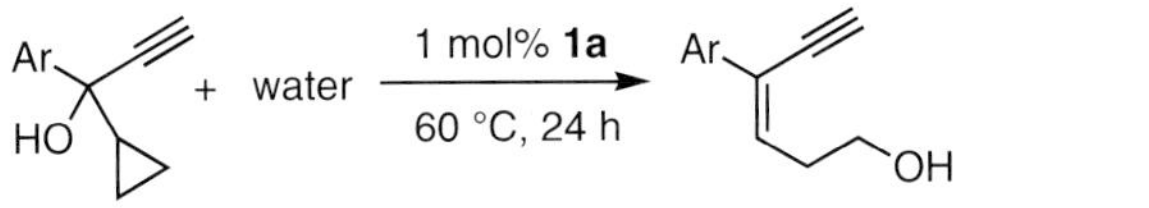
+ water
1 mol% 1a
60 °C, 24 h

(7.9)

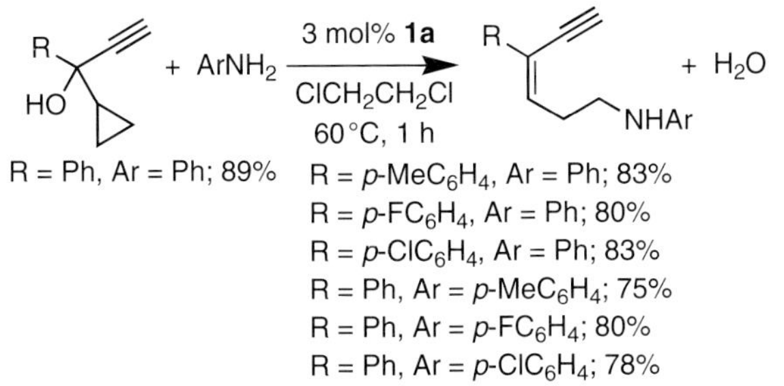
+ ArNH2
3 mol% 1a
ClCH2CH2Cl
60 °C, 1 h
+ H2O
R = Ph, Ar = Ph; 89%
R = p-MeC6H4, Ar = Ph; 83%
R = p-FC6H4, Ar = Ph; 80%
R = p-ClC6H4, Ar = Ph; 83%
R = Ph, Ar = p-MeC6H4; 75%
R = Ph, Ar = p-FC6H4; 80%
R = Ph, Ar = p-ClC6H4; 78%

Scheme 7.34 Selective preparation of tri-substituted conjugated enynes.

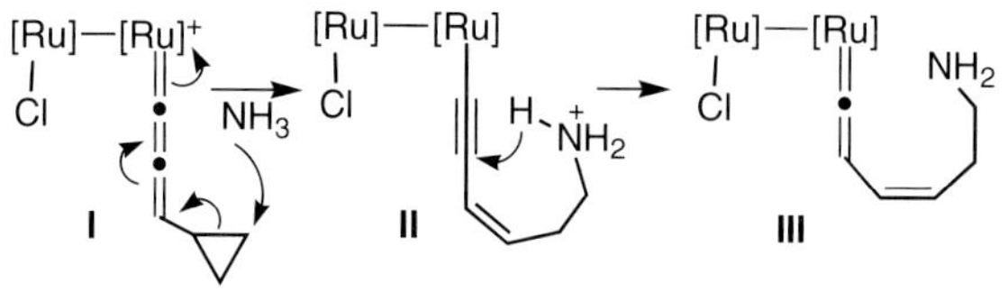

Scheme 7.35 Density functional theory calculation at the B3LYP/LANL2DZ level of theory for the model reaction.

By taking into consideration the product structure, the nucleophilic attack should occur on the ε-carbon of a cyclopropane of the intermediate allenylidene complex. In order to know whether this process followed by ring-opening is energetically favorable or not, a density functional theory calculation was performed at the B3LYP/LANL2DZ level of theory for the model reaction [CpRuCl(μ_2-SMe)$_2$RuCp(=C=C=CH(CHCH$_2$CH$_2$))]$^+$ (**I**; Cp = η^5-C$_5$H$_5$) with NH$_3$ (Scheme 7.35). As a result, the expected nucleophilic attack of NH$_3$ on the ε-carbon of a cyclopropane ring connected to an allenylidene moiety was shown to occur easily, giving the corresponding ruthenium-alkynyl complex (**II**). This is followed by the smooth transfer of one of the protons on the nitrogen atom to an alkynyl moiety to give the corresponding vinylidene complex (**III**). Thus, the reaction pathway shown in Scheme 7.35 is energetically favorable and reasonable.

To account for highly selective formation of (*E*)-enynes, a reaction pathway as shown in Scheme 7.36 was proposed. The steric repulsion between a phenyl group and a cyclopropyl moiety might give predominant formation of **IV** which is susceptible to nucleophilic attack by aniline on the cyclopropane ring to give (*E*)-enynes with complete selectivity. It is well known that the interaction of the cyclopropyl bonding orbitals with the carbon p orbital of the carbocation imposes a preference for the bisected conformation (the eclipsed conformation) of the cyclopropylmethyl cation rather than the perpendicular conformation [43]. A similar stabilization between cyclopropyl bonding orbitals and the p orbital of the γ–carbon in the allenylidene ligand may also occur in **IV**, as shown in Scheme 7.36.

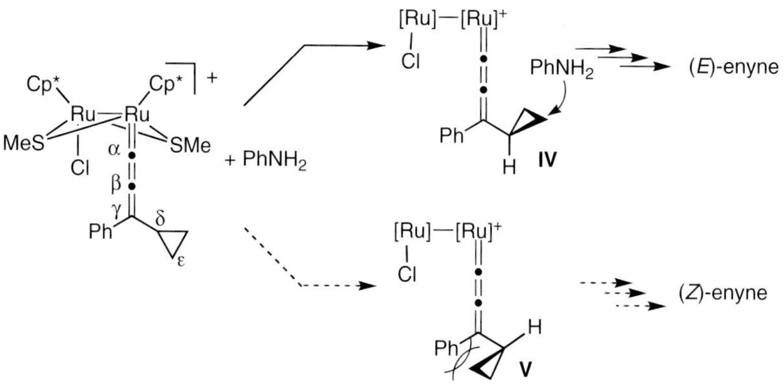

Scheme 7.36 Reaction pathway to account for highly selective formation of (*E*)-enynes.

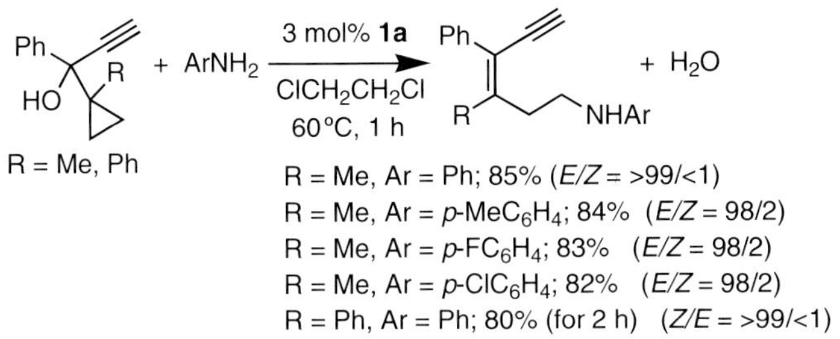

Scheme 7.37 shows a reaction:

Ph−≡−R(cyclopropyl with HO, R) + ArNH₂ →(3 mol% **1a**, ClCH₂CH₂Cl, 60 °C, 1 h) Ph−C≡C−C(=CR)(CH₂CH₂NHAr) + H₂O

R = Me, Ph

R = Me, Ar = Ph; 85% (E/Z = >99/<1)
R = Me, Ar = p-MeC$_6$H$_4$; 84% (E/Z = 98/2)
R = Me, Ar = p-FC$_6$H$_4$; 83% (E/Z = 98/2)
R = Me, Ar = p-ClC$_6$H$_4$; 82% (E/Z = 98/2)
R = Ph, Ar = Ph; 80% (for 2 h) (Z/E = >99/<1)

Scheme 7.37 Selective preparation of tetra-substituted conjugated enynes.

For synthetic purpose, the selective preparation of tetrasubstituted enynes was also investigated in the reactions of 1-cyclopropyl-2-propyn-1-ols bearing a substituent at the α-position in a cyclopropane ring with anilines (Scheme 7.37). As expected, the corresponding tetrasubstituted enynes were obtained in high yields with almost complete selectivity.

7.7
Preparation of Dicationic Chalcogenolate-Bridged Diruthenium Complexes and Their Dual Catalytic Activity

In addition to the catalytic activity of chalcogenolate-bridged diruthenium complexes, which are monocationic species, the corresponding dicationic chalcogenolate-bridged diruthenium complexes, directly prepared from neutral diruthenium complexes and 2 molar equiv of silver trifluoromethanesulfonate (AgOTf), were found to show novel catalytic activity [44]. Thus, reactions of propargylic alcohols with acetone afforded the corresponding hexadienones in moderate to good yields, in sharp contrast to the formation of γ-ketoalkynes when neutral or monocationic diruthenium complexes were employed as catalysts [21].

Treatment of methanethiolate-bridged diruthenium complex **1a** with 2 molar equiv of AgOTf at room temperature gave the corresponding dicationic thiolate-bridged diruthenium complex [Cp*Ru(μ$_2$-SMe)$_2$RuCp*(OH$_2$)](OTf)$_2$ (**5a**) in a high yield (Scheme 7.38). Similarly, the dicationic selenolate- and tellurolate-bridged diruthenium complexes [Cp*Ru(μ$_2$-YMe)$_2$RuCp*(OH$_2$)](OTf)$_2$ (**5b**, Y = Se; **5c**, Y = Te) were prepared in good yields (Scheme 7.38). The molecular structure of **5a** was determined

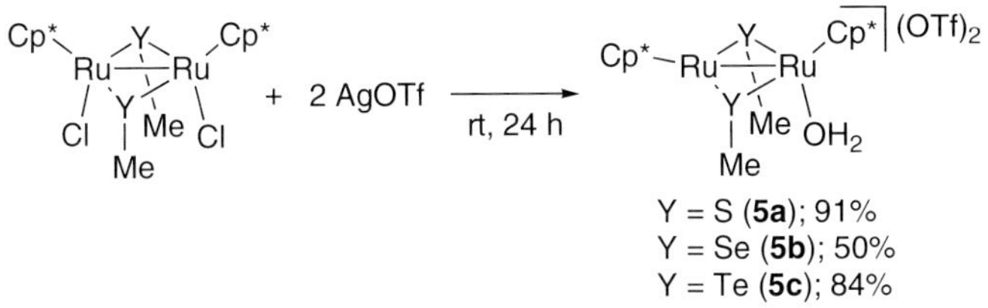

Y = S (**5a**); 91%
Y = Se (**5b**); 50%
Y = Te (**5c**); 84%

Scheme 7.38 Preparation of dicationic chalcogenolate-bridged diruthenium complexes.

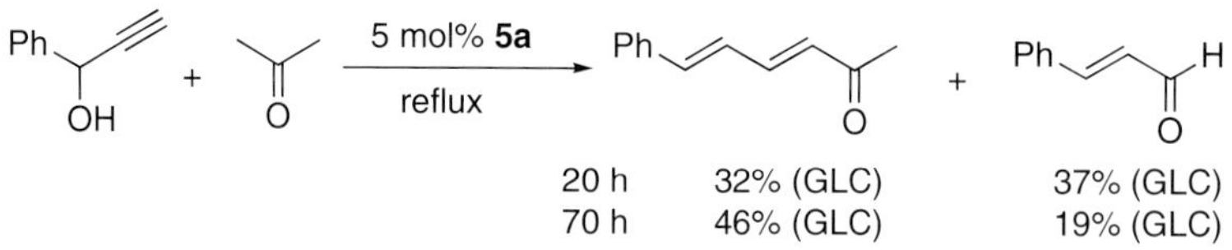

Scheme 7.39 Reaction of propargylic alcohol with acetone catalyzed by dicationic diruthenium complex.

	20 h	32% (GLC)	37% (GLC)
	70 h	46% (GLC)	19% (GLC)

by X-ray analysis. The bond length of Ru–Ru (2.61 Å) in **5a** is shorter than that of the Ru–Ru bond in **1a** (2.84 Å).

Reactions of propargylic alcohol with acetone in the presence of a catalytic amount of **5a** at reflux temperature gave a conjugated hexadienone together with cinnamaldehyde (Scheme 7.39). Prolonged reaction time increased the yield of the hexadienone by sacrificing the yield of cinnamaldehyde. For comparison, the catalytic activities of other dicationic diruthenium complexes were investigated in the reaction of propargylic alcohol with acetone at reflux temperature. Use of dicationic selenolate- and tellurolate-bridged diruthenium complexes (**5b** and **5c**) afforded the hexadienone in moderate yields (Scheme 7.40), in contrast to the result with neutral or monocationic tellurolate-bridged diruthenium complexes which did not promote propargylic alkylation of propargylic alcohols with acetone.

In order to obtain some information on the reaction mechanism, the reaction of propargylic alcohol with acetone in the presence of a catalytic amount of **5a** was monitored. The result indicated that the catalytic formation of the hexadienone proceeded via the initial isomerization of propargylic alcohol to cinnamaldehyde followed by aldol condensation between the produced aldehyde and acetone, and then dehydration. In fact, heating of propargylic alcohol in the presence of a catalytic amount of **5a** gave only cinnamaldehyde (Scheme 7.41), and the separate reaction of cinnamaldehyde with acetone in the presence of a catalytic amount of **5a** gave the hexadienone

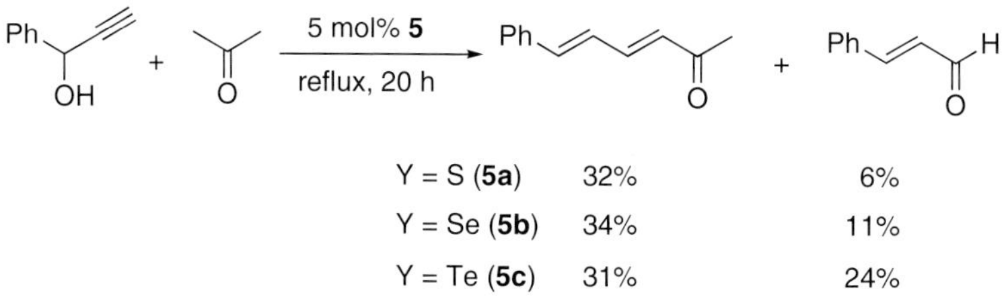

Y = S (**5a**)	32%	6%
Y = Se (**5b**)	34%	11%
Y = Te (**5c**)	31%	24%

Scheme 7.40 Reactions of propargylic alcohol with acetone catalyzed by dicationic chalcogenolate-bridged diruthenium complexes.

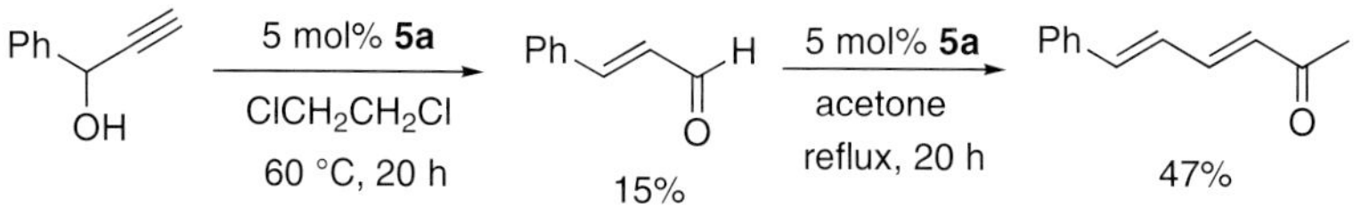

Scheme 7.41 Reaction of cinnamaldehyde with acetone.

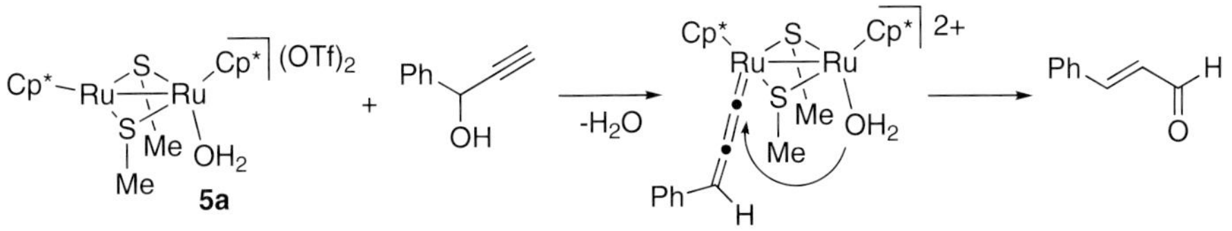

7.8 Other Catalytic Reactions via Allenylidene Complexes as Key Intermediates

Scheme 7.42 Proposed reaction pathway for cinnamaldehyde via an allenylidene complex.

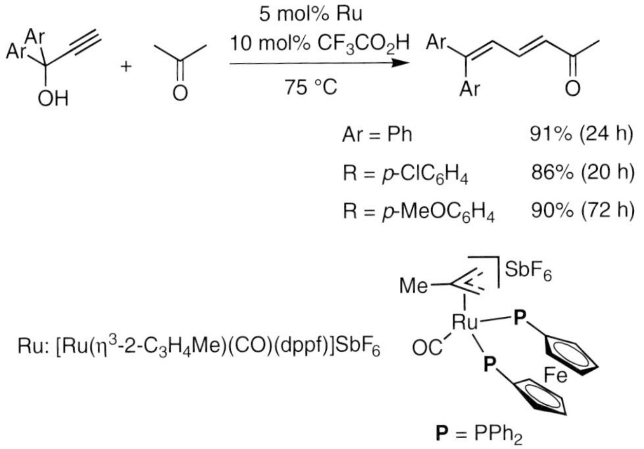

Scheme 7.43 Monoruthenium complex-catalyzed reactions of propargylic alcohols with acetone.

(Scheme 7.41). It seems to be reasonable to presume that the isomerization of propargylic alcohol to cinnamaldehyde proceeds via an intramolecular nucleophilic attack of coordinated water on an electropositive α-carbon of the allenylidene ligand (Scheme 7.42). Then, dicationic diruthenium complexes work as Lewis acids to promote the aldol condensation between cinnamaldehyde and acetone. Thus, the dual catalytic activity of dicationic chalcogenolate-bridged diruthenium complexes is essential to promote the present novel reaction between propargylic alcohols and acetone.

Quite recently, Gimeno and coworkers reported the same type of coupling reaction between propargylic alcohols and ketones using the mononuclear ruthenium complex [Ru(η^3-2-C$_3$H$_4$Me)(CO)(dppf)]SbF$_6$ and trifluoroacetic acid (Scheme 7.43) [45]. They also proposed that the attack of water on the α-carbon of the allenylidene ligand is a key step in producing the corresponding aldehydes [46], which react with ketones to give the final products.

7.8
Other Catalytic Reactions via Allenylidene Complexes as Key Intermediates

After our discovery of the chalcogenolate-bridged diruthenium complex-catalyzed propargylic substitution reactions of propargylic alcohols with a variety of nucleophiles via allenylidene intermediates, as described in the previous sections, some

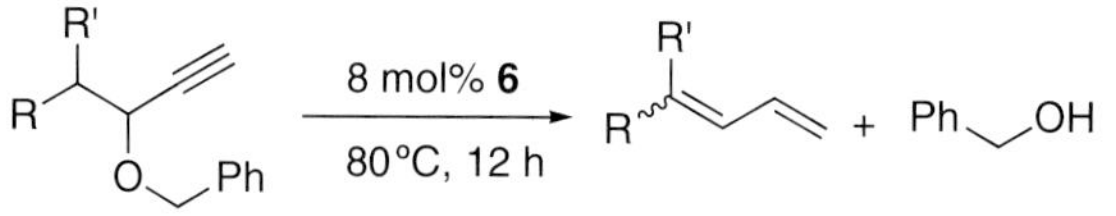

7 Allenylidene Complexes in Catalysis

Scheme 7.44 Monoruthenium complex-catalyzed bond fission reactions of propargylic ethers.

Reaction conditions: 8 mol% **6**, 80 °C, 12 h

[TpRu(PPh$_3$)(MeCN)$_2$]PF$_6$ **6**

R = Ph, R' = H	72%
R = $^nC_{12}H_{25}$, R' = H	78%
R = Ph, R' = Me	78%

mononuclear ruthenium complexes have been reported to promote the catalytic organic transformation via allenylidene complexes. Especially, Liu and coworkers have found interesting catalytic reactions using a monoruthenium complex [TpRu(PPh$_3$)(MeCN)$_2$]PF$_6$ (Tp=tris(1-pyrazolyl)borate) (**6**), typical results being shown in this section.

In 2002, Liu and coworkers reported bond-fission reactions of propargylic ethers in the presence of a catalytic amount of the monoruthenium complex at 80 °C to give the corresponding conjugated dienes in good yields as a mixture of E/Z isomers (Scheme 7.44) [47]. Although no intermediates including allenylidene complexes were detected in this reaction system, the reaction pathway via such intermediates was proposed, as shown in Scheme 7.45. The first step is the dealkoxylation from the propargylic ether to give the allenylidene complex. Then, the attack of the generated alcohol on the α-carbon atom of the allenylidene complex, followed by isomerization

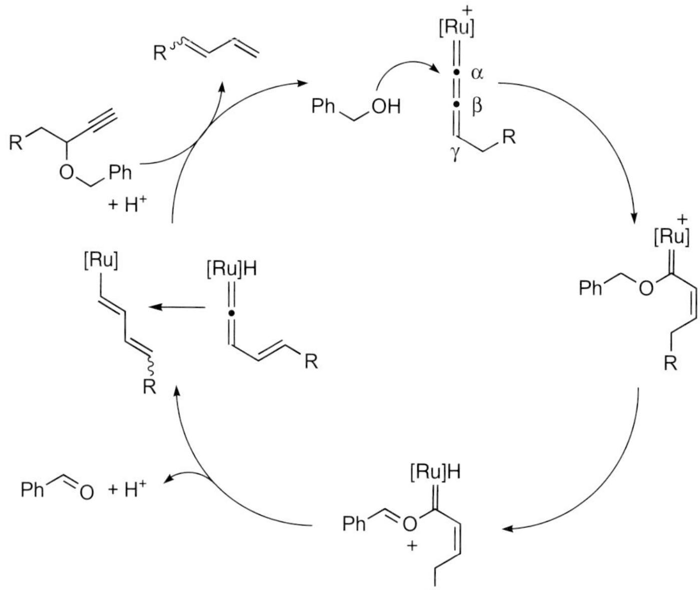

[Ru] = TpRu(PPh$_3$)

Scheme 7.45 Proposed reaction pathway for the bond fission reactions of propargylic ethers.

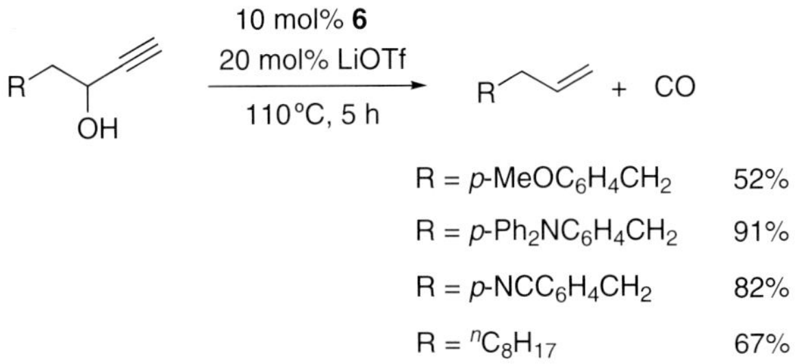

7.8 Other Catalytic Reactions via Allenylidene Complexes as Key Intermediates

Scheme 7.46 Monoruthenium complex-catalyzed decarbonylation of propargylic alcohols.

R = p-MeOC$_6$H$_4$CH$_2$	52%
R = p-Ph$_2$NC$_6$H$_4$CH$_2$	91%
R = p-NCC$_6$H$_4$CH$_2$	82%
R = ^{n}C$_8$H$_{17}$	67%

of the produced carbene, led to the formation of the conjugated diene. In fact, from cyclic propargylic ethers, the corresponding conjugated dienes bearing a formyl moiety were produced (Equation 7.10).

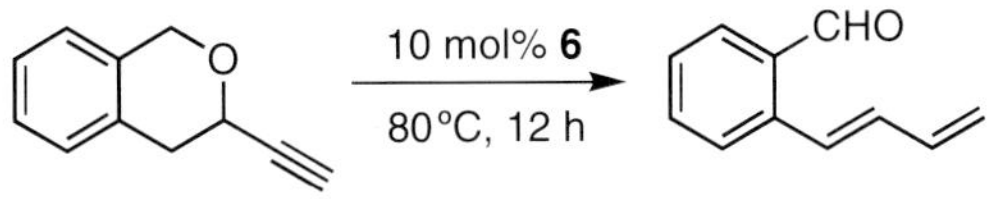

$$(7.10)$$

They also reported the selective cleavage of the carbon–carbon triple bond of propargylic alcohols [48]. Reaction of propargylic alcohols in the presence of a catalytic amount of **6** and LiOTf at 110 °C gave the corresponding terminal alkenes in good yields together with carbon monoxide (Scheme 7.46). The presence of a catalytic amount of LiOTf is necessary to promote this catalytic decarbonylation. Although no intermediates including allenylidene complexes were detected in this catalytic reaction, they proposed Scheme 7.47 where the attack of OH$^-$ anion on the

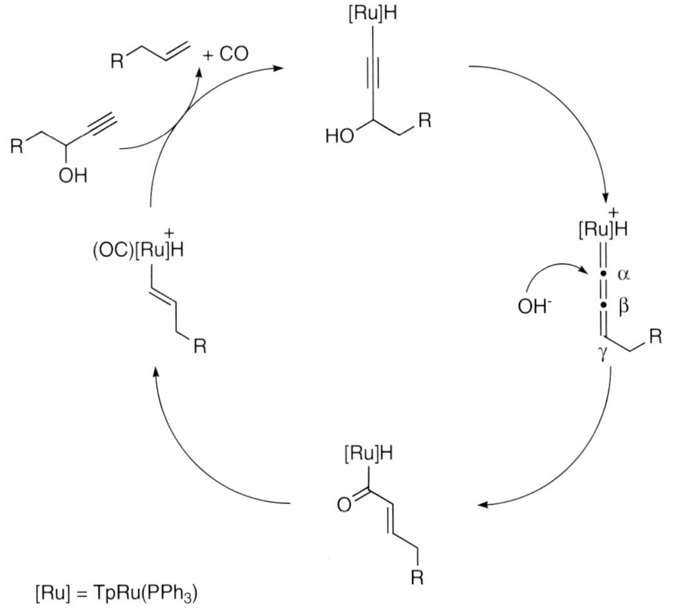

[Ru] = TpRu(PPh$_3$)

Scheme 7.47 Proposed reaction pathway for decarbonylation of propargylic alcohols.

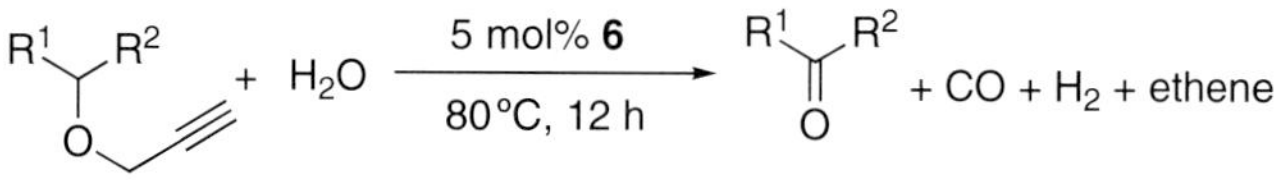

Scheme 7.48 Monoruthenium complex-catalyzed decarbonylation of propargylic ethers.

$R^1 = Ph$, $R_2 = {}^nC_4H_9$ — 90%
$R^1 = Ph$, $R_2 = {}^nC_6H_{13}$ — 91%
$R^1 = p\text{-MeOC}_6H_4$, $R_2 = {}^nC_5H_{11}$ — 97%
$R^1 = p\text{-FC}_6H_4$, $R_2 = {}^nC_5H_{11}$ — 84%

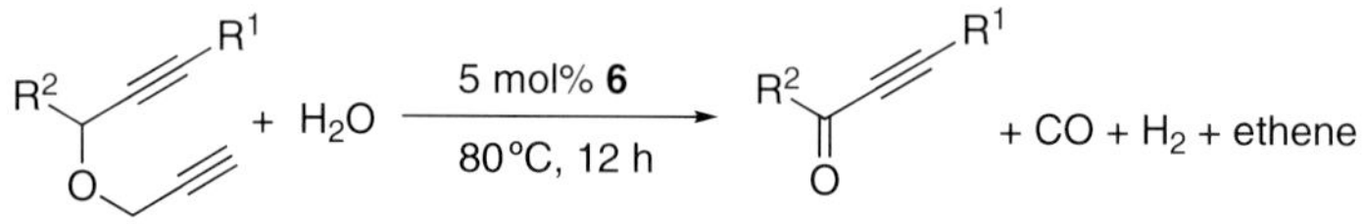

Reaction: propargylic ether + H_2O $\xrightarrow{5\text{ mol\% }\mathbf{6},\ 80°C,\ 12\text{ h}}$ ketone + CO + H_2 + ethene

$R^1 = Ph$, $R_2 = {}^nC_4H_9$ — 84%
$R^1 = {}^nC_6H_{13}$, $R_2 = Ph$ — 79%
$R^1 = {}^nC_4H_9$, $R_2 = Ph$ — 82%
$R^1 = {}^nC_6H_{13}$, $R_2 = Et$ — 82%

α-carbon atom of the generated allenylidene complexes is a key step giving the corresponding ruthenium-acyl complexes followed by decarbonylation. In fact, Bianchini and coworkers previously reported a stoichiometric reaction of an allenylidene complex bearing PNP ligand with water to give the corresponding terminal alkene and carbon monoxide (Equation 7.11) [49]. More recently, Liu and coworkers reported the transformation of aryl and alkynyl propargylic ethers to aryl and alkynyl ketones [50]. Reactions of aryl and alkynyl propargylic ethers with water in the presence of a catalytic amount of the monoruthenium complex at 80 °C gave the corresponding aryl and alkynyl ketones together with carbon monoxide, dihydrogen, and ethene (Scheme 7.48).

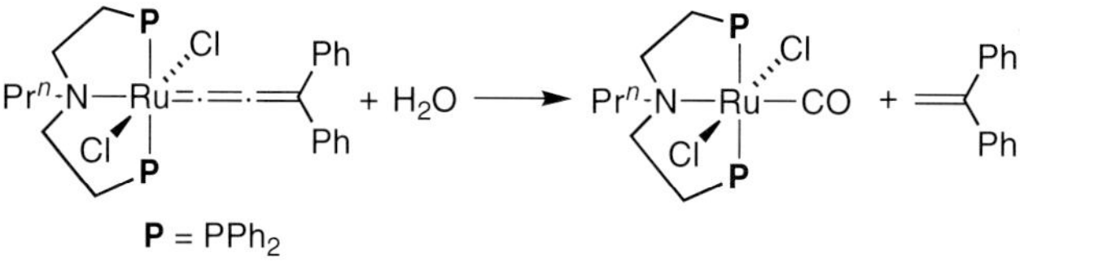

$$\text{[Pr}^n\text{-N-Ru=C=C(Ph)_2 complex]} + H_2O \longrightarrow \text{[Pr}^n\text{-N-Ru-CO complex]} + \text{CH}_2\text{=C(Ph)}_2 \qquad (7.11)$$

$P = PPh_2$

Results of catalytic reactions in which allenylidene complexes work as catalytic precursors [51] are not included in this chapter.

7.9
Conclusion

This chapter is mainly a summary of our recent findings on the thiolate-bridged diruthenium complex-catalyzed novel organic transformation of propargylic alcohols.

All reactions proceeded via ruthenium-allenylidene complexes as key intermediates. The propargylic substitution reaction described may substitute and/or compensate for the well-known Nicholas reaction. Our results on the propargylation of aromatic compounds with propargylic alcohols bearing an internal alkyne group [52] and the propargylic reduction of propargylic alcohols bearing an internal alkyne moiety [53] catalyzed by cationic thiolate-bridged diruthenium complexes are not included in this chapter because the reactions did not seem to proceed via ruthenium-allenylidene intermediates. In addition, sequential reactions via allenylidene complexes as key intermediates are not described [54]. We have a feeling that our findings have stimulated the development of many other transition metal-catalyzed propargylic substitution reactions of propargylic alcohol, bearing not only a terminal alkyne moiety but also an internal alkyne moiety, with a variety of nucleophiles to give the corresponding propargylic substituted products [42, 55]. The development of an asymmetric version of the catalytic reaction via allenylidene intermediates [56] as well as a novel organic transformation via cumulenylidene complexes [57] awaits future work.

References

1 For reviews, see: (a) Werner, H. (1997) *Chemical Communications*, 903; (b) Touchard, D. and Dixneuf, P.H. (1998) *Coordination Chemistry Reviews*, **178–180**, 409; (c) Bruce, M.I. (1998) *Chemical Reviews*, **98**, 2797; (d) Cadierno, V., Gamasa, M.P. and Gimeno, J. (2001) *European Journal of Inorganic Chemistry*, 571; (e) Puerta, M.C. and Valerga, P. (1999) *Coordination Chemistry Reviews*, **193–195**, 977; (f) Selegue, J.P. (2004) *Coordination Chemistry Reviews*, **248**, 1543; (g) Winter, R.F. and Zális, S. (2004) *Coordination Chemistry Reviews*, **248**, 1565; (h) Rigaut, S., Touchard, D. and Dixneuf, P.H. (2004) *Coordination Chemistry Reviews*, **248**, 1585; (i) Bruce, M.I. (2004) *Coordination Chemistry Reviews*, **248**, 1603; (j) Cadierno, V., Gamasa, M.P. and Gimeno, J. (2004) *Coordination Chemistry Reviews*, **248**, 1627; (k) Fischer, H. and Szesni, N. (2004) *Coordination Chemistry Reviews*, **248**, 1659.

2 For recent reviews, see: (a) Bruneau, C. (2004) *Ruthenium Catalysts and Fine Chemistry* (eds C. Bruneau and P.H. Dixneuf), Springer, Berlin, Heidelberg, p. 125; (b) Bruneau, C. and Dixneuf, P.H. (2006) *Angewandte Chemie-International Edition*, **45**, 2176.

3 For recent reviews, see: (a) Grubbs, R.H. and Trnka, T.M. (2004) *Ruthenium in Organic Synthesis* (ed. S.I. Murahashi), Wiley-VCH, Weinheim, Chapter 6, pp. 153–177; (b) Grubbs, R.H., Wenzel, A.G. and Chatterjee, A.K. (2007) *Comprehensive Organometallic Chemistry III*, Vol. 11 (eds R.H. Crabtree and D.M.P. Mingos), Elsevier, Amsterdam, p. 179; (c) Connon, S.J. and Blechert, S. (2004) *Ruthenium Catalysts and Fine Chemistry* (eds C. Bruneau and P.H. Dixneuf), Springer, Berlin, Heidelberg, p. 93; (d) Mulzer, J., Ohler, E. and Gaich, T. (2007) *Comprehensive Organometallic Chemistry III*, Vol. 11 (eds R.H. Crabtree and P.H. Dixneuf), Elsevier, Amsterdam, p. 207; (e) Mori, M. and Kitamura, T. (2007) *Comprehensive Organometallic Chemistry III*, Vol. 11 (eds R.H. Crabtree and P.H. Dixneuf), Elsevier, Amsterdam, p. 271.

4 For recent reviews, see: (a) Valyaev, D.A., Semeikin, O.V. and Ustynyuk, N.A. (2004) *Coordination Chemistry Reviews*, **248**, 1679; (b) Werner, H. (2004) *Coordination Chemistry Reviews*, **248**, 1693.

5 For reviews, see: (a) Bruneau, C. and Dixneuf, P.H. (1999) *Accounts of Chemical Research*, **32**, 311;

(b) McDonald, F.E. (1999) *Chemistry – A European Journal*, **5**, 3103; (c) Trost, B.M. (2002) *Accounts of Chemical Research*, **35**, 695; (d) Katayama, H. and Ozawa, F. (2004) *Coordination Chemistry Reviews*, **248**, 1703; (e) Fischmeister, C., Bruneau, C. and Dixneuf, P.H. (2004) *Ruthenium in Organic Synthesis* (ed. S-.I. Murahashi), Wiley-VCH, Weinheim, pp. 189–217; (f) Miki, K., Uemura, S. and Ohe, K. (2005) *Chemistry Letters*, **34**, 1068; (g) Trost, B.M., Frederiksen, M.U. and Rudd, M.T. (2005) *Angewandte Chemie-International Edition*, **44**, 6630; (h) Varela, J.A. and Saá C. (2006) *Chemistry – A European Journal*, **12**, 6450.

6 (a) Fischer, E.O., Kalder, H.-J., Frank, A., Kohler, H. and Huttner, G. (1976) *Angewandte Chemie (International Edition in English)*, **15**, 623; (b) Berke, H. (1976) *Angewandte Chemie (International Edition in English)*, **15**, 624.

7 Selegue, J.P. (1982) *Organometallics*, **1**, 217.

8 (a) Cadierno, V., Conejero, S., Gamasa, M.P., Gimeno, J., Pérez-Carreno, E. and García-Granda, S. (2001) *Organometallics*, **20**, 3175; (b) Cadierno, V., Conejero, S., Gamasa, M.P., Gimeno, J., Favello, L.R. and Llusar, R.M. (2002) *Organometallics*, **21**, 3716; (c) Cadierno, V., Conejero, S., Gamasa, M.P. and Gimeno, J. (2002) *Organometallics*, **21**, 3837; (d) Cadierno, V., Gamasa, M.P., Gimeno, J., Pérez-Carreno, E. and García-Granda, S. (2003) *Journal of Organometallic Chemistry*, **670**, 75; (e) Cadierno, V., Conejero, S., Gamasa, M.P. and Gimeno, J. (2003) *Dalton Transactions*, 3060.

9 (a) Nicholas, K.M. (1987) *Accounts of Chemical Research*, **20**, 207, and references therein; (b) Caffyn, A.J.M. and Nicholas, K.M. (1995) *Comprehensive Organometallic Chemistry II*, Vol. 12 (eds E.W. Abel, F.G.A. Stone and G. Wilkinson), Pergamon, New York, Chapter 7, p. 1; (c) Green, J.R. (2001) *Current Organic Chemistry*, **5**, 809; (d) Teobald, B.J. (2002) *Tetrahedron*, **58**, 4133.

10 See for example: Yen, Y.-S., Lin, Y.-C., Huang, S.-L., Liu, Y.-H., Sung, H.-L. and Wang, Y. (2005) *Journal of the American Chemical Society*, **127**, 18037.

11 Trost, B.M. and Flygare, J.A. (1992) *Journal of the American Chemical Society*, **114**, 5476.

12 (a) The exact structure of these complexes by X-ray crystallographic analysis was not known at this stage; Dev, S., Imagawa, K., Mizobe, Y., Cheng, G., Wakatsuki, Y., Yamazaki, H. and Hidai, M. (1989) *Organometallics*, **8**, 1232; (b) Hidai, M. and Mizobe, Y. (2005) *Canadian Journal of Chemistry*, **83**, 358.

13 Nishibayashi, Y. and Uemura, S. (2006) *Current Organic Chemistry*, **10**, 135.

14 For reviews, see for example: (a) Tsuji, J. (1995) *Palladium Reagents and Catalysts*, John Wiley & Sons, New York, p. 290; (b) Trost, B.M. and Van Vranken, D.L. (1996) *Chemical Reviews*, **96**, 395; (c) Trost, B.M. and Lee, C. (2000) *Catalytic Asymmetric Synthesis* (ed. I. Ojima), Wiley-VCH, New York, Chapter 8E; (d) Nishibayashi, Y. and Uemura, S. (2007) *Comprehensive Organometallic Chemistry III*, Vol. 11 (eds R.H. Crabtree and P.H. Dixneuf), Elsevier, p. 75.

15 (a) Nishibayashi, Y., Wakiji, I. and Hidai, M. (2000) *Journal of the American Chemical Society*, **122**, 11019; (b) Nishibayashi, Y., Milton, M.D., Inada, Y., Yoshikawa, M., Wakiji, I., Hidai, M. and Uemura, S. (2005) *Chemistry – A European Journal*, **11**, 1433.

16 The structure of all the complexes has been definitely clarified by X-ray crystallographic analysis and reported in the following papers; (a) Nishibayashi, Y., Imajima, H., Onodera, G., Hidai, M. and Uemura, M. (2004) *Organometallics*, **23**, 26; (b) Nishibayashi, Y., Imajima, H., Onodera, G., Inada, Y., Hidai, M. and Uemura, S. (2004) *Organometallics*, **23**, 5100.

17 Inada, Y., Nishibayashi, Y., Hidai, M. and Uemura, S. (2002) *Journal of the American Chemical Society*, **124**, 15172.

18 Kondo, T., Kanda, Y., Baba, A., Fukuda, K., Nakamura, A., Wada, K., Morisaki, Y. and

Mitsudo, T. (2002) *Journal of the American Chemical Society*, **124**, 12960.

19 (a) Milton, M.D., Onodera, G., Nishibayashi, Y. and Uemura, S. (2004) *Organic Letters*, **6**, 3993; (b) Onodera, G., Matsumoto, H., Milton, M.D., Nishibayashi, Y. and Uemura, S. (2005) *Organic Letters*, **7**, 4029.

20 Cadierno, V., García-Garrido, S.E. and Gimeno, J. (2006) *Advanced Synthesis and Catalysis*, **348**, 101.

21 Nishibayashi, Y., Wakiji, I., Ishii, Y., Uemura, S. and Hidai, M. (2001) *Journal of the American Chemical Society*, **123**, 3393.

22 Nakamura, E., Yoshikai, N. and Yamanaka, M. (2002) *Journal of the American Chemical Society*, **124**, 7181.

23 Ammal, S.C., Yoshikai, N., Inada, Y., Nishibayashi, Y. and Nakamura, E. (2005) *Journal of the American Chemical Society*, **127**, 9428.

24 Cadierno, V., Díez, J., García-Garrido, S.E. and Gimeno, J. (2004) *Chemical Communications*, 2716.

25 Caffyn, A.J.M. and Nicholas, K.M. (1993) *Journal of the American Chemical Society*, **115**, 6438.

26 Nishibayashi, Y., Imajima, H., Onodera, G. and Uemura, S. (2005) *Organometallics*, **24**, 4106.

27 Nishibayashi, Y., Onodera, G., Inada, Y., Hidai, M. and Uemura, S. (2003) *Organometallics*, **22**, 873.

28 Inada, Y., Nishibayashi, Y. and Uemura, S. (2005) *Angewandte Chemie-International Edition*, **44**, 7715.

29 Nishibayashi, Y., Yoshikawa, M., Inada, Y., Hidai, M. and Uemura, S. (2004) *Journal of Organic Chemistry*, **69**, 3408.

30 (a) Nishibayashi, Y., Yoshikawa, M., Inada, Y., Hidai, M. and Uemura, S. (2002) *Journal of the American Chemical Society*, **124**, 11846; (b) Inada, Y., Yoshikawa, M., Milton, M.D., Nishibayashi, Y. and Uemura, S. (2006) *European Journal of Organic Chemistry*, 881.

31 (a) Bustelo, E. and Dixneuf, P.H. (2005) *Advanced Synthesis and Catalysis*, **347**, 393; (b) Fischmeister, C., Toupet, L. and Dixneuf, P.H. (2005) *New Journal of Chemistry*, **29**, 765; (c) Bustelo, E. and Dixneuf, P.H. (2007) *Advanced Synthesis and Catalysis*, **349**, 933.

32 Nishibayashi, Y., Inada, Y., Hidai, M. and Uemura, S. (2002) *Journal of the American Chemical Society*, **124**, 7900.

33 Baya, M., Buil, M.L., Esteruelas, M.A., López, A.M., Oñate, E. and Rodríguez, J.R. (2002) *Organometallics*, **21**, 1841.

34 Nishibayashi, Y., Inada, Y., Hidai, M. and Uemura, S. (2003) *Journal of the American Chemical Society*, **125**, 6060.

35 Mann, M., Muller, C. and Tyrrell, E. (1998) *Journal of the Chemical Society-Perkin Transactions 1*, 1427.

36 For example: Pelter, A., Smith, K. and Brown, H.C. (1988) *Borane Reagents*, Academic Press, London.

37 Miyaura and co-workers have reported the rhodium- or iridium-catalyzed *trans*-hydroboration of terminal alkynes, where hydroboration of the vinylidene complexes is considered as a key reaction path; Ohmura, T., Yamamoto, Y. and Miyaura, N. (2000) *Journal of the American Chemical Society*, **122**, 4990.

38 Onodera, G., Nishibayashi, Y. and Uemura, S. (2006) *Organometallics*, **25**, 35.

39 Organoboranes have been known to work as good radical precursors. For example, see: (a) Brown, H.C. and Negishi, E. (1971) *Journal of the American Chemical Society*, **93**, 3777; (b) For recent reviews, see: Ollivier, C. and Renaud, P. (2001) *Chemical Reviews*, **101**, 3415; (c) Schaffner, A.-P. and Renaud, P. (2004) *European Journal of Organic Chemistry*, 2291.

40 (a) Rigaut, S., Maury, O., Touchard, D. and Dixneuf, P.H. (2001) *Chemical Communications*, 373; (b) Rigaut, S., Monnier, F., Mousset, F., Touchard, D. and Dixneuf, P.H. (2002) *Organometallics*, **21**, 2654; (c) Rigaut, S., Costuas, K., Touchard, D., Saillard, J.-Y., Golhen, S. and Dixneuf, P.H. (2004) *Journal of the American Chemical Society*, **126**, 4072.

41 Yamauchi, Y., Onodera, G., Sakata, K., Yuki, M., Miyake, Y., Uemura, S. and

Nishibayashi, Y. (2007) *Journal of the American Chemical Society*, **129**, 5175.

42 Some metal salts and Brønsted acid were reported to work as catalysts for the propargylic substitution reactions, see: (a) Georgy, M., Boucard, V. and Campagne, J.-M. (2005) *Journal of the American Chemical Society*, **127**, 14180; (b) Zhan, Z., Yu, J., Liu, H., Cui, Y., Yang, R., Yang, W. and Li, J. (2006) *Journal of Organic Chemistry*, **71**, 8298; (c) Sanz, R., Martínez, A., Áivarez-Gutiérrez, J.M. and Rodríguez, F. (2006) *Euroepan Journal of Organic Chemistry*, 1383.

43 (a) For a review of cyclopropylmethyl cations, see: Richey, H.G., Jr. (1972) *Carbonium Ions*, Vol. III (eds G.A. Olah and P.V.R. Schleyer), Wiley-Interscience, New York, Chapter 25; (b) Childs, R.F., Kostyk, M.D., Lock, C.J.L. and Mahendran, M. (1990) *Journal of the American Chemical Society*, **112**, 8912.

44 Onodera, G., Matsumoto, H., Nishibayashi, Y. and Uemura, S. (2005) *Organometallics*, **24**, 5799.

45 Cadierno, V., Díez, J., García-Garrido, S.E. and Gimeno, J. (2006) *Advanced Synthesis and Catalysis*, **348**, 2125.

46 (a) Alvarez, P., Bassetti, M., Gimeno, J. and Mancini, G. (2001) *Tetrahedron Letters*, **42**, 8467; (b) Suzuki, T., Tokunaga, M. and Wakatsuki, Y. (2002) *Tetrahedron Letters*, **43**, 7531.

47 Yeh, K.-L., Liu, B., Lo, C.-Y., Huang, H.-L. and Liu, R.-S. (2002) *Journal of the American Chemical Society*, **124**, 6510.

48 Datta, S., Chang, C.-L., Yeh, K.-L. and Liu, R.-S. (2003) *Journal of the American Chemical Society*, **125**, 9294.

49 Bianchini, C., Peruzzini, M., Zanobini, F., Lopez, C., Rios, I. and Romerosa, A. (1999) *Chemical Communications*, 443.

50 Shen, H.-C., Su, H.-L., Hsueh, Y.-C. and Liu, R.-S. (2004) *Organometallics*, **23**, 4332.

51 Maddock, S.M. and Fin, M.G. (2001) *Angewandte Chemie-International Edition*, **40**, 2138.

52 Nishibayashi, Y., Inada, Y., Yoshikawa, M., Hidai, M. and Uemura, S. (2003) *Angewandte Chemie-International Edition*, **42**, 1495.

53 Nishibayashi, Y., Shinoda, A., Miyake, Y., Matsuzawa, H. and Sato, M. (2006) *Angewandte Chemie-International Edition*, **45**, 4835.

54 (a) Nishibayashi, Y., Yoshikawa, M., Inada, Y., Milton, M.D., Hidai, M. and Uemura, S. (2003) *Angewandte Chemie-International Edition*, **42**, 2681; (b) Milton, M.D., Inada, Y., Nishibayashi, Y. and Uemura, S. (2004) *Chemical Communications*, 2712; (c) Nishibayashi, Y., Yoshikawa, M., Inada, Y., Hidai, M. and Uemura, S. (2004) *Journal of the American Chemical Society*, **126**, 16066.

55 (a) Matsuda, I., Komori, K. and Itoh, K. (2002) *Journal of the American Chemical Society*, **124**, 9072; (b) Sherry, B.D., Radosevich, A.T. and Toste, F.D. (2003) *Journal of the American Chemical Society*, **125**, 6076; (c) Luzung, M.R. and Toste, F.D. (2003) *Journal of the American Chemical Society*, **125**, 15760; (d) Evans, P.A. and Lawler, M.J. (2006) *Angewandte Chemie-International Edition*, **45**, 4970.

56 Matsuzawa, H., Miyake, Y. and Nishibayashi, Y. (2007) *Angewandte Chemie-International Edition*, **46**, 6488.

57 Yamauchi, Y., Yuki, M., Tanabe, Y., Miyake, Y., Inada, Y., Uemura, S. and Nishibayashi, Y. (2008) *Journal of the American Chemical Society*, in press.

8
Ruthenium Allenylidenes and Indenylidenes as Catalysts in Alkene Metathesis

Raluca Malacea and Pierre H. Dixneuf

8.1
Introduction

Alkene metathesis has become an important and unavoidable catalytic reaction that has brought a revolution in synthetic methodology with applications not only in organic synthesis but also in supramolecular materials and polymer science [1]. This is due, following the elucidation of the catalytic mechanism by Chauvin [2], to the discovery of well-defined and efficient alkylidene-metal RCH=M catalysts, such as the molybdenum Schrock catalyst [3] and the ruthenium Grubbs catalysts $RuX_2(=CHR)L_2$ [4]. The latter, especially, tolerate functional groups and have found many applications in both organic syntheses and polymer science. The usefulness of the RCH=M-based catalysts in alkene metathesis has led us to question the possible involvement of the homologs vinylidene RCH=C=M and allenylidene $R_2C=C=C=M$ complexes as precatalysts for this catalytic reaction.

The first vinylidene in catalysis was reported as early as 1985. An *in situ* generated vinylidene intermediate [$(OC)_4W=C=CHPh$] was shown to initiate the polymerization of phenylacetylene [5]. Since 1992, the discovery of the Grubbs-type catalyst $RuX_2(=CHR)L_2$ [4] has promoted the use of related vinylidenes Ru=C=CHR in alkene metathesis, and the first ruthenium-vinylidene promoted ROMP reaction was disclosed with $RuCl_2(=C=CHR)(PCy_3)_2$ catalyst in 1996 [6]. Since then a variety of ruthenium-vinylidenes have been discovered, some of them were generated *in situ* directly from terminal alkynes [7] and were efficiently applied in ring opening metathesis polymerization (ROMP), ring closing metathesis (RCM) and cross-metathesis reactions [8]. The resulting applications of metal-vinylidenes in alkene metathesis have been the topic of recent reviews [9–11].

Among the $R_2C(=C)_n=Ru$ homologs promoting alkene metathesis the most recent discoveries deal with the allenylidene-ruthenium and related pre-catalysts. This chapter is devoted to the class of ruthenium-allenylidene metathesis pre-catalysts, their intramolecularly rearranged indenylidene catalysts, and their use in various alkene metathesis reactions for which they show good to excellent catalytic activity and selectivity.

8.2
Propargyl Derivatives as Alkene Metathesis Initiator Precursors: Allenylidenes, Indenylidenes and Alkenylalkylidenes

In 1998 it was revealed that allenylidene-ruthenium complexes, arising simply from propargylic alcohols, were efficient precursors for alkene metathesis [12]. This discovery first initiated a renaissance in allenylidene metal complexes as possible alkene metathesis precursors, then it was observed and demonstrated that allenylidene-ruthenium complexes rearranged into indenylidene-ruthenium intermediates that are actually the real catalyst precursors. The synthesis of indenylidene-metal complexes and their efficient use in alkene metathesis are now under development. The interest in finding a convenient source of easy to make alkene metathesis initiators is currently leading to investigation of other routes to initiators from propargylic derivatives.

The objective of this chapter is to report on these various aspects: allenylidenes in alkene metathesis, their transformation into indenylidenes, alkene metathesis with indenylidene complexes, other propargylic derivatives as alkene metathesis initiators and their application in alkene metathesis.

8.2.1
Allenylidene-Ruthenium Complexes as Alkene Metathesis Catalyst Precursors: the First Evidence

The observation by Selegue in 1982 [13] that 16 electron ruthenium(II) intermediates could activate terminal propargyl alcohols into ruthenium-allenylidene complexes, via 3-hydroxyvinylidene-metal intermediates, showed not only that these allenylidene complexes were stable toward the action of the released water but, especially, that it could be an excellent way to generate allenylidene-metal complexes from easily accessible sources, the propargyl alcohols (Equation 8.1).

$$\text{Cp(PR}_3)_2\text{RuCl} + \text{HC} \equiv \text{CCPh}_2\text{OH} \xrightarrow{\text{NaPF}_6} \left[\text{Ru}=\text{C}=\text{C}\begin{matrix}\text{H}\\\text{CPh}_2\text{OH}\end{matrix}\right]\text{PF}_6 \longrightarrow \left[\text{Cp(PR}_3)_2\text{Ru}=\text{C}=\text{C}=\text{CPh}_2\right]\text{PF}_6 \quad (8.1)$$

This general route to a variety of allenylidene-metal complexes from propargyl alcohols, was demonstrated soon after for ruthenium(II) complexes and for a large set of metal complexes [14–17].

This evidence of easy stoichiometric formation of ruthenium-allenylidene led Trost, 10 years later, to propose for the first time such an intermediate in a ruthenium (II)-catalyzed transformation of functional propargyl alcohols [18]. Since 2000 Nishibayashi et al. [19–23] have developed a set of catalytic propargylations, with

8.2 Propargyl Derivatives as Alkene Metathesis Initiator Precursors

C–C and carbon–heteroatom bond formation, based on allenylidene-ruthenium intermediates arising from propargyl alcohols activated by binuclear systems [Cp*RuCl(μ_2-SR)$_2$RuCp*(X)], (See Chapter 7). Arene-ruthenium(II) systems have also been used to promote the direct propargylation with propargylic alcohols of C- and O- pronucleophiles, although via a metal stabilized propargylic cation M(η-R-C≡C-CR$_2^+$) [24–26]. In parallel, allenylidene-metal catalysts were suggested as intermediates for several catalytic transformations: the dimerization of tin hydrides [27], the transetherification of vinyl ether [28], hydrogen transfer reaction [29] and even atom transfer radical polymerization [30].

In the field of alkene metathesis ruthenium-allenylidene precursors have made, since 1998, an important contribution to catalysis [12, 31, 32], for the formation of cycles and macrocycles via RCM, ROMP and acyclic diene metathesis (ADMET) polymerization.

The first evidence for allenylidene-metal alkene metathesis precursors was initiated by the synthesis of ionic, 18 electrons (arene)ruthenium-allenylidenes containing a bulky and electron-donating phosphine [12, 31, 32]. They were produced by abstraction of a halide from RuCl$_2$(PR$_3$)(arene) in the presence of propargylic alcohol and a non-coordinating salt [12] and then initial reaction with AgX derivatives, followed by propargylic alcohol addition, that allowed the easy exchange of the counter anion (X$^-$) [31]. Allenylidene variations were easily obtained by changing the nature of the propargylic alcohols [32] (Scheme 8.1).

Alkene metathesis, promoted by the allenylidene–ruthenium complexes, was revealed in the RCM of diallyl tosylamide [12]. The first studies showed some significant influences [31, 32].

1. The most electron-donating and bulky phosphine led to the highest catalyst efficiency: PCy$_3$ > P ^{i}Pr$_3$ ≫ PPh$_3$.
2. The nature of the counter anion was strongly related to the catalyst activity and corresponded to the sequence TfO$^-$ ≫ PF$_6^-$ = BPh$_4^-$ ≫ BF$_4^-$. Actually, the triflate allowed catalysis at room temperature and the BF$_4^-$ anion inhibited the reaction, this inhibition being neutralized by the addition of HBF$_4$.

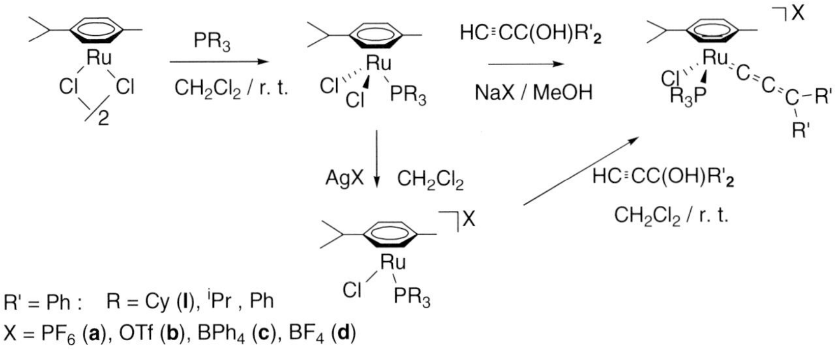

Scheme 8.1 Initial pathways for the synthesis of allenylidene-ruthenium alkene metathesis catalysts.

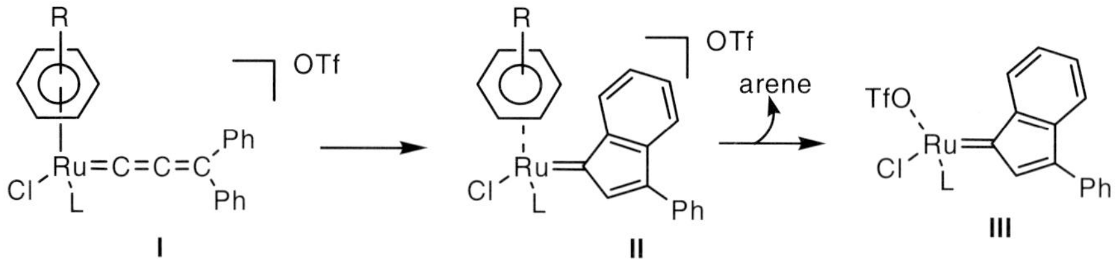

Scheme 8.2 Transformation of allenylidene-ruthenium into active indenylidene-ruthenium catalyst.

3. The activity was higher with *p*-cymene, as coordinated arene, than with hexamethylbenzene which leads to a more stable ruthenium-arene bond.
4. 3-3-diarylallenylidene-ruthenium complexes were shown to be efficient but the most simple 3,3-diphenylallenylidene-ruthenium derivative was shown to lead to the most active catalysts.

Now, it has been shown [33, 34] that allenylidene-metal precursors **I** generate the indenylidene-metal intermediate **III** which is the real catalyst precursor (Scheme 8.2). Thus, we now understand that to generate the active species **III**, the *para*-cymene ligand is more easily displaced from the ruthenium site and the triflate, which interacts weakly with the ruthenium-allenylidene, favors the formation of the indenylidene ligand and arene displacement.

8.2.2
Allenylidene-Ruthenium Complexes in RCM, Enyne Metathesis and ROMP

8.2.2.1 RCM Reactions
The allenylidene-ruthenium(arene) catalyst precursors **I** have been used for the synthesis of macrocycles by the RCM reaction and were revealed as active as the first generation Grubbs catalyst RuCl$_2$(=CHPh)(PCy$_3$)$_2$ [35], depending on the nature of the diene functional groups and macrocycle size [32] (Scheme 8.3). These macrocycle syntheses show that the allenylidene ruthenium catalysts **I** offer functional group tolerance.

Fluorinated α-aminophosphonates containing two pendent alkene chains with catalyst **Ia** led to six- or seven-membered cyclic phosphonates as potential antibacterial reagents [36] (Scheme 8.4).

8.2.2.2 Enyne Metathesis
The allenylidene-ruthenium complexes **I** also catalyze the enyne metathesis to alkenylcycloalkenes with a 1,3-diene structure. Initial studies showed the transformation of simple enynes with ether function [37] (Scheme 8.5). This reaction was significantly accelerated by initial catalyst photochemical activation, which is now understood to favor the rearrangement of the allenylidene- into the active indenylidene-ruthenium moiety and arene displacement.

Alternatively, fluorinated aminoesters with the enyne structure were transformed into a variety of vinyl cyclic amino acid derivatives and related bicyclic derivatives via Diels–Alder and oxidation reactions [38] (Scheme 8.6).

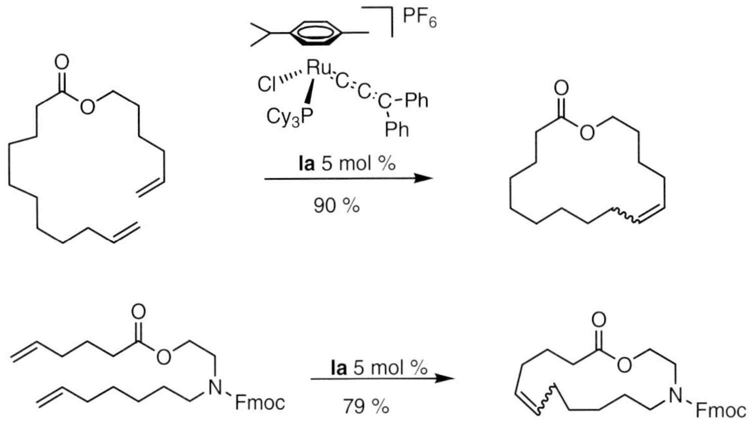

Scheme 8.3 Macrocycle syntheses with alkene metathesis allenylidene-ruthenium catalyst **Ia**.

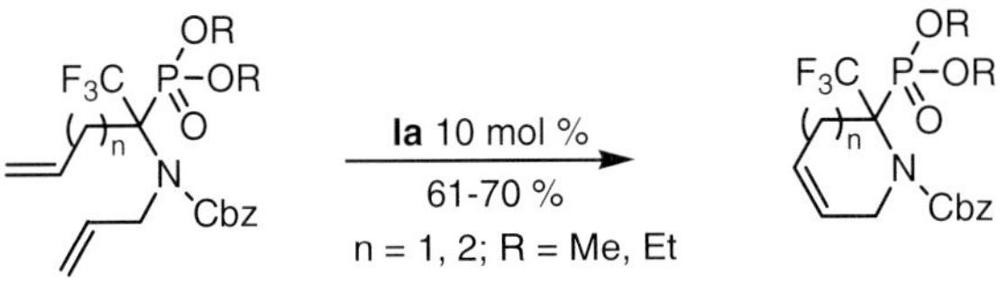

Scheme 8.4 RCM reaction leading to cyclic fluorinated aminophosphonates.

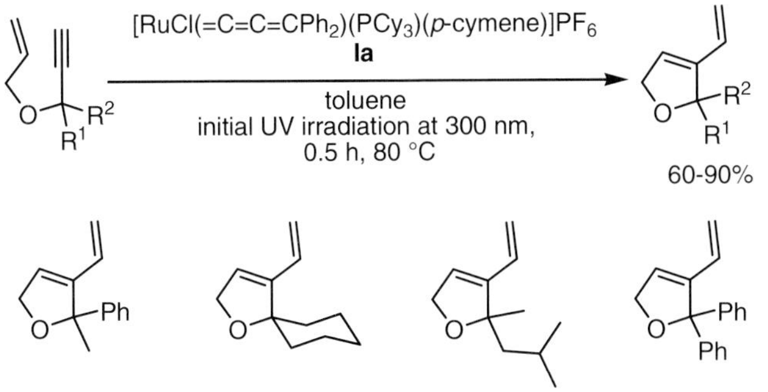

Scheme 8.5 Enyne metathesis catalyzed by the photochemically activated ruthenium allenylidene precursor **Ia**.

8.2.2.3 ROMP Promoted by Allenylidene Complexes

Allenylidene-ruthenium complex **Ib** readily promotes the ROMP of norbornene, much faster than the precursor $RuCl_2(PCy_3)(p\text{-cymene})$ [39] (Table 8.1, entry 1). The ROMP of cyclooctene requires heating at 80 °C (5 min), however a pre-activation of the catalyst allows the polymerization to take place at room temperature. The activation consists, for example, in a preliminary heating at 80 °C or UV irradiation of the catalyst before addition of the cyclic alkene, conditions under which rearrangement into indenylidene and arene displacement take place [39] (Table 8.1, entries 2,3). The arene-free allenylidene complexes, the neutral $RuCl_2(=C=C=CPh_2)$

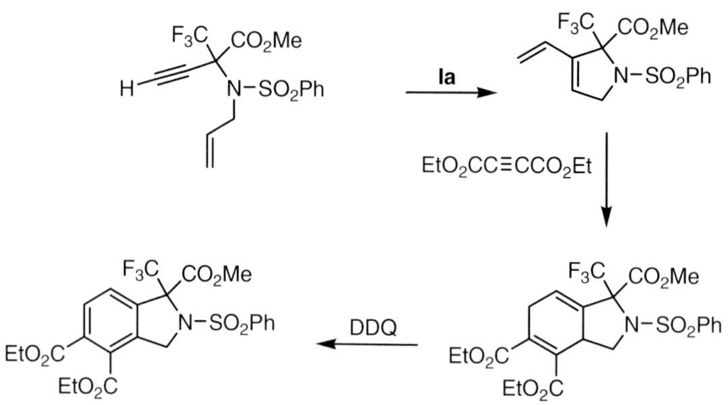

Scheme 8.6 Enyne metathesis in the synthesis of fluorinated bicyclic aminoesters.

$(PCy_3)_2(dmso)$ and the ionic $[RuCl(=C=C=CPh_2)(PCy_3)(dmso)_2]OTf$ also promote the ROMP of norbornene but their efficiency is lower than that of complexes **I** [40] (Table 8.1, entries 4,5).

8.2.3
Indenylidene-Ruthenium Complexes: the Alkene Metathesis Catalytic Species from Allenylidene Ruthenium Complexes

8.2.3.1 The First Evidence
The evidence that ruthenium-allenylidenes were easy to make and efficient alkene metathesis precursors motivated several groups to design new allenylidene metal complexes and to explore their impact on alkene metathesis. Nolan first reported

Table 8.1 ROMP of cycloalkenes with ruthenium allenylidene catalysts.

Monomer	Catalyst	Ratio[a]	Conditions	Yield (%)	Mn[b] ($\times 10^{-3}$)	PDI[c]
Norbornene	$[RuCl(=C=C=CPh_2)(L)$ (p-cymene)]OTf **Ib**	1000	5 min, 22 °C	90	198	1.8
Cyclooctene	$[RuCl(=C=C=CPh_2)(L)$ (p-cymene)]OTf **Ib**	1000	5 min, 80 °C	90	267	1.4
Cyclooctene	$[RuCl(=C=C=CPh_2)(L)$ (p-cymene)]OTf **Ib**	150	UV 2 h, 22 °C	99	143	1.8
Norbornene	$RuCl_2(=C=C=CPh_2)(L)_2$ (dmso)	300	4 h, 60 °C	56	16	4.1
Norbornene	$[RuCl(=C=C=CPh_2)(L)$ $(dmso)_2]OTf$	300	4 h, 60 °C	70	37	3.9

$L = PCy_3$.
[a] monomer/catalyst molar ratio;
[b] Mn: molecular weight (number average) (g mol^{-1});
[c] polydispersity of the polymer.

the synthesis of [RuCl(=C=C=CPh$_2$)(NHC)(p-cymene)]PF$_6$ **V** upon reaction of RuCl$_2$(NHC)(p-cymene) **IV** with propargylic alcohol [41] (Equation 8.2). Both complexes **IV** and **V** showed catalytic activity for RCM of diallylmalonate.

$$\text{(8.2)}$$

Then Nolan reported the synthesis of RuCl$_2$(=C=C=CPh$_2$)(PCy$_3$)(L) **VI** (L = PCy$_3$) and **VII** (L = IMes). On treatment of [RuCl$_2$(p-cymene)]$_2$ with 2 equiv of PCy$_3$ in the presence of propargyl alcohol, complex **VI** was obtained and on addition of the carbene IMes it afforded the allenylidene complex **VII** (Scheme 8.7) [42]. The allenylidene complexes **VI** and **VII**, explored as alkene metathesis catalysts for RCM reaction, offered disappointing activity, suggesting again that the allenylidene ligand is not an alkene metathesis initiator, even on displacement of a PCy$_3$ ligand as in the Grubbs catalyst.

However, the reaction of RuCl$_2$(PPh$_3$)$_4$ with only HC≡CCPh$_2$OH directly led to the indenylidene complex **VIII**, which gave, after phosphine exchange with PCy$_3$, the indenylidene complex **IX** [42].

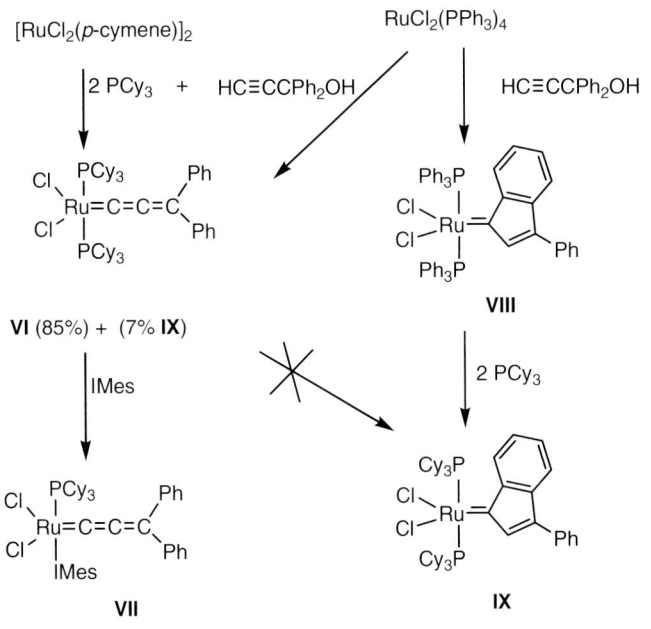

Scheme 8.7 Synthesis of allenylidene and indenylidene complexes with the Grubbs catalyst structure.

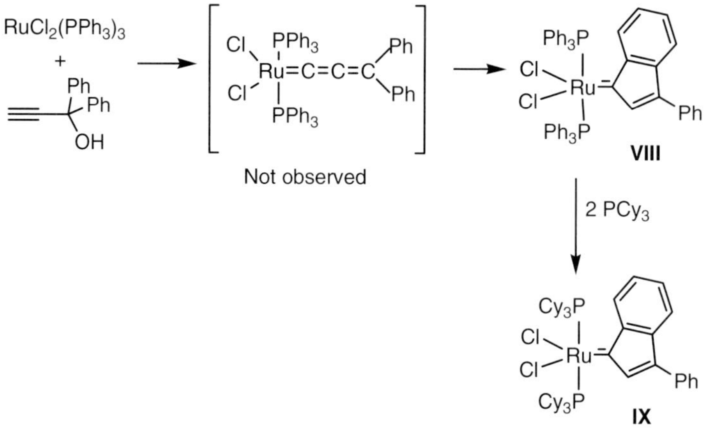

Scheme 8.8 Direct formation of alkene metathesis ruthenium-indenylidene catalyst **IX**.

At the same time, Fürstner et al. [43–44] also observed that RuCl$_2$(PPh$_3$)$_3$ on reaction with HC≡CCPh$_2$OH did not lead to the expected allenylidene complex but to the same indenylidene complex **VIII** (Scheme 8.8) [45].

Moreover, Fürstner showed that whereas complex **VIII** was inactive for RCM reaction, its PCy$_3$ analog **IX** was, in contrast, very active in a variety of RCM reactions. The latter is now commercially available and currently used in alkene metathesis (see Section 8.3 for further applications).

These observations indicate that when the metal complex is electron-rich, the allenylidene-metal complexes are stable (**VI** and **VII**), even on heating or protonation [42]. However, with less electron-rich systems (e.g., PPh$_3$ ligands instead of PCy$_3$ or NHC) the corresponding allenylidene complex was never observed, to the profit of the indenylidene complex **VIII**. These results suggested that the allenylidene-ruthenium complex is a transient species that rearranges into the indenylidene complex **VIII**, as was observed for a C$_5$ cumulenylidene [48].

Presumably, the complex **IX** is expected to generate the same intermediate RuCl$_2$(=CH$_2$)(PCy$_3$)$_2$ as the Grubbs catalysts after the first cycle of a terminal diene RCM reaction.

Other closely related ruthenium-allenylidene were made and evaluated in alkene metathesis [32]. Werner et al. [49] also produced allenylidene complexes of analogous structure to that of the Grubbs catalyst, but containing hemilabile phosphine such as complex **X** (Scheme 8.9). However, the Ru—O bond may be too stable to initiate the rearrangement into indenylidene, the coordination of alkene and to become a catalyst.

Le Gendre and Moise [50] produced an allenylidene ruthenium complex analogous to **I** but with a titanium(IV)-containing phosphine such as **XI** (Scheme 8.9). Its rearrangement into indenylidene was not observed and its catalyst activity remained moderate.

The neutral NHC complex **XII**, analogous to **V** (Equation 8.2) but with almost orthogonal linked NHC and arene ligands, was considered as a potential allenylidene-

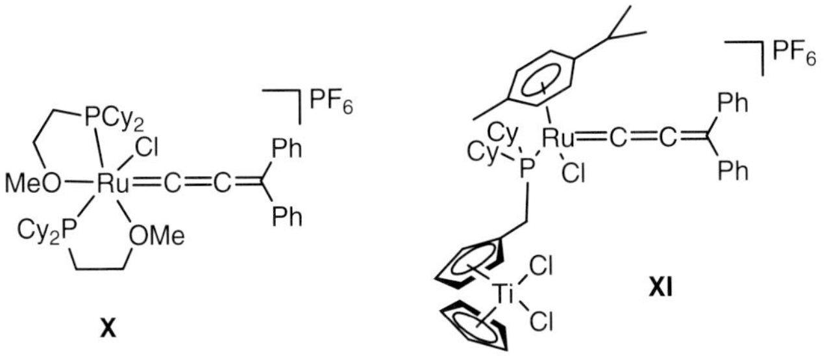

Scheme 8.9 Allenylidene-ruthenium complexes.

ruthenium complex **XIII**, but the latter or its related indenylidene complex could not be isolated [51] (Scheme 8.10). The resulting *in situ* prepared complex appeared to be an alkene metathesis catalyst showing the crucial role of the propargyl alcohol as alkene metathesis initiator, although its structure was not elucidated and its activity was lower than that of non-chelating arene complexes **I**.

8.2.3.2 The Intramolecular Allenylidene to Indenylidene Rearrangement Demonstration

Kinetic studies of diallyltosylamide RCM reaction monitored by NMR and UV/VIS spectroscopy showed that thermal activation of the catalyst precursors **Ia** and **Ib** (25–80 °C) led to the *in situ* formation of a new species which could not be identified but appeared to be the active catalytic species [52]. Attempts to identify this thermally generated species were made in parallel by protonation of the catalysts **I**. Indeed, the protonation of allenylidene-ruthenium complex **Ia** by HBF$_4$ revealed a significant increase in catalyst activity in the RCM reaction [31, 32]. The influence of the addition of triflic acid to catalyst **Ib** in the ROMP of cyclooctene at room temperature (Table 8.2, entries 1,3) was even more dramatic. For a cyclooctene/ruthenium ratio of 1000 the TOF of ROMP with **Ib** was 1 min^{-1} and with **Ib** and 5 equiv. of TfOH it reached 950 min^{-1} [33].

The stoichiometric protonation of complex **Ib** with TfOH at low temperature was followed by ^{1}H, ^{13}C NMR spectroscopy [34]. At −40 °C, the protonation of the

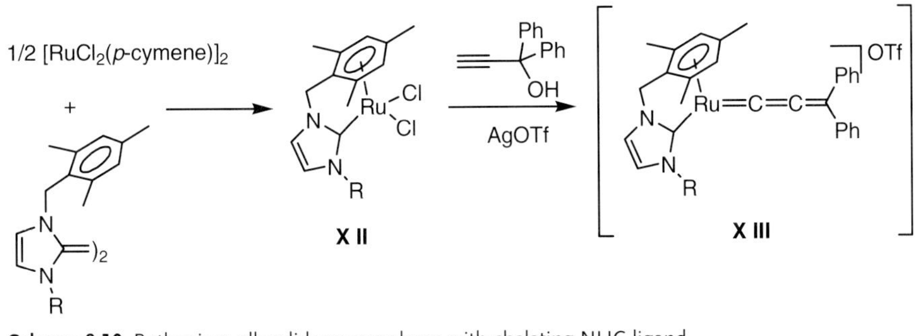

Scheme 8.10 Ruthenium allenylidene complexes with chelating NHC ligand.

Table 8.2 Acid-promoted cyclooctene polymerization with catalysts I at room temperature[a].

Entry	Catalyst	Ratio[b]	Acid[c] (amount(eq))	Time	Yield (%)	$10^{-3} \times M_n$	PDI	TOF (min^{-1})
1	Ib	1000	—	15 h	95	143	1.9	1
2	Ib	1000	HBF$_4$ (5)	1 min	92	224	1.7	920
3	Ib	1000	HOTf (5)	1 min	95	238	1.6	950
4	Ib	10 000	HOTf (5)	5 min	97	387	1.5	1.940
5	Ib	100 000	HOTf (100)	5 min	88	857	1.4	17.600

[a] 4.5×10^{-3} mol of cyclooctene in 2.5 mL of PhCl.
[b] [cyclooctene]/[Ru].
[c] Related to ruthenium complex.

allenylidene-ruthenium complex **Ib** with 1.2 equiv of TfOH led to the initial formation of the characterized alkenylcarbyne complex **XIV** by protonation of the allenylidene C$_2$ carbon atom (Scheme 8.11).

This carbyne was shown not to be the RCM active species. At $-20\,°C$ it rearranged spontaneously into the indenylidene complex **XV** with release of TfOH. This intramolecular transformation corresponds to the electrophilic *ortho*-substitution of one phenyl group by the electrophilic carbyne carbon of **XIV**. The carbene complex **XV** was identified as the species thermally formed *in situ* from the catalyst precursors **Ia,b** in the range 25–80 °C.

These observations constituted the first evidence for intramolecular rearrangement of allenylidene into indenylidene ligand, although the first isolation by

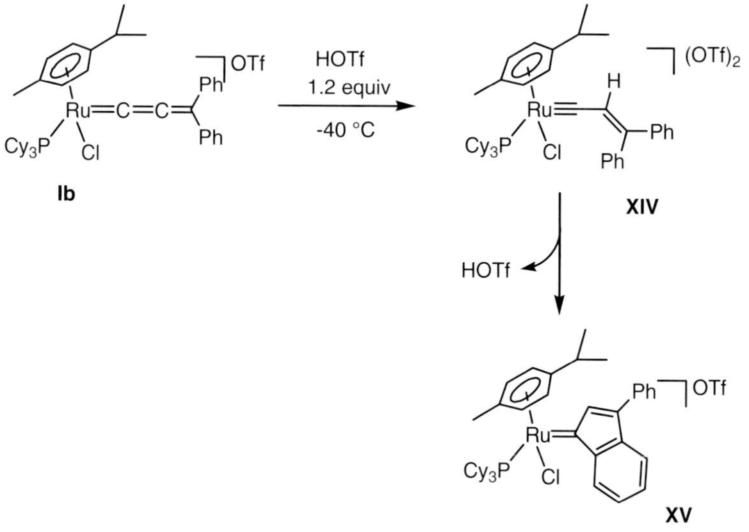

Scheme 8.11 Formation of indenylidene ruthenium complex accelerated by the protonation of ruthenium allenylidene complex.

Nolan [42] and Fürstner [43] of indenylidene-ruthenium complex strongly supported the $L_nRu=C=C=CPh_2$ rearrangement. The $L_nRu = RuCl_2(PPh_3)_2$ moiety is electrophilic enough to promote this electrophilic rearrangement, whereas it does not take place with the more electron-rich $RuCl_2(PCy_3)_2$ moiety. For the $[Ru(Cl)(PCy_3)(p$-cymene)](OTf) moiety the corresponding allenylidene **Ib** is stable, but slowly gives the indenylidene at room temperature. The advantage of the acid-promoted transformation **Ib** → **XV** is that it is complete even at $-20\,°C$ whereas at room temperature without acid addition this transformation is only partial, the allenylidene complex **Ib** being a reservoir of the indenylidene active catalytic species, freed either thermally or on protonation.

This complex **XV** was proved to be an active alkene metathesis catalyst even at $0\,°C$ or room temperature. The *in situ* generated catalyst **XV** catalyzed the ROMP of cyclooctene at room temperature at high cyclooctene/Ru ratio (Table 8.2 entries 4,5) reaching TOF of more than $17\,000\,min^{-1}$.

Although the *in situ* generation and use of **XV** is convenient, the presence of acid also favors slow decomposition of **XV**. It was thus suitable to isolate catalyst **XV** by filtration of its acidic solution on basic alumina [34]. Thus the indenylidene-ruthenium(arene) complexes containing various phosphines were isolated (Scheme 8.12).

8.2.3.3 Applications of Isolated Indenylidene-Ruthenium Complexes in ROMP

The catalytic activities for ROMP of cyclooctene of isolated indenylidene complexes **XV–XVII** were evaluated in chlorobenzene (Table 8.3).

Catalyst **XV** promoted the cyclooctene ROMP with an extremely low catalyst loading (30 000 : 1) reaching a TOF of $9200\,min^{-1}$ (Table 8.3, entries 1,2) but its activity decreased at a ratio 100 000 : 1. The catalyst containing P^iPr_3 **XVI** showed excellent activity whereas, as expected, complex **XVII** with the poor electron donor and less bulky phosphine PPh_3 showed low activity.

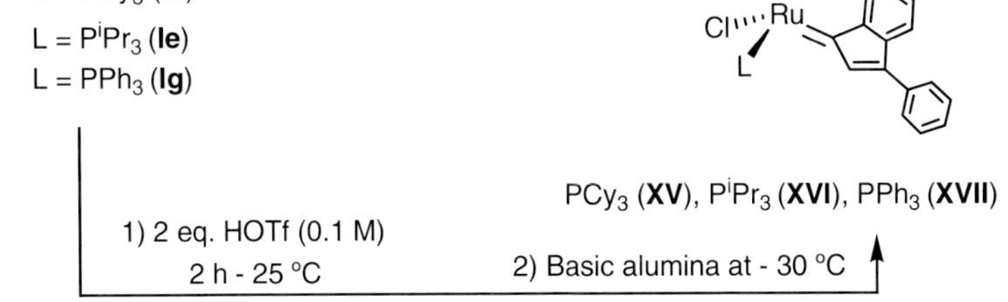

Scheme 8.12 Synthesis of ruthenium indenylidene complexes.

Table 8.3 Cyclooctene polymerization catalyzed by isolated indenylidene complexes **XV–XVII** at room temperature[a].

Entry	Catalyst	Ratio[b]	Time	Yield (%)	$10^{-3} \times M_n$	TOF (min^{-1})
1	XV	10 000	1 min	99	549	9900
2	XV	30 000	3 min	92	754	9200
4	XVI	10 000	2 min	93	621	4650
5	XVII	10 000	16 h	42	387	4

[a] 4.5×10^{-3} mol of cyclooctene in 2.5 mL of PhCl.
[b] [monomer]/[Ru].

Catalyst **XV** was also able to polymerize cyclopentene at extremely low loading (cyclopentene: Ru = 10 000 : 1). Whereas at room temperature after initial gelification, due to polymerization, a decrease in viscosity was observed, at low temperature, −40 °C for 1 h, 98% of cyclopentene polymer could be obtained.

8.2.3.4 Indenylidene-Ruthenium(arene) Catalyst in Diene and Enyne RCM

The isolated catalyst **XV** was evaluated in the RCM reaction of dienes and enynes at room temperature. The RCM of dienes easily takes place, even for disubstituted alkene bonds (Table 8.4). For enynes, the formation of vinylcycloalkenes takes place more slowly but in high yield [34].

8.2.4
Propargylic Ethers as Alkene Metathesis Initiator Precursors: Generation of Alkenyl Alkylidene-Ruthenium Catalysts

In the early syntheses of alkenyl alkylidene-ruthenium catalysts, the first generation of Grubbs catalyst, it was observed that propargyl chloride could be a convenient source of the vinylcarbene initiator [53] with respect to the previous one arising from activation of cyclopropene [4] (Equation 8.3). In this synthesis the alkylidene hydrogen atom arises from the ruthenium hydride.

$$Ru(H)(H_2)Cl(PCy_3)_2 + H\!\!\equiv\!\!\!-\!\!\overset{Cl}{\underset{}{\diagdown}}\!\!\!\!\diagup \longrightarrow \underset{PR_3}{\overset{PR_3}{Cl-Ru\overset{Cl}{=}\!\!\!\diagdown}} \quad (8.3)$$

Recently, another general transformation of propargylic derivatives into alkenyl alkylidene ligand and alkene metathesis initiator was observed. It corresponds to the activation of mixed propargylic alkyl ethers by hydride transfer from the α-carbon ether alkyl group to the propargylic moiety. It was revealed by the study of RuCl(OTf)(PCy$_3$)(arene) complexes, the precursor of 16 electron ruthenium reactive species by triflate displacement. It was shown to transform an enyne into its RCM product, thus suggesting that an alkene metathesis catalyst was generated *in situ* in this reaction (Equation 8.4) [54].

Table 8.4 RCM and enyne metathesis reactions promoted by catalyst XV[a].

Substrate	Product	Time	Yield (%)
Ts-N(allyl)(allyl)	Ts-N pyrroline	1 min	99
Ts-N(crotyl)(allyl)	Ts-N pyrroline	30 min	95
Ts-N(methallyl)(allyl)	Ts-N methylpyrroline	16 h	65
cyclohexyl propargyl allyl ether	vinyl dihydrofuran spirocyclohexane	90 min	93
fluorenyl propargyl allyl ether	vinyl dihydrofuran spirofluorene	24 h	95
geranyl propargyl allyl ether	vinyl dihydrofuran geranyl	24 h	78

[a] 2.5 mL of PhCl [monomer]/[Ru] = 50, room temperature [34].

$$\underset{\text{R}^1\ \text{R}^2}{\text{propargyl allyl ether}} \xrightarrow[\text{CH}_2\text{Cl}_2,\ \text{RT, 1 h}]{[\text{RuCl}(p\text{-cymene})(\text{PCy}_3)]\text{TfO}\ (2\ \text{mol\%})} \underset{95\%}{\text{R}^1\ \text{R}^2\ \text{vinyl dihydrofuran}} \quad (8.4)$$

The stoichiometric interaction of an enyne and [RuCl(PCy$_3$)(pcymene)]B(Ar^F)$_4$ **XVIIIa** containing a bulky non-coordinating anion B(ArF)$_4^-$ showed by NMR at −30 °C the formation of the alkenyl alkylidene ruthenium complex and acrolein. This formation could be understood by the initial formation of a vinylidene intermediate and transfer of a hydride from the oxygen α-carbon atom to the electrophilic vinylidene carbon, as a retroene reaction step (Scheme 8.13) [54].

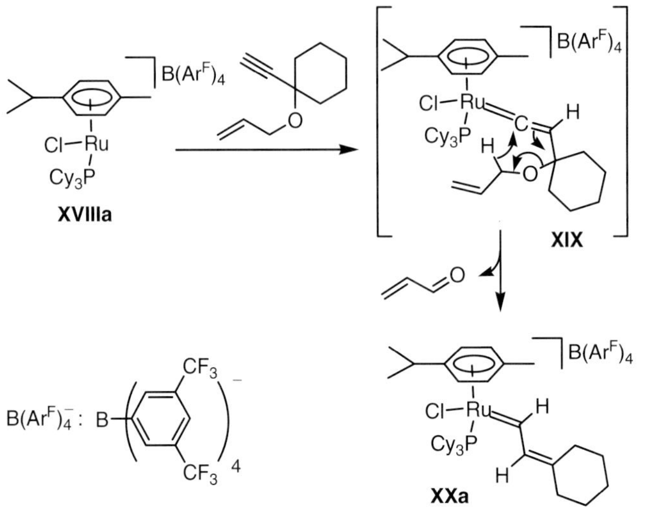

Scheme 8.13 Formation of ruthenium alkenylcarbene from propargylic enyne.

A similar hydride transfer was observed by Ipaktschi in the rearrangement of a 3-dialkylaminovinylidene-tungsten derivative [55].

The stoichiometric reaction of propyl propargylic ether with the similar ruthenium precursor **XVIIIb** (X: OTf) led to the formation of the same isolable carbene-ruthenium complex **XXb** and propanal (Scheme 8.14).

Under these conditions the catalyst precursor **XXb** containing the triflate anion could be isolated and evaluated in metathesis. The catalyst **XXb** underwent efficient RCM reaction of dienes and enynes (Table 8.5) [54].

This stoichiometric reaction constitutes a new contribution to vinylidene chemistry and a novel method to generate alkenylcarbene ligand from simple propargyl alkyl ethers rather than via activation of cyclopropenes [4] or by stoichiometric activation of butadiene [6]. When linked to a suitable metal-ligand moiety this carbene constitutes an alkene metathesis initiator.

A variety of alkenylalkylidene ruthenium complexes have been obtained by a quite different route: the reaction of alkylidene-ruthenium complex with a functional alkyne via a [2 + 2] cycloaddition [56].

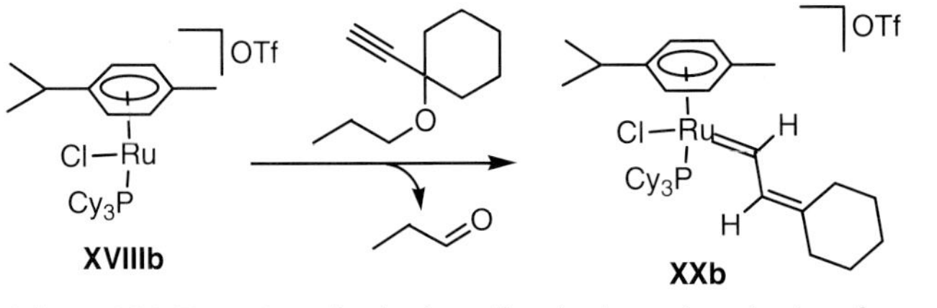

Scheme 8.14 Formation of ruthenium alkenylcarbene via activation of propyl propargyl ether.

Table 8.5 Diene and enyne RCM reactions with 2 mol% of complex XXb at room temperature.

Substrate	Product	Time	Conv (%)	TOF (h^{-1})
(cyclohexyl propargyl allyl ether)	(spiro product)	1 h	99	49
Ts–N(diallyl)	Ts–N (pyrroline)	15 min	95	190
Ph,Ph diallyl ether	Ph,Ph dihydropyran	4 h	99	12
EtO$_2$C, CO$_2$Et diene	EtO$_2$C, EtO$_2$C cyclopentene	2 h	99	25
(diester diene)	(macrocyclic diester)	1 h	99	49

8.3 Indenylidene-Ruthenium Catalysts in Alkene Metathesis

8.3.1 Preparation of Indenylidene-Ruthenium Catalysts

Two observations initiated a strong motivation for the preparation of indenylidene-ruthenium complexes via activation of propargyl alcohols and the synthesis of allenylidene-ruthenium intermediates. The first results from the synthesis of the first indenylidene complexes **VIII** and **IX** without observation of the expected allenylidene intermediate [42–44] (Schemes 8.7 and 8.8), and the initial evidence that the well-defined complex **IX** was an efficient catalyst for alkene metathesis reactions [43–44]. The second observation concerned the direct evidence that the well-defined stable allenylidene ruthenium(arene) complex **Ib** rearranged intramolecularly into the indenylidene-ruthenium complex **XV** via an acid-promoted process [22, 23] (Scheme 8.11) and that the in situ prepared [33] or isolated [34] derivatives **XV** behaved as efficient catalysts for ROMP and RCM reactions.

The NHC ligand-containing indenylidene complexes **XXI**, **XXII** and **XXIII**, **XXIV** were readily prepared by reaction of the NHC ligand with complex **VIII** and **IX**, respectively [57] (Scheme 8.15).

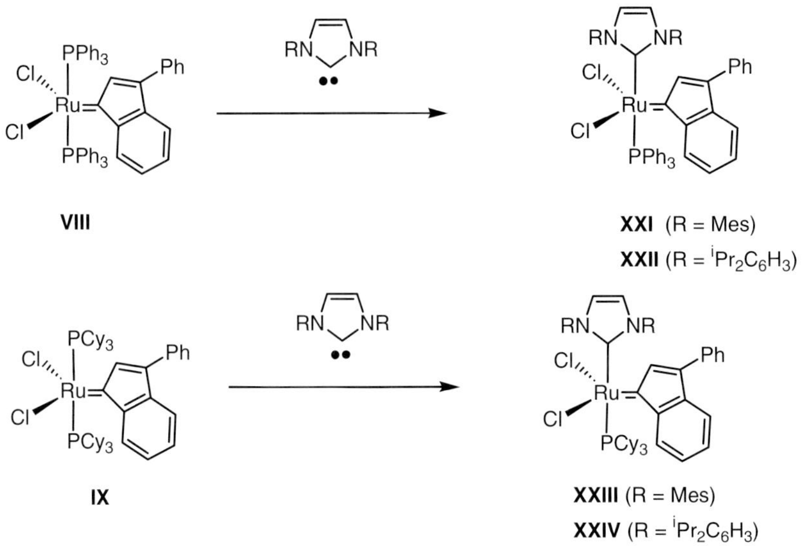

Scheme 8.15 Preparation of indenylidene ruthenium complexes containing a NHC ligand.

The binuclear indenylidene complex **XXV** was readily obtained on reaction of complex **IX** with [RuCl$_2$(p-cymene)]$_2$ (Equation 8.5) [58].

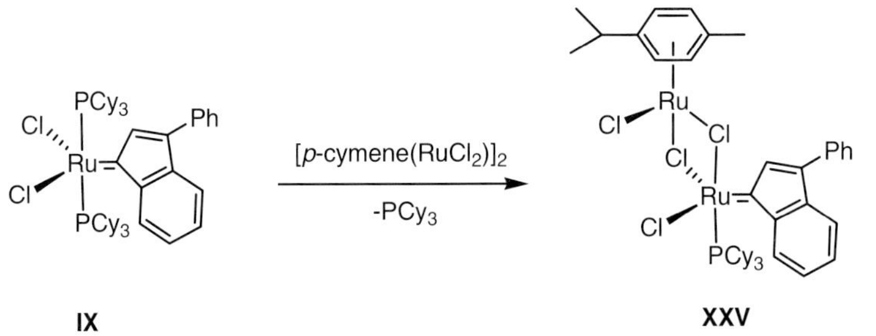

(8.5)

Similarly to the third generation Grubbs catalyst preparation, the related catalyst **XXVI** was obtained on addition of pyridine to precursor **IX** (Equation 8.6) [59].

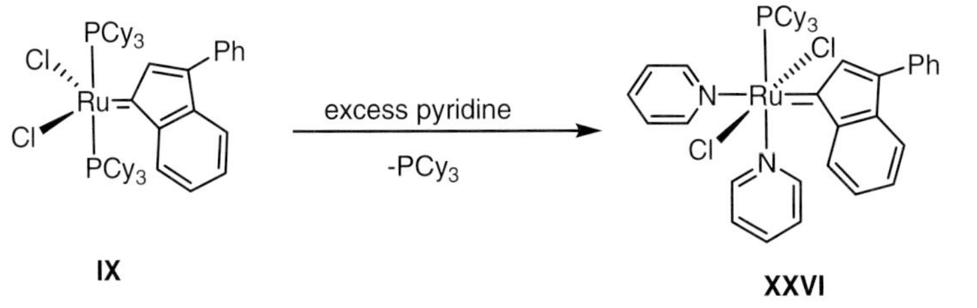

(8.6)

8.3 Indenylidene-Ruthenium Catalysts in Alkene Metathesis

Recently, the reaction of the Phoban phosphine with precursor **VIII** was reported to give a very stable catalyst precursor **XXVII** (Equation 8.7) [60].

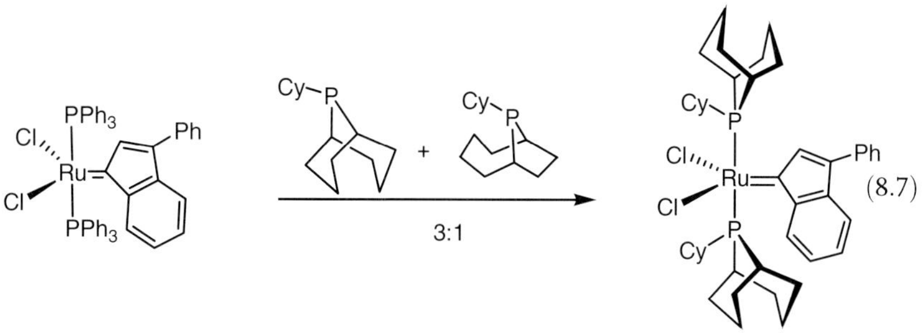

(8.7)

The indenylidene catalysts **XXVIII a–d**, containing a bidentate ligand, were obtained on reaction of complex **IX** or **XXI** with the corresponding iminophenol thalium derivative (Scheme 8.16) [61].

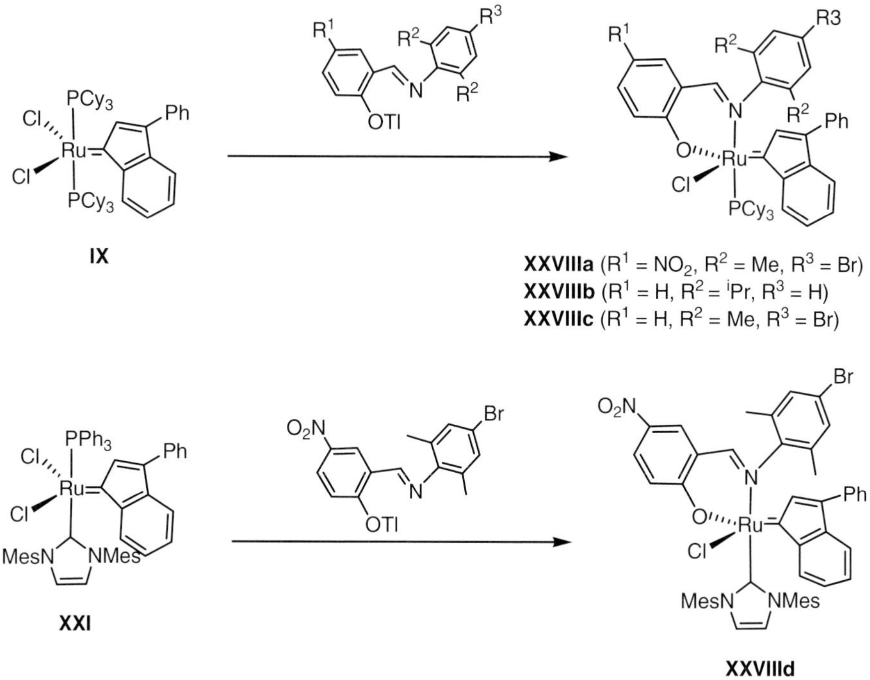

XXVIIIa (R^1 = NO_2, R^2 = Me, R^3 = Br)
XXVIIIb (R^1 = H, R^2 = ^{i}Pr, R^3 = H)
XXVIIIc (R^1 = H, R^2 = Me, R^3 = Br)

Scheme 8.16 Preparation of indenylidene complexes with bidentate iminophenolate ligand.

8.3.2
Ruthenium Indenylidene Complexes in Alkene Metathesis

The first efficient application of a well-defined ruthenium indenylidene complex in metathesis was described in 1999 by Fürstner for the total synthesis of a cyclic prodigiosin derivative, a potential lead compound for the development of immunosuppressive agents. The RCM using the ruthenium indenylidene complex **IX** (10 mol%) as precatalyst leads to the transformation of the N-protonated diene into the desired macrocycle in 65% yield (Equation 8.8) [43].

$$\text{(8.8)}$$

Actually, applications of indenylidene-ruthenium complexes for alkene metathesis were reported before, at a time when the action mode of their ruthenium allenylidene precursors was not known. These complexes catalyzed a variety of RCM reactions of dienes and enynes [31, 32, 47] (see Section 8.2.2).

The same catalyst **IX** was used in the RCM of a functional diene to give, in 87% yield, a tetrahydroazepine derivative which is a precursor in the synthesis of the metabolite (-)-Balanol (Equation 8.9) [44].

$$\text{(8.9)}$$

Starting from complex **IX**, Fürstner developed a homobimetallic phenylindenylidene complex **XXV** (Equation 8.5), and both of these were used in the cyclization of medium-sized rings by RCM. A series of examples is presented which shows that indenylidene complexes are as good as or superior to the classical Grubbs first generation catalyst in terms of yield, reaction rate, and tolerance towards different functional groups (Scheme 8.17) [58].

Catalyst **IX** (5 mol %) was also used for RCM or self-metathesis of different alkene derivatives. It allowed the preparation of macrocycles: the nonylprodigiosin precursors and their analogs or the efficient self-metathesis of multifunctional alkene (Scheme 8.18) [62].

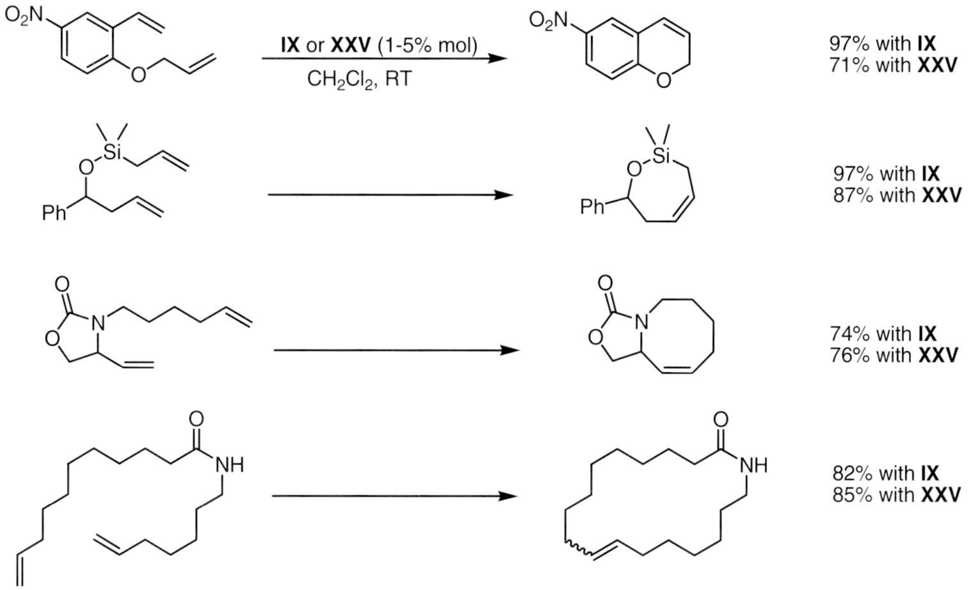
O2N
IX or XXV (1-5% mol)
CH2Cl2, RT
97% with IX
71% with XXV
Si
Ph
97% with IX
87% with XXV
74% with IX
76% with XXV
NH
82% with IX
85% with XXV

Scheme 8.17 Ring closing metathesis of dienes with ruthenium-indenylidenes **IX** and **XXV**.

The ruthenium indenylidene **IX** catalyzed the ring closing metathesis of a homoallyl ether diene to give a bicyclic compound, which is an intermediate for the synthesis of different marine natural products (Equation 8.10) [63].

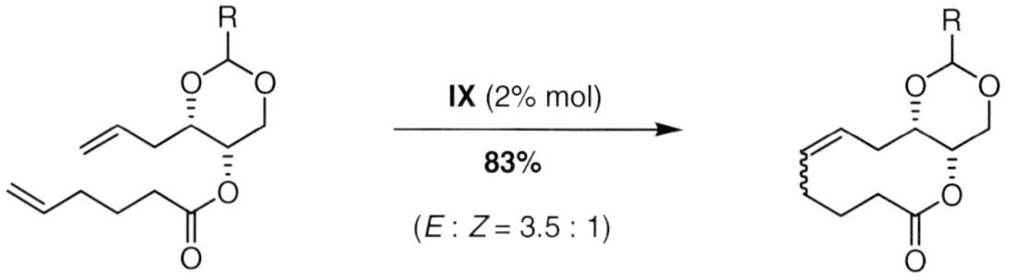
R
IX (2% mol)
83%
(E : Z = 3.5 : 1)

R = *p*MeO-C_6H_4

(8.10)

The same reaction (RCM) was used as the key step for the formation of a family of potent herbicidal 10-membered lactones. An important aspect from the preparative point of view is the control of stereochemical outcome of the RCM by the choice of catalyst. Thus, the use of the ruthenium indenylidene complex **IX** always leads to the corresponding (*E*)-alkenes, whereas the second generation of Grubbs catalyst bearing a N-heterocyclic carbene ligand affords the isomeric (*Z*)-olefin with good selectivity (Scheme 8.19) [64].

During the same period, Nolan showed also that indenylidene complexes are active catalyst precursors in the RCM of dienes. The reactions were performed on the NMR scale and moderate to good yields were obtained for diethyl diallylmalonate,

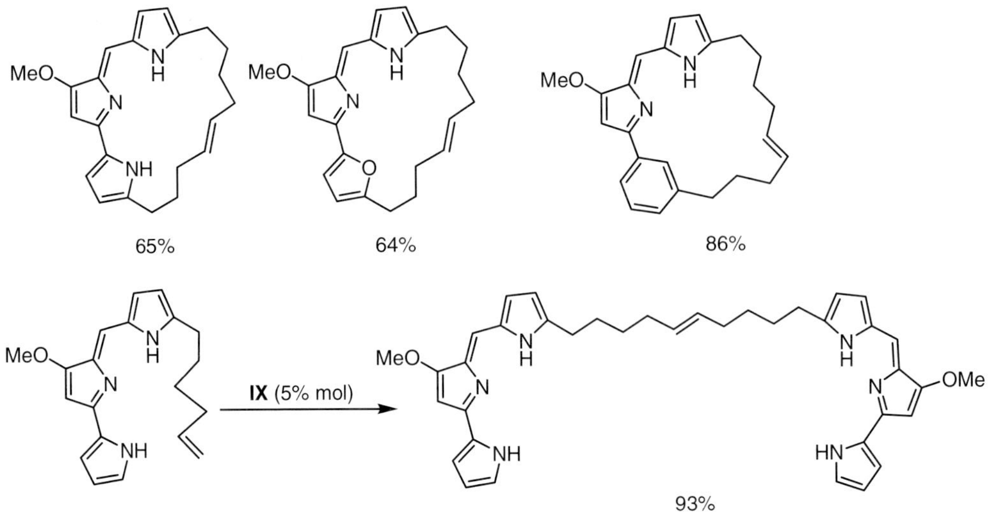

Scheme 8.18 Nonylprodigiosin analogs synthesized by RCM and self-metathesis with catalyst **IX**.

diallyltosylamine and diethyl bis(2-methylallyl)malonate (Scheme 8.20). It is to be pointed out that these ruthenium indenylidene complexes offer higher thermal stability as compared to the related alkylidene complexes [57].

The pyridine-containing ruthenium-based complex **XXVI** developed by Nolan [59] from the indenylidene complex **IX**, promoted the RCM of various dienes (Equation 8.6). Kinetic studies were carried out and showed that, despite a rapid initiation, the presence of pyridine in the reaction mixture has a negative effect on the stability of the active species and only moderate catalytic conversions were obtained [59] (Scheme 8.21).

(Arene)ruthenium-indenylidene complex **XV** (Scheme 8.11), developed by Dixneuf et al. [33, 34] was also a very efficient catalyst for RCM of dienes and enyne metathesis (Table 8.4).

One of the latest compounds of this class is the phoban-indenylidene complex **XXVII**, synthesized by Forman et al. in 2006 [60]. This robust catalyst was tested in self-metathesis and ethenolysis reactions of methyl oleate, giving rise to significantly

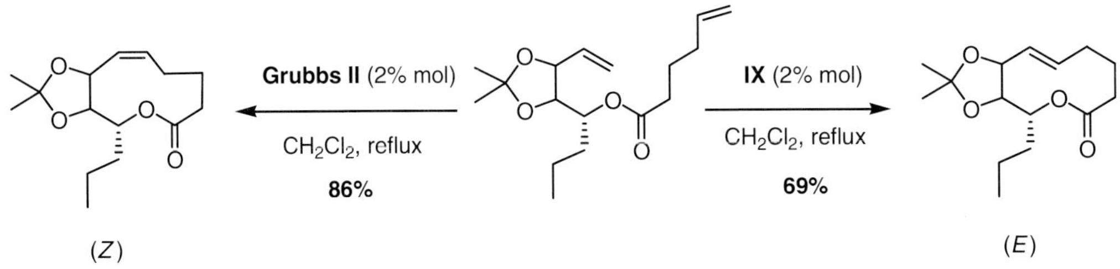

Scheme 8.19 Ring closing metathesis stereochemistry depending on the catalyst nature.

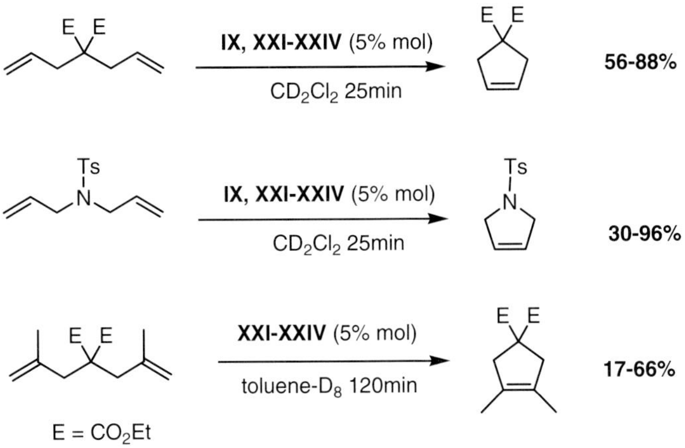

8.3 Indenylidene-Ruthenium Catalysts in Alkene Metathesis

Scheme 8.20 Ring closing metathesis of simple dienes with ruthenium-indenylidenes **IX, XXI-XXIV**.

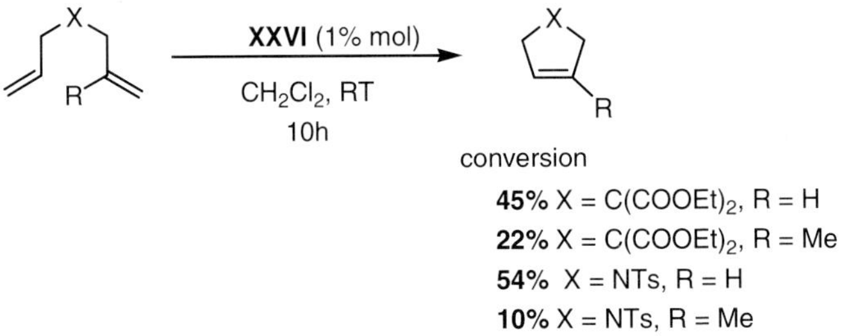

Scheme 8.21 RCM reactions with pyridine indenylidene catalyst **XXVI**.

higher conversions than Grubbs first generation catalyst (Scheme 8.22). For the self-metathesis reactions of methyl oleate to 9-octadecene and dimethyl 9-octadecene-1,18-dioate, 50% conversion was reached for a substrate to catalyst ratio of 40 000 : 1, and for the ethenolysis reaction of methyl oleate to 1-decene and methyl 9-decenoate, 70% conversion was reached for a substrate to catalyst ratio of 10 000 : 1 [60].

8.3.3
Polymerization with Ruthenium Indenylidene Complexes

The ruthenium indenylidene complex **XV**, prepared *in situ* from RuCl(η^6-*p*-cymene)(=C=C=CPh$_2$)(PCy$_3$)][CF$_3$SO$_3$] **Ib** with different acids showed very high activity in the cyclooctene polymerization (Table 8.2) [33].

The isolated complex **XV** and its derivatives having P*i*Pr$_3$ (**XVI**) and PPh$_3$ (**XVII**) instead of PCy$_3$ were also tested in cyclooctene polymerization at room temperature. Very good yields were obtained with **XV** and **XVI** (Table 8.3) [34].

The ruthenium indenylidene Schiff base complexes **XXVIIIa** and **XXVIIId**, synthesized by Verpoort, were evaluated in atom-transfer radical polymerization of methyl methacrylate. The polymerization was initiated by ethyl 2-bromo-2-methyl-

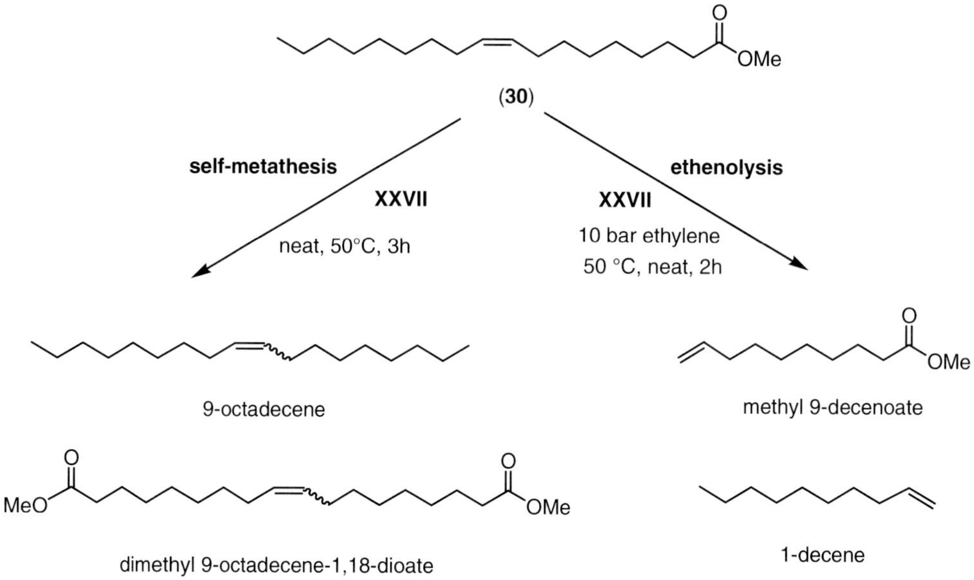

272 | *8 Ruthenium Allenylidenes and Indenylidenes as Catalysts in Alkene Metathesis*

Scheme 8.22 Self-metathesis and ethenolysis of methyloleate with Ru-indenylidene **XXVII**.

propionate at 85 °C and the results showed that the indenylidene complex **XXVIIIa** is moderately active for the conversion of MMA, but the polymer is produced in a well-controlled manner and possesses a very narrow molecular-weight distribution (MWD) ($M_w/M_n = 1.19$). Complexes **XXVIIIa** and **XXVIIId** promoted the ROMP of typical low-strain cyclic olefins, namely cyclooctene and cyclopentene (Table 8.6).

Table 8.6 ROMP of low-strained cyclic olefins using catalysts **XXVIIIa**, **XXVIIId** in toluene at RT.

Substrate	Catalyst	Monomer/Catalyst	Time	Conversion (%)	$10^3 \times M_n$	M_w/M_n
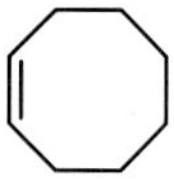 cyclooctene	XXVIIIa	10 000	17 h	53	89	1.79
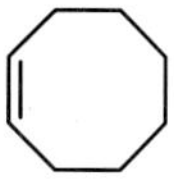 cyclooctene	XXVIIId	10 000	15 min	100	121	1.60
 cyclopentene	XXVIIIa	5000	17 h	41	20	1.78
 cyclopentene	XXVIIId	5000	17 h	78	32	1.55

Table 8.7 ATRP of styrene 170 °C, 17 h.

Catalyst	Yield (%)	$10^{-3} \times M_n$	M_w/M_n
VIII	37	17.0	1.30
IX	44	20.5	1.32
XXI	71	30.6	1.30
XXIII	75	29.9	1.33
XXVIIIb	30	13.8	1.30
XXVIIIa	56	23.8	1.12

Complex **XXVIIIa** and especially complex **XXVIIId** proved to be highly active catalysts for the ROMP of these olefins [61].

Indenylidene compounds **VIII, IX, XXI, XXIII, XXVIIIa** and **XXVIIIb** act as atom transfer radical polymerization catalysts for the polymerization of methyl methacrylate and styrene in high yields and with good control (Table 8.7). The catalytic activity can be dramatically improved by transforming the complexes into cationic species by treatment with $AgBF_4$ [61].

Nitrile rubber polymers, having lower molecular weight have been prepared by metathesis of nitrile butadiene rubber with ruthenium indenylidene complexes [65].

8.3.4
Other Catalytic Reactions Promoted by Indenylidenes

Another application of ruthenium indenylidene complexes was the atom transfer radical addition of carbon tetrachloride to vinyl monomers reported by Verpoort [61]. This Kharasch reaction afforded good yields for all substrates tested, especially with the catalyst **VIII** (Equation 8.11, Table 8.8).

$$\underset{R''}{\overset{R'}{=}} + Cl_3C-X \xrightarrow[\text{80°C, 17h}]{\textbf{VIII, IX, XXI, XXIII, XXVIIIa, XXVIIIb} (0.33\% \text{ mol})}{\text{toluene}} Cl_3C\underset{R''}{\overset{X}{\underset{R'}{\bigvee}}}$$

(8.11)

The nature of the chlorinated reagent is crucial for promoting the Kharasch addition reaction (Equation 8.11). The results showed that carbon tetrachloride could be added to various olefins in a regioselective way. Under these reaction conditions, no polymerization products were detected. In contrast, when chloroform was used as the halide source the methyl methacrylate and styrene conversions reached only 33% and 40% with the best performing system (**VIII**), and a significant fraction of polymers was observed [61].

The indenylidene complexes **IX** and **XXVIIIc** were also reported to promote the addition of different carboxylic acids to terminal alkynes to give enol esters, the Markovnikov addition product being the major product with, in some cases, the competing catalytic dimerization of terminal alkynes [61].

Table 8.8 ATRA of carbon tetrachloride to various olefins catalyzed by ruthenium indenylidene complexes.

Substrate	Yield (%)					
	VIII	IX	XXI	XXIII	XXVIIIb	XXVIIIa
Me, COOMe (methyl methacrylate)	94	73	79	70	58	70
iBu, COOMe	88	76	78	74	60	73
COOMe	78	59	76	55	47	38
nBu, COOMe	83	60	63	56	40	44
styrene	87	72	75	69	67	83
CN	42	12	15	9	23	30
C$_6$H$_{13}$	56	18	30	12	11	40

8.4
Conclusion

The above results show that allenylidene-ruthenium complexes, easily obtained from propargylic alcohols, are good precursors for catalytic alkene metathesis. It was demonstrated that they operate via their intramolecularly rearranged indenylidene-ruthenium complexes. These indenylidene complexes can be easily modified by ligand exchange to offer a variety of stable but active catalysts for alkene metathesis for both RCM and ROMP reactions. They constitute useful tools for the alkene metathesis catalyst box of which the content is constantly increasing, to the advantage of catalyst stability, activity, recycling and multiple applications.

References

1 Grubbs, R.H. (2003) *Handbook of metathesis*, Wiley VCH, Weinheim.

2 Hérisson, J.L. and Chauvin, Y. (1971) *Makromolekulare Chemie*, **141**, 161.

Chauvin, Y. (1973) *Comptes Rendus de L'Academie des Sciences, Paris*, **276**, 169. Chauvin, Y., Commereuc, D. and Cruypelinck, D. (1976) *Makromolekulare Chemie*, **177**, 2637.

3 Schrock, R.R., Murdzek, J.S., Bazan, G.C., Robbins, J., Di Mare, M. and O'Regan, M. (1990) *Journal of the American Chemical Society*, **112**, 3875 Schrock, R.R. (2004) *Journal of Molecular Catalysis A-Chemical*, **213**, 21.

4 Nguyen, S.T., Johnson, L.K., Grubbs, R.H. and Ziller, J.W. (1992) *Journal of the American Chemical Society*, **114**, 3974.

5 Landon, S.J., Shulman, P.M. and Geoffroy, G.L. (1985) *Journal of the American Chemical Society*, **107**, 6739.

6 Schwab, P., Grubbs, R.H. and Ziller, J.W. (1996) *Journal of the American Chemical Society*, **118**, 100.

7 Sémeril, D., Bruneau, C. and Dixneuf, P.H. (2001) *Helvetica Chimica Acta*, **84**, 3335. Sémeril, D., Bruneau, C. and Dixneuf, P.H. (2002) *Advanced Synthesis and Catalysis*, **344**, 585.

8 Katayama, H. and Ozawa, F. (1998) *Chemistry Letters*, 67. Louie, J. and Grubbs, R.H. (2001) *Angewandte Chemie-International Edition*, **40**, 247.

9 Bruneau, C. and Dixneuf, P.H. (2006) *Angewandte Chemie-International Edition*, **45**, 2176.

10 Katayama, H. and Ozawa, F. (2004) *Coordination Chemistry Reviews*, **248**. 1703.

11 Dragutan, V. and Dragutan, I. (2004) *Platinum Metal Reviews*, **48**, 148.

12 Fürstner, A., Picquet, M., Bruneau, C. and Dixneuf, P.H. (1998) *Chemical Communications*, 1315.

13 Selegue, J.P. (1982) *Organometallics*, **1**, 217.

14 Rigaut, S., Monnier, F., Mousset, F., Touchard, D. and Dixneuf, P.H. (2002) *Organometallics*, **21**, 2654.

15 Baya, M., Buil, M.L., Esteruelas, M.A., López, A.M., Oñate, E. and Rodríguez J.R. (2002) *Organometallics*, **21**, 1841.

16 Bustelo, E., Jiménez-Tenorio, M., Mereiter, K., Puerta, M.C. and Valerga, P. (2002) *Organometallics*, **21**, 1903.

17 Cadierno, V., Conejero, S., Gamasa, M.P., Gimeno, J., Falvello, L.R. and Llusar, R.M. (2002) *Organometallics*, **21**, 3716.

18 Trost, B.M. and Flygare, J.A. (1992) *Journal of the American Chemical Society*, **114**, 5476.

19 Nishibayashi, Y., Wakiji, I. and Hidai, M. (2000) *Journal of the American Chemical Society*, **122**, 11019.

20 Nishibayashi, Y., Inada, Y., Hidai, M. and Uemura, S. (2003) *Journal of the American Chemical Society*, **125**, 6060.

21 Nishibayashi, Y., Inada, Y., Hidai, M. and Uemura, S. (2002) *Journal of the American Chemical Society*, **124**, 7900.

22 Nishibayashi, Y., Inada, Y., Hidai, M. and Uemura, S. (2003) *Journal of the American Chemical Society*, **125**, 6060.

23 Nishibayashi, Y., Milton, M.D., Inada, Y., Yoshikawa, M., Wakiji, I., Hidai, M. and Uemura, S. (2005) *Chemistry – A European Journal*, **11**, 1453. Inada, Y., Nishibayashi, Y. and Uemura, S. (2005) *Angewandte Chemie-International Edition*, **44**, 7715. Nishibayashi, Y., Milton, M.D., Inada, Y., Yoshikawa, M. and Uemura, S. (2006) *European Journal of Organic Chemistry*, 881.

24 Bustelo, E. and Dixneuf, P.H. (2005) *Advanced Synthesis and Catalysis*, **347**, 393.

25 Fischmeister, C., Toupet, L. and Dixneuf, P.H. (2005) *New Journal of Chemistry*, **29**, 765.

26 Bustelo, E. and Dixneuf, P.H. (2007) *Advanced Synthesis and Catalysis*, **349**, 933.

27 Maddock, S.M. and Finn, M.G. (2001) *Angewandte Chemie-International Edition*, **40**, 2138.

28 Saoud, M., Romerosa, A., Mañas Carpio, S., Gonsalvi, L. and Peruzzini, M. (2003) *European Journal of Inorganic Chemistry*, 1614.

29 Yeh, K.-L., Liu, B., Lo, C.-Y., Huang, H.-L. and Liu, R.-S. (2002) *Journal of the American Chemical Society*, **124**, 6510.

30 Opstal, T. and Verpoort, F. (2003) *Polymer Bulletin*, **50**, 17.

31 Picquet, M., Touchard, D., Bruneau, C. and Dixneuf, P.H. (1999) *New Journal of Chemistry*, **23**, 141.

32 Fürstner, A., Liebl, M., Lehmann, C.W., Picquet, M., Kunz, R., Bruneau, C., Touchard, D. and Dixneuf, P.H. (2000) *Chemistry – A European Journal*, **6**, 1847.

33 Castarlenas, R. and Dixneuf, P.H. (2003) *Angewandte Chemie-International Edition*, **42**, 4524.

34 Castarlenas, R., Vovard, C., Fischmeister, C. and Dixneuf, P.H. (2006) *Journal of the American Chemical Society*, **128**, 4079.

35 Schwab, P., Grubbs, R.H. and Ziller, J.W. (1996) *Journal of the American Chemical Society*, **118**, 100.

36 Osipov, S.N., Artyushin, O.I., Kolomiets, A.F., Bruneau, C., Picquet, M. and Dixneuf, P.H. (2001) *European Journal of Inorganic Chemistry*, 3891.

37 Picquet, M., Bruneau, C. and Dixneuf, P.H. (1998) *Chemical Communications*, 2249.

38 Sémeril, D., Le Nôtre, J., Bruneau, C., Dixneuf, P.H., Kolomiets, A.F. and Osipov, S. (2001) *New Journal of Chemistry*, **25**, 16.

39 Castarlenas, R., Sémeril, D., Noels, A.F., Demonceau, A. and Dixneuf, P.H. (2002) *Journal of Organometallic Chemistry*, **663**, 235.

40 Alaoui Abdallaoui, I., Sémeril, D. and Dixneuf, P.H. (2002) *Journal of Molecular Catalysis A: Chemical*, **182–183**, 577.

41 Jafarpour, L., Huang, J., Stevens, E.D. and Nolan, S.P. (1999) *Organometallics*, **18**, 3760.

42 Schanz, H.-J., Jafarpour, L., Stevens, E.D. and Nolan, S.P. (1999) *Organometallics*, **18**, 5187.

43 Fürstner, A., Grabowski, J. and Lehmann, C.W. (1999) *Journal of Organic Chemistry*, **64**, 8275.

44 Fürstner, A. and Thiel, O.R. (2000) *Journal of Organic Chemistry*, **65**, 1738.

45 The activity of the catalyst was not due to the expected allenylidene [46,47] but to its rearranged indenylidene complex [43,44].

46 Harlow, K.J., Hill, A.F. and Wilton-Ely, J.D.E.T. (1999) *Journal of the Chemical Society-Dalton Transactions*, 285.

47 Fürstner, A., Hill, A.F., Liebl, M. and Wilton-Ely, J.D.E.T. (1999) *Chemical Communications*, 601.

48 Touchard, D., Haquette, P., Daridor, A., Toupet, L. and Dixneuf, P.H. (1994) *Journal of the American Chemical Society*, **116**, 11157.

49 Jung, S., Brandt, C.D. and Werner, H. (2001) *New Journal of Chemistry*, **25**, 1101.

50 Le Gendre, P., Picquet, M., Richard, P. and Moise, C. (2002) *Journal of Organometallic Chemistry*, **643–644**, 231.

51 Cetinkaya, B., Demir, S., Özdemir, I., Toupet, L., Sémeril, D., Bruneau, C. and Dixneuf, P.H. (2003) *Chemistry – A European Journal*, **9**, 2323.

52 Bassetti, M., Centola, F., Sémeril, D., Bruneau, C. and Dixneuf, P.H. (2003) *Organometallics*, **22**, 4459.

53 Wilhelm, T.E., Belderrain, T.R., Brown, S.N. and Grubbs, R.H. (1997) *Organometallics*, **16**, 3867.

54 Castarlenas, R., Eckert, M. and Dixneuf, P.H. (2005) *Angewandte Chemie-International Edition*, **44**, 2576.

55 Ipaktschi, J., Mohsseni-Ala, J., Dülmer, A., Loschen, C. and Frenkin, G. (2005) *Organometallics*, **24**, 977. Ipaktschi, J., Rooshenas, P. and Dülmer, A. (2005) *Organometallics*, **24**, 6239.

56 Fürstner, A., Davies, P.W. and Lehmann, C.W. (2005) *Organometallics*, **24**, 4065.

57 Jafarpour, L., Schanz, H.-J., Stevens, E.D. and Nolan, S.P. (1999) *Organometallics*, **18**, 5416. Fürstner, A., Thiel, O.R., Ackermann, L., Schanz, H.-J. and Nolan, S.P. (2000) *Journal of Organic Chemistry*, **65**, 2204. Nolan, S.P. and Huang, J. (2000) EP1115491A1.

58 Fürstner, A., Guth, O., Düffels, A., Seidel, G., Liebl, M., Gabor, B. and Mynott, R. (2001) *Chemistry – A European Journal*, **7**, 481.

59 Clavier, H., Petersen, J.L. and Nolan, S.P. (2006) *Journal of Organometallic Chemistry*, **691**, 5444.

60 Forman, G.S., Bellabarba, R.M., Tooze, R.P., Slawin, A.M.Z., Karch, R. and Winde,

R. (2006) *Journal of Organometallic Chemistry*, **691**, 5513.

61 Opstal, T. and Verpoort, F. (2003) *Angewandte Chemie-International Edition*, **42**, 2876. Opstal, T. and Verpoort, F. (2003) *New Journal of Chemistry*, **25**, 257. Opstal, T. and Verpoort, F. (2002) *Synlett*, **6**, 935. Drozdzak, R., Allaert, B., Ledoux, N., Dragutan, I., Dragutan, V. and Verpoort, F. (2005) *Advanced Synthesis and Catalysis*, **347**, 1721.

62 Fürstner, A., Grabowski, J., Lehmann, C.W., Kataoka, T. and Nagai, K. (2001) *ChemBioChem*, **2**, 60.

63 Fürstner, A. and Schlede, M. (2002) *Advanced Synthetic Catalysis*, **344**, 657.

64 Fürstner, A., Radkowski, K., Wirtz, C., Goddard, R., Lehmann, C.W. and Mynott, R. (2002) *Journal of the American Chemical Society*, **124**, 7061.

65 Guerin, F (2005). EP172091A1.

9
Rhodium and Group 9–11 Metal Vinylidenes in Catalysis

Sean H. Wiedemann and Chulbom Lee

9.1
Introduction

Most pioneering studies of vinylidene-mediated catalysis were concerned with transition metals from Groups 6 and 8 of the periodic table (Chapters 5, 6, 10). These endeavors led to the recognition that unique and valuable products can result from the modification of alkynes via their vinylidene complexes. Examples of Group 9–11 transition metal vinylidene-mediated catalysis have increased in number recently. Reported examples share some common features.

1. All metal vinylidenes described herein are derived from alkynes. While alkyne-to-vinylidene interconversion typically occurs via the 1,2-shift of a hydrogen atom, the corresponding migration of heavier main group heteroatoms is also possible.

2. None of the methods described in this chapter utilize a pre-formed metal vinylidene as an active catalyst precursor. The occurrence of metal vinylidene intermediates is instead inferred on the basis of product structure, isotopic labeling experiments, and computational studies.

3. Often, selectivity for a vinylidene-mediated pathway is heavily dictated by substrate structure. It is especially true in the case of hetero-atom substituted alkynes that π-alkyne/vinylidene rearrangement is driven by a reduction in steric interactions at the metal center.

Vinylidene-mediated reactions involving rhodium and iridium are discussed separately from those involving Groups 10 and 11 transition metals. The reactions of Group 9 metal vinylidenes are more numerous and have more in common with one another. Extensive stoichiometric organometallic literature aids in the understanding of these processes. In contrast, reactions of Groups 10 and 11 metal vinylidenes are more scattered and often controversial.

Metal Vinylidenes and Allenylidenes in Catalysis: From Reactivity to Applications in Synthesis
Edited by Christian Bruneau and Pierre Dixneuf
Copyright © 2008 WILEY-VCH Verlag GmbH & Co. KGaA, Weinheim
ISBN: 978-3-527-31892-6

Tables of results presented in this chapter are excerpted from literature data in an effort to provide only the most meaningful and illustrative examples. Also, for the sake of brevity, deuterium labeling experiments are generally not depicted. The interested reader is directed to the primary literature.

9.2 Rhodium and Iridium Vinylidenes in Catalysis

9.2.1 Introduction

The ability to harness alkynes as effective precursors of reactive metal vinylidenes in catalysis depends on rapid alkyne-to-vinylidene interconversion [1]. This process has been studied experimentally and computationally for [$MCl(PR_3)_2$] (M = Rh, Ir, Scheme 9.1) [2]. Starting from the π-alkyne complex **1**, oxidative addition is proposed to give a transient hydridoacetylide complex (**3**) which can undergo intramolecular 1,3-H-shift to provide a vinylidene complex (**5**). Main-group atoms presumably migrate via a similar mechanism. For iridium, intermediates of type **3** have been directly observed [3]. Section 9.3 describes the use of an alternate alkylative approach for the formation of rhodium vinylidene intermediates bearing two carbon-substituents (alkenylidenes).

When desired vinylidene-mediated pathways are not sufficiently favorable, Group 9 metal catalysts can access a set of typical side-reaction pathways. Alkyne dimerization to give conjugated enynes or higher oligomers is often observed. Polysubstituted benzenes resulting from [2 + 2 + 2] alkyne cyclotrimerization are also common coproducts. Fortunately, the selectivity of rhodium and iridium catalysts can often be modulated by the variation of spectator ligands.

The reactions discussed in this section are divided into three classes based on the manner in which the incipient vinylidenes undergo further bond-forming reactions. Some of the most significant rhodium vinylidene-mediated synthetic methods reported to date are those that involve vinylidene π-bonds participating in pericyclic processes. These reactions permit a complex carbon framework to be assembled through unique bond disconnections. This subject has been reviewed [4]. Anti-Markovnikov hydrofunctionalization reactions catalyzed by Group 9 metal complexes are also discussed in this section. Finally, a small, but growing, number of rhodium-catalyzed multi-component coupling reactions have been discovered which apparently occur via one or more metal vinylidene intermediates.

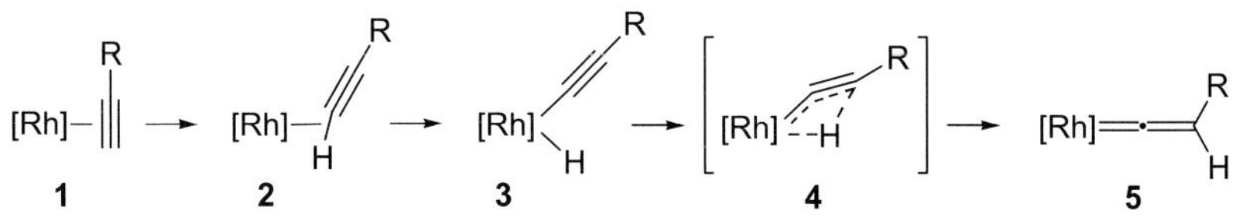

Scheme 9.1 Alkyne-to-vinylidene conversion.

9.2.2
Carbocyclization/Pericyclic Reactions

In the course of research on catalysis by synthetic metalloporphyrins, Tagliatesta and coworkers reported a unique cyclodimerization reaction of aryl acetylenes to give 2-aryl naphthalenes (**6**, Scheme 9.2) [5]. Ru(CO) and RhCl complexes effectively promote this reaction, with the latter catalyst (**8**) providing generally superior yields and selectivities for a small range of substrates (Table 9.1). As a synthetic method, Tagliatesta's cyclodimerization reaction is most remarkable for its efficiency. Yields of up to 78% were observed at a substrate/catalyst ratio of 5700 : 1. Successful recycling of recovered catalyst was also demonstrated.

The authors' proposed mechanism for dimerization involves initial formation of metal vinylidene complex **9** via 1,2-H-migration. A second molecule of arylacetylene acts as a dienophile in a formal [4 + 2] Diels–Alder cycloaddition with **9**. A subsequent

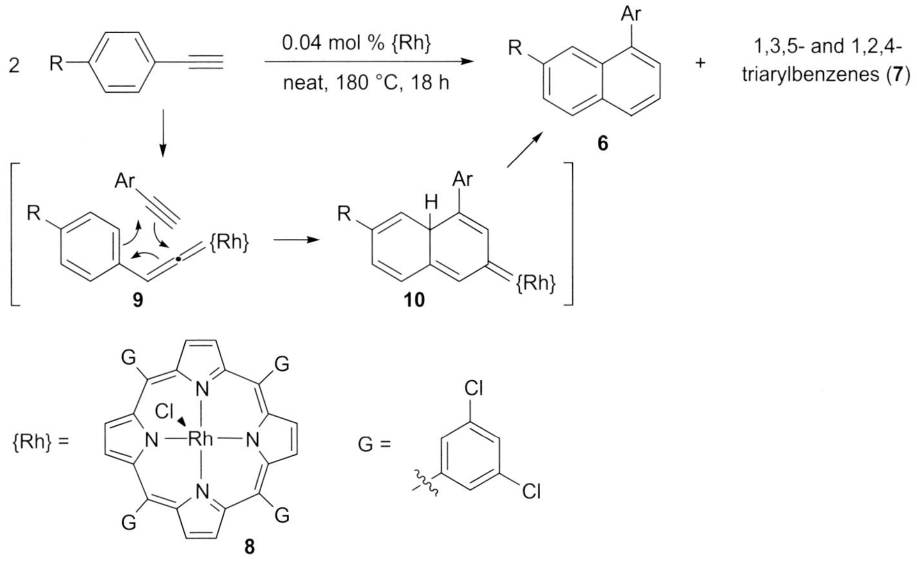

Scheme 9.2 Cyclodimerization of arylacetylenes with a rhodium porphyrin catalyst.

Table 9.1 Cyclodimerization of phenylacetylenes.

Entry	R	Yield 6 (%)	Yield 7 (%)
1	H	17	25
2	Cl	1	98
3	OMe	49	14
4	Me	69	30
5	m-OMe	78[a]	21

[a] 1 : 1 mixture of regioisomers.

1,4-H-shift is invoked for the formation of **6** and regeneration of catalyst **8**. The proposed mechanism is unusual insofar as the π-bonds of electroneutral alkynes and arenes seldom participate in Diels–Alder reactions. The intermediacy of metal vinylidenes is supported by the failure of internal alkynes to dimerize under the reported conditions. More importantly, mechanistic restrictions imposed by the porphyrin ligand set severely restrict conceivable alternative mechanisms.

Another rhodium vinylidene-mediated reaction for the preparation of substituted naphthalenes was discovered by Dankwardt in the course of studies on 6-endo-dig cyclizations of enynes [6]. The majority of his substrates (not shown), including those bearing internal alkynes, reacted via a typical cationic cycloisomerization mechanism in the presence of alkynophilic metal complexes. In the case of silylalkynes, however, the use of [Rh(CO)$_2$Cl]$_2$ as a catalyst unexpectedly led to the formation of predominantly 4-silyl-1-silyloxy naphthalenes (**12**, Scheme 9.3). Clearly, a distinct mechanism is operative. The author's proposed catalytic cycle involves the formation of Rh(I) vinylidene intermediate **14** via 1,2-silyl-migration. A nucleophilic addition reaction is thought to occur between the enol-ether and the electrophilic vinylidene α-position of **14**. Subsequent H-migration would be expected to provide the observed product. Formally a 6π-electrocyclization process, this type of reaction is promoted by W(0)- and Ru(II)-catalysts (Chapters 5 and 6).

Dankwardt's vinylidene-mediated reaction apparently operates concurrently with cationic cycloisomerization to give 3-silyl-1-silyloxy naphthalenes (e.g., **13**). From his data, a direct comparison can be made of the effect of different metal complexes and silyl groups on selectivity for a vinylidene-mediated reaction pathway (Table 9.2). At least in this instance, Rh(I) is more "vinylidene friendly" than Pt(II). Iwasawa and coworkers [7], in an isolated related report, also obtained high selectivity for silyl-shifted products in the presence of a Rh(I)-catalyst, albeit one with a substantially different ligand set from that employed by Dankwardt.

Lee and coworkers envisaged engaging metal vinylidene intermediates derived from simple enynes (**16**) in formal intramolecular 4π-cycloaddition (Scheme 9.4) [8].

Whereas the thermally allowed 6π-pericyclic reactions of metal vinylidenes described above engage the vinylidene C=C π-bond, hope for success in a thermally

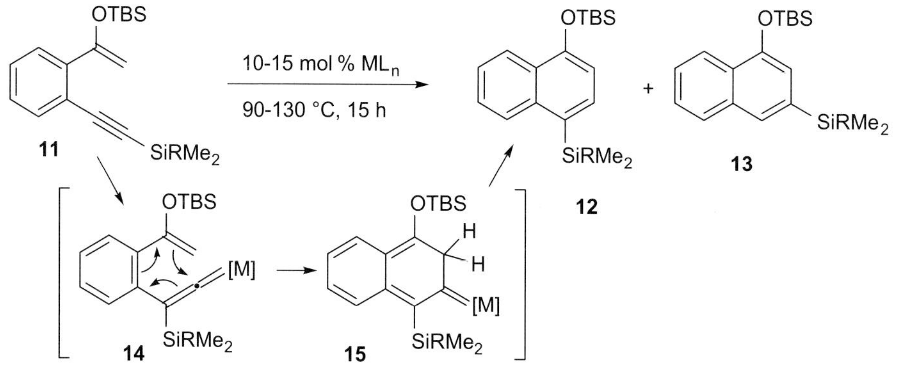

Scheme 9.3 Rh(I)- and Pt(II)-catalyzed carbocyclization.

Table 9.2 Cycloisomerization of silylated enynes.

Entry	R	ML_n	Solvent	Temp (°C)	Yield (%)	12 : 13
1	Me	$[Rh(CO)_2Cl]_2$	None	90	75	4.9 : 1
2	Me	$PtCl_2$			77	1 : 1.9
3	Ph	$[Rh(CO)_2Cl]_2$	PhMe	130	80	5.2 : 1
4	Ph	$PtCl_2$			68	1.3 : 1
5	*t*Bu	$[Rh(CO)_2Cl]_2$	PhMe	130	91	7.4 : 1
6	*t*Bu	$PtCl_2$			49	1 : 3.7

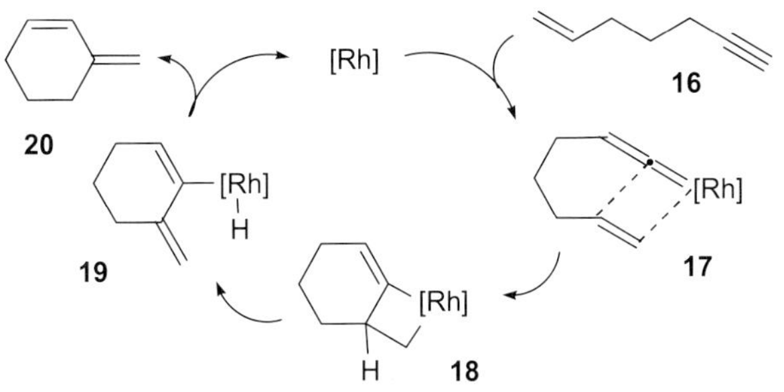
[Rh]
16
20
[Rh]
H
19
[Rh]
17
[Rh]
H
18

Scheme 9.4 Cycloisomerization of 1,6-enynes via rhodium vinylidene intermediates.

forbidden [2 + 2]-cycloaddition lay in exploiting the highly polarized M=C π-bond. Vinylidene/olefin cyclization of intermediate **17** would lead to rhodacycle **18**, which could break down via β-hydride elimination/reductive elimination to give a methylene cyclohexene (**20**). This overall transformation, employing a Rh(I)/phosphine catalyst mixture, was discovered by Grigg et al. and later fully described by Lee and coworkers (Equation 9.1) [9]. Mechanistic data gathered from isotopic labeling experiments support the proposed reaction mechanism for enyne cycloisomerization.

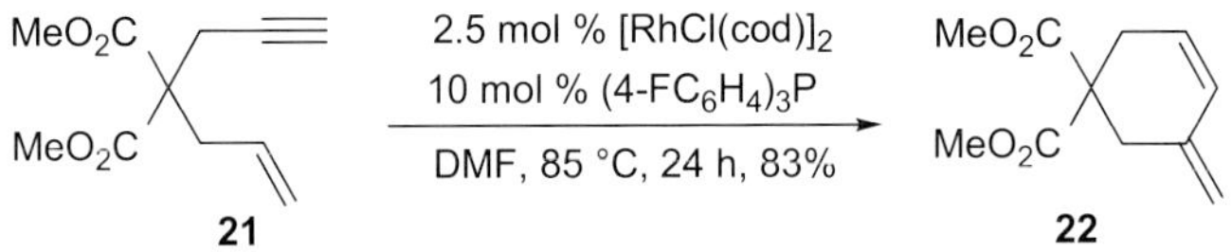
MeO_2C
MeO_2C
21
2.5 mol % $[RhCl(cod)]_2$
10 mol % $(4\text{-}FC_6H_4)_3P$
DMF, 85 °C, 24 h, 83%
MeO_2C
MeO_2C
22

(9.1)

Under optimized conditions, cycloisomerizations of a number of functionalized hept-1-en-6-ynes took place in good-to-excellent yields (Table 9.3). Heteroatom substitution was tolerated both within the tether and on its periphery. Alkynyl silanes and selenides underwent rearrangement to provide cyclized products in moderate yield (entries 6 and 7). One example of seven-membered ring formation was reported (entry 5). Surprisingly, though, substitution was not tolerated on the alkene moiety of the reacting enyne. The authors surmize that steric congestion retards the desired [2 + 2]-cycloaddition reaction to the point that side reactions, such as alkyne dimerization, become dominant.

Table 9.3 Rh(I)-catalyzed cycloisomerization of enynes.

Entry[a]	Substrate	Product	Yield (%)
1	TBDPSO-ene-yne	TBDPSO-diene	80
2	BnO-decalin-yne	BnO-octahydronaphthalene-ene	90
3	BnO-decalin-yne-allyl	BnO-octahydronaphthalene-diene	99
5[b]	Ph-O-Si-yne	Ph-O-Si-diene (7-membered)	77
6[c]	Si-O-C$_5$H$_{11}$-yne	Si-O-C$_5$H$_{11}$-diene	51
7[c]	BnO, BnO, SePh-yne	BnO, BnO, SePh-cyclohexadiene	64

[a] Conditions: See Equation 9.1.
[b] 20 mol % RhCl(Ph$_3$P)$_3$, DMF, 85 °C, 24 h.
[c] 20 mol % RhCl(Ph$_3$P)$_3$, PhMe, 125 °C, 1–3 d.

Lee and coworkers went on to investigate the action of their enyne cycloisomerization catalyst on N-propargyl enamines (Equation 9.2) [10].

$$\underset{23}{\text{Bz-N-cyclopentenyl-propargyl}} \xrightarrow[\text{DMF, 25 °C, 24 h, 95\%}]{\substack{5 \text{ mol \% [RhCl(C}_2\text{H}_4)_2]_2 \\ 25 \text{ mol \% (4-FC}_6\text{H}_4)_3\text{P} \\ 1 \text{ equiv DABCO}}} \underset{24}{\text{Bz-N-bicyclic diene}} \quad (9.2)$$

In that report, electron-rich olefins were found to react with Rh(I) vinylidenes differently from the manner in which electron-neutral olefins do (Scheme 9.5). A nucleophilic addition reaction, rather than a pericyclic ring closure, was evident. Because the reaction is strongly promoted by organic base, proton-shuttling is believed to account for the conversion of zwitterionic intermediate **26** to the observed

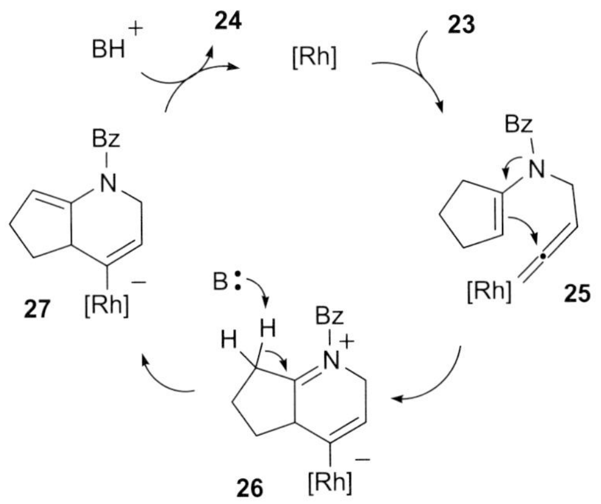

9.2 Rhodium and Iridium Vinylidenes in Catalysis | 285

Scheme 9.5 Proposed mechanism of Rh(I)-catalyzed N-propargyl enamine cycloisomerization.

product (24). N-Propargyl enamines with various substitution patterns underwent Rh(I)-catalyzed cycloisomerization in the presence of DABCO at 25 °C (Table 9.4). These conditions are especially mild, and the stabilized enamine starting materials are easy to handle.

Table 9.4 Rh(I)-catalyzed cycloisomerization of N-propargyl enamines.

Entry[a]	Substrate	Product	Yield (%)
1	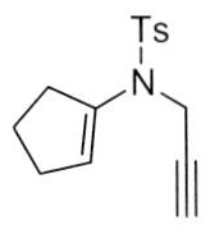 cyclopentenyl-N(Ts)-CH$_2$C≡CH	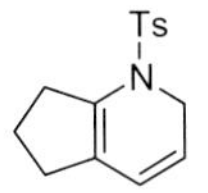 Ts-N bicyclic dihydropyridine	83
2	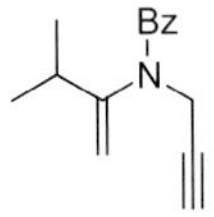 iPr-C(=CH$_2$)-N(Bz)-CH$_2$C≡CH	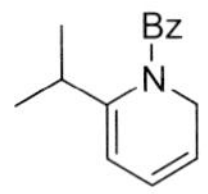 iPr-substituted N(Bz) dihydropyridine	81
3	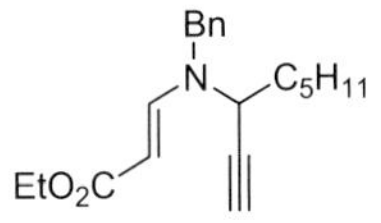 EtO$_2$C-CH=CH-N(Bn)-CH(C$_5$H$_{11}$)-C≡CH	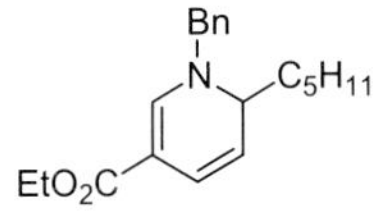 EtO$_2$C, Bn, C$_5$H$_{11}$ dihydropyridine	69
4	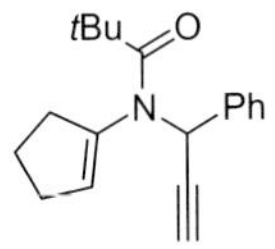 cyclopentenyl-N(CO$_2$tBu)-CH(Ph)-C≡CH	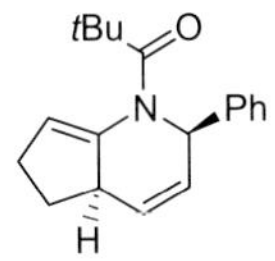 tBuO$_2$C-N, Ph bicyclic dihydropyridine	99
5[b]	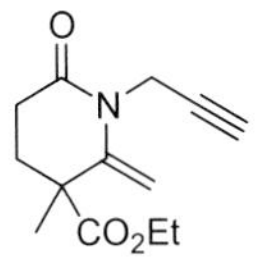 piperidinone with =CH-CO$_2$Et and N-CH$_2$C≡CH	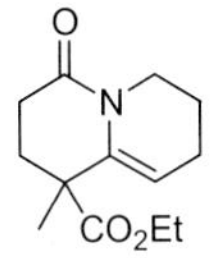 bicyclic enamide with CO$_2$Et	83

[a] Conditions: See Equation 9.2.
[b] Reaction post-treated with H$_2$.

Substrates bearing a propargylic stereocenter exhibited perfect 1,4-stereoinduction (entry 4). No specific model was proposed to account for different product olefin isomers (endocyclic versus exocyclic) obtained from different substrate structures.

This reaction constitutes a rare example of the use of enamines in transition metal catalysis and a unique method for the synthesis of six-membered aza-heterocycles.

Uemura and coworkers discovered another unique rhodium vinylidene-mediated cycloisomerization reaction [11]. They found that in the presence of an electron-rich Rh(I)-complex, $[RhCl(iPr_3P)_2]_2$, (*Z*)-hexa-3-en-1,5-diynes bearing an alkyl substituent at one terminus undergo cycloisomerization to give allylbenzenes (Equation 9.3).

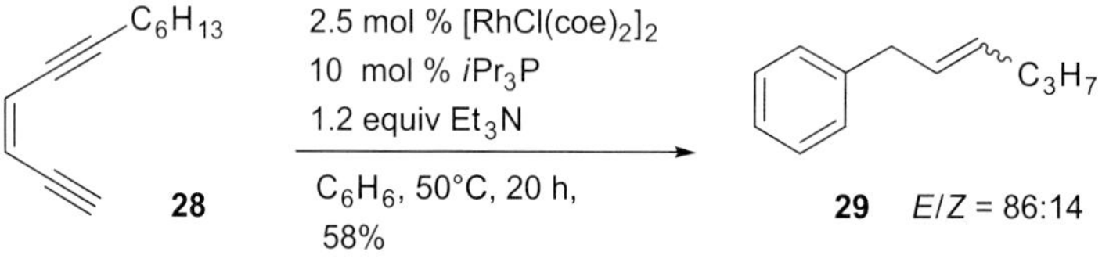
C6H13
28
2.5 mol % [RhCl(coe)2]2
10 mol % iPr3P
1.2 equiv Et3N
C6H6, 50°C, 20 h,
58%
C3H7
29 E/Z = 86:14

(9.3)

This reaction may be viewed mechanistically as involving a variant of the Myers cycloaromatization, in which a metal vinylidene takes the place of an allene (Scheme 9.6). Instead of producing a *C,C*-diradical, the electrocyclization step of Uemura's reaction gives a monomeric Rh(II)-intermediate bearing a carbon-centered radical (**31**). According to the proposed mechanism, a 1,5-hydride shift in **31** leads to another intermediate (**32**) bearing a radical that can combine with the Rh (II)-center to even its electron count. The resulting rhodacyclohexene intermediate (**33**) can break down by β-hydride elimination/reductive elimination to afford the observed allylbenzene product (**29**). A deuterium labeling experiment confirmed the expected 1,2-H-shift. Interestingly, the addition of a hydride donor (1,4-cyclohexadiene) had no effect on the cycloaromatization reaction of **28**. In contrast, the use a

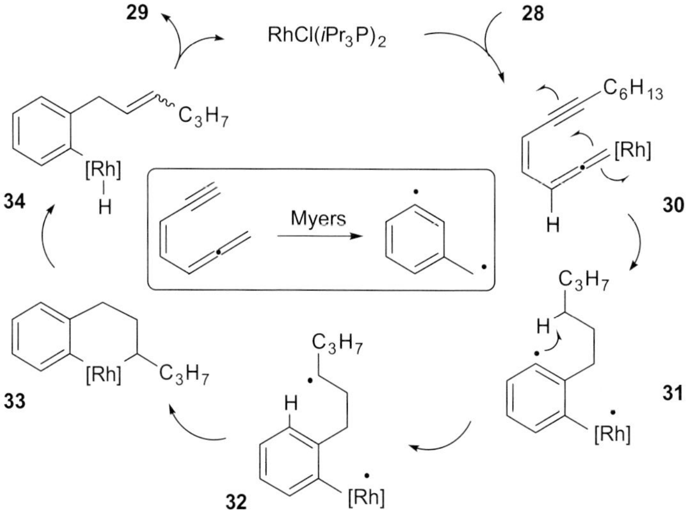
29
RhCl(iPr3P)2
28
C6H13
[Rh]
H
30
C3H7
[Rh]
H
34
Myers
C3H7
H
31
[Rh]
C3H7
H
[Rh]
32
[Rh]
C3H7
33

Scheme 9.6 Proposed catalytic cycle for cycloaromatization of 3-en-1,5-diynes.

Table 9.5 Rh(I)-catalyzed cycloaromatization of enediynes.

Entry[a]	Enediyne	Product	Yield (%)
1	R = (CH₂-allyl)	phenethyl + styryl benzene	68
2	R = (CH₂-allyl), cyclohexenyl substrate	cyclohexyl-phenethyl + cyclohexenyl-styryl	64
3	R = CH₂-SiMe₂iPr	PhCH₂CH₂-SiMe₂-CH=CMe₂	69[b]
4	o-(C₆H₁₃-alkynyl)(propargyl)benzene	naphthyl-CH₂CH=CH-C₃H₇, E/Z=86:14	46

[a] Conditions: See Equation 9.3.
[b] Reaction carried out at 80 °C with slow addition of enediyne over 8 h in the presence of 25 mol % dimethyl maleate.

hydride donor was essential in a related stoichiometric metalla-Myers-type process mediated by ruthenium [12].

A number of substituted allylarenes were synthesized under mild conditions using Et₃N as an additive (Table 9.5). Curiously, only products of endocyclic β-hydride elimination (from the homobenzylic position) were observed, though mixtures of E/Z isomers were obtained from substrates bearing aliphatic chains (entry 4, Equation 9.3). In entry 3, the substrate cannot undergo a 1,5-hydride shift; hence, the product resulting from a 1,6-hydride shift was obtained.

In a full account of their work on the cycloaromatization, Uemura and coworkers reported an alternate reaction pathway that is available when a mechanism involving β-hydride elimination is not accessible (Equation 9.4) [13].

$$\text{35 (enediyne-SiMe}_3\text{)} \xrightarrow[\text{Et}_3\text{N, C}_6\text{H}_6\text{, 80 °C}]{\substack{2.5 \text{ mol \% [RhCl(coe)}_2]_2 \\ 10 \text{ mol \% } i\text{Pr}_3\text{P}}} \text{36 (bicyclic-SiMe}_2\text{)} \quad (9.4)$$

8 h slow addition, 59%

Bicyclic products were obtained from C–C-reductive elimination of a presumed rhodacycle intermediate (see **33**). Both of Uemura's reactions constitute significant contributions to the catalysis literature because they involve (i) vinylidenes, (ii) catalysis of a pericyclic process, and (iii) C–H bond activation.

9.2.3
Anti-Markovnikov Hydrofunctionalization

Although ruthenium and Group 6 metal catalysts are commonly employed for anti-Markovnikov alkyne hydrofunctionalization (Chapter 10), some interesting rhodium- and iridium-catalyzed methods have also been reported. These can be divided into three groups based on the nature of the incoming functional group: (i) additions of nitrogen and oxygen nucleophiles, (ii) hydroboration and (iii) alkyne dimerization. Each of these reactions is known to occur either in the presence of other transition metals or through catalytic mechanisms that do involve metal vinylidene intermediates. Vinylidene-mediated processes generally offer atypical chemo- or stereo-selectivity.

The Trost group has disclosed two methods of rhodium vinylidene-mediated hydrofunctionalization: (i) intramolecular hydroalkoxylation of alkynols and (ii) intramolecular hydroamination of 2-ethynylanilines. The endo-dig-cyclization of homo- and bishomopropargyl alcohols is an attractive method for the formation of five- and six-membered cyclic enol ethers, the latter of which may be viewed as glycal donors for polysaccharide synthesis (Scheme 9.7) [14].

Tungsten- and molybdenum-catalyzed methods involving vinylidene intermediates have been described for this transformation (Chapter 5). The use of rhodium provides some advantages in terms of catalyst turnover and selectivity. The catalyst formed *in situ* from a RhCl source and a fluorinated triarylphosphine promotes the cyclization of a variety of alkynols (Table 9.6).

The same catalyst mixture functions in several other vinylidene-mediated processes. Good yields were obtained in the present case using low catalyst loadings but high ligand:metal ratios. Excess ligand was found to suppress undesired alkyne dimerization. Compared to other catalysts, Trost's Rh(I)-system is especially suitable for use in 6-endo-cyclizations (entries 3 and 4). Even heteroatom-substituted substrates, including those capable of undergoing 5-endo-ring closure (entry 3), were cyclized to dihydropyrans in good yield.

Alkynyl anilines are simple and convenient starting materials for the preparation of indoles. Unprotected, unfunctionalized 2-ethynyl aniline can be cycloisomerized to indole via molybdenum vinylidene-mediated catalysis [15]. Unlike $(Et_3N)Mo(CO)_5$,

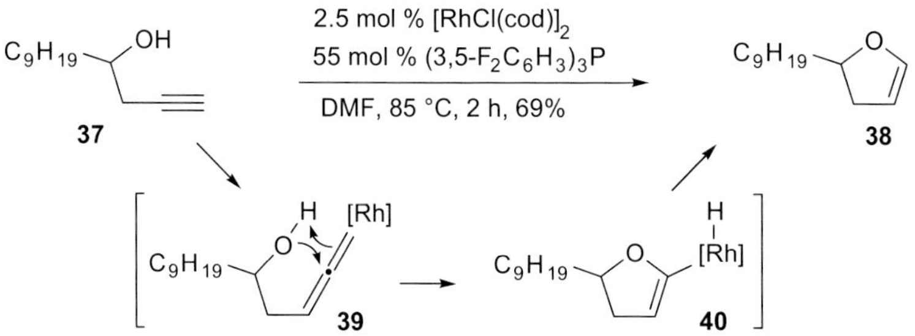

Scheme 9.7 Cyclization of homopropargylic alcohols.

Table 9.6 Rh(I)-catalyzed hydroalkoxylation.

Entry	Substrate	Product	Yield (%)
1	C_9H_{19}, H_3C-, OH, alkyne	C_9H_{19}, H_3C- dihydrofuran	62
2	Ar-CH(OH)-CH_2-C≡CH; Ar = 3,4-(MeO)$_2$C$_6$H$_3$	Ar-dihydrofuran	74
3[a]	BocHN-CH(OBn)-CH$_2$-CH(OH)-CH$_2$-C≡CH	BocHN-OBn dihydropyran	67
4	NHTs substrate with OH and alkyne	NHTs dihydropyran	52

[a]Conditons: 7.5 mol % RhCl[(4-FC$_6$H$_4$)$_3$P]$_3$, 45 mol %(4-FC$_6$H$_4$)$_3$P.

though, Rh-catalysts need not be activated by photoirradiation. Furthermore, Trost's catalyst system tolerates a wide variety of functional groups, providing substituted indoles in good yields, even at low catalyst loadings (Table 9.7) [16]. Electron-withdrawing substituents promote the desired reaction and enable the use of other nucleophiles. For example, a 2-ethynylphenol was cycloisomerized to the corresponding benzofuran (entry 6). As expected for a vinylidene-mediated transformation, internal alkynes did not participate in hydroamination.

Table 9.7 Rh(I)-catalyzed indole synthesis *via* hydroamination.

Entry[a]	Substrate	Product		Yield (%)
1	R-C$_6$H$_3$(NH$_2$)(C≡CH)	R-indole (NH)	R=Me	72
2			R=Cl	88
3			R=CN	89
4	Cl-C$_6$H$_3$(NHR)(C≡CH)	Cl-indole (NR)	R=Bn	73
5			R=PMB	74
6	OHC-C$_6$H$_2$(OMe)(OH)(C≡CH)	OHC-benzofuran-OMe		52

[a]1 mol % [RhCl(cod)]$_2$, 4 mol % (4-FC$_6$H$_4$)$_3$P, DMF, 85 °C, 2 h

Hydroboration and hydrosilylation reactions of alkynes, when they do not proceed spontaneously, can be catalyzed by numerous transition metals [17]. Metal vinylidene-mediated processes uniquely provide (Z)-alkenes via "trans-addition". In 2000, Miyaura and coworkers discovered that both Rh(I)- and Ir(I)-complexes supported by bulky electron-rich phosphine ligands catalyze the hydroboration of *tert*-butylacetylene to give alkenylboronate **41** with ≥95:5 Z/E selectivity (Equation 9.5) [18].

$$
\begin{array}{c}
\text{1.5 mol \% [MCl(cod)]}_2 \\
\text{6 mol \% }i\text{Pr}_3\text{P, 5 equiv Et}_3\text{N} \\
\hline
\text{C}_6\text{H}_{12}\text{, 25 °C, 2 h} \\
\text{M = Rh} \quad 69\% \quad Z/E\ 95{:}5 \\
\text{M = Ir} \quad 71\% \quad Z/E\ 97{:}3
\end{array}
\tag{9.5}
$$

(with 1.2 equiv alkyne + HBpin → (Z)-**41** + (E)-**41**)

This reaction occurs rapidly at room temperature using a small excess of alkyne and either pinacol- (HBpin) or catecholborane (HBcat). When an excess of borane was used, the Z/E ratio of the products was slowly eroded, eventually attaining a thermodynamic distribution of isomers. Equilibration presumably occurs via addition/elimination of excess Rh–H. Miyaura's method provides a useful synthetic complement to known cis-hydroboration methods. Under optimized conditions, good yields and high stereoselectivity (≥90:10) were achieved for a variety of alkenylboronates (Table 9.8). The best selectivities were generally obtained with the use of catecholborane and Et$_3$N as an additive. As in related reactions, the presence of base seems to suppress undesired reaction pathways.

According to a deuterium-labeling experiment, Miyuara's hydroboration is actually a 1,1-addition process with concomitant 1,2-H-shift, rather than a true trans-addition. The olefin geometry of the product boronate is presumably determined during an insertion of boron or hydrogen into the α-position of rhodium vinylidene intermediate **43** (Table 9.8).

$$
\text{C}_4\text{H}_9{-}{\equiv}{-}\text{H} + \text{Et}_3\text{SiH} \xrightarrow[\text{PhMe, 55 °C, 24 h, 91\%}]{\text{0.01 mol \% RhCl(PPh}_3)_3} \text{C}_4\text{H}_9\text{CH=CHSiEt}_3
\tag{9.6}
$$

1.02 equiv   **44** dr >95:5

Rh(I)/R$_3$P complexes also catalyze (Z)-selective hydrosilylation of alkynes (Equation 9.6) [19]. Although Miyaura's hydroboration and this reaction bear superficial similarities to one another, rhodium vinylidenes are not part of the generally accepted mechanism in the latter case.

Many Rh(I)-complexes are capable of dimerizing or oligomerizing alkynes to some degree. Seemingly small changes in reaction conditions can affect the stereo- and regio-selectivity of dimerization. The formation of (Z)-head-to-head dimers, however, can be indicative of metal vinylidene intermediates. This correlation was observed for Ru(II)-catalyzed dimerizations (Chapter 10) and also holds true for the Rh(I)- and Ir(I)-catalyzed processes described herein.

The group of Lin and coworkers has studied the dimerization of ethynylarenes through Rh(III) vinylidene-mediated catalysis (Scheme 9.8) [20]. Active Rh(III)-

Table 9.8 Rh(I)-Catalyzed trans-hydroboration.

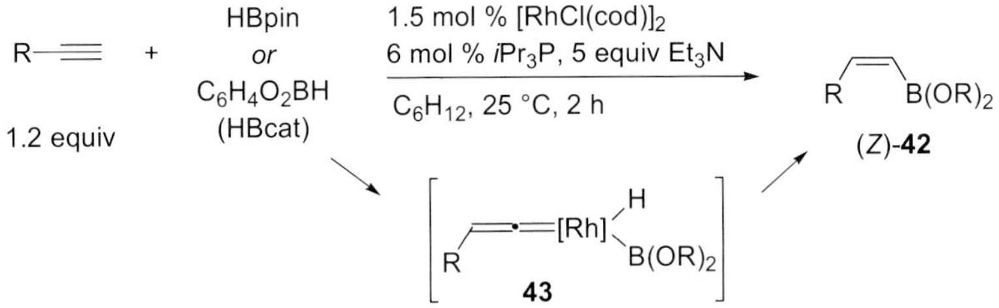

$$R\!-\!\!\equiv\ +\ \begin{array}{c}\text{HBpin}\\ \text{or}\\ \text{C}_6\text{H}_4\text{O}_2\text{BH}\\ \text{(HBcat)}\end{array}\ \xrightarrow[\text{C}_6\text{H}_{12},\ 25\ ^\circ\text{C},\ 2\ \text{h}]{\text{1.5 mol \% [RhCl(cod)]}_2\ \ 6\ \text{mol \%}\ i\text{Pr}_3\text{P, 5 equiv Et}_3\text{N}}\ \ R\diagup\!\!=\!\!\diagdown B(OR)_2$$

1.2 equiv, (Z)-42

$$\left[\begin{array}{c}R\diagup\!\!=\!\!=[Rh]\diagdown_{B(OR)_2}^{H}\\ 43\end{array}\right]$$

Entry	Alkyne	Borane	Yield (%)
1	C$_8$H$_{17}$—≡	HBpin	81
		HBcat	79
2	TBSO\—≡	HBpin	59
		HBcat	70
3	TMS—≡	HBpin	59
		HBcat	70
4	Ph—≡	HBpin	67
		HBcat	60

catalysts were formed *in situ* by the action of MeI on a Rh(I)-precatalyst. Under these conditions, some alkyne methylation was observed, but, remarkably, by careful control of the stoichiometry, arylpropyne byproducts could be minimized (<5%). Two distinct alkyne dimerization modes were observed, depending on the solvent. In THF, (*E*)-head-to-head dimers were obtained in good-to-excellent yield, while in MeOH, (*Z*)-head-to-head dimers were formed selectively. The latter reaction takes place at room temperature with quite low catalyst loading. Dimers of electron-poor ethynylarenes were furnished with the greatest efficiency and selectivity (Table 9.9).

In most cases, none of the following were observed: (*E*)-head-to-head dimers, head-to-tail dimers, or trimers. High regio- and stereo-selectivities set this method apart from Rh(I)-catalyzed dimerization methods.

Based on a series of stoichiometric organometallic reactions, a catalytic cycle for (*Z*)-head-to-head dimerization was proposed (Scheme 9.9).

When, and only when, all reaction constituents are combined, a vinylidene complex (**49**) forms. The observed enyne product is generated by coupling of the σ-acetylide of **49** with its vinylidene moiety, followed by protodemetallation. The integrity of kinetically controlled product ratios is safeguarded by the avoidance of Rh–H intermediates. (*Z*)-Selectivity was rationalized on the basis of minimizing

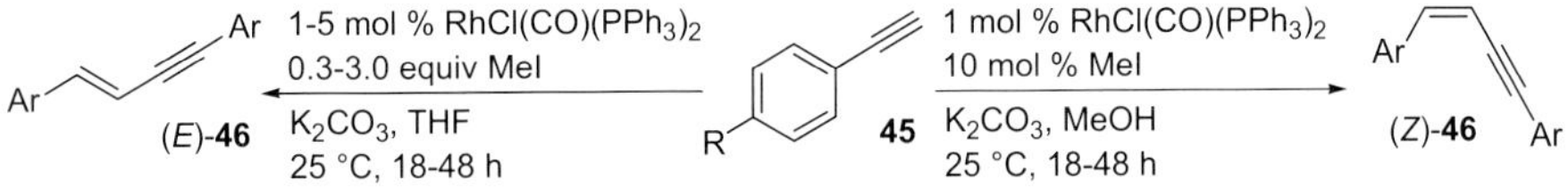

$$\text{Ar}\diagdown\!\!=\!\!\diagup^{\text{Ar}}\ \xleftarrow[\text{K}_2\text{CO}_3,\ \text{THF};\ 25\ ^\circ\text{C, 18-48 h}]{\text{1-5 mol \% RhCl(CO)(PPh}_3)_2\ \ 0.3\text{-}3.0\ \text{equiv MeI}}\ R\!-\!\!\!\!\bigcirc\!\!\!\!-\!\!\equiv\ \xrightarrow[\text{K}_2\text{CO}_3,\ \text{MeOH};\ 25\ ^\circ\text{C, 18-48 h}]{\text{1 mol \% RhCl(CO)(PPh}_3)_2\ \ 10\ \text{mol \% MeI}}\ \text{Ar}\diagup\!\!=\!\!\diagdown_\text{Ar}$$

(*E*)-**46**   **45**   (*Z*)-**46**

Scheme 9.8 Solvent-dependent selectivity in Rh(III)-catalyzed Dimerization of terminal alkynes.

Table 9.9 Rh(III)-catalyzed dimerization of 4-substituted phenylacetylenes.

Entry[a]	R	Yield (Z)-46 in MeOH (%)	Yield (E)-46 in THF (%)
1	H	63[b]	75
2	CN	82	54
3	CF_3	91	90
4	F	25	39
5	Br	90	88

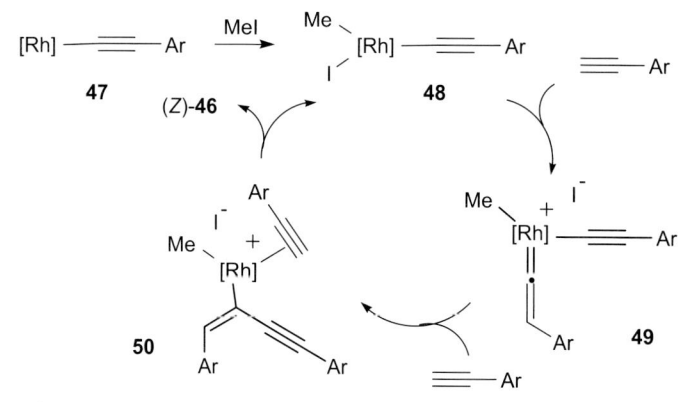

[a]Conditions: See Scheme 9.8.
[b]E/Z 13:82.

Scheme 9.9 Proposed catalytic cycle for dimerization of alkynes in MeOH.

ligand–ligand steric interactions during insertion (49 → 50). Overall, this work is significant as a unique example of well-characterized Rh(III)-vinylidene-mediated catalysis. The origin of the profound solvent effect on selectivity remains to be determined.

Miyaura and coworkers have disclosed an iridium-catalyzed alkyne dimerization reaction whose stereochemical outcome is determined by ligand structure (Equation 9.7) [21].

R_3P	Time	Yield (%)	E:Z
Ph_3P	6 h	83	94:6
nPr_3P	24 h	70	8:91

(9.7)

The catalyst prepared from PPh_3 and a simple Ir(I) salt, [IrCl(cod)]$_2$, promotes (E)-selective head-to-head dimerization in good yield, while the combination of [IrCl(cod)]$_2$ with an electron-rich phosphine (PPr_3) affords (Z)-enynes with reduced efficiency, but comparable selectivity. The results of [IrCl(cod)]$_2$/PPr_3-catalyzed dimerization reactions carried out according to Scheme 9.10 are summarized in Table 9.10.

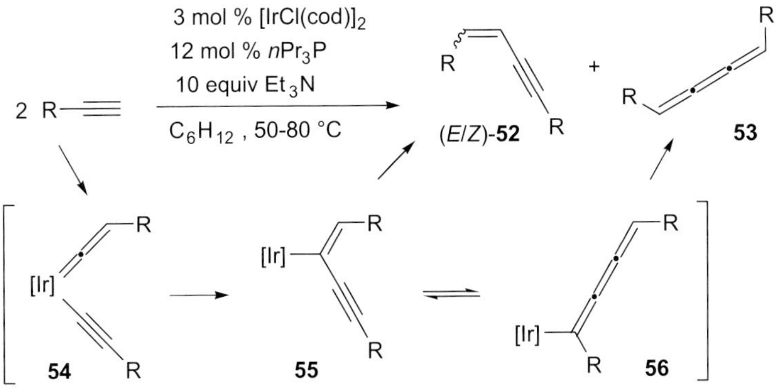

Scheme 9.10 Ir(I)-catalyzed alkyne dimerization leading to butatriene derivatives.

Table 9.10 Ir(I)-catalyzed alkyne dimerization.

Entry	R	Time (h)	Yield (%)	(E)-52:(Z)-52:53
1	Ph$_2$MeSi	20	62	11:89:0
2	4-MeC$_6$H$_4$	24	67	24:76:0
3	2-MeC$_6$H$_4$	24	78	13:87:0
4	Me$_2$(TBSO)C	36	71	0:22:78
5	tBu	18	74	0:7:93

Reaction outcomes are hardly affected by the electronic character of the alkyne substrates. On the other hand, large alkyne substituents favor (Z)-enyne formation, up to a certain threshold (entries 1–3). The most sterically encumbered alkynes are converted into (Z)-butatrienes (entries 4 and 5).

The need for a base additive in this reaction implies the intermediacy of acetylide complexes (Scheme 9.10). As in the Rh(III)-catalyzed reaction, vinylidene acetylide 54 undergoes α-insertion to give the vinyl-iridium intermediate 55. A [1,3]-propargyl/allenyl metallatropic shift can give rise to the cumulene intermediate 56. The individual steps of Miyaura's proposed mechanism have been established in stoichiometric experiments. In the case of (E)-selective head-to-head dimerization, vinylidene intermediates are not invoked. The authors argue that electron-rich phosphine ligands affect stereoselectivity by favoring alkyne C–H oxidative addition, a step often involved in vinylidene formation.

One more example of iridium vinylidene-mediated alkyne dimerization catalysis emerged from the study of diamidonaphthalene-bridged diiridium complexes (e.g., 58) [22]. Complexes of this type readily react with internal trimethylsilylalkynes to give vinylidene complexes via 1,2-silyl migration. Complex 58 catalyzes stereoselective head-to-head dimerization and trimerization of terminal alkynes to give mixtures of (Z)-butenyne (52) and (Z,E)-hexa-3,5-diene-1-yne (57) products (Table 9.11). While chemoselectivity was variable depending on temperature and the alkyne substituent, regio- and stereo-selectivity were very high.

Table 9.11 Diiridium complex **58**-catalyzed alkyne dimerization.

Entry	R	Temp (°C)	Conversion (%)	52:57
1	Ph	50	90	45:55
2	Ph	90	100	27:73
3	tBu	60	100	46:54
4	TMS	40	100	100:0
5	TES	40	100	100:0

The authors presume that transient vinylidene intermediates are involved in the observed reactions. This hypothesis is consistent with the observed (Z)-selectivity of dimerization and the unusual facility with which complexes of type **58** form isolable vinylidenes. Intermetallic cooperation in bimetallic complexes has been recognized to facilitate vinylidene-mediated processes [23]; however, the operative mechanism in the present case remains unknown.

9.2.4
Multi-Component Coupling

Using established principles of late-transition metal catalysis, several research groups have engineered multi-component coupling reactions from the basis set of known Group 9 metal vinylidene-mediated reactions. In 2004, Jun and coworkers described a new method for the synthesis of enones via rhodium vinylidene-mediated hydrative dimerization of alkynes (Table 9.12) [24].

Optimized reaction conditions call for the use of Wilkinson's catalyst in conjunction with the organocatalyst 2-amino-3-picoline (**60**) and a Brønsted acid. Jun and coworkers have demonstrated the effectiveness of this catalyst mixture for a number of reactions including hydroacylation and C–H bond functionalization [25]. Whereas, in most cases, the Lewis basic pyridyl nitrogen of the cocatalyst acts to direct the insertion of rhodium into a bond of interest, in this case the opposite is true – the pyridyl nitrogen directs the attack of cocatalyst onto an organorhodium species (Scheme 9.11). Hydroamination of the vinylidene complex **61** by 3-amino-2-picoline gives the chelated amino-carbene complex **62**, which is in equilibrium with σ-bound hydrido-rhodium tautomers **63** and **64**.

Jun's proposed mechanism was probed by a deuterium labeling experiment using N-methyl co-catalyst **66** (Equation 9.8). Hydridoimidoyl complexes (e.g., **64**), which are implicated by Jun in C–C bond formation, cannot form when secondary amine **66** is used as a cocatalyst. Instead, C–H reductive elimination from a complex

Table 9.12 Rh(I)-catalyzed hydrative dimerization of alkynes.

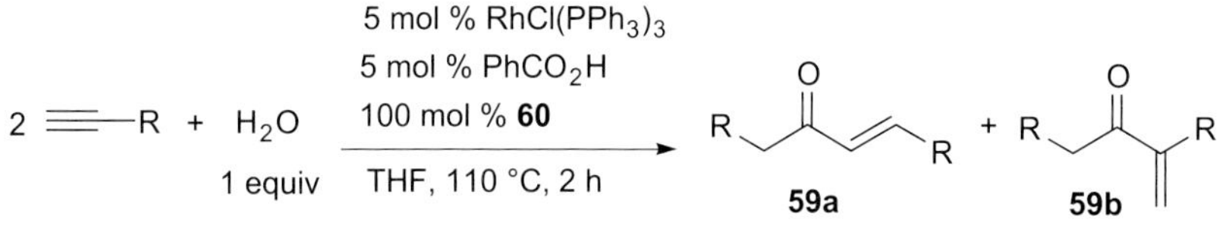

$$2\ \equiv\!\!-R\ +\ H_2O \xrightarrow[\text{THF, 110 °C, 2 h}]{\begin{array}{c}\text{5 mol \% RhCl(PPh}_3)_3\\ \text{5 mol \% PhCO}_2\text{H}\\ \text{100 mol \% }\mathbf{60}\end{array}} R\!\!-\!\!\overset{O}{\underset{}{\|}}\!\!-\!\!\diagup\!\!\diagdown\!\!\!_R\ +\ R\!\!-\!\!\overset{O}{\underset{}{\|}}\!\!\diagdown\!\!\overset{R}{\underset{\|}{}}$$

1 equiv   **59a**   **59b**

Entry	Alkyne	Yield (%)	59a : 59b
1	≡—C$_6$H$_{13}$	82	78 : 22
2	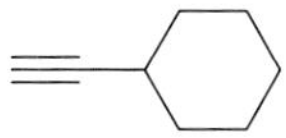 ≡—cyclohexyl	66	75 : 25
3	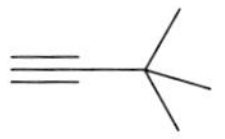 ≡—C(CH$_3$)$_3$	59	0 : 100

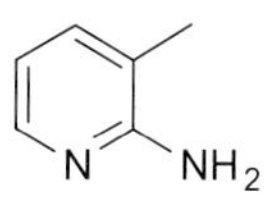

60: 3-methyl-2-aminopyridine

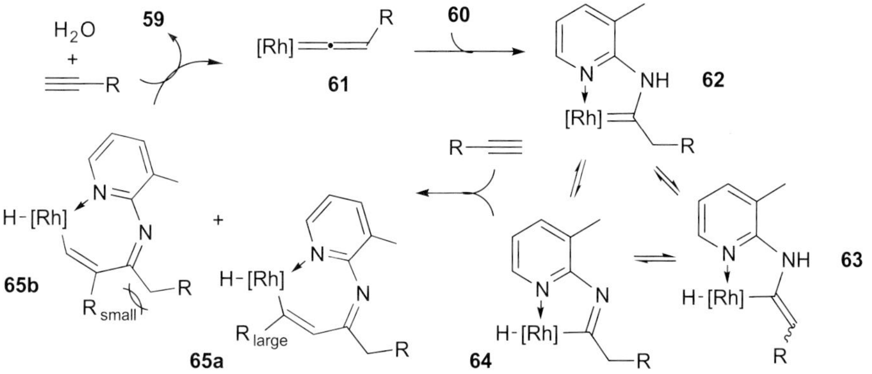

Scheme 9.11 Proposed mechanism for hydrative dimerization of terminal alkynes.

analogous to intermediate **64** leads to the formation of enamine **68**. The near-statistical distribution of deuterium between the vinylic positions of **68** is consistent with the proposed addition/elimination mechanism.

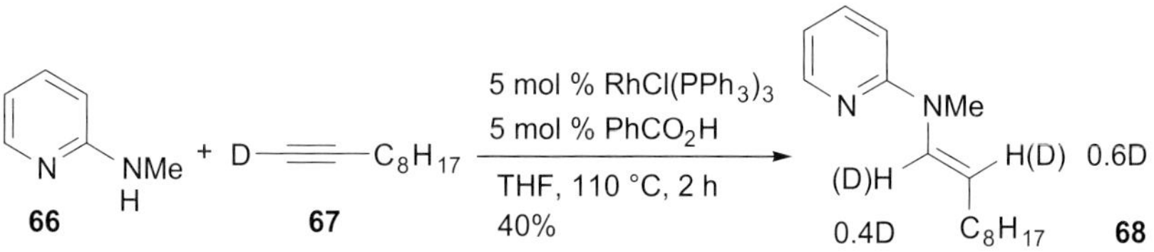

$$\underset{\mathbf{66}}{\text{PyNMeH}} + \text{D}\!\!-\!\!\equiv\!\!-\!\!\underset{\mathbf{67}}{C_8H_{17}} \xrightarrow[\substack{\text{THF, 110 °C, 2 h}\\40\%}]{\substack{\text{5 mol \% RhCl(PPh}_3)_3\\ \text{5 mol \% PhCO}_2\text{H}}} \underset{\mathbf{68}}{\text{PyNMe-C(=CH(D)0.4D)-C}_8\text{H}_{17}\ \text{H(D) 0.6D}}$$

(9.8)

Both head-to-head and head-to-tail isomers are formed from Jun's hydrative dimerization reaction. Regioselectivity is controlled by steric interactions during migratory insertion (**64** → **65**). In the final steps of the proposed catalytic cycle,

Table 9.13 Rh(I)-Catalyzed cyclodimerization of alkynes with allylamine.

$$2 \; \equiv\!\!-R \; + \; \overset{}{\diagup\!\!\diagdown}\!\!-NH_2 \xrightarrow[\text{THF, 70 °C, 30 h}]{\text{5 mol \% RhCl(PPh}_3)_3 \;\; \text{6 mol \% NH}_4BF_4} \; \underset{69}{\text{pyrroline}}$$

2.0 equiv

Entry	Alkyne	Yield (%)
1	≡—CH₂CH₂C(O)OMe	70
2	≡—CH₂CH₂—O—(tetrahydropyranyl)	64
3	≡—CH₂CH₂CH₂—OTBS	66
4	≡—CH₂CH₂CH₂—CN	71
5	≡—cyclohexyl	58
6	≡—phenyl	61

reductive elimination and imine hydrolysis lead to enone formation and regeneration of 3-amino-2-picoline (**60**). Unexpectedly, substoichiometric loading of organocatalyst **60** did not prove practical, nor were reactions involving functionalized alkynes successful.

In 2006, Chatani and coworkers published a rhodium vinylidene-mediated alkyne cyclodimerization incorporating allylamine (Table 9.13) [26]. Although RhCl(PPh₃)₃ alone was a competent catalyst for the stereospecific formation of (*E*)-3-alkylidene-3,4-dihydro-2*H*-pyrroles (**69**), the addition of an ammonium salt helped to suppress undesired alkyne homocoupling. Under optimized conditions, a variety of functionalized alkynes underwent cyclodimerization with allyl- or crotylamine in good yield.

Chatani's proposed mechanism bears some similarity to that of Jun's reaction (Scheme 9.12). They both begin with hydroamination of the C=C π-bond of a rhodium vinylidene. The resultant aminocarbene complexes (**71** and **62**) are each in equilibrium with two tautomers. The conversion of **71** to imidoyl-alkyne complex **74** involves an intramolecular olefin hydroalkynylation. Intramolecular *syn*-carbometallation of intermediate **74** is thought to be responsible for ring closure and the apparent stereospecificity of the overall reaction. In the light of the complexity of Chatani and coworkers' mechanism, the levels of chemoselectivity that they achieved should be considered remarkable. For example, 5-endo-cyclization of intermediate **72** was not observed, though it has been for more stabilized rhodium aminocarbenes bearing pendant olefins [27].

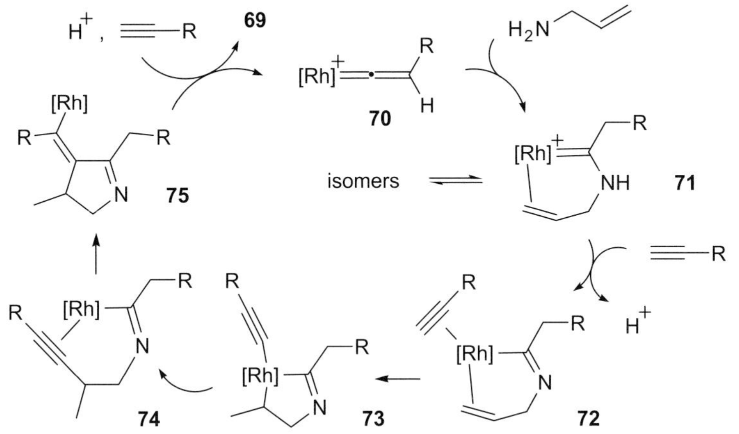

Scheme 9.12 Rh(I)-catalyzed coupling of terminal alkynes with allylamine – Catalytic cycle.

In 2006, Lee and coworkers reported a multi-component cyclization reaction combining aryl or vinyl boronic acids with 1,5-enynes (Scheme 9.13) [28].

The concept behind this reaction can be traced to the work of Werner and coworkers who demonstrated the clean α-insertion of phenyl and other organic groups into rhodium vinylidenes to give vinyl-rhodium complexes (Scheme 9.14) [29].

The addition of Rh-Csp^2 σ-bonds to enones is a well-studied process. Lee's reaction is predicated on the idea that rhodium-catalyzed conjugate addition of boronic acids to enones can be "interrupted" by 1,1-insertion into an alkyne. Thanks to the high reactivity of rhodium toward alkynes and the effects of tethering, a partly intramo-

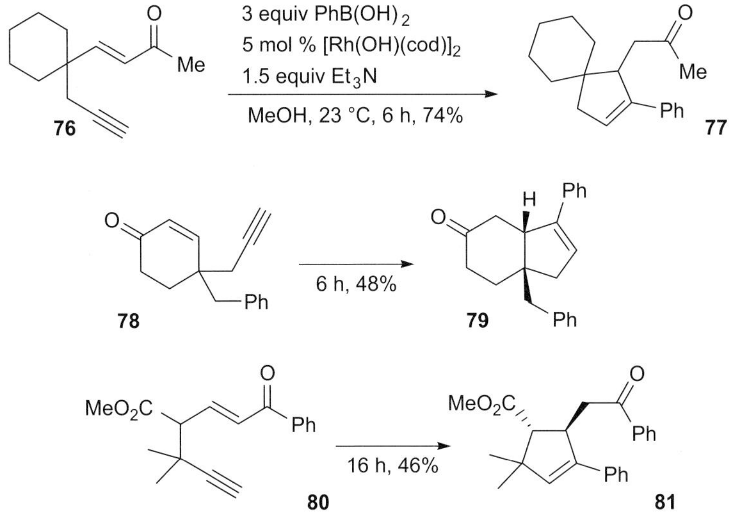

Scheme 9.13 Rh(I)-catalyzed arylative cyclization of 1,5-enynes.

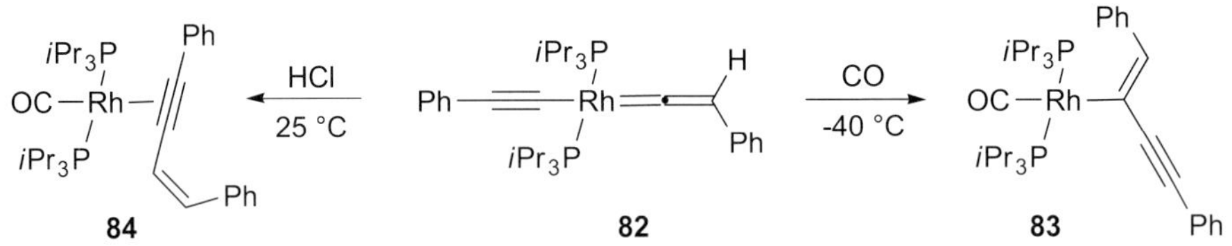

Scheme 9.14 Oxidation state-dependent stereoselectivity in α-insertions into a Rh vinylidene.

lecular three-component coupling was achieved in preference to two-component coupling.

Alkynyl enones with varying degrees and patterns of substitution were cyclized in moderate-to-good yields under a mild set of reaction conditions (Scheme 9.13). A hydroxylic solvent (MeOH) and Rh-OH were employed to facilitate the transmetallation of boronic acids, and Et_3N was added to favor the formation of vinylidenes. Both (E)- and (Z)-alkenyl boronic acids and aryl boronic acids, some bearing reactive functional groups, were efficiently coupled (Table 9.14). Relative 1,2-stereocontrol was demonstrated (80 → 81), but cyclization was limited to the formation of five-membered rings. Substrates entirely lacking substitution along the backbone of the alkynyl enone were not described, implying that the Thorpe–Ingold effect contributes to some degree to the success of reported substrates.

The proposed reaction mechanism is shown in Scheme 9.15. Starting from the phenyl-rhodium complex 87, alkyne rearrangement is expected to furnish the phenyl-vinylidene complex 88. Migration of a phenyl ligand onto the vinylidene moiety of 88 must occur such that the vinyl Rh–C bond and the enone tether of the resultant complex (89) attain a cis-relationship to one another. Intramolecular conjugate

Table 9.14 Rh(I)-Catalyzed arylative and alkenylative cyclization of enynes.

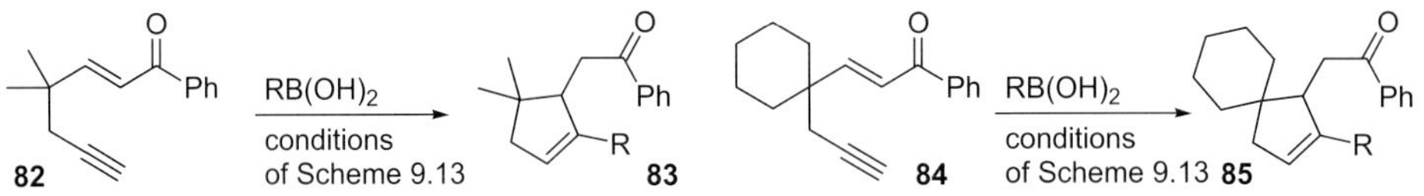

Entry	Enone	RB(OH)$_2$		Time (h)	Yield (%)
1	82	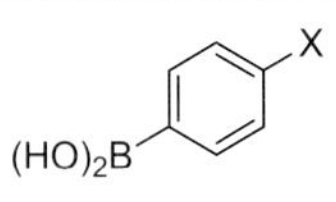 (HO)$_2$B-Ar-X	X=OMe	1.5	67
2	82		X=OH	0.5	62
3	82		X=CHO	8	49
4	82	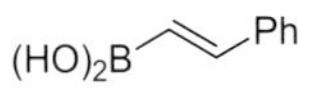 (HO)$_2$B-CH=CH-Ph		0.2	69
5	84	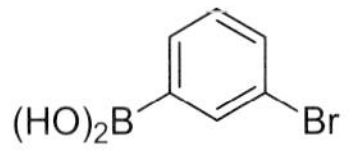 (HO)$_2$B-Ar-Br		2	63
6	84	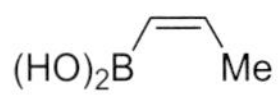 (HO)$_2$B-CH=CH-Me		0.2	73

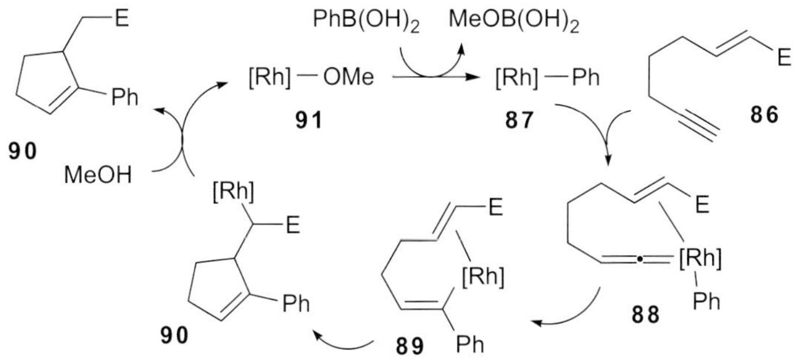

Scheme 9.15 Proposed catalytic cycle for arylative and alkenylative cyclization of enynes.

addition can thereby proceed to provide cyclized products after protodemetallation. A deuterium labeling study confirmed the expected 1,2-H-migration.

Notably, the proposed stereoselectivity of α-insertion seems to run opposite to that observed for some related processes (*vide supra*). Such systems differ from the present system, though, because they involve an octahedral Rh(III) center, whereas intermediate **88** is a Rh(I)-square-planar complex. In Werner's original stoichiometric studies, the same mechanistic dichotomy was evident (Scheme 9.14) – square planar Rh(I)-complexes underwent cisoid insertion in the absence of other factors and transoid addition in the presence of an oxidant (HX). In intermediate **88**, coordination of the alkene moiety to the rhodium center may also play a role in directing insertion.

9.3
Rhodium Alkenylidenes in Catalysis

The metal vinylidene intermediates discussed elsewhere in this chapter are limited to a single carbon-substituent on account of the 1,2-migration process by which they form from terminal alkynes. Alkenylidenes—vinylidenes bearing two carbon-substituents—are formed by nucleophilic addition of the β-carbon of a metal acetylide to an electrophile (Scheme 9.16) [30].

When the electrophile is an alkyl halide, a C–C σ-bond is forged; thus, alkenylidene formation is irreversible. Vinylidene formation by 1,2-migration, on the other hand, is generally reversible. Because of this contrast, alkenylidenes can offer access to new catalytic reaction manifolds, in addition to unique molecular architecture.

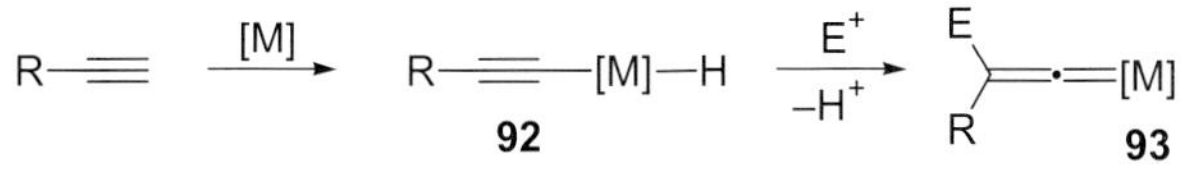

Scheme 9.16 Formation of metal alkenylidenes.

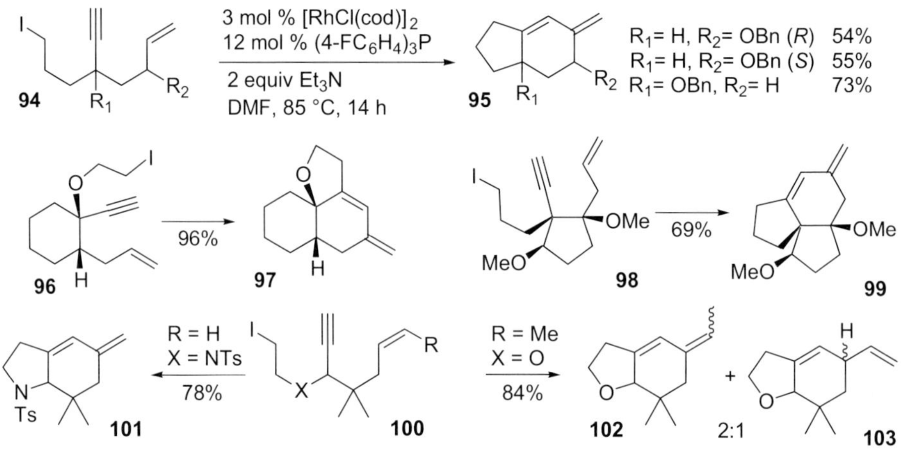

Scheme 9.17 Rh(I)-catalyzed double cyclization of iodoenynes.

The Lee group originated rhodium alkenylidene-mediated catalysis by combining acetylide/alkenylidene interconversion with known metal vinylidene functionalization reactions [31]. Thus, the first all-intramolecular three-component coupling between alkyl iodides, alkynes, and olefins was realized (Scheme 9.17). Prior to their work, such tandem reaction sequences required several distinct chemical operations. The optimized reaction conditions are identical to those of their original two-component cycloisomerization of enynes (see Section 9.2.2, Equation 9.1) except for the addition of an external base (Et$_3$N). Various substituted [4.3.0]-bicyclononene derivatives were synthesized under mild conditions. Oxacycles and azacycles were also formed. The use of DMF as a solvent proved essential; reactions in THF afforded only enyne cycloisomerization products, leaving the alkyl iodide moiety intact.

Double cyclization of iodoenynes is proposed to occur through a Rh(I)-acetylide intermediate **106**, which is in equilibrium with vinylidene **105** (Scheme 9.18). Organic base deprotonates the metal center in the course of nucleophilic displacement and removes HI from the reaction medium. Once alkenylidene complex **107** is generated, it undergoes [2 + 2]-cycloaddition and subsequent breakdown to release cycloisomerized product **110** in the same fashion as that discussed previously (Scheme 9.4). Deuterium labeling studies support this mechanism.

The reaction of iodoenyne **100** serves as a demonstration of the unique reactivity of alkenylidenes (Scheme 9.17) [32]. In simple rhodium vinylidene-mediated enyne cycloisomerization, substrates bearing internal olefins are not reactive; only decomposition and dimerization byproducts are observed. In contrast, substrate **100** (R = Me, X = O) undergoes three-component coupling in high yield. This result may be rationalized by supposing that alkenylidene intermediate **107**, because it is formed irreversibly, represents a catalyst resting state. Relatively few reaction pathways may be available to intermediate **107**, allowing it to persist long enough to engage a sluggish substrate, such as an internal olefin. Unselective β-hydride

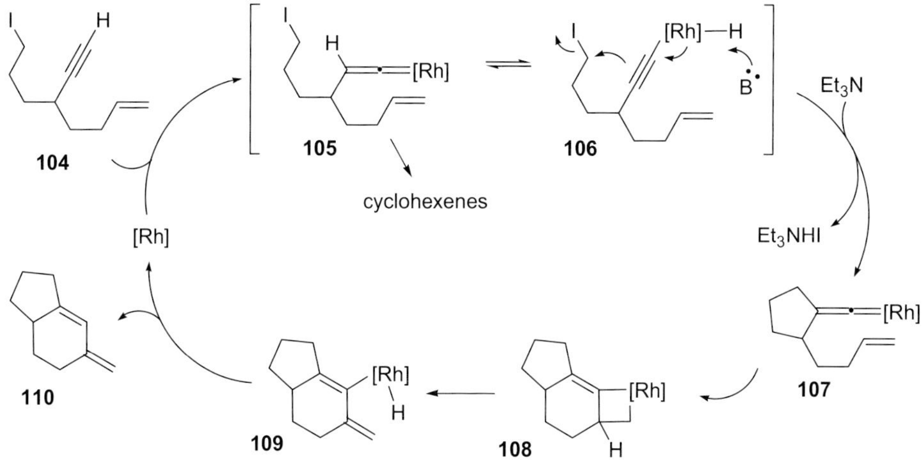

Scheme 9.18 Rh(I)-catalyzed double cyclization of iodoenynes – Proposed catalytic cycle.

elimination from the resulting rhodacyclobutane intermediate (see **108**, Scheme 9.18) produces a mixture of product regioisomers (**102**, **103**).

Lee and coworkers went on to show that the concept of alkenylidene formation and functionalization by a single catalyst can be applied to other transformations. Under conditions similar to those reported by Trost and coworkers for vinylidene-mediated catalytic intramolecular hydroalkoxylation (see Section 9.2.3), alcohol **111** was transformed into a mixture of enol ethers with moderate selectivity for three-component coupling (Equation 9.9).

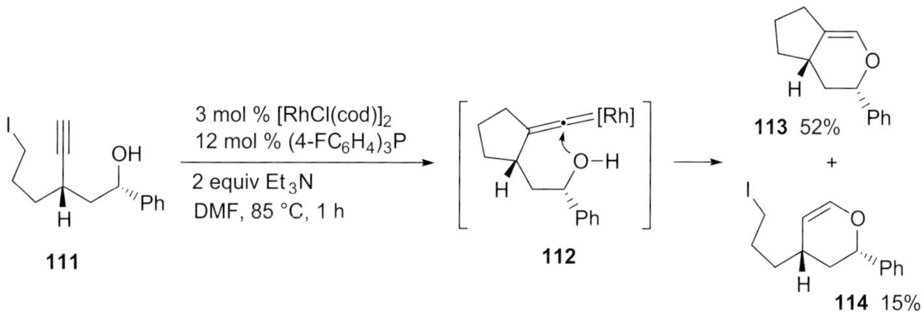

(9.9)

Like alcohols, arenes can attack the electrophilic α-position of metal vinylidenes (see Section 9.4.6). Substrate **115** was transformed into tetracycle **117** in high yield, presumably via 6π-electrocyclization and subsequent rearomatization (Equation 9.10). To date, no intermolecular examples of metal alkenylidene-mediated catalysis have come to light. The extension of Lee's alkylative approach to catalysis by other metals may prove fruitful in this regard.

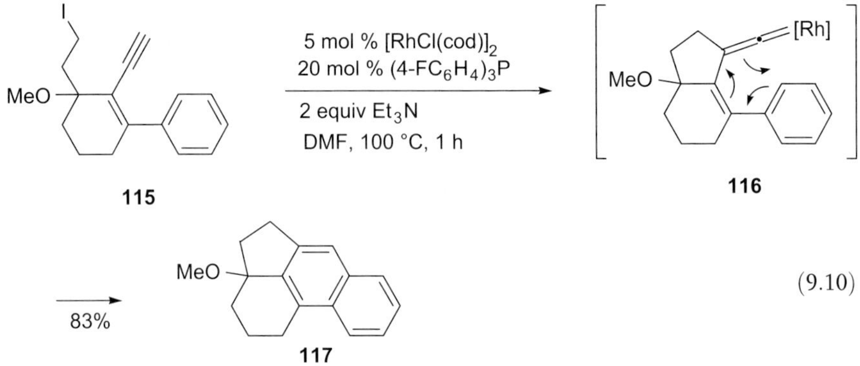

302 | 9 Rhodium and Group 9–11 Metal Vinylidenes in Catalysis

(9.10)

9.4
Group 10 and 11 Metal Vinylidenes in Catalysis

9.4.1
Introduction

The metals of Groups 10 and 11 are significantly less well-studied than those of Groups 8 and 9 in vinylidene-mediated catalysis. Relevant stoichiometric organometallic chemistry has been described [33]. Why, then, is there a general paucity of catalysis? Each of the aforementioned metals has a preferred mode of reactivity with alkynes. Palladium typically catalyzes cross-coupling and cis-1,2-addition reactions with alkynes. Platinum and gold are best known for activating alkynes as electrophiles in cation- or carbene-mediated processes. Finally, copper acetylide chemistry has been studied for more than a century. To elicit vinylidene-type reactivity from the metals of Groups 10 and 11, special conditions are required under which other, more typical, reaction manifolds are inaccessible. The discoveries described herein were mostly serendipitous. Note that late-transition metal vinylidenes are invoked as intermediates in some reactions that occur on heterogeneous catalyst surfaces [34]. Such species are often difficult to characterize and their study is beyond the scope of this work.

9.4.2
Nickel Vinylidenes in Catalysis

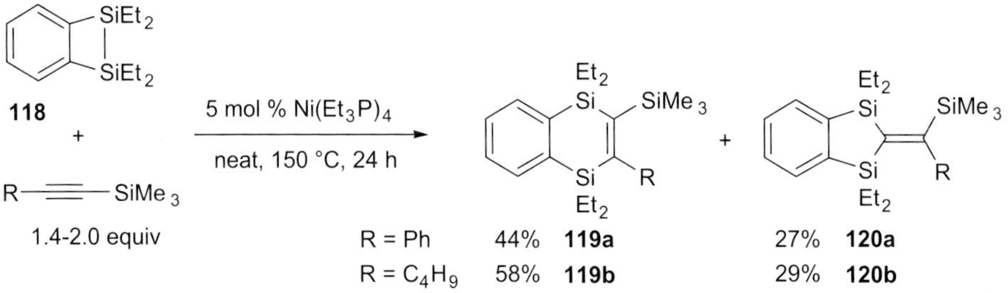

(9.11)

Ishikawa and coworkers have studied the unique reactivity of strained cyclic disilanes (Equation 9.11) [35]. Transition metals, especially those of Group 10, readily insert into the Si–Si bond of disilacyclobutene **118** and can catalyze the addition of that bond across a variety of unsaturated acceptors. In the case of Ni(0)-catalyzed reactions of **118** with trimethylsilyl alkynes, insertion was found to occur both in a 1,2- and in a 1,1-fashion. The latter of these pathways implies a 1,2-silyl-migration, presumably occurring at the metal center. A nickel vinylidene intermediate was therefore proposed, though efforts to prove its existence were inconclusive. Similar vinylidene intermediates have been proposed by Ishikawa and coworkers to account for migrations observed in related palladium- and platinum-catalyzed reactions [36]. To date, no computational evidence has been presented to support these reaction mechanisms. Moreover, monomeric nickel vinylidene complexes have not been characterized outside of matrix isolation [37]. Given the unique properties of disilacyclobutenes, the significance of this work in the context of metal vinylidene-mediated catalysis remains to be established.

9.4.3
Palladium Vinylidenes in Catalysis

The sole report of homogeneous palladium catalysis invoking vinylidene intermediates comes from the laboratory of Buono and coworkers [38]. They discovered a reaction unique to air-stable palladium catalysts **122**, which form from the self-assembly of secondary phosphine oxides with Pd(II) (Equation 9.12).

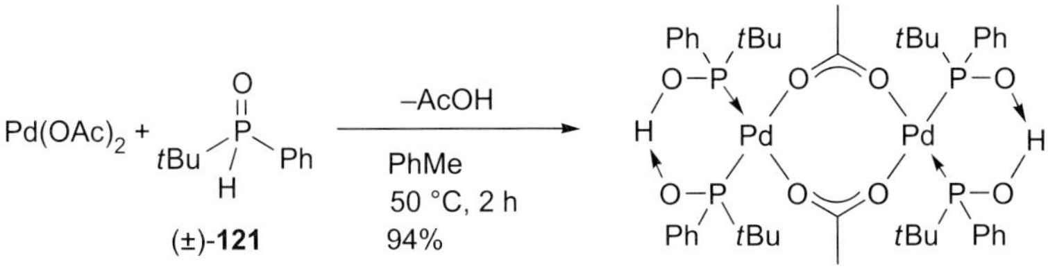

$$Pd(OAc)_2 + \underset{(\pm)\text{-}121}{\overset{O}{\underset{tBu\diagup\,\,\,\backslash Ph}{\overset{\|}{P}}}} \xrightarrow[\substack{PhMe \\ 50\,°C,\,2\,h \\ 94\%}]{-AcOH} (\pm)\text{-}tBu\text{-}\mathbf{122} \quad (9.12)$$

Under mild conditions, norbornene and phenylacetylene were selectively coupled to give alkylidene cyclopropane **123** (Scheme 9.19).

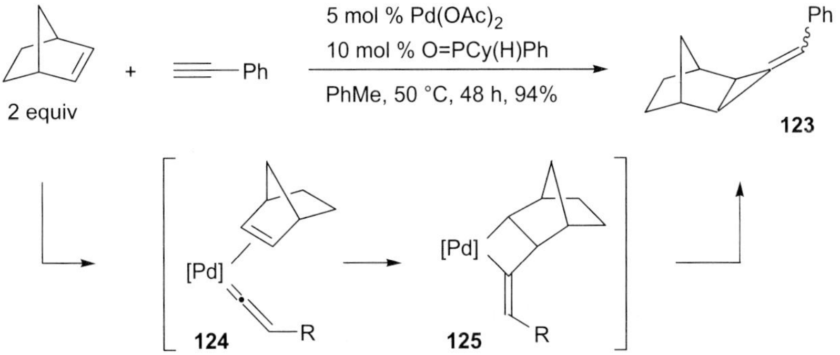

Scheme 9.19 Palladium-catalyzed alkylidenecyclopropane formation.

Table 9.15 Pd(II)-catalyzed cyclopropanation of strained olefins.

Entry[a]	Alkene	≡—R		Product	Temp (°C)	Time (h)	Yield (%)
1	norbornene deriv.	X=H	CH$_2$SO$_2$Ph	cyclopropanated product	25	48	66
2		X=H	CH$_2$OAc		25	50	73
3		X=H	C(CH$_3$)$_2$OH		50	36	75
4		X=CH$_2$OAc	Ph		50	36	58
5	benzonorbornadiene	X=CH$_2$	Ph		25	60	84
6		X=O	Ph		50	48	56

[a]Conditions: See Scheme 9.19.

Using in situ-generated catalyst Cy-**122**, no starting material self-condensation or other major side reactions were detected during cyclopropanation. Buono's reaction is tolerant of functionalized alkynes, but only bicyclo-[2.2.1]-heptenes and related strained olefins undergo cyclopropanation (Table 9.15). The successful use of substrates bearing allylic and propargylic acetates illustrates the atypical reactivity of catalysts **122**. No products arising from metal-alkylidene-mediated cyclopropanation, in entry 2, or palladium π-allyl formation, in entry 4, were observed.

Using optically pure **121**, available from resolution by HPLC, the authors demonstrated a rare example of asymmetric catalysis involving metal vinylidenes (Equation 9.13). The synthesis of **126**, enriched in a single olefin geometrical isomer, was achieved in 59% ee.

$$\text{norbornene} + \text{HC≡C-Ph} \xrightarrow[\text{PhMe, 50 °C, 60 h, 53\%}]{\substack{5 \text{ mol \% Pd(OAc)}_2 \\ 10 \text{ mol \% (−)-}\mathbf{121}}} \mathbf{126} \quad 59\% \text{ ee} \tag{9.13}$$

The authors' proposed mechanism, outlined in Scheme 9.19, was tested using a deuterium-labeling experiment. H-migration consistent with initial formation of a Pd-vinylidene was observed. The key intermediate of Buono's mechanism is a palladacyclobutane (**125**) resulting from [2 + 2]-cycloaddition. Direct C—C reductive elimination from intermediate **125** proceeds to give highly strained products (**123**), despite the apparent availability of a β hydride elimination pathway [39].

9.4.4
Platinum Vinylidenes in Catalysis

Yamamoto and coworkers unexpectedly discovered a vinylidene-mediated reaction during their studies of platinum-catalyzed enyne cycloisomerization (Scheme 9.20) [40].

Whereas, in the presence of PtBr$_2$, internal alkynes of type **127** were cleanly transformed to vinyl naphthalenes (**129**), similar terminal alkynes were converted primarily to indenes (**131**). The authors quickly discovered that the observed C—H insertion reaction is very substrate specific. Only tertiary benzylic C—H bonds could be

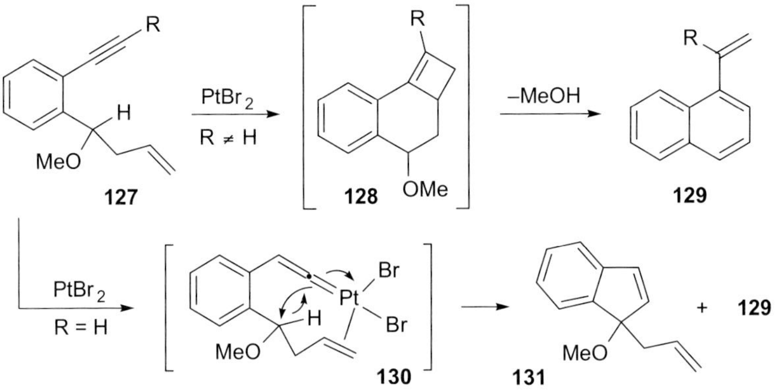

Scheme 9.20 Substitution-dependent modes of Pt-catalyzed enyne cycloisomerization.

activated and, curiously, only substrates bearing an allyl group at the benzylic position were active. Changing the tether length between the arene and olefin resulted in complete suppression of desired reactivity. Substrates varying in their substitution at other positions were screened (Table 9.16). Alteration of the test substrate **127** did not result in significantly improved yields, though excellent selectivity for C—H insertion over enyne cycloisomerization was achieved (entries 7 and 8).

The intermediacy of a platinum vinylidene in Yamamoto's reaction was supported by the results of isotopic labeling studies. DFT calculations were used to further probe the proposed reaction mechanism. In contrast to the prevailing model of alkyne/vinylidene interconversion for Rh(I)-catalysts, direct $C_\alpha \rightarrow C_\beta$ 1,2-H-migration is implicated in the formation of vinylidene **130**. Direct C—H insertion via a single

Table 9.16 PtBr$_2$-catalyzed formation of indenes.

Entry[a]	Substrate		Yield (%)	131:129
1	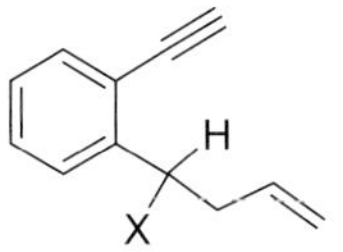	X=OMe	61	90:10
2		X=OBn	52	95:5
3[b]		X=Ph	40	ND
5	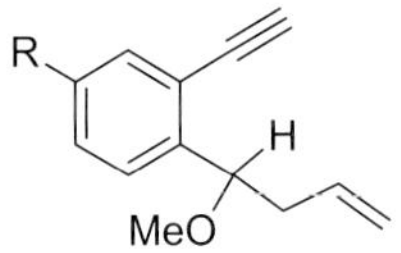	R=OMe	20	100:0
6		R=CF$_3$	36	59:41
7	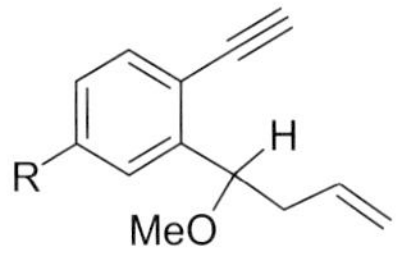	R=OMe	65	100:0
8		R=F	50	100:0

[a] 5 mol% PtBr$_2$, MeCN, 120 °C, 12–24 h.
[b] 10 mol% PtBr$_2$.

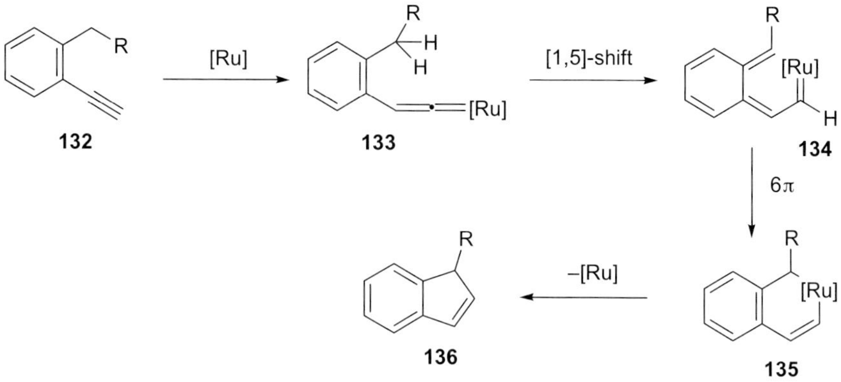

Scheme 9.21 Ru-catalyzed formation of indenes.

transition-state accounts for the conversion of **130** to the observed product (**131**). Near-perfect square planar geometry at platinum is maintained throughout the reaction, thanks to the orientation of the olefin moiety, which evidently acts as a directing group in a fashion typical for C−H functionalization reactions.

Recently, a general catalytic method for the conversion of 2-alkyl-1-ethynylbenzenes to indenes was disclosed by the group of Liu [41]. Their proposed mechanism involves the stepwise insertion of a ruthenium vinylidene into a benzylic C−H bond (Scheme 9.21).

The reactions of Yamamoto and Liu are sufficiently dissimilar that different modes of catalysis may indeed be operative.

9.4.5
Copper Vinylidenes in Catalysis

Following reports of efficient Cu(I)-catalyzed alkyne/azide cycloaddition on solid phase and in solution by Meldal [42] and Sharpless [43], respectively, the formerly obscure Huisgen reaction soared to prominence as a versatile tool for covalent chemical ligation. The so-called "click" reaction can be catalyzed by a number of copper sources in a variety of media (Equation 9.14).

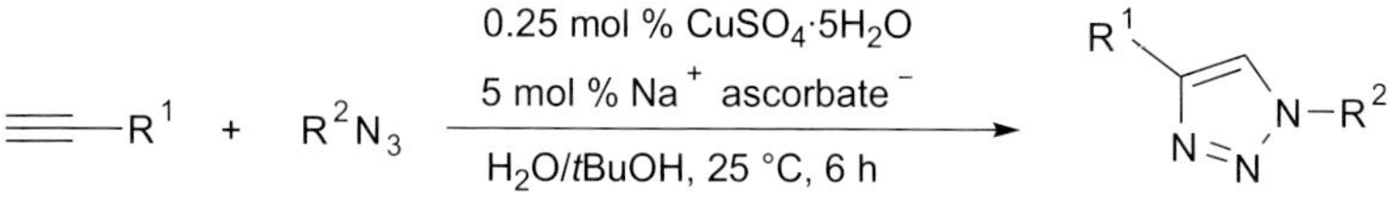

$$\equiv\!-R^1 \; + \; R^2N_3 \xrightarrow[\text{H}_2\text{O}/t\text{BuOH, 25 °C, 6 h}]{\substack{\text{0.25 mol \% CuSO}_4\cdot 5\text{H}_2\text{O} \\ \text{5 mol \% Na}^+ \text{ascorbate}^-}} \underset{\mathbf{137}}{R^1\!\!-\!\!\underset{N=N}{\underset{|}{C}}\!\!=\!\!\underset{}{\overset{}{C}}\!\!-\!\!N\!\!-\!\!R^2} \qquad (9.14)$$

On the basis of DFT calculations, a catalytic cycle involving a copper vinylidene intermediate has been proposed (Scheme 9.22) [44]. The reaction is initiated by copper acetylide (**138**) formation. Sharpless and coworkers next invoke an unusual [3 + 3]-cycloaddition that would be forbidden by orbital symmetry, were it not stepwise. Coordination of an azide to complex **138** generates a zwitterionic complex (**139**). Internal nucleophilic attack of the acetylide moiety of **139** on the electrophilic

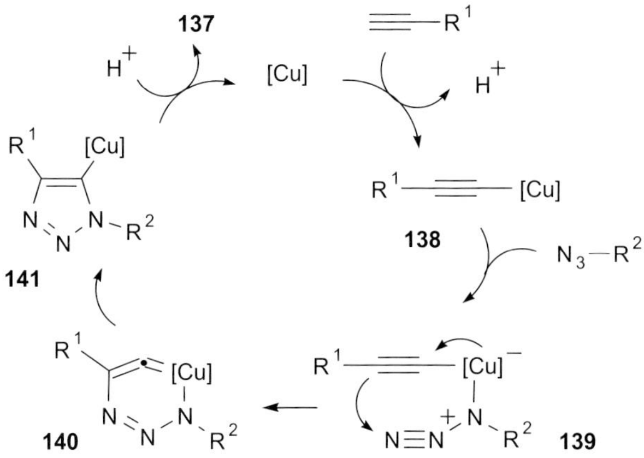

Scheme 9.22 Cu-catalyzed alkyne/azide coupling.

terminus of its azide ligand results in the neutralization of charge and formation of a strained, cyclic copper vinylidene complex (**140**).

Metallacycle **140** rapidly expels product by α-migration of an amido ligand, and then protodemetallation of the resulting copper triazolide complex (**141**). This mechanism adequately accounts for the regioselectivity and reduced activation barrier observed for the Cu(I)-catalyzed Huisgen reaction. The copper vinylidene formation step (**139** → **140**) proceeds via a mechanism similar to that invoked by Lee and coworkers for alkenylidene formation (see Section 9.3). It has been suggested that this step may occur in a dimeric Cu_2L_2 system, giving a less strained eight-membered metallacycle [45]. Given the present rarity of copper vinylidenes in the chemical literature, only further study will permit lingering mechanistic uncertainties to be resolved.

9.4.6
Gold Vinylidenes in Catalysis

Thanks to a recent renaissance in gold catalysis, new gold-mediated transformations are being discovered with ever-increasing frequency. Some of these discoveries have been attributed to the action of gold vinylidenes. Fürstner and coworkers uncovered one such example while screening catalysts for intramolecular alkyne hydroarylation (Scheme 9.23) [46].

Whereas $InCl_3$ was found to efficiently catalyze Friedel-Crafts-type cyclization of chloroalkynes (**142**) to give 10-chlorophenanthrenes (**143**), AuCl induced cycloisomerization with concomitant 1,2-halogen-migration to give 9-halo-phenanthrenes (**144**). The latter transformation can be carried out via tungsten [47] and rhodium (see Section 9.3) vinylidene-mediated catalysis. DFT calculations, likewise, support a mechanism involving 6π-electrocyclization of a Au(I)-vinylidene intermediate for Fürstner's reaction [48]. An alternate mechanism involving Freidel–Crafts-type hydroarylation, followed by 1,2-halogen migration, was ruled out computationally.

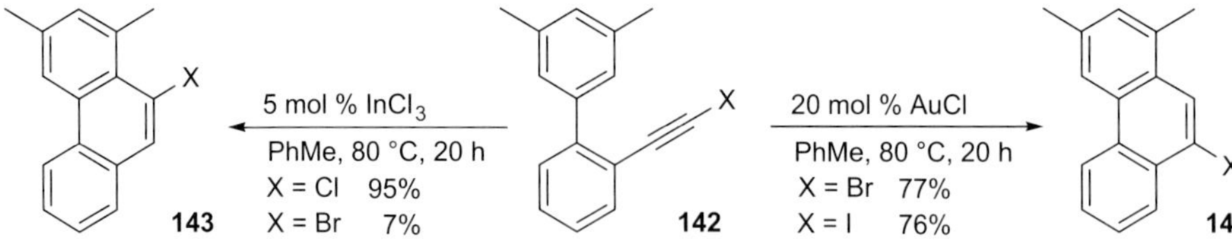

Scheme 9.23 Metal-dependent cyclization modes in intramolecular alkyne hydroarylation.

Relative to Au(III), in general, and In(III), in this case, Au(I) is more inclined to react through vinylidene intermediates due, in part, to its linear geometry, which offers little steric impediment to incoming nucleophiles.

Vinylidene-mediated cyclization or coupling reactions involving 1,2-migration of a heavy p-block element (i.e. X, SiR_3, SnR_3) are synthetically useful because (i) the required functionalized alkynes are readily available and (ii) the products of 1,2-migration are amenable to further functionalization. Accordingly, Gevorgyan and coworkers focused on the use of functionalized alkynes in their gold-catalyzed method for the synthesis of pyrrole-fused heterocycles (Equation 9.15) [49].

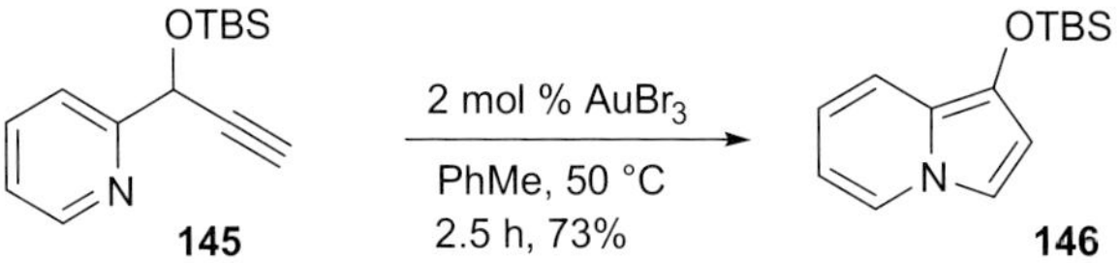

(9.15)

Mild conditions discovered for the cyclization of propargyl pyridine **145** were applied to other substituted and elaborated heterocycles, giving indolizidine-type products in good-to-excellent yields (Table 9.17).

Entry 3 constitutes the first reported example of a metal vinylidene-mediated reaction involving a 1,2-germyl migration. The formation of germylated product is extremely rapid, in agreement with the high reactivity of alkynylgermanes. Both Au(III)- and Au(I)-salts catalyze the observed reaction; therefore, a Au(I)-vinylidene mediated reaction mechanism is favored (Scheme 9.24).

Intramolecular nucleophilic attack on the α-position of the vinylidene complex **148** affords a zwitterionic species (**149**). Next, a formal 1,3-H-shift must occur before release of the observed product (**146**). Based on the results of an isotope labeling study, this process is believed to occur via two discrete 1,2-H-migration events (**149** → **150** → **147**). Interestingly, no back-migration of heteroatom substituents is observed [50].

In 2000, Hashmi and coworkers reported that certain alkynyl furans (**151**) undergo rapid cycloisomerization to give bicyclic phenols (**152**) in the presence of $AuCl_3$ at room temperature (Equation 9.16) [51]. A number of late-transition metal catalysts promote this transformation [52]. Echavarren and coworkers have studied the Pt-catalyzed variant [53], which is believed to proceed via a mechanism involving Pt-cyclopropylcarbene intermediates [54].

Table 9.17 $AuBr_3$-catalyzed synthesis of pyrrole-containing heterocycles.

Entry[a]	Substrate	Product	Group	Time (h)	Yield (%)
1	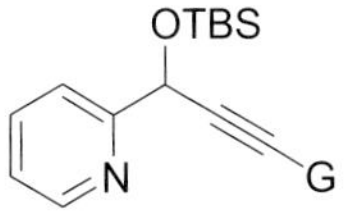	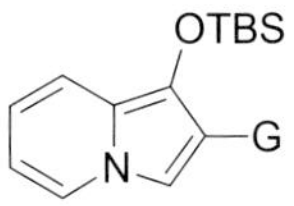	G = $SiMe_3$	1.5	63
2			G = $SnBu_3$	0.5	64
3[b]			G = $GeMe_3$	0.5	92
4	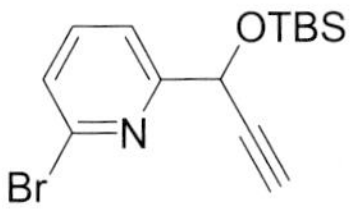	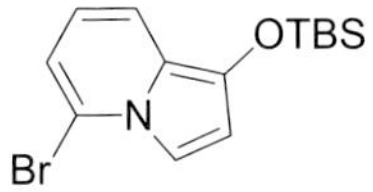		1.5	62
5	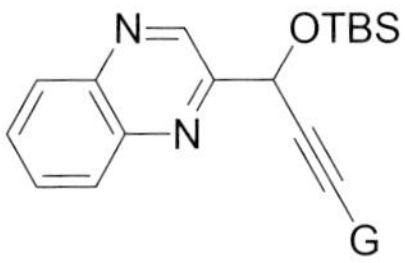	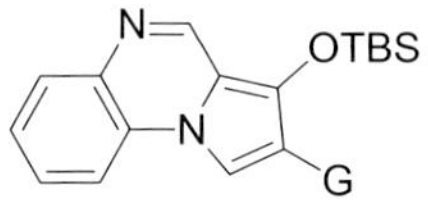	G = H	2.0	94
6			G = $SiMe_3$	4.0	78
7	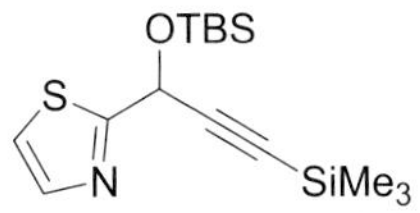	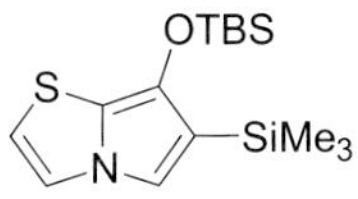		3.5	56

[a]Conditions: See Equation 9.15.
[b]Reaction run at 25 °C.

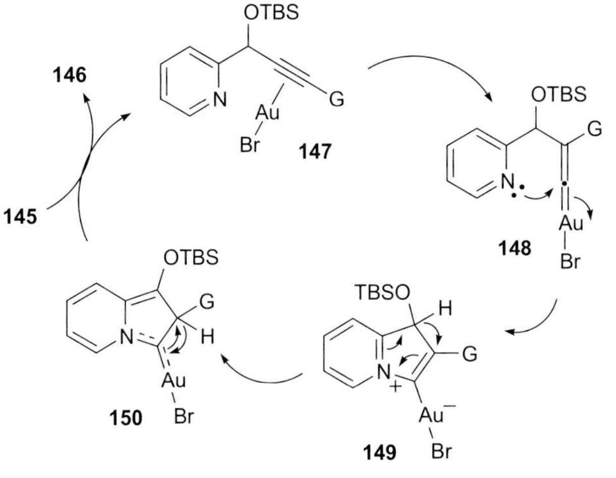

146
147
148
149
150
145

Scheme 9.24 Proposed catalytic cycle for pyrrole-containing heterocycle formation.

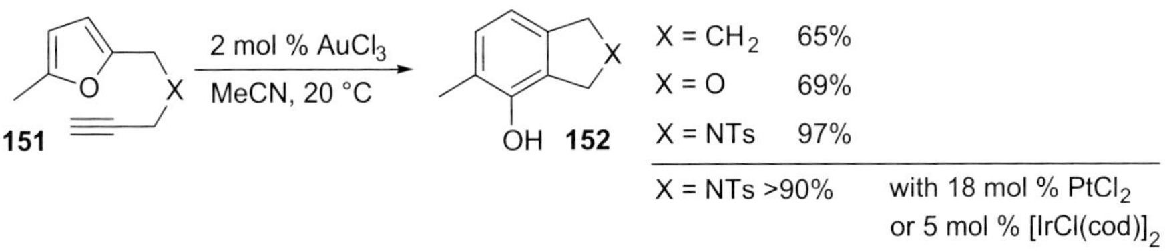

2 mol % AuCl3
MeCN, 20 °C
151
152
X = CH2 65%
X = O 69%
X = NTs 97%
X = NTs >90% with 18 mol % PtCl2
or 5 mol % [IrCl(cod)]2

(9.16)

There are some experimental [55] and computational [56] hints to the effect that vinylidene intermediates may be involved in the Au(III) system. At this time, it is unclear whether all of the metals that catalyze furan/alkyne cycloisomerization operate by the same mechanism.

9.5
Conclusion

The literature available on the subject of Group 9 transition metal vinylidene-mediated catalysis has reached a level of development such that the rational design of new transformations is now possible. Simple metal halide precursors accompanied by either bulky trialkylphosphines or electron-poor triarylphosphines readily form reactive vinylidenes in the presence of alkynes. The use of excess ligand or organic base additives can suppress undesired side reactions. Under these conditions, polar solvents and functionalized substrates are well tolerated. Testing new concepts in Group 9 transition metal vinylidene formation or elaboration has become relatively straightforward. Important synthetic methods are expected to proceed from both completely unprecedented reactions and intermolecular variants of existing reactions.

The prospects for vinylidene-mediated catalysis involving transition metals from Groups 10 and 11 are less certain. Continuing robust interest in the area of gold catalysis is sure to provide new revelations about the control of gold π-alkyne/vinylidene interconversion. The use of ligand-controlled catalysts, as opposed to simple metal salts, may aid in these efforts. Our ability to predict which conditions will favor vinylidene-mediated reactivity will increase as the body of knowledge on the subject grows.

9.6
Note Added in Proof

After submission of the original manuscript, additional relevant papers were published.

Buono and coworkers developed a platinum analog of their Pd-catalyzed cyclopropanation method (Section 9.4.3). Their new catalyst, $Pt(\eta^2$-acetato$)$-$\{[(R)(Ph)PO]_2H\}$, was successfully applied to reactions of heteroatom-containing alkenes [57].

Chatani and coworkers published an efficient method for the Rh(I)-catalyzed anti-Markovnikov hydroamination of terminal alkynes using either primary or secondary amines [58]. This reactivity had been observed earlier in the course of their studies on hydrative alkyne dimerization (Equation 9.8).

Two recent publications feature metal vinylidenes functioning as 1,3-dipole equivalents, as in the Cu-catalyzed Huisgen cyclization (Section 9.4.5). Fürstner and coworkers described intramolecular Diels-Alder reactions of unactivated dienynes catalyzed via a proposed [4+3]-diene/copper vinylidene cycloaddition [59].

Bertrand and coworkers invoke a retroiminoene-type mechanism for the formation of gold vinylidenes in an interesting crosscoupling of enamines and terminal alkynes to give allenes [60]. This reaction constitutes both a new reaction manifold for metal vinylidenes and a new retron for stereochemically defined allenes.

References

1. Wakatsuki, Y. (2004) *Journal of Organometallic Chemistry*, **689**, 4092–4109.
2. Grotjahn, D.B., Zeng, X., Cooksy, A.L., Kassel, W.S., DiPasquale, A.G., Zakharov, L.N. and Rheingold, A.L. (2007) *Organometallics*, **26**, 3385–3402.
3. Höhn, A. and Werner, H. (1990) *Journal of Organometallic Chemistry*, **382**, 255–272.
4. Varela, J.A. and Saá, C. (2006) *Chemistry – A European Journal*, **12**, 6450–6456.
5. Elakkari, E., Floris, B., Galloni, P. and Tagliatesta, P. (2005) *European Journal of Organic Chemistry*, 889–894.
6. Dankwardt, J.W. (2001) *Tetrahedron Letters*, **42**, 5809–5812.
7. Miura, T., Murata, H., Kiyota, K., Kusama, H. and Iwasawa, N. (2004) *Journal of Molecular Catalysis A-Chemical*, **213**, 59–71.
8. Kim, H. and Lee, C. (2005) *Journal of the American Chemical Society*, **127**, 10180–10181.
9. Grigg, R., Stevenson, P. and Worakun, T. (1988) *Tetrahedron*, **44**, 4967–4972.
10. Kim, H. and Lee, C. *Journal of the American Chemical Society*, **2006**, **128**, 6336–6337. Correction: Kim, H. and Lee, C. (2006) *Journal of the American Chemical Society*, **128**, 10629.
11. Ohe, K., Kojima, M., Yonehara, K. and Uemura, S. (1996) *Angewandte Chemie-International Edition in English*, **35**, 1823–1825.
12. Wang, Y. and Finn, M.G. (1995) *Journal of the American Chemical Society*, **117**, 8045–8046.
13. Manabe, T., Yanagi, S.I., Ohe, K. and Uemura, S. (1998) *Organometallics*, **17**, 2942–2944.
14. Trost, B.M. and Rhee, Y.H. (2003) *Journal of the American Chemical Society*, **125**, 7482–7483.
15. McDonald, F.E. and Chatterjee, A.K. (1997) *Tetrahedron Letters*, **38**, 7687–7690.
16. Trost, B.M. and McClory, A. (2007) *Angewandte Chemie-International Edition in English*, **46**, 2074–2077.
17. Trost, B.M. and Ball, Z.T. (2005) *Synthesis*, 853–887.
18. Ohmura, T., Yamamoto, Y. and Miyaura, N. (2000) *Journal of the American Chemical Society*, **122**, 4990–4991.
19. Ojima, I., Clos, N., Donovan, R.J. and Ingallina, P. (1990) *Organometallics*, **9**, 3127–3133.
20. Lee, C.C., Lin, Y.C., Liu, Y.H. and Wang, Y. (2005) *Organometallics*, **24**, 136–143.
21. Ohmura, T., Yorozuya, S., Yamamoto, Y. and Miyaura, N. (2000) *Organometallics*, **19**, 365–367.
22. Jiménez, M.V., Sola, E., Lahoz, F.J. and Oro, L.A. (2005) *Organometallics*, **24**, 2722–2729.
23. Ammal, S.C., Yoshikai, N., Inada, Y., Nishibayashi, Y. and Nakamura, E. (2005) *Journal of the American Chemical Society*, **127**, 9428–9438.
24. Park, Y.J., Kwon, B.I., Ahn, J.A., Lee, H. and Jun, C.H. (2004) *Journal of the American Chemical Society*, **126**, 13892–13893.
25. Jun, C.H. and Lee, J.H. (2004) *Pure and Applied Chemistry*, **76**, 577–587.
26. Fukumoto, Y., Kinashi, F., Kawahara, T. and Chatani, N. (2006) *Organic Letters*, **8**, 4641–4643.
27. Tan, K.L., Bergman, R.G. and Ellman, J.A. (2002) *Journal of the American Chemical Society*, **124**, 3202–3203.
28. Chen, Y. and Lee, C. (2006) *Journal of the American Chemical Society*, **128**, 15598–15599.

29 Werner, H. (2004) *Coordination Chemistry Reviews*, **248**, 1693–1702.
30 Davies, S.G. and Scott, F. (1980) *Journal of Organometallic Chemistry*, **188**, C41–C42.
31 Joo, J.M., Yuan, Y. and Lee, C. (2006) *Journal of the American Chemical Society*, **128**, 14818–14819.
32 Joo, J.M. and Lee, C. unpublished results.
33 Werner, H., Ilg, K., Lass, R. and Wolf, J. (2002) *Journal of Organometallic Chemistry*, **661**, 137–147.
34 Stacchiola, D., Molero, H. and Tysoe, W.T. (2001) *Catalysis Today*, **65**, 3–11.
35 Naka, A., Okazaki, S., Hayashi, M. and Ishikawa, M. (1995) *Journal of Organometallic Chemistry*, **499**, 35–41.
36 Naka, A., Okada, T., Kunai, A. and Ishikawa, M. (1997) *Journal of Organometallic Chemistry*, **547**, 149–156.
37 Kline, E.S., Kafafi, Z.H., Hauge, R.H. and Margrave, J.L. (1987) *Journal of the American Chemical Society*, **109**, 2402–2409.
38 Bigeault, J., Giordano, L. and Buono, G. (2005) *Angewandte Chemie-International Edition in English*, **44**, 4753–4757.
39 Johnson, T.H. and Cheng, S.-S. (1979) *Journal of the American Chemical Society*, **101**, 5277–5280.
40 Bajracharya, G.B., Pahadi, N.K., Gridnev, I.D. and Yamamoto, Y. (2006) *Journal of Organic Chemistry*, **71**, 6204–6210.
41 Odedra, A., Datta, S. and Liu, R.-S. (2007) *Journal of Organic Chemistry*, **72**, 3289–3292.
42 Tornøe, C.W., Christensen, C. and Meldal, M. (2002) *Journal of Organic Chemistry*, **67**, 3057–3064.
43 Rostovtsev, V.V., Green, L.G., Fokin, V.V. and Sharpless, K.B. (2002) *Angewandte Chemie-International Edition in English*, **41**, 2596–2599.
44 Himo, F., Lovell, T., Hilgraf, R., Rostovtsev, V.V., Noodleman, L., Sharpless, K.B. and Fokin, V.V. (2005) *Journal of the American Chemical Society*, **127**, 210–216.
45 Bock, V.D., Hiemstra, H. and van Maarseveen, J.H. (2005) *European Journal of Organic Chemistry*, 51–68.
46 Mamane, V., Hannen, P. and Fürstner, A. (2004) *Chemistry – A European Journal*, **10**, 4556–4575.
47 Miura, T. and Iwasawa, N. (2002) *Journal of the American Chemical Society*, **124**, 518–519.
48 Soriano, E. and Marco-Contelles, J. (2006) *Organometallics*, **25**, 4542–4553.
49 Seregin, I.V. and Gevorgyan, V. (2006) *Journal of the American Chemical Society*, **128**, 12050–12051.
50 Shen, H.-C., Pal, S., Lian, J.-J. and Liu, R.-S. (2003) *Journal of the American Chemical Society*, **125**, 15762–15763.
51 Hashmi, A.S.K., Frost, T.M. and Bats, J.W. (2000) *Journal of the American Chemical Society*, **122**, 11553–11554.
52 Hashmi, A.S.K., Frost, T.M. and Bats, J.W. (2001) *Organic Letters*, **3**, 3769–3771.
53 Martín-Matute, B., Cárdenas, D.J. and Echavarren, A.M. (2001) *Angewandte Chemie-International Edition in English*, **40**, 4754–4757.
54 Martín-Matute, B., Nevado, C., Cárdenas, D.J. and Echavarren, A.M. (2003) *Journal of the American Chemical Society*, **125**, 5757–5766.
55 Hashmi, A.S.K., Weyrauch, J.P., Kurpejović, E., Frost, T.M., Miehlich, B., Frey, W. and Bats, J.W. (2006) *Chemistry – A European Journal*, **12**, 5806–5814.
56 Rabaâ, H., Engels, B., Hupp, T. and Hashmi, A.S.K. (2007) *International Journal of Quantum Chemistry*, **107**, 359–365.
57 Bigeault, J., Giordano, L., de Riggi, I., Gimbert, Y. and Buono, G. (2007) *Organic Letters*, **9**, 3567–3570.
58 Fukumoto, Y., Asai, H., Shimizu, M. and Chatani, N. (2007) *Journal of the American Chemical Society*, **129**, 13792–13793.
59 Fürstner, A. and Stimson, C.C. (2007) *Angewandte Chemie International Edition*, **46**, 8845–8849.
60 Lavallo, V., Frey, G.D., Kousar, S., Donnadieu, B. and Bertrand, G. (2007) *Proceedings of the National Academy of Sciences*, **104**, 13569–13573.

10
Anti-Markovnikov Additions of O-, N-, P-Nucleophiles to Triple Bonds with Ruthenium Catalysts

Christian Bruneau

10.1
Introduction

A variety of metal complexes are able to promote the Markovnikov addition of nucleophiles to alkynes via Lewis acid-type activation of triple bonds (Equation 10.1) [1]. Starting from terminal alkynes, the anti-Markovnikov addition to form the other vinylic regioisomer (Equation 10.2) is less common and requires selected catalysts. A mechanistic possibility to explain this regioselectivity, corresponding to the addition of the nucleophile at the less substituted carbon of the C≡C triple bond, is the formation of a metal vinylidene intermediate featuring a highly reactive electrophilic C_α carbon atom as in an organic heteroallene [2–4]. This type of activation, followed by addition of nucleophiles (mainly O-nucleophiles) has been performed with W, Cr, Mo, Rh and Ru precursors. In this chapter, we will focus on ruthenium-catalyzed anti-Markovnikov nucleophilic addition to terminal alkynes, even though the evidence for a ruthenium vinylidene intermediate has not always been demonstrated. The activity of the other metals in the same type of addition is discussed in Chapters 5 (Cr, Mo, W) and 9 (Rh).

$$H\!-\!\!\equiv\!\!-R \xrightarrow{L_nM} H\!-\!\!\equiv\!\!-R \xrightarrow{Nu-H} \underset{H}{\overset{H}{\diagup}}\!\!=\!\!\underset{Nu}{\overset{R}{\diagdown}} \qquad (10.1)$$

Markovnikov addition

$$H\!-\!\!\equiv\!\!-R \xrightarrow{L_nM} L_nM\!=\!C\!=\!C\overset{R}{\underset{H}{\diagdown}} \xrightarrow{Nu^-} \underset{LnRu}{\overset{Nu}{\diagup}}\!\!=\!\!\underset{H}{\overset{R}{\diagdown}} \xrightarrow{H^+} \underset{H}{\overset{Nu}{\diagup}}\!\!=\!\!\underset{H}{\overset{R}{\diagdown}} \qquad (10.2)$$

anti-Markovnikov addition

Metal Vinylidenes and Allenylidenes in Catalysis: From Reactivity to Applications in Synthesis
Edited by Christian Bruneau and Pierre Dixneuf
Copyright © 2008 WILEY-VCH Verlag GmbH & Co. KGaA, Weinheim
ISBN: 978-3-527-31892-6

10.2
C—O Bond Formation

10.2.1
Addition of Carbamic Acids: Synthesis of Vinylic Carbamates and Ureas

The first example of anti-Markovnikov addition of O-nucleophiles to terminal alkynes was the catalytic addition of ammonium carbamates, generated *in situ* from secondary amines and carbon dioxide, to terminal alkynes, which regioselectively produced vinylcarbamates (Scheme 10.1) [5]. It was also the first time that a metal vinylidene was suggested as an active intermediate in catalysis [5].

Success was obtained with $Ru_3(CO)_{12}$ as catalyst precursor [6], but the most efficient catalysts were found in the $RuCl_2$(arene)(phosphine) series. These complexes are known to produce ruthenium vinylidene species upon reaction with terminal alkynes under stoichiometric conditions, and thus are able to generate potential catalysts active for anti-Markovnikov addition [7]. Similar results were obtained by using $Ru(\eta^4\text{-cyclooctadiene})(\eta^6\text{-cyclooctatriene})/PR_3$ as catalyst precursor [8]. (Z)-Dienylcarbamates were also regio- and stereo-selectively prepared from conjugated enynes and secondary aliphatic amines (diethylamine, piperidine, morpholine, pyrrolidine) but, in this case, $RuCl_2$(arene)(phosphine) complexes were not very efficient and the best catalyst precursor was Ru(methallyl)$_2$(diphenylphosphinoethane) [9] (Scheme 10.1).

The formation of vinylcarbamates is restricted to secondary amines and also to terminal alkynes, which is in line with the formation of a metal vinylidene intermediate. It is noteworthy that even starting from secondary amines, the presence of a hydroxy group in propargylic alcohols drove the reaction towards the formation of β-keto carbamates, resulting from initial Markovnikov addition of the carbamate anion to the triple bond followed by intramolecular transesterification [10]. The proposed general catalytic cycle which applies for the formation of vinylic carbamates is shown in Scheme 10.2.

A catalytic reaction also took place under similar conditions with primary aliphatic amines but it led to the formation of symmetrical ureas and the best results were

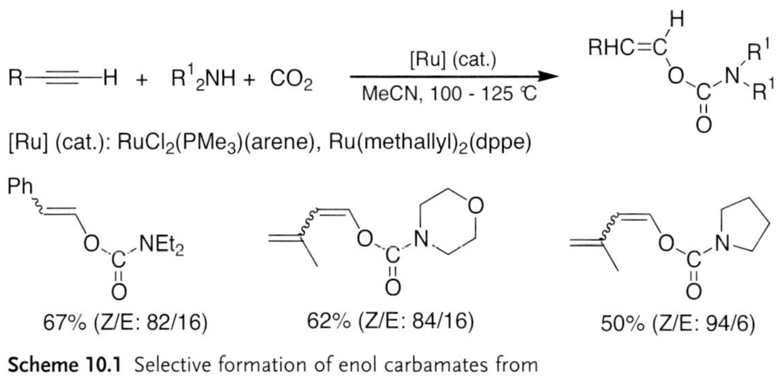

Scheme 10.1 Selective formation of enol carbamates from secondary amines, CO_2, and terminal alkynes.

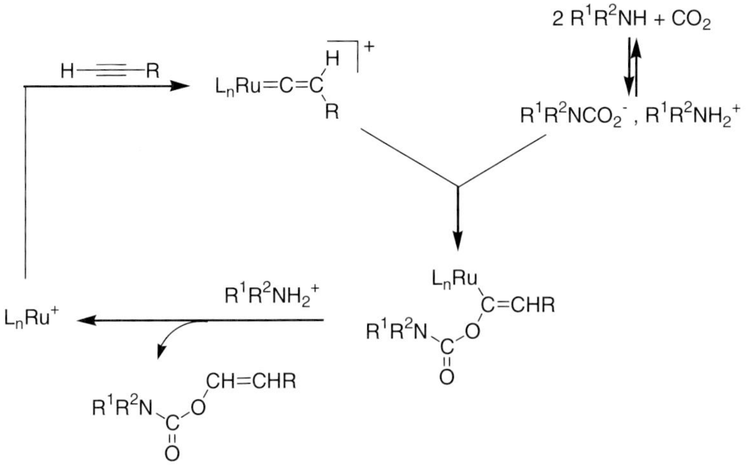

Scheme 10.2 A plausible mechanism for vinylic carbamate formation.

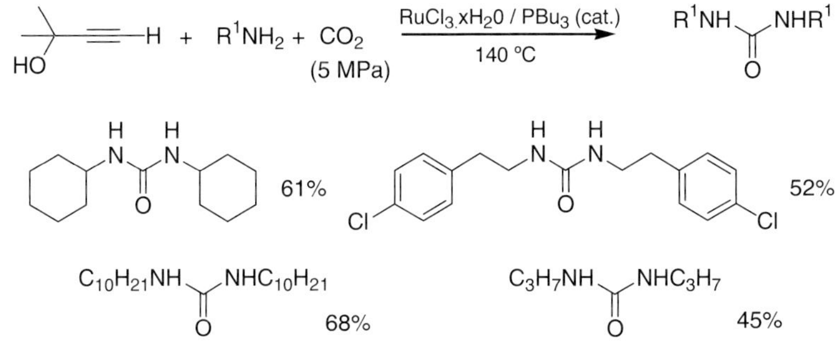

Scheme 10.3 Formation of ureas from primary amines, CO_2, and terminal alkynes.

obtained in the presence of a propargylic alcohol as the alkyne, and $RuCl_3 \cdot 3H_2O/3$ PBu_3 or $RuCl_2$(hexamethylbenzene)(PMe_3) as metal catalyst (Scheme 10.3) [11].

From simple terminal alkynes, the catalytic system generated in this case is also thought to proceed via a ruthenium vinylidene active species and is very efficient for the formal elimination of water by formation of an organic adduct (Equation 10.3) [12].

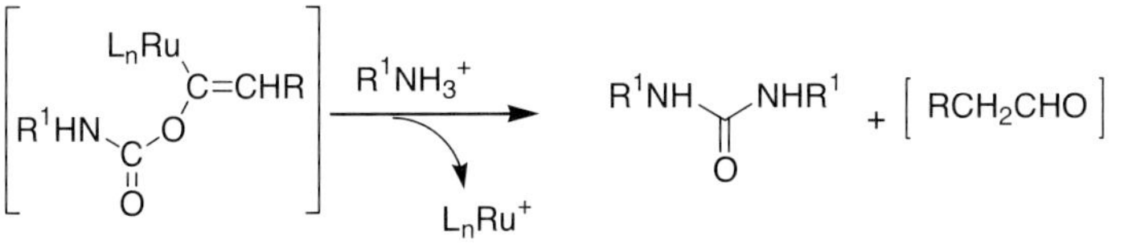

10.2.2
Addition of Carboxylic Acids: Synthesis of Enol Esters

The Markovnikov addition of carboxylic acids to terminal alkynes to produce geminal enol esters has been carried out with a variety of efficient ruthenium precursors such as $Ru_3(CO)_{12}$ [13], $Ru(cod)_2/PR_3$ [14], $RuCl_2(PR_3)(arene)$ [15, 16], or $[Ru(O_2CH)(CO)_2(PR_3)]_2$ [17]. In contrast, some π-allyl ruthenium complexes containing a chelating diphosphine ligand were the first metal complexes which favored the anti-Markovnikov addition of carboxylic acids to terminal alkynes to form (Z)- and (E)-enol esters with high regio- and stereo-selectivity [18–20] according to Equation 10.4. The major stereoisomers were always the (Z)-enol esters as the result of a formal trans-addition of the carboxylic acid to the triple bond.

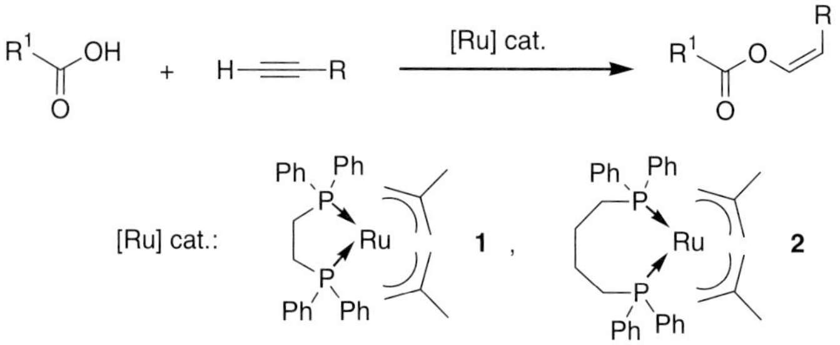

$$R^1\text{COOH} + H\text{≡}R \xrightarrow{\text{[Ru] cat.}} R^1\text{COO-CH=CHR} \quad (10.4)$$

[Ru] cat.: Ru(methallyl)₂(dppe) **1**, Ru(methallyl)₂(dppb) **2**

For this selective addition, the best catalyst precursors were Ru(methallyl)₂(dppe) **1** and Ru(methallyl)₂(dppb) **2**. The choice of the appropriate complex depended on the steric demand of both the alkyne and the carboxylic acid. A large variety of carboxylic acids and alkynes have been used, including N-protected amino acids, α-hydroxy acids, and functionalized alkynes such as enynes, diynes and propargylic ethers [21–23] (Scheme 10.4). The addition took place under mild conditions and carboxylic acids of

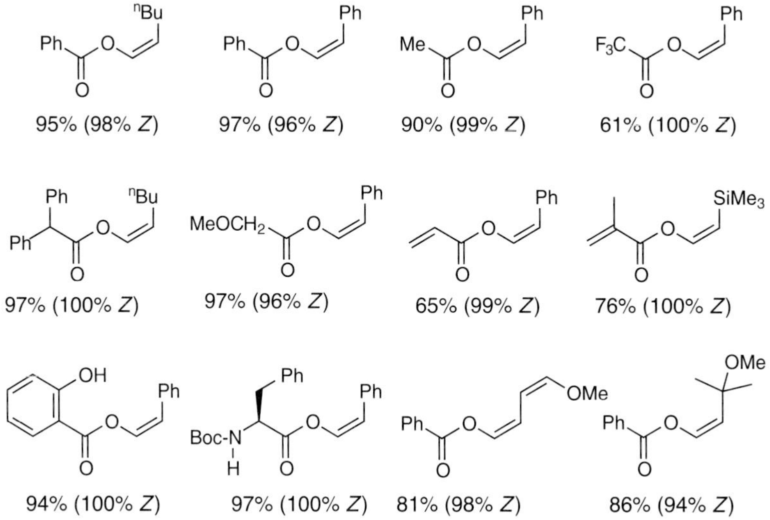

- 95% (98% Z)
- 97% (96% Z)
- 90% (99% Z)
- 61% (100% Z)
- 97% (100% Z)
- 97% (96% Z)
- 65% (99% Z)
- 76% (100% Z)
- 94% (100% Z)
- 97% (100% Z)
- 81% (98% Z)
- 86% (94% Z)

Scheme 10.4 Selected examples of (Z)-enol esters.

low pK_a, such as CF_3CO_2H and $CHCl_2CO_2H$, provided high reactivity and regioselectivity at low temperature (0–20 °C).

The regioselective anti-Markovnikov addition of benzoic acid to phenylacetylene has also been carried out with success in the presence of ruthenium complexes containing a tris(pyrazolyl)borate (Tp) ligand (RuCl(Tp)(cod) **3**, RuCl(Tp)(tmeda) **4**, RuCl(Tp)(pyridine)$_2$**5**) with a stereoselectivity in favor of the (*E*)-enol ester isomer [24]. More recently, new catalyst precursors derived from [RuCl$_2$(*p*-cymene)]$_2$, such as RuCl$_2$(triazol-5-ylidene)(*p*-cymene) **6** and RuCl(*p*-cymene)(*o*-Ph-(triazol-5-ylidene) **7** [25], or from RuCl$_2$(PPh$_3$)$_3$**8** [26], and the three-component catalytic system easily generated *in situ* from [RuCl$_2$(*p*-cymene)]$_2$/P(*p*-C$_6$H$_4$Cl)$_3$/DMAP [27] have revealed their potential to perform the anti-Markovnikov addition of a variety of carboxylic acids to phenylacetylene and terminal aliphatic alkynes. In contrast, ruthenium vinylidene complexes such as **9** have been reported as active catalysts for the addition of carboxylic acids to alkynes, but in most cases they favored the Markovnikov addition [28] (Scheme 10.5).

The σ-enynyl complex Ru(Tp)[PhC=C(Ph)C≡CPh](PMe^{i}Pr$_2$) **10** efficiently catalyzed the regioselective cyclization of α,ω-alkynoic acids involving an anti-Markovnikov intramolecular addition to give unsaturated lactones [29] (Equation 10.5).

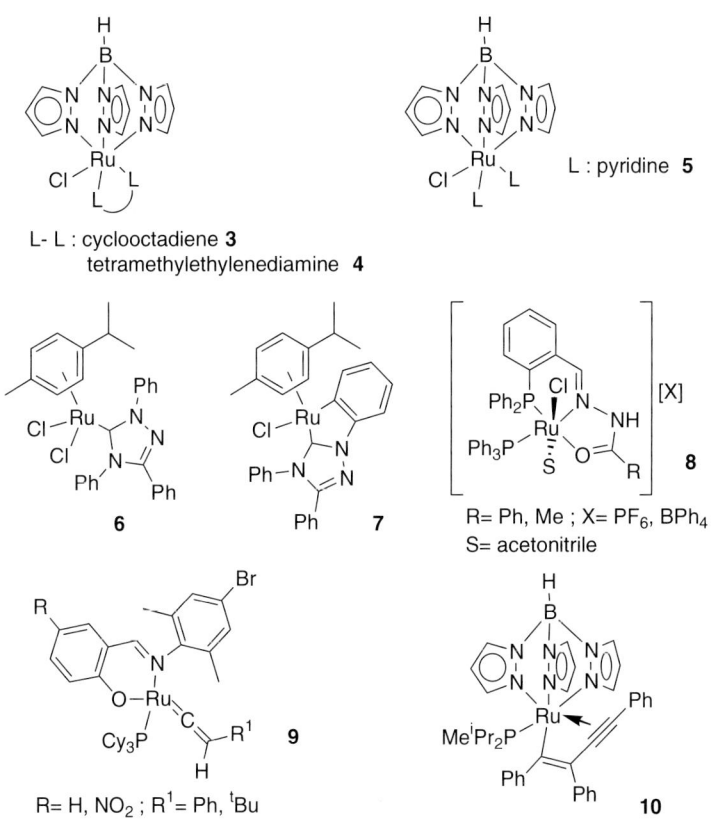

Scheme 10.5 Some ruthenium catalysts leading to anti-Markovnikov addition of carboxylic acids to terminal alkynes.

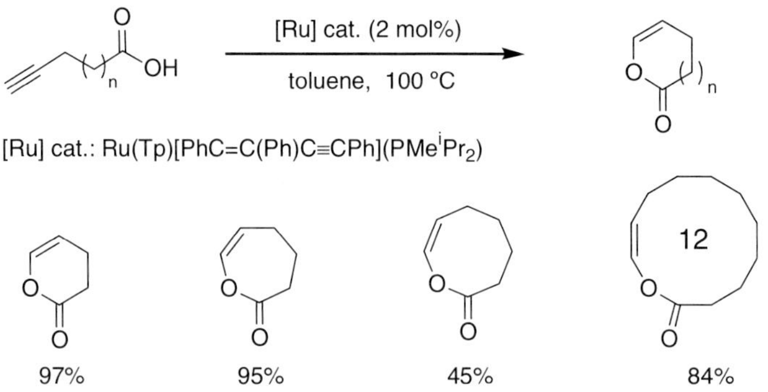

[Ru] cat. (2 mol%), toluene, 100 °C

[Ru] cat.: Ru(Tp)[PhC=C(Ph)C≡CPh](PMe^{i}Pr$_2$)

(10.5)

97%   95%   45%   84% (12)

Whereas the Markovnikov addition of carboxylic acids to propargylic alcohols produces β-ketoesters, resulting from intramolecular transesterification [30, 31], the addition to propargylic alcohols in the presence of Ru(methallyl)$_2$(dppe) 1 at 65 °C leads to hydroxylated alk-1-en-1-yl esters via formation of a hydroxy vinylidene intermediate [32, 33]. The stereoselectivities are lower than those obtained from non-hydroxylated substrates. These esters, which are protected forms of aldehydes, can easily be cleaved under thermal or acidic conditions to give conjugated enals, corresponding to the formal isomerization products of the starting alcohols (Scheme 10.6).

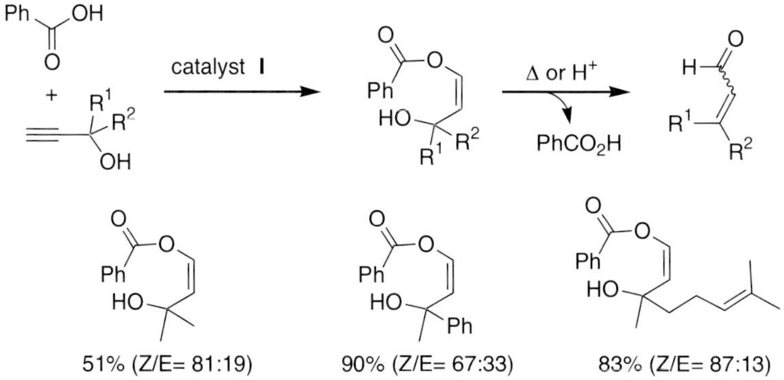

51% (Z/E= 81:19)   90% (Z/E= 67:33)   83% (Z/E= 87:13)

Scheme 10.6 Regioselective formation of 3-hydroxylated enol esters resulting from anti-Markovnikov addition.

10.2.3
Addition of Water: Synthesis of Aldehydes

The Lewis acid-catalyzed addition of water to terminal alkynes leads to ketones following Markovnikov's rule [34]. The first selective catalytic formation of aldehydes was reported by Tokunaga and Wakatsuki who used RuCl$_2$(C$_6$H$_6$)(PPh$_2$(C$_6$F$_5$)) + 3 PPh$_2$(C$_6$F$_5$) 11 or [RuCl$_2$(C$_6$H$_6$)]$_2$ associated to 8 equiv of the water-soluble ligand P(3-C$_6$H$_5$SO$_3$Na)$_3$ (TPPTS) 12 in alcohol at 65–100 °C (Equation 10.6) [35].

$$R\!\!\equiv\ +\ H_2O \xrightarrow{\text{catalyst}} R\!\!-\!\!CH_2\!\!-\!\!CHO \qquad (10.6)$$

With these catalytic systems, a variety of linear aliphatic terminal alkynes were transformed into aldehydes with good selectivity but phenylacetylene and *tert*-butylacetylene showed low reactivity. In addition, these catalytic systems suffered from high catalyst and phosphine loadings. The efficiency, regioselectivity of the addition, and substituent tolerance were improved by using RuCl(Cp)(phosphine)$_2$ or RuCl(Cp)(diphenylphosphinomethane) **13** as catalyst precursor [36], which made possible the preparation of aldehydes from bulky aliphatic alkynes (*tert*-BuC≡CH), aromatic alkynes (PhC≡CH), diynes (HC≡C(CH$_2$)$_6$C≡CH) and functional terminal alkynes (NC(CH$_2$)$_3$C≡CH, PhCH$_2$O(CH$_2$)$_2$C≡CH, ...) in more than 80% yield at 100 °C. The mechanism of this reaction was investigated in detail by isolation of intermediates, deuterium-labeling experiments and DFT calculations [37]. The most probable catalytic cycle first involves protonation of a ruthenium(II)-π-alkyne species to give a Ru(IV)-vinylidene intermediate via a Ru(IV)-vinyl species. The nucleophilic addition of water to the α-carbon of the vinylidene ligand followed by reductive elimination affords the aldehyde (Scheme 10.7).

It is noteworthy that the indenyl complex RuCl(η^5-C$_9$H$_7$)(PPh$_3$)$_2$ **14** provides an efficient catalyst precursor for the anti-Markovnikov hydration of terminal alkynes in aqueous media, especially in micellar solutions with either anionic (sodium dodecylsulfate (SDS)) or cationic (hexadecyltrimethylammonium bromide (CTAB)) surfactants [38]. This system can be applied to the hydration of propargylic alcohols to selectively produce β-hydroxyaldehydes, whereas RuCl(Cp)(PMe$_3$)$_2$ gives α,β-unsaturated aldehydes (the Meyer Schuster rearrangement products)(Scheme 10.8) [39].

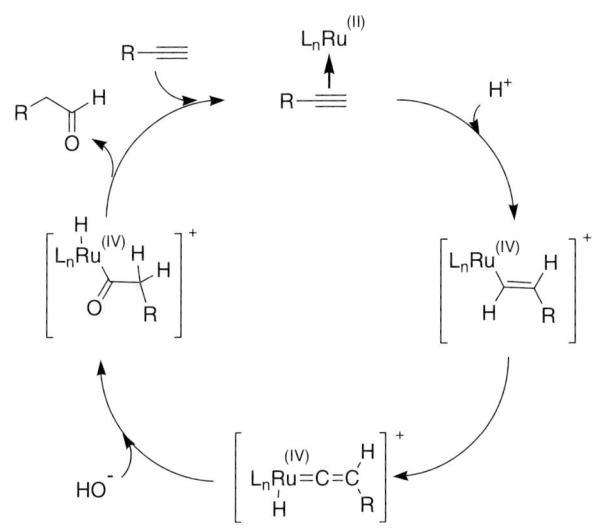

Scheme 10.7 A mechanism proposed for the anti-Markovnikov hydration of alkynes.

Scheme 10.8 Two hydration processes involving ruthenium vinylidene intermediates.

The formation of β-hydroxyaldehydes from propargylic alcohols has also been observed in aqueous media in the presence of a catalytic amount of water-soluble ruthenium sulfophthalocyanine complex and the heterogeneous ruthenium hydroxyapatite catalyst [40].

Recently, a new class of supramolecular CpRu-containing catalysts for hydration of alkynes has emerged. These catalysts are based on the supramolecular self-assembly of monodentate ligands through hydrogen bonding association, as shown in Scheme 10.9 [41–43]. The remarkable activity of catalytic systems such as **15–17**

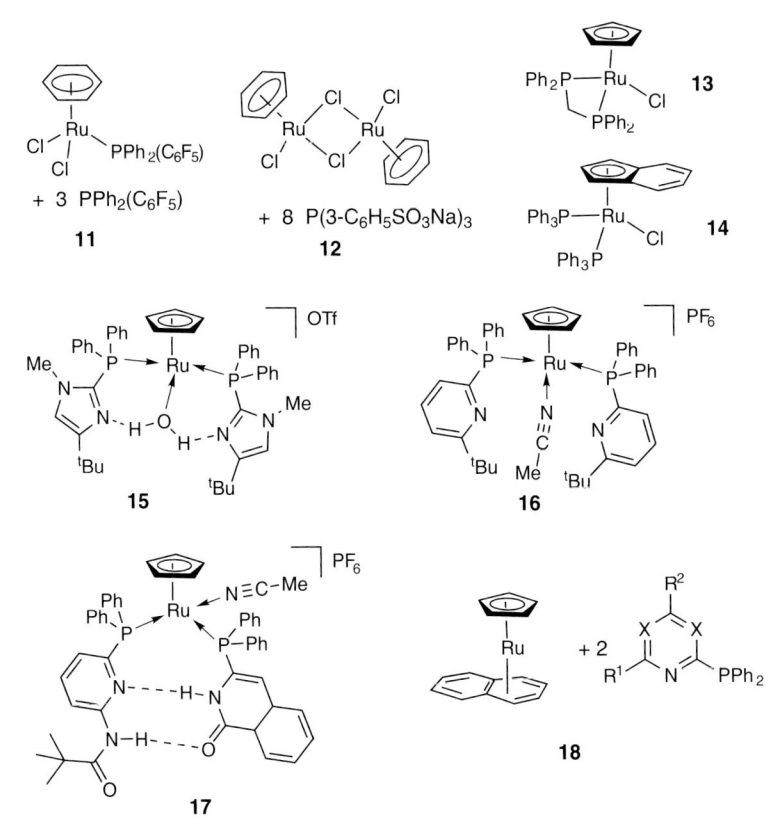

Scheme 10.9 Ruthenium catalysts for anti-Markovnikov hydration of alkynes.

in terms of reactivity and selectivity is based on the principle of cooperative catalysis between the metal center and the functional groups of the ligands. In complexes such as **17**, the interacting monophosphine ligands behave as a diphosphine with large bite angle. Based on the same concept, very active catalysts have been formed upon treatment of the air-stable ruthenium(naphthalene)(cyclopentadienyl) hexafluorophosphate with 2 equiv of pyridine-phosphine or triazine-phosphine ligand **18**. With this system, very high efficiencies and selectivities were obtained for a wide range of functional terminal alkynes at moderate temperature (45–65 °C) [44].

10.2.4
Addition of Alcohols: Synthesis of Ethers and Ketones

10.2.4.1 Intermolecular Addition: Formation of Unsaturated Ethers and Furans

The ruthenium-catalyzed direct addition of saturated aliphatic alcohols to non-activated alkynes remains a challenge. Only allyl alcohol has been successfully involved in the intermolecular addition to phenylacetylene to produce an ether and the enal resulting from Claisen rearrangement (Equation 10.7) [24]. Thus, in refluxing toluene, in the presence of a catalytic amount of RuCl(tris(pyrazolyl)borate)(pyridine)$_2$, a 1 : 1 mixture of allyl β-styryl ether and 2-phenylpent-4-enal was obtained in 72% overall yield.

$$\text{Ph}-\!\!\!\equiv\ +\ \diagup\!\!\!\diagdown\text{OH}\ \xrightarrow[\text{toluene, 111 °C}]{\text{RuCl(Tp)(pyridine)}_2}\ \text{Ph}\diagdown\!\!\!\diagup\text{O}\diagdown\!\!\!\diagup\ +\ \text{Ph}\diagdown\!\!\!\diagup\!\!\!\diagdown\!\!\!\diagup\text{O} \quad (10.7)$$

Tp: tris(pyrazolyl)borate

The isomerization of terminal epoxyalkynes into furans catalyzed by RuCl(Tp)(PPh$_3$)(MeCN) in the presence of Et$_3$N as a base at 80 °C in 1,2-dichloroethane is explained by a related intramolecular nucleophilic addition of the oxygen atom of the epoxide to the α-carbon atom of a ruthenium vinylidene intermediate, as shown by deuteration in the 3-position of the furan (Scheme 10.10) [45]. This reaction is specific for terminal alkynes and tolerates a variety of functional groups (ether, ester, acetal, tosylamide, nitrile).

10.2.4.2 Intermolecular Addition with Rearrangement: Formation of Unsaturated Ketones

Trost's group has shown that another selective reaction involving C–O bond formation followed by rearrangement and C–C bond formation occurred when Cp-containing ruthenium complexes were used as catalytic precursors. With RuCl(Cp)(PPh$_3$)$_2$ in the presence of NH$_4$PF$_6$, an additive known to facilitate chloride abstraction from the metal center, the addition of allylic alcohols to terminal alkynes afforded unsaturated ketones [46, 47]. It has been shown that the key steps are the

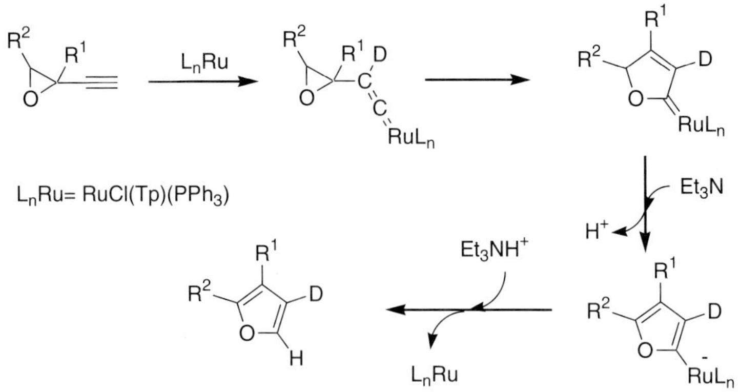

Scheme 10.10 Cycloisomerization of ethynyloxiranes.

$R^1 = H$; $R^2 = C_7H_{15}$ (84%), $C_{12}H_{25}$ (91%)
$R^1 = Me$; $R^2 = CH_2OH$ (86%), CH_2OCOPh (81%), CH_2OBn (91%)
$R^1 = CH_2NTsCH_2Ph$; $R^2 = H$ (89%)

nucleophilic addition of the allylic alcohol to a ruthenium vinylidene species followed by formation of an allyl-metal intermediate via sigmatropic rearrangement (Scheme 10.11) [47].

From α-substituted allylic alcohols, the formation of β,γ-unsaturated ketones is favored, whereas conjugated enones are obtained from simple allyl alcohol [46]. This transformation of terminal alkynes via coupling with allylic alcohol and formation of a C−C bond with atom economy has been applied to the synthesis and modification of natural compounds such as rosefuran and steroids [48, 49].

The preparation of optically active ruthenium vinylidene complexes with the objective of performing the asymmetric version of this reaction has been attempted but has led to moderate enantioselectivities [50, 51].

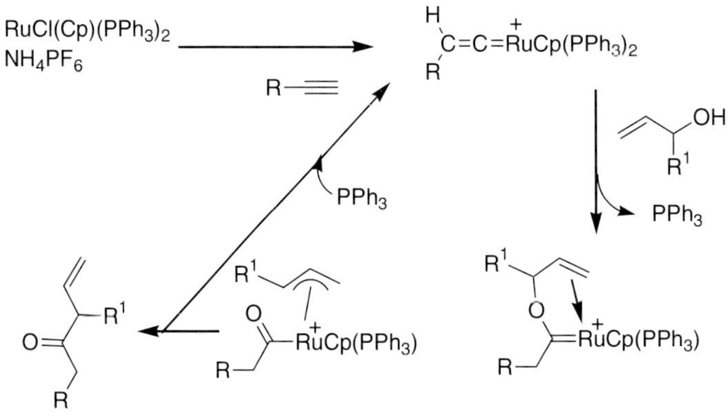

Scheme 10.11 Formation of unsaturated ketones via anti-Markovnikov addition of allylic alcohols to terminal alkynes.

10.2.4.3 Intramolecular Addition: Formation of Cyclic Enol Ethers and Lactones from Pent-4-yn-1-ols and But-3-yn-1-ols

Homopropargylic alcohols (but-3-ynols) as well as propargylic epoxides, are suitable products to form cyclic ruthenium alcoxycarbenes upon intramolecular nucleophilic addition of the nucleophilic O atom to the electrophilic α-carbon of the ruthenium vinylidene species. The cyclization of acetylenic alcohols to enol ethers resulting from anti-Markovnikov intramolecular addition has been widely reported with Group 6 metal catalysts (see Chapter 5). This cyclization is much less common with ruthenium catalysts. The sole efficient catalytic system is based on $RuCl(Cp)(p\text{-}FC_6H_4)_3P)_2$ (5 mol%), $(p\text{-}FC_6H_4)_3P$ (20 mol%), Bu_4NPF_6 (15 mol%), N-hydroxysuccinimide sodium salt (50 mol%) in dimethylformamide at 85 °C. Under these conditions, pent-4-yn-1-ols were completely transformed and selectively converted into dihydropyrans within 25 h (Scheme 10.12) [52].

From the same substrates, a different reaction took place when the electron-deficient phosphine was replaced by the electron-rich $(p\text{-}MeOC_6H_4)_3P$ phosphine. In the presence of this slightly modified catalytic system, the recovery of the organic ligand as a lactone was made possible by oxidation of the intermediate cyclic alkoxycarbene with N-hydroxysuccinimide, a mild oxidant which did not destroy the catalyst (Scheme 10.13) [52].

Both oxidative cyclization and cycloisomerization were applied to a variety of substrates, including sugar derivatives, the only restriction to the formation of lactones was the presence of a tertiary alcohol functionality.

Scheme 10.14 rationalizes the divergent behavior of the two catalytic systems in these selective transformations of pent-1-yn-ols. The presence of phosphine ligands promotes the formation of ruthenium vinylidene species which are key intermediates in both reactions. The more electron-rich $(p\text{-}MeOC_6H_4)_3P$ phosphine favors the formation of a cyclic oxacarbene complex which leads to the lactone after attack of the N-hydroxysuccinimide anion on the carbenic carbon. In contrast, the more labile electron-poor $(p\text{-}FC_6H_4)_3P$ phosphine is exchanged with the N-hydroxysuccinimide anion and makes possible the formation of an anionic ruthenium intermediate which liberates the cyclic enol ether after protonation.

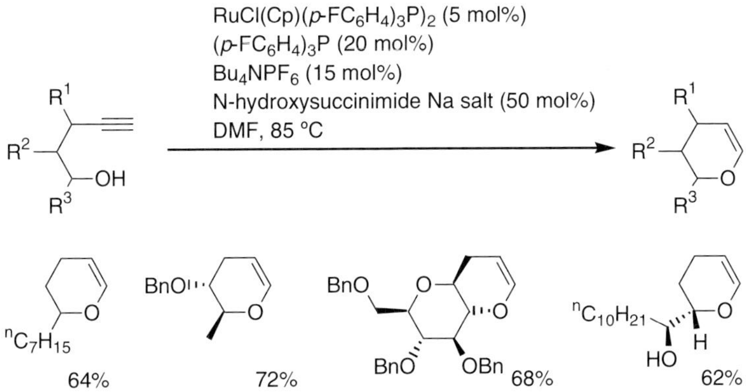

Scheme 10.12 Preparation of cyclic enol ethers from pent-4-yn-1-ols.

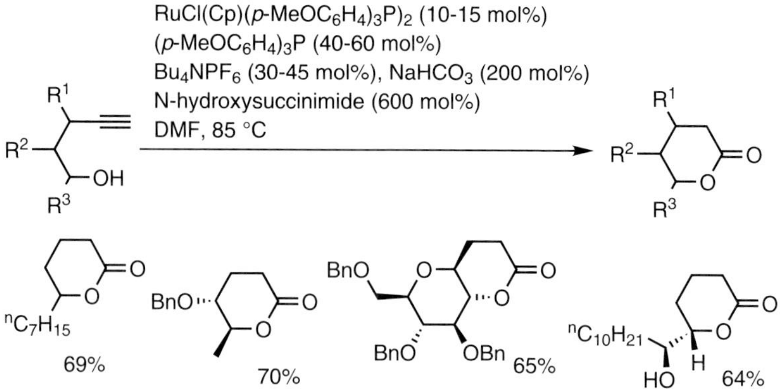

Scheme 10.13 Preparation of δ-lactones from pent-4-yn-1-ols.

Starting from but-3-yn-1-ols (homopropargylic alcohols), a similar oxidative cyclization leading to butyrolactones was observed. The best catalytic system reported up to now is based on RuCl(C$_5$H$_5$)(cod), tris(2-furyl)phosphine as ancillary ligand, NaHCO$_3$ as a base, in the presence of Bu$_4$NBr or Bu$_4$NPF$_6$, and N-hydroxysuccinimide

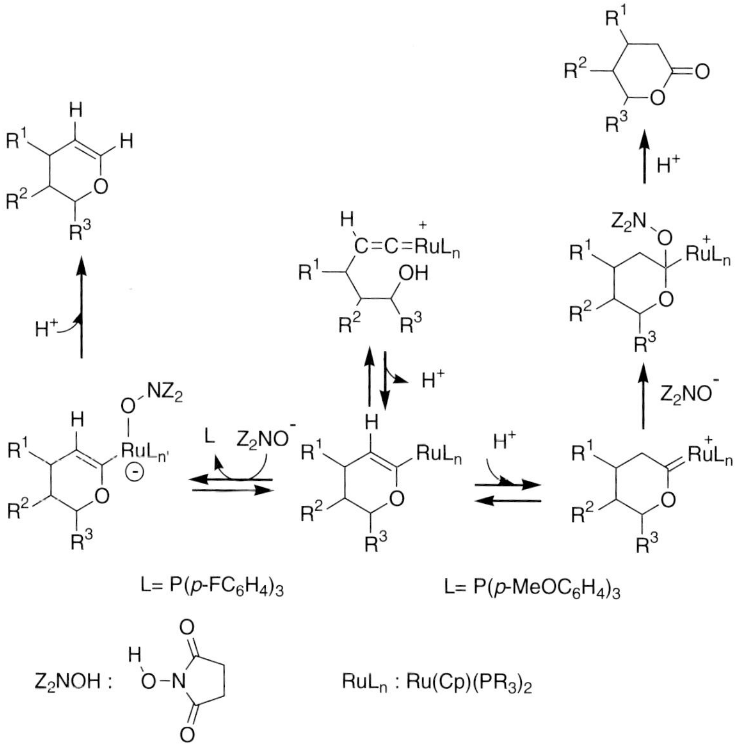

Scheme 10.14 Mechanistic proposal for cyclic enol ether and lactone formation based on a common ruthenium vinylidene intermediate.

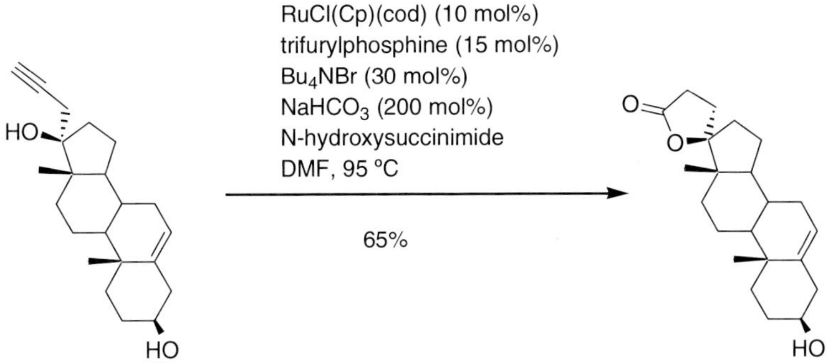

Scheme 10.15 Lactonization of a steroidal homopropargylic alcohol structure.

as the oxidant in DMF–water at 95 °C as illustrated by a steroidal compound in Scheme 10.15 [53].

10.3
Formation of C—N Bonds via Anti-Markovnikov Addition to Terminal Alkynes

10.3.1
Addition of Amides to Terminal Alkynes

The direct addition of secondary amides to terminal alkynes has been successfully carried out in the presence of a ruthenium catalyst generated *in situ* from Ru(methallyl)$_2$(cod) (2 mol%) as metal source, P^{n}Bu$_3$ (6 mol%) and dimethylaminopyridine (4 mol%), in toluene at 100 °C [54]. Under these conditions the (*E*)-enamides were stereoselectively formed from a variety of cyclic amides and ureas (Scheme 10.16).

When P^{n}Bu$_3$ is replaced by the diphosphine ligand Cy$_2$PCH$_2$PCy$_2$ in the presence of water as additive, the addition takes place with the same regioselectivity but with the reverse stereoselectivity and the (*Z*)-enamides are obtained preferentially [54].

10.3.2
Formation of Nitriles *via* Addition of Hydrazines to Terminal Alkynes

The formation of a ruthenium vinylidene is proposed as the key intermediate in the regioselective addition of hydrazine to terminal alkynes [55]. This reaction, which proceeds via addition of the primary amino group of a 1,1-disubstituted hydrazine followed by deamination, provides an unprecedented access to a variety of aromatic and aliphatic nitriles. The tris(pyrazolyl)borate complex RuCl(Tp)(PPh$_3$)$_2$ gave the best catalytic activity in the absence of any chloride abstractor (Scheme 10.17).

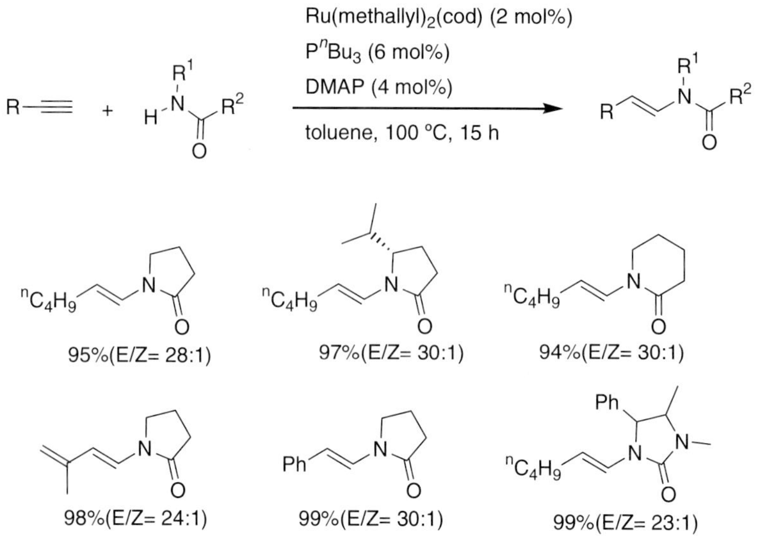

Scheme 10.16 Enamides from anti-Markovnikov addition of amides to alkynes.

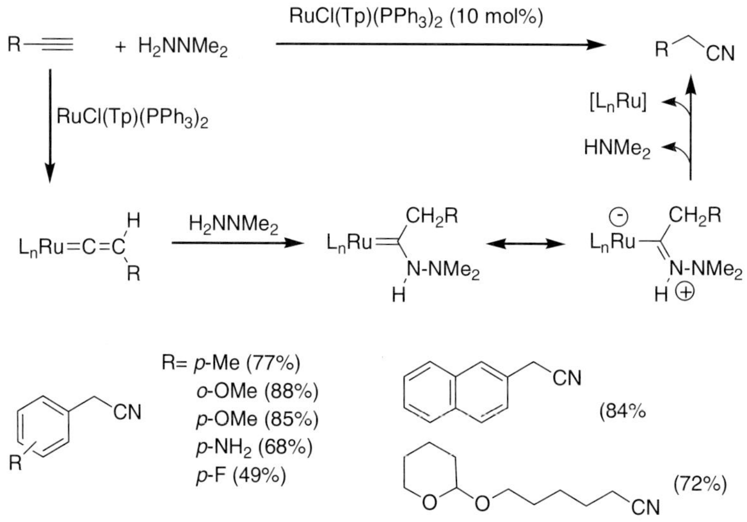

Scheme 10.17 Nitriles from addition of hydrazine to terminal alkynes.

10.4
Hydrophosphination: Synthesis of Vinylic Phosphine

The addition of secondary phosphines HPR_2 to prop-2-ynols in the presence of RuCl(C_5Me_5)(cod) or RuCl(C_5Me_5)(PPh$_3$)$_2$ provides the first regio- and stereo-selective direct hydrophosphination of propargylic alcohols and leads to bifunctional

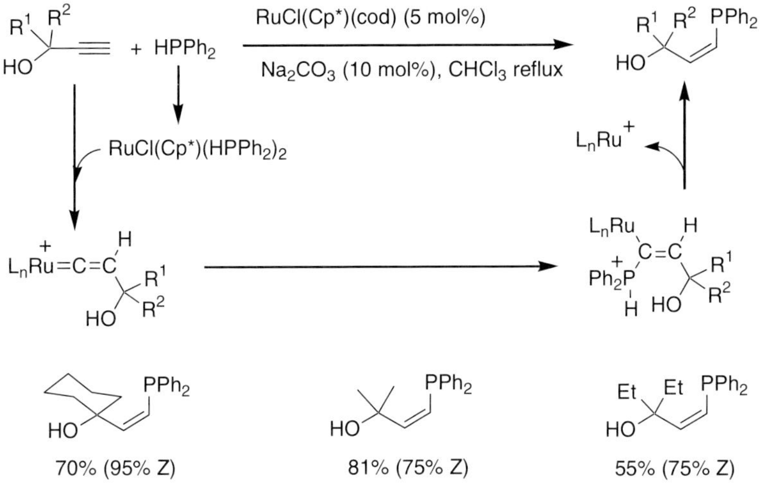

10.5 C–C Bond Formation: Dimerization of Terminal Alkynes

Scheme 10.18 Preparation of 3-hydroxyprop-1-enyl phosphines.

(Z)-olefins [56]. Indeed, the reaction of tertiary propargylic alcohols with diphenylphosphine in the presence of Na_2CO_3 in refluxing $CHCl_3$ leads to 3-diphenylphosphinoprop-2-enols in good yields and high stereoselectivity in favor of the (Z)-isomer (Scheme 10.18). The Z-isomer with observed OH···P interaction is the kinetic product, which readily isomerizes into the E-isomer over silica.

It is noteworthy that the presence of a phosphine ligand is required to trigger the formation of the hydroxyvinylidene species, which is the key intermediate in this regioselective hydrophosphination reaction. This is possible starting from $RuCl(Cp^*)(PPh_3)_2$ or from $RuCl(Cp^*)(HPPh_2)_2$, which is easily generated from substitution of the labile cod ligand in $RuCl(Cp^*)(cod)$.

10.5
C–C Bond Formation: Dimerization of Terminal Alkynes

Terminal alkynes can undergo several types of interaction with ruthenium centers. In addition to the formation of ruthenium vinylidene species, a second type of activation provides alkynyl ruthenium complexes via oxidative addition.

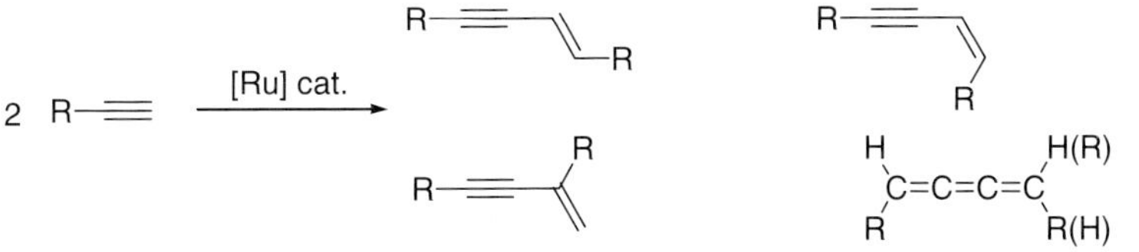

$$\text{(10.8)}$$

When these two types of coordination take place at the same metal center, the migration of the alkynyl ligand to the C_α-carbon atom of the vinylidene can occur, to form enynyl intermediates which, upon protonation by the terminal alkyne, lead to

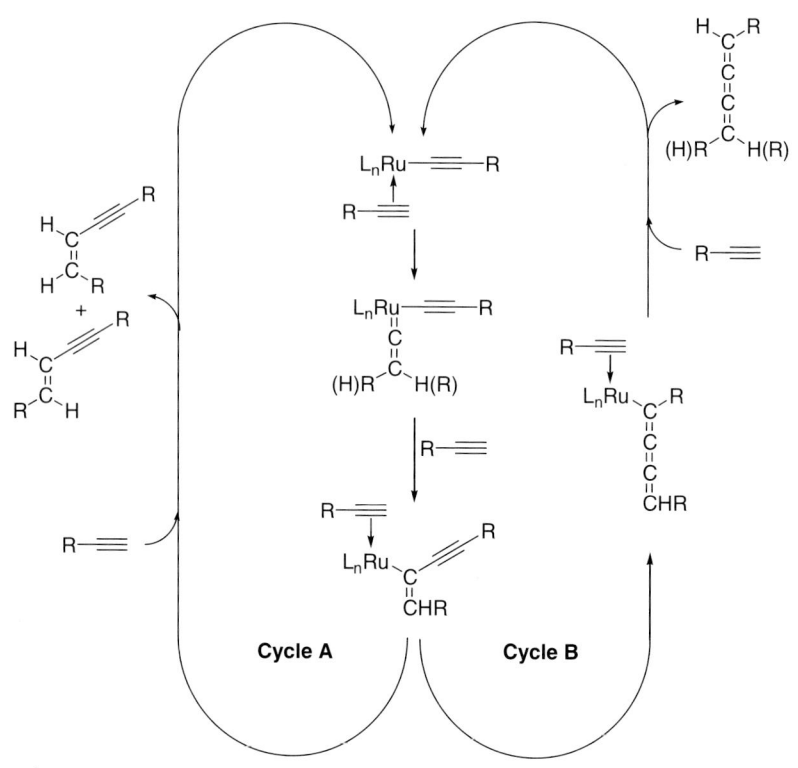

Scheme 10.19 Proposed catalytic cycles for terminal alkynes dimerization.

the formation of enynes corresponding to alkyne dimerization (Equation 10.8 – Scheme 10.19, cycle A), as already mentioned with Group 9 metals (Chapter 9). In special cases, the rearrangement of the enynyl ligand to an allenylidenyl ligand can occur and the formation of the butatriene dimer is observed (Scheme 10.19, cycle B).

Thus, ruthenium complexes containing a bulky electron-donating polydentate nitrogen ligand, such as a tris(pyrazolyl)borate (Tp) **19** [57–59], a polypodal phosphorus ligand like P(CH$_2$CH$_2$PPh$_2$)$_3$**20**, **21** [60, 61], P(CH$_2$CH$_2$PPh$_2$)$_3$**22** [62], a pentamethylcyclopentaniedyl **23** [63] or an indenyl **24** [64] ligand are efficient catalysts for the selective head-to-head dimerization of terminal alkynes to enynes (Scheme 10.20).

Most of these catalytic systems are able to dimerize either aromatic alkynes, such as phenylacetylene derivatives, or aliphatic alkynes, such as trimethylsilylacetylene, tert-butylacetylene and benzylacetylene. The stereochemistry of the resulting enynes depends strongly on both the alkyne and the catalyst precursor. It is noteworthy that the vinylidene ruthenium complex RuCl(Cp*)(PPh$_3$)(=C=CHPh) catalyzes the dimerization of phenylacetylene and methylpropiolate with high stereoselectivity towards the (E)-enyne [65, 66], and that head-to-tail dimerization is scarcely favored with this catalyst. It was also shown that the metathesis catalyst RuCl$_2$(P-Cy$_3$)$_2$(=CHPh) reacted in refluxing toluene with phenylacetylene to produce a

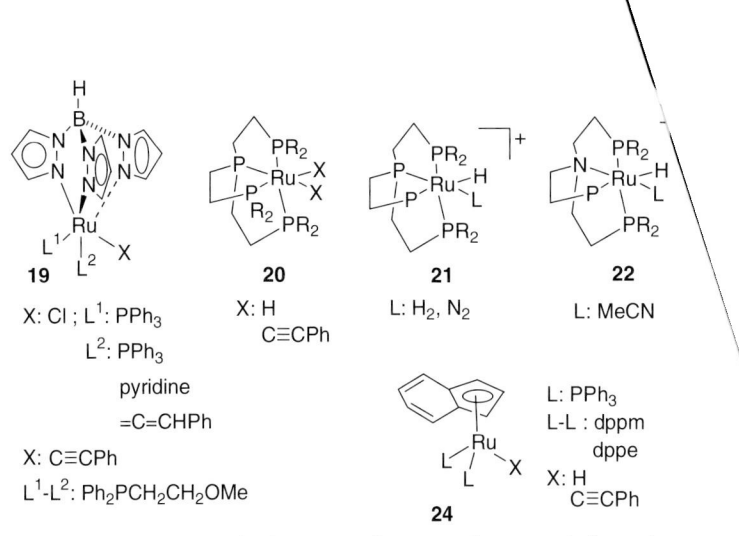

Scheme 10.20 Examples of ruthenium catalysts active for terminal alkynes dimerization.

ruthenium vinylidene species, which promoted the regioselective dimerization of phenylacetylene to (E)-1,4-diphenylbutenyne [67]. The catalytic system based on [RuCl$_2$(p-cymene)]$_2$ operating at room temperature in acetic acid leads to high stereoselectivity in favor of the (E)-enynes from aromatic alkynes without formation of enol acetates [68]. On the other hand, complex 22 [62], RuCl$_2$(iPr$_3$P)$_2$(=C=CHPh) [69], and a dinuclear bis(pentamethylcyclopentadienyl)diruthenium complex bearing bridging thiolates [70], have revealed very good catalytic efficiency for the production of the (Z)-1,4-enyne structure.

The head-to-head dimerization with formation of a butatriene derivative was very scarcely observed as the main catalytic route (Scheme 10.19, cycle B). Nevertheless, this was the case with benzylacetylene in the presence of RuH$_3$Cp*(PCy$_3$) as catalyst precursor in tetrahydrofuran at 80 °C which gave more than 95% of (Z)-1,4-dibenzylbutatriene [66], and with tert-butylacetylene with two efficient catalytic systems capable of generating zero-valent ruthenium species, RuH$_2$(PPh$_3$)$_3$(CO) and Ru(cod)(cot) in the presence of an excess of triisopropylphosphine, which led to (Z)-1,4-di-tert-butylbutatriene as the major compound [71–73].

Finally, it can be noted that some cross-dimerization of terminal alkynes with internal alkynes, where ruthenium vinylidene intermediates are postulated, have also been reported [74, 75].

10.6
Conclusion

The anti-Markovnikov addition of carbamates to terminal alkynes was introduced as the first example of catalytically active metal vinylidene in 1986. The development of this concept to other O-nucleophiles followed immediately and carboxylic acids, water and allylic alcohols were used to produce the corresponding addition products. The intermediates, which formally corresponds to [...]ile, has also been investigated extensively and [...] enynes and butatrienes. The catalytic systems [...]ophiles, and are being improved continuously. [...] volving metal vinylidenes is increasing [76], and [...] transformations are now appearing [77].

L: PPh_3
PCy_3
PMe_3

[...] Yus, M.
, 3079.
.H. (1999)
ch, 32, 311.
Organometallic

af, P.H. (2006)
ernational Edition,

P.H. and Lécolier, S.
Letters, 27, 6333.
euf, P.H. (1986) Journal
ociety. Chemical
, 790.
ki, Y., Bruneau, C. and
(1989) Journal of Organic
54, 1518.
, T., Hori, Y., Yamakawa, Y. and
abe, Y. (1987) Tetrahedron Letters, 28,
/.
Höfer, J., Doucet, H., Bruneau, C. and Dixneuf, P.H. (1991) Tetrahedron Letters, 32, 7409.

10 Bruneau, C. and Dixneuf, P.H. (1987) Tetrahedron Letters, 28, 2005.
11 Fournier, J., Bruneau, C., Dixneuf, P.H. and Lécolier, S. (1991) Journal of Organic Chemistry, 56, 4456.
12 Bruneau, C. and Dixneuf, P.H. (1992) Journal of Molecular Catalysis, 74, 97.
13 Rotem, M. and Shvo, Y. (1983) Organometallics, 2, 1689.
14 Mitsudo, T., Hori, Y., Yamazaki, Y. and Watanabe, Y. (1987) Journal of Organic Chemistry, 52, 2230.
15 Bruneau, C., Neveux, M., Kabouche, Z., Ruppin, C. and Dixneuf, P.H. (1991) Synlett, 755.
16 Bruneau, C. and Dixneuf, P.H. (1997) Journal of the Chemical Society. Chemical Communications, 507.
17 Neveux, M., Seiller, B., Hagedorn, F., Bruneau, C. and Dixneuf, P.H. (1993) Journal of Organometallic Chemistry, 451, 133.
18 Doucet, H., Höfer, J., Bruneau, C. and Dixneuf, P.H. (1993) Journal of the Chemical Society. Chemical Communications, 850.
19 Doucet, H., Martin-Vaca, B., Bruneau, C. and Dixneuf, P.H. (1995) Journal of Organic Chemistry, 60, 7247.
20 Dixneuf, P.H., Bruneau, C. and Dérien, S. (1998) Pure and Applied Chemistry, 70, 1065.
21 Doucet, H., Höfer, J., Derrien, N., Bruneau, C. and Dixneuf, P.H. (1996) Bulletin de la Societe Chimique de France, 133, 939.
22 Doucet, H., Derrien, N., Kabouche, Z., Bruneau, C. and Dixneuf, P.H. (1997) Journal of Organometallic Chemistry, 551, 151.
23 Kabouche, A., Kabouche, Z., Bruneau, C. and Dixneuf, P.H. (1999) Journal of Chemical Research (M), 1247.
24 Gemel, C., Trimmel, G., Slugovc, C., Kremel, S., Mereiter, K., Schmid, R. and Kirchner, K. (1996) Organometallics, 15, 3998.
25 Melis, K., Samulkiewicz, P., Rynkowski, J. and Verpoort, F. (2002) Tetrahedron Letters, 43, 2713.
26 Pelagatti, P., Bacchi, A., Balordi, M., Bolano, S., Calbiani, F., Elviri, L., Gonsalvi, L., Pelizzi, C., Peruzzini, M. and Rogolino,

D. (2006) *European Journal of Inorganic Chemistry*, 2422.
27 Goossen, L.J., Paetzold, J. and Koley, D. (2003) *Chemical Communications*, 706.
28 Opstal, T. and Verpoort, F. (2002) *Synlett*, 935.
29 Jimenez Tenorio, M., Puerta, M.C., Valerga, P., Moreno-Dorado, F.J., Guerra, F.M. and Massanet, G.M. (2001) *Chemical Communications*, 2324.
30 Devanne, D., Ruppin, C. and Dixneuf, P.H. (1988) *Journal of Organic Chemistry*, 53, 925.
31 Bruneau, C., Kabouche, Z., Neveux, M., Seiller, B. and Dixneuf, P.H. (1994) *Inorganica Chimica Acta*, 222, 154.
32 Picquet, M., Bruneau, C. and Dixneuf, P.H. (1997) *Journal of the Chemical Society. Chemical Communications*, 1201.
33 Picquet, M., Fernandez, A., Bruneau, C. and Dixneuf, P.H. (2000) *European Journal of Organic Chemistry*, 2361.
34 Hintermann, L. and Labonne, A. (2007) *Synthesis*, 8, 1121.
35 Tokunaga, M. and Wakatsuki, Y. (1998) *Angewandte Chemie-International Edition*, 37, 2867.
36 Suzuki, T., Tokunaga, M. and Wakatsuki, Y. (2001) *Organic Letters*, 3, 735.
37 Tokunaga, M., Suzuki, T., Koga, N., Fukushima, T., Horiuchi, A. and Wakatsuki, Y. (2001) *Journal of the American Chemical Society*, 123, 11917.
38 Alvarez, P., Bassetti, M., Gimeno, J. and Mancini, G. (2001) *Tetrahedron Letters*, 42, 8467.
39 Suzuki, T., Tokunaga, M. and Wakatsuki, Y. (2002) *Tetrahedron Letters*, 43, 7531.
40 d'Alessandro, N., Di Deo, M., Bonetti, M., Tonucci, L., Morvillo, A. and Bressan, M. (2004) *European Journal of Inorganic Chemistry*, 810.
41 Grotjahn, D.B., Incarvito, C.D. and Rheingold, A.L. (2001) *Angewandte Chemie-International Edition*, 40, 3884.
42 Grotjahn, D.B. and Lev, D.A. (2004) *Journal of the American Chemical Society*, 126, 12232.
43 Chevallier, F. and Breit, B. (2006) *Angewandte Chemie-International Edition*, 45, 1599.
44 Labonne, A., Kribber, T. and Hintermann, L. (2006) *Organic Letters*, 8, 5853.
45 Lo, C.-Y., Guo, H., Lian, J.-J., Shen, F.-M. and Liu, R.-S. (2002) *Journal of Organic Chemistry*, 67, 3930.
46 Trost, B.M., Dyker, G. and Kulawiec, R.J. (1990) *Journal of the American Chemical Society*, 112, 7809.
47 Trost, B.M. and Kulawiec, R.J. (1992) *Journal of the American Chemical Society*, 114, 5579.
48 Trost, B.M. and Flygare, J.A. (1994) *Journal of Organic Chemistry*, 59, 1078.
49 Trost, B.M., Kulawiec, R.J. and Hammes, A. (1993) *Tetrahedron Letters*, 34, 587.
50 Trost, B.M., Vidal, B. and Thommen, M. (1999) *Chemistry – A European Journal*, 5, 1055.
51 Nishibayashi, Y., Takei, I. and Hidai, M. (1997) *Organometallics*, 16, 3091.
52 Trost, B.M. and Rhee, Y.H. (2002) *Journal of the American Chemical Society*, 124, 2528.
53 Trost, B.M. and Rhee, Y.H. (1999) *Journal of the American Chemical Society*, 121, 11680.
54 Goossen, L.J., Rauhaus, J.E. and Deng, G. (2005) *Angewandte Chemie-International Edition*, 44, 4042.
55 Fukumoto, Y., Dohi, T., Masaoka, H., Chatani, N. and Murai, S. (2002) *Organometallics*, 21, 3845.
56 Jérôme, F., Monnier, F., Lawicka, H., Dérien, S. and Dixneuf, P.H. (2003) *Chemical Communications*, 696.
57 Slugovc, C., Mereiter, K., Zobetz, E., Schmid, R. and Kirchner, K. (1996) *Organometallics*, 15, 5275.
58 Slugovc, C., Doberer, D., Gemel, C., Schmid, R., Kirchner, K., Winkler, B. and Stelzer, F. (1998) *Monatshefte für Chemie*, 129, 221.
59 Pavlik, S., Gemel, C., Slugovc, C., Mereiter, K., Schmid, R. and Kirchner, K. (2001) *Journal of Organometallic Chemistry*, 617–618, 301.

60 Bianchini, C., Peruzzini, M., Zanobini, F., Frediani, P. and Albinati, A. (1991) *Journal of the American Chemical Society*, **113**, 5453.

61 Bianchini, C., Frediani, P., Masi, D., Peruzzini, M. and Zanobini, F. (1994) *Organometallics*, **13**, 4616.

62 Chen, X., Xue, P., Sung, H.H.Y., Williams, I.D., Peruzzini, M., Bianchini, C. and Jia, G. (2005) *Organometallics*, **24**, 4330.

63 Yi, C.S. and Liu, N. (1996) *Organometallics*, **15**, 3968.

64 Bassetti, M., Marini, S., Tortorella, F., Cadierno, V., Diez, J., Pilar Gamasa, M. and Gimeno, J. (2000) *Journal of Organometallic Chemistry*, **593–594**, 292.

65 Yi, C.S. and Liu, N. (1997) *Organometallics*, **16**, 3910.

66 Yi, C.S. and Liu, N. (1999) *Synlett*, 281.

67 Melis, K., De Vos, D., Jacobs, P. and Verpoort, F. (2002) *Journal of Organometallic Chemistry*, **659**, 159.

68 Bassetti, M., Pasquini, C., Raneri, A. and Rosato, D. (2007) *Journal of Organic Chemistry*, **72**, 4558.

69 Katayama, H., Nakayama, M., Nakano, T., Wada, C., Akamatsu, K. and Ozawa, F. (2004) *Macromolecules*, **37**, 13.

70 Matsuzaka, H., Takagi, Y., Ishii, Y., Nishio, M. and Hidai, M. (1995) *Organometallics*, **14**, 2153.

71 Wakatsuki, Y., Yamazaki, H., Kumegawa, N., Satoh, T. and Satoh, J.Y. (1991) *Journal of the American Chemical Society*, **113**, 9604.

72 Wakatsuki, Y. and Yamazaki, H. (1995) *Journal of Organometallic Chemistry*, **500**, 349.

73 Yamazaki, H. (1976) *Journal of the Chemical Society. Chemical Communications*, 841.

74 Katayama, H., Yari, H., Tanaka, M. and Ozawa, F. (2005) *Chemical Communications*, 4336.

75 Shirakawa, E., Nakayama, K., Morita, R., Tsuchimoto, T., Kawakami, Y. and Matsubara, T. (2006) *Bulletin of the Chemical Society of Japan*, **79**, 1963.

76 Trost, B.M. and McClory, A. (2008) *Chemistry an Asian Journal*, **3**, 164.

77 Kim, H., Goble, S.D. and Lee, C. (2007) *Journal of the American Chemical Society*, **129**, 1030.

Index

a

acetylene-vinylidene tautomerization   134ff, 152f
ω-acetylenic silyl enol ether   173
acetylethynyl complex   103ff
acyclic diene metathesis (ADMET)   253f
acyl complexes   15, 161, 182f
– dehydration   15
cis-acyl-2-ethynylcyclopropanes   182f
alcohols   321ff
aldol condensation   242f
aldehydes   318ff
– β-hydroxyaldehydes   319ff
alkenyl-carbene complexes   70ff
alkoxy-carbene complexes   74
alkynyl-allenylidene resonance   70
σ-alkynyl complex   73ff
σ-allenyl complexes   71ff, 76
allenylidene complexes   11f, 42f, 61ff, 109f, 132, 151ff, 217ff
– addition of nucleophiles   11f
– alkenyl-amino-allenylidene complexes   85ff
– allenylidene to indenylidene rearrangement   259ff
– amino-allenylidene complexes   83ff
– as catalysts in metathesis   251ff
– as intermediates   78
– bridging allenylidene complexes   69
– C7-bridged   89
– C-C coupling reactions   79f
– chiral allenylidene complexes   77
– cyclization reactions   81ff
– Diels-Alder reactions   11, 86, 236ff
– deprotonation   89
– dienyl-allenylidene complexes   85
– diheterocyclizations   88
– dinuclear allenylidene complexes   65ff
– electrophilic additions to Cβ   70
– enyne metathesis   254f
– heteroatom-substituted allenylidene complexes   64
– heterocyclizations   87
– intramolecular cyclization   221
– nucleophilic additions to Cα or Cγ   71ff
– polyunsaturated allenylidene complexes   85
– reactions   80ff, 219ff
– reactions with alcohols   219f
– reactions with alkynes   80, 84
– reactions with amides   219f
– reactions with amines   219f
– reactions with carbodiimides   87
– reactions with carbon monoxide   87
– reactions with dinucleophiles   87
– reactions with hydroxylamines   87
– reactions with imines   83
– reactions with pyrazoles   87
– reactions with thiols   219f
– reactions with ynamines   85
– reactivity   69ff, 81ff
– reduction   89
– ring closing metathesis (RCM)   254
– ring closing reactions   88, 254
– ring opening metathesis polymerization (ROMP)   255f
– solid state structure   99
– substitution reactions   219ff
– synthesis   62ff
allenylidene-ene reactions   236ff
allenyl-vinylidene complex   83
allylic alcohol   322
alkene metathesis   251ff
alkenes   209ff, 245ff, 251ff
alkenyl esters   318
alkenylalkylidenes   252ff, 262ff
alkenylcarbene-metal complex   163ff
alkenylstannane   165f

alkenylvinylidenes 28
cis-1-alkenyl-2-ethynylcyclopropanes 184
alkylation 8
1,2-alkylidene shift 200ff
alkynals 211
alkyne complexes 4ff, 327ff
– redox rearrangements 4f
alkyne coupling 45
alkyne dimerization 281, 290ff, 327ff
– hydrative 294f
– Ir(I)-catalyzed 292f
– Rh(I)-catalyzed cyclodimerization 281, 296f
– Rh(III)-catalyzed 291
1-alkynes 2ff, 313ff, 327ff
– 1,1-functionalization 297ff
– isomerization to vinylidenes 3ff, 144ff, 152f, 280, 327
– addition of nucleophiles 313ff, 327f
alkyne-vinylidene tautomerization 144ff, 152f, 280, 299, 305
alkynol compounds 186f
ω-alkynol cyclization 165f
alkynol cycloisomerization 142ff, 153
bis-alkynyl silyl enol ether 177
alkynylalkenyl complex 119
alkynylimines 178
alkynylsilanes 282f, 302f
– silyl migration 282f, 302f, 308f
– dimerization 292f
alkynyl-metal complexes 8ff, 186, 313ff
allenynes 187
allylic alcohol 227f
amides 325
aminocarbene complexes 34f
andirolactone 267
azabutadiene-2-ethynyl complexes 116ff
3-azadienes 202ff
azaphosphacarbene complexes 75
azetidinylidene complexes 35f, 39, 82, 159ff
azidoalkynyl complexes 78
azonia-butadienyl complex 74ff, 82
azulene 233f

b

back donation 132
barrier of rotation of vinylidene ligands 132f
benzofuranylidene complex 182f
benzopyranylidene complex 178f
benzylidene tungsten complex 160
BINAP 228
biscarbyne complexes 141ff
bond cleavage 245f
bond fission reactions 244ff
Brønsted acid 239, 294
butadiyne 103ff

butadiynyl complex 104ff
butatriene-Rh 79
butatrienylidene complexes 4f, 100ff, 114ff
– reactions with nucleophiles 115ff
– synthesis 103ff
– X-ray structure analyses 100ff
butatrienyl(dihydrido) complex 119
3-butynol 162, 165f, 170
3-butynylamine 165f

c

carbamates 314ff
carbene complex 99ff, 163, 217
– cyclic carbene complex 163, 170f
carbocyclization, see cyclization
carbonyl ylide 182f
carboxylic acids 316ff
carbyne complexes 11ff, 26, 260f
– deprotonation 11ff
catalysis 193ff, 217ff, 251ff, 279ff, 313ff
– asymmetric 304
– homocoupling reaction 238f
– metal-allenylidene complexes 217ff, 251ff
– metal-vinylidene complexes 193ff, 251ff
– alkene metathesis 251ff
– nucleophilic additions 313ff
catalyst precursors 252ff, 319f
C-C coupling reactions 79f
– allenylidene complexes 79f
– alkyne dimerization 327ff
chiral compounds 228ff
chiraphos 222
chromenone derivatives 231ff
chromium 72ff, 110ff, 120, 159ff
C-H bond activation 287, 294f, 304f
cinnamaldehyde 242
Claisen rearrangement 235f
"click" reaction, Huisgen reaction 306
C-N bond formation 325ff
copper vinylidene complexes in catalysis 307f
cordycepin 167
coronene derivatives 201f
C-O bond formation 314ff
C-P bond formation 326f
cumulenylidene complexes 99ff, 217, 258f
– synthesis 103
cyclization 146ff, 165ff, 170f, 175ff, 193ff, 221ff
– catalytic 193ff
– cyclocarbonylation 208ff
– cyclodecarbonylation 207f
– electrocyclization 178ff
– endo- and exo 146ff, 175ff
– endo-selective 176
– hydrative cyclization 211ff

– intramolecular 221ff
– of alkynals 211
– of 3-azadienes 202ff
– of 3,5-dien-1-ynes 197ff, 202
– of 1,5-enynes 211ff
– of enynyl epoxides 204ff
– of 3-en-1-ynes 195ff
– of ethynylbenzenes 202
– of cis-ethynyl-2-vinyloxiranes 203f
– stoichiometric 193ff
– tandem cyclization 177
– umpolung cyclization 212
cycloaddition reactions 74, 81ff, 164f, 182f, 209ff, 231ff
– [1+2] cycloaddition reaction 149
– [2+2] cycloaddition reaction 82ff, 148ff, 160, 160f
– [3+2] cycloaddition reaction 182f, 186f
– of alkenes and enynes 209ff
– of allenylidene complexes 74, 81ff
– of 1,3-diketones 231ff
– ruthenium-catalyzed 209ff, 231
cycloaromatization 196ff, 286ff, 301
– Myers 286f
– via 6π-electrocyclization 301, 307
cyclobutylidene complexes 29, 85, 161f
cyclocarbonylations 208ff
– catalytic 208ff
– via addition of ruthenium vinylidene intermediates 208ff
– of 1,1'-bis(silylethynyl)ferrocene 208f
cyclodecarbonylation 207f
cycloisomerizations 148ff, 167ff, 323ff
– endo- and exo 148ff
cycloheptatrienes 184
cyclopropanation 172, 303f
cyclopropane 240f
cyclopropyl groups 239f

d
decarbonylation 245ff
DFT-calculations 305f, 319
dicyclohexylcarbodiimide 159
Diels-Alder reaction 86, 180f, 281
– Rh-catalyzed
dienyl-carbene complex 164
2,3-dihydrofuran 165
dihydropyran 167ff
dihydropyridinium complexes 82
dimerization 209f, 328f
1,4-diynes 77
diynyl complex 104ff

e
electrocyclization 178ff, 182f, 195ff

electron density distribution 130f
electron reservoirs 142
electronic structure 100ff, 129ff
electrophiles 7, 25ff, 217
electrophilic attack 102, 131f, 152
enediynes 194f
enol esters 315ff
enynes 19, 77, 195, 209ff, 211ff, 222f, 241, 253ff, 327ff
– metathesis 253ff
– Pt-catalyzed cycloisomerization 304ff
– Rh(I)-catalyzed cycloisomerization 282ff, 300f
– Rh(I)-catalyzed cycloisomerization with N-propargyl enamines 284ff
– tetrasubstituted 241
σ-enynyl complexes 76, 317f
enynyl epoxides 204ff
epoxyalkynes 170
ethenolysis 272f
ethers 321
– cyclic enol eters 323ff
– lactones, see lactones
– unsaturated 321
β-ethynyl α,β-unsaturated carbonyl compounds 178f
o-ethynylphenylcabonyl compounds 178ff
ethynylquinoline complexes 116ff
o-ethynylstyrene 178
cis-ethynyl-2-vinyloxiranes 203f

f
1,1'-bis(silylethynyl)ferrocene 208f
ferrocenylmethylamine 115
Fischer carbene complexes 130, 159, 165ff, 217
– cyclic Fischer carbene complexes 165ff
Friedel-Crafts reaction 307
furans 233f
– unsaturated 321
furanylidene complexes 186f
2-furyl-carbene complexes 180f

g
glycosylation 168f
gold vinylidene complexes in catalysis 307ff
Grubbs catalyst 251

h
1,2-halogen shift 6, 307
heptahexaenylidene complex 113ff, 123f
– reaction with dimethylamine 114
– synthesis 113
hetero-Claisen rearrangement 115
hetero-Cope rearrangement 115

hexadienones 241ff
hexapentaene complexes 80ff, 121ff, 123
Huisgen reaction 306
hydrazines 325f
hydride migration 136
hydrido-alkynyl-intermediate 135ff, 139
hydroalkoxylation 288f, 301
hydroamination 288f, 294ff
hydroboration 238f, 290f
1,2-hydrogen shift 2ff, 135
1,3-hydrogen shift 3, 137ff
1,4-hydrogen shift 105ff
hydrophosphination 326f
hydrosilylation 290
hydroxyalkynyl complexes 81
hydroxyl-carbene complexes 75

i

indenes 207f
indenylidene complexes 71ff, 251ff
– generation 265ff
– in alkene metathesis 268ff
– in polymerization 271ff
– in ROMP 261f
– in RCM 262, 268ff
– in the Kharasch reaction 273f
indoles 233f, 288f
insertion 99, 290f, 297ff, 307
π-π interaction 229
interconversion (Z/E isomeric) 151ff
iodinated vinylidene complex 178f
iodoalkyne 178
iridium 41f, 103ff, 110, 119f, 292ff
– bimetallic iridium complex 294
iron 104ff
isomerization 17
– η^2-vinyl ether complexes 17

k

ketene intermediates 205ff
γ-ketoalkynes 77
β-ketoesters 318
ketones 321ff
– unsaturated 321ff
Kharasch reaction 273f

l

lactones 317ff, 323ff
ligand substitution 30f
– in vinylidene complexes 30f

m

manganese 8ff, 103ff
anti-Markovnikov addition 313ff
metal alkynyls 9ff

– redox rearrangements 9f
– oxidative coupling 9ff
metallacyclobutene 164
metallacumulenes 100
– atomic charges 102
– bond dissociation energy 101
– bond lengths 101
– bond length variation 101
– CCC- angle 101
– charge distribution 102
– DFT calculations 101
– even-chain 101
– HOMO 102
– HOMO-LUMO-gap 102
– LUMO 102
– MCC-angle 101
– odd-chain 101
metallacyclobutane 82
metal-alkynyl complexes 7f,
– redox rearrangements 9f,
metal-allenylidene complexes, *see* allenylidene complexes
metal-assisted proton migration 144ff
metal-carbon complexes 14
metal-carbyne complexes 11f
metal-vinylidene π-bonding 133f
metal-vinylidene complexes, *see* vinylidene complexes
metal transfer in allenylidene complexes 89
metathesis 164f, 187f, 251ff
– ring closing metathesis 187f
– self-metathesis 272
multi-component coupling 294ff, 300ff
muricatacin 167

n

naphthalene derivatives 198f, 209, 233f
– formation via metal-vinylidene complexes 198ff
NHC (N-heterocyclic carbenes) 258f
Nicholas reaction 218
nickel vinylidene complexes in catalysis 302f
nitriles 325f
nucleophiles 20ff, 118, 172ff, 213f, 218f, 313ff
– N-nucleophiles 325ff
– O-nucleophiles 314ff
– P-nucleophiles 326ff
nucleophilic attack 102ff, 131f, 144ff, 180f, 313ff
nucleophilicity 103ff, 218

o

optically active compounds 228
oxacycle formation 123, 162, 169f

2-oxacyclopentadienylidene-metal complex 162f
3-oxapentadienyl ruthenium complex 123
oxepins 182f
oxidative addition 3, 16f, 207f
– of 1-alkynes 3
– of α-chlorvinylsilanes 16f

p

palladium vinylidene complexes in catalysis 303f
pentacarbonylvinylidene complexes 35ff
pentacarbonyl-metal complexes 161ff
pentatetraenylidene complexes 100ff, 109ff, 119ff
– intramolecular addition 121
– reduction 121
– synthesis 108
– thermolysis 121
– X-ray structure analyses 100ff
pentatetraenylidene ligand 108ff
η^3-pentatrienyl complexes 79
penta-1,3-diyne 108ff
pentenylidene complex 162f
phosphinovinyl complexes 35
phosphino-allenyl complexes 72ff
phosphinylation 222f
phosphonio-alkynyl complexes 73ff
phosphonio-butadienyl complexes 73ff
pinacolborane 238f
platinum vinylidene complexes in catalysis 304f
polymerization 271ff
preparation 1ff,
– of metal vinylidene complexes 1ff
propadienylidene complexes, see allenylidene complexes
propargylation 233ff
– of aromatic compounds 233ff
propargylic compounds 42ff, 221ff, 252, 318ff, 323ff
– homopropargylic alcohol 323ff
– propargylic alcohol 42ff, 252f, 323ff
– propargylic alkylated products 223f
– propargylic amines 221ff
– propargylic enamines 284ff
– propargylic ethers 221ff
– propargylic sulfides 221ff
propargylic substitution 76, 219ff
– enantioselective 229f
2-propynol 163, 326f
protonation 9
pyrans 234ff
pyranylidene complex 182
pyridine derivatives 202f
pyrroles 233f

q

quinoline 178f

r

racemization 229
reactions of metallacumulenes, see metallacumulenes
reactions of metal-allenylidene complexes, see allenylidene complexes
reactions of metal-vinylidene complexes, see vinylidene complexes
reactions proceeding via vinylidene complexes 42ff
1,4-rearrangement 106ff
redox rearrangements of vinylidene complexes 9f,
regioselectivity 102, 224, 313ff
rhenium 112ff
rhodium 41f, 110ff, 122, 279ff
– alkyne complex 6
– vinylidene complexes 279ff
ring closing metathesis (RCM) 187f, 251
ring opening 240f
– of cyclopropane rings 240f
ring opening metathesis polymerization (ROMP) 251
ruthenium 3, 6, 15, 108ff, 120ff, 193ff, 217ff, 251ff, 313ff
– alkenylalkylidenes 252ff, 262ff
– alkyne complex 6
– allenylidene complexes 217ff, 251ff
– diruthenium complexes 218, 225, 252
– indenylidene-ruthenium complexes 256ff, 265ff, 271ff
– vinylidene complexes 3ff, 193ff

s

Schrock carbene complexes 130
Schrock catalyst 251
Schrock's molybdenum alkylidene complex 187f
self-assembly 303
[3,3]-sigmatropy 182ff
silyl enol ether 173
1,1'-bis(silylethynyl)ferrocene 208f
stavudine 167
substitution reactions 219ff
– allylic substitution reactions 219, 227ff
– enantioselective allylic substitution reactions 227ff
– propargylic substitution reactions 219ff
synergistic effect 226

t

tandem cyclization 177
tautomerization 134ff, 139ff, 144ff, 148f, 152f, 184f, 224
tetrahydropyranylidene complex 168f
thiophenes 233f
titanocene vinylidene complexes 33ff, 150
trienyl carbene complex 164
triethylamine 279ff
triphenylketimine 160
trisaccharide 169f
triyne 113ff
tungsten 6, 72ff, 111ff, 120, 142ff, 159ff, 186f
– tungsten-vinylidene intermediate 144ff
– tungsten alkenyl species 146f

u

ureas 314ff

v

vinyl carbene complex 164f
vinylic compounds 314ff
– synthesis of vinylic carbamates 314ff
– synthesis of vinylic phosphines 326f
– synthesis of vinylic ureas 314ff
vinylidenes, see vinylidene complexes
vinylidene complexes 1ff, 99ff, 193ff, 217, 279ff, 313ff
– addition of electrophiles 8, 27
– addition of water 20
– adducts with other metal fragments 28ff
– alcohols 21
– alkylation 27
– as catalysts 193ff, 279ff, 302ff, 313ff
– as reaction intermediates 42ff, 197f, 313ff
– bimetallic vinylidene complexes 8f
– carbocyclization 193ff
– carbon nucleophiles 22
– chemistry of specific complexes 33ff
– coordinated phosphines 24
– coupling 24
– cycloaddition reactions 27ff, 34, 283f
– deprotonation 20
– displacement of the vinylidene ligand 30
– formation of alkenylalkylidenes 264ff
– formation of π-bonded ligands 25
– formation of carbamates 314f
– formation of cyclopropenes 23f
– formation of enol esters 316ff
– formation of ureas 314f
– from acyl complexes 15
– from alkenes 16
– from 1-alkynes 3ff
– from carbynes 11ff
– from cyclic alkynes 17
– from metal alkynyls 6ff
– from metal allenylidenes 11
– from metal-carbon complexes 14
– from metal-carbyne complexes 11ff
– from metal-cyclopropenyl complexes 15
– from vinyls 15
– halogen nucleophiles 22
– heavier analogs of metal vinylidene complexes 150f
– hydride 22
– intramolecular reactions 23ff
– intramolecular metathesis 28
– iodinated vinylidene complex 178f
– ligand substitution 30f
– metal-vinylidene complexes 3ff
– migration of SiR_3, SnR_3, SR, SeR 5f
– neutral vinylidene complexes 3ff
– nitrogen 22
– optically active 322
– oxidation 10f
– phosphorus 22
– photolytic demetallation 30f
– porphyrin vinylidene complexes 17
– preparative methods 2ff, 193f
– protonation 7ff, 26f
– reactions 20ff
– reactions at Cα 20ff, 313
– reactions at Cβ 25ff
– reactions with electrophiles 25ff
– reactions with nucleophiles 20ff, 313ff
– redox rearrangements 9f
– solid state structure 99
– stoichiometric reactions 19ff
– sulfur 21
– tautomerization 134ff, 139ff, 144ff, 148ff, 152f, 184f
– titanocene vinylidene complexes 150
– tungsten-vinylidene intermediate 144ff
– vinylidene-alkyne coupling 25
vinylidene transfer 41
vinylidene-alkyne tautomerization 139ff, 184ff
vinylvinylidene complexes 2, 17ff
– by dehydration 18
– by deprotonation 18
– from enynes 19
vinylvinylidenes, see vinylvinylidene complexes

w

Wittig type reactions 76
Wilkinson's catalyst 284f, 290, 294ff

y

ynamine 112ff